Consciousness and Quantum Mechanics

PHILOSOPHY OF MIND SERIES

SERIES EDITOR: David J. Chalmers, *Australian National University*

The Conscious Brain
Jesse Prinz

Simulating Minds
The Philosophy, Psychology, and Neuroscience of Mindreading
Alvin I. Goldman

Supersizing the Mind
Embodiment, Action, and Cognitive Extension
Andy Clark

Perception, Hallucination, and Illusion
William Fish

Phenomenal Concepts and Phenomenal Knowledge
New Essays on Consciousness and Physicalism
Torin Alter and Sven Walter

Phenomenal Intentionality
George Graham, John Tienson, and Terry Horgan

The Character of Consciousness
David J. Chalmers

The Senses
Classic and Contemporary Philosophical Perspectives
Fiona Macpherson

Attention is Cognitive Unison
An Essay in Philosophical Psychology
Christopher Mole

The Contents of Visual Experience
Susanna Siegel

Consciousness and the Prospects Of Physicalism
Derk Pereboom

Consciousness and Fundamental Reality
Philip Goff

The Phenomenal Basis of Intentionality
Angela Mendelovici

Seeing and Saying
The Language of Perception and the Representational View of Experience
Berit Brogaard

Perceptual Learning
The Flexibility of the Senses
Kevin Connolly

Combining Minds
How to Think About Composite Subjectivity
Luke Roelofs

The Epistemic Role of Consciousness
Declan Smithies

The Epistemology of Non-Visual Perception
Berit Brogaard and Dimitria Electra Gatzia

What are Mental Representations?
Edited by Joulia Smortchkova, Krzysztof Dołęga, and Tobias Schlicht

Consciousness and Quantum Mechanics
Edited by Shan Gao

Consciousness and Quantum Mechanics

Edited by

SHAN GAO

OXFORD
UNIVERSITY PRESS

OXFORD
UNIVERSITY PRESS

Oxford University Press is a department of the University of Oxford.
It furthers the University's objective of excellence in research, scholarship,
and education by publishing worldwide. Oxford is a registered trade mark of
Oxford University Press in the UK and certain other countries.

Published in the United States of America by Oxford University Press
198 Madison Avenue, New York, NY 10016, United States of America.

Cataloging-in-Publication data is on file at Library of Congress
ISBN 978–0–19–750166–5

DOI: 10.1093/oso/9780197501665.001.0001

Printed by Integrated Books International, United States of America

Contents

Acknowledgments

I would like to thank Peter Ohlin and David Chalmers for their kind support as I worked on this project, and the referees who gave helpful suggestions on how the work could best serve its targeted audience. During the editing of this book, I have been assisted by research funding from the National Social Science Foundation of China. Finally, I am deeply indebted to my parents, Qingfeng Gao and Lihua Zhao, my wife, Huixia, and my daughter, Ruiqi, for their unflagging love and support.

List of Contributors

J. Acacio de Barros (San Francisco State University)

Michel Bitbol (French National Centre for Scientific Research)

David J. Chalmers (New York University)

Shan Gao (Shanxi University)

Philip Goff (Durham University)

Stuart Hameroff (University of Arizona)

Lucien Hardy (Perimeter Institute for Theoretical Physics)

Basil J. Hiley (University of London)

Jenann Ismael (Columbia University)

Adrian Kent (University of Cambridge and Perimeter Institute for Theoretical Physics)

Peter J. Lewis (Dartmouth College)

Kelvin J. McQueen (Chapman University)

Carlos Montemayor (San Francisco State University)

Elias Okon (National Autonomous University of Mexico)

Roger Penrose (University of Oxford)

Paavo Pylkkänen (University of Helsinki and University of Skövde)

William Seager (University of Toronto)

Miguel Ángel Sebastián (National Autonomous University of Mexico)

Michael Silberstein (Elizabethtown College)

Lee Smolin (Perimeter Institute for Theoretical Physics)

Paul Skokowski (University of Oxford)

W. Mark Stuckey (Elizabethtown College)

Information on Editor

Shan Gao is Professor of Philosophy at the Research Center for Philosophy of Science and Technology, Shanxi University. He is the founder and managing editor of *International Journal of Quantum Foundations*, and the author of several books including the recent monograph *The Meaning of the Wave Function: In Search of the Ontology of Quantum Mechanics* (Cambridge University Press 2017). His research focuses on the philosophy of physics, especially the foundations of quantum mechanics. He also has interests in the philosophy of mind and the philosophy of science.

Introduction

Consciousness and quantum mechanics are two mysteries in our times. It has been conjectured that a deep connection between them may exist. The connection is bidirectional. On the one hand, an analysis of the conscious mind and psychophysical connection seems indispensable in understanding quantum mechanics and solving the notorious measurement problem (Gao, 2019).[1] Indeed, as Harvey Brown once emphasized, "The issue of psychophysical parallelism is at the heart of the problem of measurement in quantum mechanics" (Brown, 1996). On the other hand, it seems that in the end quantum mechanics, the most fundamental theory of the physical world, will be relevant to understanding consciousness and even solving the mind-body problem when assuming a naturalist view, even though we are not quite there yet (Atmanspacher, 2019). Therefore, a careful and thorough examination of possible connections between consciousness and quantum mechanics is not only necessary but also even pressing in order to unravel these two mysteries.

This book is the first volume that provides a comprehensive review and thorough analysis of intriguing conjectures about the connection between consciousness and quantum mechanics. It contains seventeen original chapters that are written by leading experts in this research field. This book is accessible to graduate students in physics and philosophy of mind. It will be of value to students and researchers in physics with an interest in the

[1] Quantum mechanics, according to its standard formulation, says that when there is no measurement, the wave function of a physical system evolves in time according to the linear and deterministic Schrödinger equation, while when a measurement is made on the system, its wave function will collapse to one branch that corresponds to the result of the measurement in a nonlinear and stochastic way. The measurement problem of quantum mechanics, in John Bell's words, is then to answer the question: "What exactly qualifies some physical systems to play the role of 'measurer'?" (Bell, 1990). Quantum mechanics is not a precise physical theory without solving the measurement problem. There are also other formulations of the measurement problem that are independent of standard quantum mechanics (see Chapter 8 of this volume and references therein), and solving the measurement problem in one way or another such as resorting to consciousness or not, will lead to alternative quantum theories, which will be discussed in several chapters of this volume.

Shan Gao, *Introduction* In: *Consciousness and Quantum Mechanics.* Edited by: Shan Gao,
Oxford University Press. © Oxford University Press 2022. DOI:10.1093/oso/9780197501665.003.0001

meaning of quantum mechanics, as well as to philosophers working on the foundations of quantum mechanics and philosophy of mind.

This book is organized into three parts in order to facilitate reading, although a few chapters do fit into more than one part. Echoing the possible bidirectional connections between consciousness and quantum mechanics, the first two parts are about the possible roles of consciousness in quantum theories. The first part is about the specific consciousness-collapse conjecture, and the second part is about other possible roles of consciousness in quantum theories, and the third part is about the possible roles of quantum mechanics in understanding consciousness.

A more detailed introduction of the three parts of this book is as follows. The first part investigates the relationship between consciousness and the collapse of the wave function. If consciousness really collapses the wave function, then this will provide both a solution to the measurement problem and a potential role for consciousness in the physical world. This well-known consciousness-collapse conjecture has a long history, but it is often dismissed as a very speculative and imprecise idea. Recently, there has been exciting new progress on developing the idea and making it more precise. One main reason is that theories that give precise and mathematically defined conditions for the presence or absence of consciousness have been developed, such as Tononi's (2008) integrated information theory, in which consciousness is quantified with a mathematical measure of integrated information.

In Chapter 1, David Chalmers and Kelvin McQueen give a comprehensive introduction to the consciousness-collapse idea and answer the usual objections to the idea. Moreover, they put forward a way of making the consciousness-collapse idea precise by exploring and evaluating dynamic principles governing how consciousness collapses the wave function. Two models are proposed. The first one is a simple consciousness-collapse model on which consciousness is entirely superposition-resistant. This model is subject to a conclusive objection arising from the quantum Zeno effect. The second model combines integrated information theory with Pearle's continuous-collapse theory, and it is not subject to the objection. The prospects of empirically testing these models and potential philosophical objections to them are also discussed. The authors conclude that the consciousness-collapse idea is a research program worth exploring.

In Chapter 2, Elias Okon and Miguel Ángel Sebastián introduce a consciousness-collapse model called the subjective-objective collapse model, evaluate the virtues of the model, and answer some possible objections and challenges related to it, such as the multiple realizability of conscious states. This model consists of an objective collapse scheme, where the collapse operator is associated with consciousness as a physical property or with a physical property that correlates with consciousness. Like Chalmers and McQueen's models, the advantage of this model is that consciousness is incorporated into quantum mechanics in a well-defined way, both mathematically and conceptually, and in a way that is fully compatible with the truth of physicalism if consciousness is physical.

In Chapter 3, J. Acacio de Barros and Carlos Montemayor investigate the role of consciousness in the consciousness-collapse quantum theories. According to them, the observer in quantum mechanics is an access conscious observer, rather than a phenomenally conscious observer, because measurements are not entirely determined by merely appearance properties of experiences, but rather by concrete interventions in an environment by a rational agent with specific goals that have unique theoretical meaning. Moreover, they argue that an access-consciousness version of panpsychism, which they call "panintentionalism," suffices, and it is better equipped to account for the role of consciousness involved in these theories than the standard one, based on phenomenal consciousness.

In objective collapse theories of quantum mechanics, the collapse of the wave function is not caused by consciousness. However, the conscious perceptions may also put constraints on these theories. In Chapter 4, Adrian Kent analyzes the perception constraints on mass-dependent collapse models. According to the previous analysis of Bassi et al., the parameters of these models consistent with known experiments imply that when a person observes a superposition of a few photons a collapse would happen in her eye within the normal perception time of 100 ms, and thus these models are consistent with our conscious perceptions. Kent notes a key problem of this analysis: the relevant processes are assumed to happen in a vacuum, rather than in cytoplasm. Moreover, he argues that when considering the existence of cytoplasm, these collapse models with parameters consistent with known experiments may not satisfy the perception constraints.

The second part of this book is about other possible roles of consciousness in quantum theories. In Chapter 5, Philip Goff argues that the reality of consciousness puts a constraint on the ontology of quantum mechanics. According to wave function monism, a popular interpretation of the ontology of quantum mechanics, fundamental physical reality consists of a complex-valued field in a high-dimensional space. Goff analyzes the question of whether the reality of consciousness can be accounted for by wave function monism. After criticizing the existing attempts to close the wave function/three-dimensional objects explanatory gap, he argues, based on an analysis of the grounding relationship, that the wave function monist has no way to account for consciousness, at least on the assumption that she can't account for three-dimensional objects. Goff's argument does not assume either a materialist or a non-materialist view of consciousness.

In Chapter 6, Peter J. Lewis argues that the word "experience" should not appear as a primitive in the formulation of quantum theory, just as Bell argues that the word "measurement" should not so appear. The psychophysical connection is not something that philosophers and physicists can posit at their convenience; neuroscience constrains the connections between brain structures and experience, whether or not the latter is reducible to the former. But he cautions that, while it is relatively easy to tell whether a use of "measurement" in a discussion of quantum mechanics is good or bad, it is not so easy to tell whether a use of "experience" is good or bad.

In Chapter 7, Jenann Ismael addresses a more general question than the other papers in the volume. The paper is not addressed to the connection between consciousness and quantum mechanics specifically, but to the broader question of whether physics should steer clear of the mind completely. Ismael makes a case that discussion of the mind is both legitimate and essential (as she says: "Physics doesn't stop at the surface of the skin"), but that one can do it while excluding the aspects of mentality that give rise to the mind-body problem. If consciousness has no functional or causal role of its own in the physical world, then physics will not know the difference between the conscious state and its physical basis, and thus consciousness is irrelevant to physics and we need not worry about the mind-body problem in physics. If consciousness itself has a causal role in the physical world as in the consciousness-collapse theories that treat it as a physical primitive, it will indeed become something that matters to physics. But then it also becomes something that is characterizable in terms of its physical role, effectively providing a solution to the Hard part of the mind-body problem.

In Chapter 8, Shan Gao defends his new mentalistic formulation of the measurement problem and argues that it is more appropriate than Maudlin's original formulation. Moreover, he argues that the solutions to the measurement problem need to care about the minds of observers, e.g., they need to assume a certain form of psychophysical connection, and their validity depends on our scientific and philosophical understandings of the conscious mind.

In Chapter 9, Paul Skokowski examines the role of human belief within an Everett no-collapse version of quantum mechanics. He considers the claim that an observer of a measurement resulting in a superposition ends up being deceived about her own perceptual beliefs. Skokowski argues that, upon examination of the neural vehicles that comprise the belief eigenstates of the observer, and the intentional contents of these states, the observer will not, in fact, have the deceptive belief claimed by this interpretation of quantum mechanics.

In Chapter 10, Michael Silberstein and W. M. Stuckey offer a new realist psi-epistemic, principle-based account of quantum mechanics and a neutral monist account of experience. Recent gedanken experiments and theorems in quantum mechanics, such as new iterations on Wigner's friend and delayed choice, have led many people to claim that quantum mechanics is not compatible with determinate and intersubjectively consistent experience (what some call absoluteness of observed events). They show that jettisoning wavefunction realism in favor of a principle-based account and conceiving of consciousness as qualia in favor of neutral monism, allows one to uphold the absoluteness of observed events, deflate the hard problem of consciousness, and deflate the measurement problem, all without giving up free will (i.e., no superdeterminism), locality, or the completeness of quantum mechanics. Their account requires no invocation of relative states (e.g., outcomes being relative to branches, conscious observers, etc.) and requires no "hybrid models" such as claims about "subjective collapse." They provide a take on quantum mechanics that yields a single world wherein all the observers (conscious or otherwise) agree about determinate and definite outcomes.

In Chapter 11, Michel Bitbol argues that phenomenology provides a possible way of understanding quantum mechanics and consciousness and further solving the measurement problem and the mind-body problem. According to phenomenology, a philosophical discipline that favors a first-person approach of any ontological and epistemological issues, consciousness is neither something nor a property of something, but the flux of the self-splitting of what there is into subjective existence and its objective targets, and physical systems and processes are nothing more than objects of consciousness. This supports the neo-Bohrian approaches to quantum mechanics such as QBism and participatory realism, according to which the symbols of quantum mechanics are primarily used by agents to assign probabilistic weights to the outcomes of experiments so that such agents can make consistent bets, and the insuperable dependence of these symbols on our situation and experience indirectly reveals the nature of reality so that our knowledge of it can only be participatory rather than representational, predictive rather than descriptive.

In Chapter 12, Lucien Hardy investigates the possibility that when humans are used to decide the settings at each end in a Bell experiment, we might expect to see a violation of quantum mechanics in agreement with the relevant Bell inequality. He argues that this result is well motivated when assuming superdeterminism and mind–body dualism, and if it is confirmed, it will be tremendously significant for our understandings of quantum mechanics and consciousness. Moreover, he also discusses in detail how we can perform such a Bell experiment based on current technologies.

The third part of this book is about quantum approaches to consciousness. In Chapter 13, Roger Penrose argues that the human ability to achieve conscious understanding is a non-computational process, and this requires something beyond current physical theory, an effect of gravitation on quantum mechanics, in supplying a physical basis for "the collapse of the wave function," denoted by OR. OR events are what allow a firm classical reality to arise from a quantum reality having a somewhat weaker ontological status. When appropriately orchestrated, these "proto-conscious" OR events become genuine conscious processes according to the Orch-OR proposal. Moreover, from the principles of relativity theory, it can be deduced that OR, and therefore Orch-OR, can have a certain "retro-active" effect, which may explain how conscious decisions can act within a very small fraction of a second, in contradiction with a conclusion frequently made, on the basis of various experiments, that such acts must be necessarily unconscious. According to Penrose, this provides an explanation for some puzzling related effects found by Benjamin Libet in the 1970s.

In Chapter 14, Stuart Hameroff gives an up-to-date and comprehensive review of the Penrose-Hameroff "Orch OR" theory. The theory attributes consciousness to "orchestrated" quantum computations in microtubules inside brain neurons, which terminate by Penrose objective reduction (OR), a process in the fine scale structure of the universe that introduces phenomenal experience and non-computability. The theory suggests that microtubules (1) encode memory and process information, (2) orchestrate quantum vibrational superpositions (qubits) of pi electron resonance dipoles within tubulin that unify, entangle, and (3) evolve to meet Orch OR threshold for full, rich conscious experience, most likely (4) in dendrites and soma of cortical layer 5 pyramidal neurons, and (5) selection of microtubule states that regulate axonal firings and behavior, and (6) "retroactivity" inherent in OR and Orch OR can resolve issues in quantum mechanics, free will, and Libet's famous "backward time referral." Hameroff concludes that Orch OR has explanatory power, and is testable and falsifiable.

In Chapter 15, Basil J. Hiley and Paavo Pylkkänen propose that quantum theory implies a radically new notion of matter that has not been properly understood before David Bohm's groundbreaking work. Bohm proposed that the fundamental particles of physics (such as electrons) are not merely pushed around mechanically by classical forces but are also able to respond to information. Information is thus assumed to be an objective commodity that can exist independently of the human mind and that actively guides or instructs physical processes. This notion of active information also applies in computational, biological and psychological phenomena, thus helping us to understand how the mental and physical sides of reality are related. It may even help us to understand the nature of conscious experience. The latter part of this chapter considers the deeper mathematical and physical background of quantum theory and suggests that we need to revise our basic assumptions about quantum objects, such as the role of the wave function.

In Chapter 16, William Seager argues that the goal of interpretation of a theory such as quantum mechanics is intelligibility, which aims to show how the world described by a theory can be made intuitively clear. He identifies three kinds of intelligibility: mundane, mathematical, and metaphysical, and notes that the mismatch between high mathematical intelligibility and low mundane intelligibility of quantum mechanics motivates the search for its interpretations, which attempt to provide metaphysical intelligibility. Moreover, Seager considers several such interpretations and argues that Bohm's view, which puts mentality or some basic kind of proto-consciousness as the bearer of intrinsic information, as a fundamental feature of the world, may be the best way to provide a metaphysically intelligible basis for quantum mechanics.

In Chapter 17, Lee Smolin proposes an approach to the question of how consciousness fits into the physical world in the context of a relational and realist completion of quantum mechanics called the causal theory of views. In this theory, the "beables" are the information available at each event from its causal past, and a causal universe is composed of a set of partial views of itself. Smolin proposes that conscious perceptions are aspects of some views. Concretely speaking, only those views that are novel, in the sense that they are not duplicates of the view of any event in the event's own causal past, are the physical correlates of conscious experience, and to be conscious a view must also be maximal, in the sense of being the smallest composite not being part of a larger entangled state. This gives a restricted form of panpsychism defined by a physically based selection principle that selects which views have experiential aspects, and explains why consciousness always involves awareness of a bundled grouping of qualia that define a momentary self.

Notwithstanding these new insightful thoughts about possible deep connections between consciousness and quantum mechanics, maybe we are still far away from the final answer. But this is just the impetus to do the research. I really hope this book will inspire more researchers to join the search for the ultimate reality of the universe.

References

Atmanspacher, H. (2019). Quantum Approaches to Consciousness. In *The Stanford Encyclopedia of Philosophy* (Winter 2019 Edition), Edward N. Zalta (ed.), https://plato.stanford.edu/archives/win2019/entries/qt-consciousness/.

Bell, J. S. (1990). Against "measurement," *Physics World* 3, 33–40.

Brown H. R. (1996). Mindful of quantum possibilities, *British Journal for the Philosophy of Science* 47, 189–199.

Gao, S. (2019). The measurement problem revisited, *Synthese* 196, 299–311.

Tononi, G. (2008). Consciousness as integrated information: A provisional manifesto. *Biological Bulletin* 215, 216–242.

PART I

CONSCIOUSNESS AND WAVE FUNCTION COLLAPSE

1

Consciousness and the Collapse of the Wave Function

David J. Chalmers and Kelvin J. McQueen

1. Introduction

One of the hardest philosophical problems arising from contemporary science is the problem of quantum reality. What is going on in the physical reality underlying the predictions of quantum mechanics? It is widely accepted that quantum-mechanical systems are describable by a wave function. The wave function need not assign definite position, momentum, and other definite properties to physical entities. Instead it may assign a superposition of multiple values for position, momentum, and other properties. When one measures these properties, however, one always obtains a definite result. On a common picture, the wave function is guided by two separate principles. First, there is a process of evolution according to the Schrödinger equation, which is linear, deterministic, and constantly ongoing. Second, there is a process of collapse into a definite state, which is nonlinear, nondeterministic, and happens only on certain occasions of measurement.

This picture is standardly accepted at least as a basis for empirical predictions, but it has been less popular as a story about the underlying physical reality. The biggest problem is the *measurement problem* (see Albert (1992); Bell (1990)). On this picture, a fundamental measurement-collapse principle says that collapses happen when and only when a measurement occurs. But on the face of it, the notion of "measurement" is vague and anthropocentric,

*Authors are listed in alphabetical order and contributed equally. We owe thanks to audiences starting in 2013 at Amsterdam, ANU, Cambridge, Chapman, CUNY, Geneva, Göttingen, Helsinki, Mississippi, Monash, NYU, Oslo, Oxford, Rio, Tucson, and Utrecht. These earlier presentations have occasionally been cited, so we have made some of them available at consc.net/qm. For feedback on earlier versions, thanks to Jim Holt, Adrian Kent, Kobi Kremnizer, Oystein Linnebo, and Trevor Teitel. We are grateful to Maaneli Derakhshani and Philip Pearle for their help with the mathematics of collapse models, and especially to Johannes Kleiner, who coauthored section 5 on quantum integrated information theory.

David J. Chalmers and Kelvin J. McQueen, *Consciousness and the Collapse of the Wave Function* In: *Consciousness and Quantum Mechanics*. Edited by: Shan Gao, Oxford University Press. © Oxford University Press 2022.
DOI: 10.1093/oso/9780197501665.003.0002

and is inappropriate to play a role in a fundamental specification of reality. To make sense of quantum reality, one needs a much clearer specification of the underlying dynamic processes.

Another of the hardest philosophical problems arising from contemporary science is the mind-body problem. What is the relation between mind and body, or more specifically, between consciousness and physical processes? By consciousness, what is meant is phenomenal consciousness, or subjective experience. A system is conscious when there is something it is like to be that system, from the inside. A mental state is conscious when there is something it is like to be in that state.

There are many aspects to the problem of consciousness, including the core problem of why physical processes should give rise to consciousness at all. One central aspect of the problem is the *consciousness-causation problem*: how does consciousness play a causal role in the physical world? It seems obvious that consciousness plays a causal role, but it is surprisingly hard to make sense of what this role is and how it can be played.

There is a long tradition of trying to solve the consciousness-causation problem and the quantum measurement problem at the same time, by saying that measurement is an act of consciousness, and that consciousness plays the role of bringing about wave function collapse. The locus classicus of this consciousness-collapse thesis is Eugene Wigner's 1961 "Remarks on the Mind-Body Question." There are traces of the view in earlier work by von Neumann (1955) and London and Bauer (1939).[1] In recent years the approach has been pursued by Henry Stapp (1993) and others.

The central motivations for the consciousness-collapse view come from the way it addresses these problems. Where the problem of quantum reality is concerned, the view provides one of the few interpretations of quantum mechanics that takes the standard measurement-collapse principle at face value. Other criteria for measurement may be possible, but understanding measurement in terms of consciousness has a number of motivations. First, it

[1] It is clear that von Neumann (1955) endorses a measurement-collapse interpretation, and he says (p. 418) that subjective perception is "related" to measurement, but he does not clearly identify measurement with conscious perception. In his discussion of observed systems (I), measuring instruments (II), and "actual observer" (III), he says "the boundary can just as well be drawn between I and II+III as between I+II and III." This suggests neutrality on whether the collapse process is triggered by measuring devices or by conscious observers. He also says that the boundary is "arbitrary to a very large extent" (p. 420), which is not easy to reconcile with the fact that different locations for collapse are empirically distinguishable in principle, as we discuss in section 6. London and Bauer (1939, section 11) say more clearly: "We note the essential role played by the consciousness of the observer in this transition from the mixture to the pure case. Without his effective intervention, one would never obtain a new psi function" (although see French (2020) for an alternative reading).

provides one of the few non-arbitrary criteria for when measurement occurs. Second, it is arguable that our core pretheoretical concept of measurement is that of measurement by a conscious observer. Third, the consciousness-collapse view is especially well-suited to save the central epistemological datum that ordinary conscious observations have definite results. Fourth, understanding measurement as consciousness provides a potential solution to the consciousness-causation problem: consciousness causes collapse.

Despite these motivations, the consciousness-collapse view has not been popular among contemporary researchers in the foundations of physics. Some of this unpopularity may stem from the popularity of the view in unscientific circles: for example, popular treatments by Capra (1975) and Zukav (1979) link the view to Eastern religious traditions. More substantively, the view is frequently set aside in the literature on the basis of *imprecision* and on the basis of *dualism*.

The objection from imprecision is stated succinctly by Albert (1992, pp. 82–3):

> How the physical state of a certain system evolves (on this proposal) depends on whether or not that system is conscious; and so in order to know precisely how things physically behave, we need to know precisely what is conscious and what isn't. What this "theory" predicts will hinge on the precise meaning of the word conscious; and that word simply doesn't have any absolutely precise meaning in ordinary language; and Wigner didn't make any attempt to make up a meaning for it; so all this doesn't end up amounting to a genuine physical theory either.

We think that the force of this objection is limited. Of course it is true that 'conscious' in ordinary language is highly ambiguous and imprecise, but it is easy to disambiguate the term and make it more precise. Philosophers have distinguished a number of meanings for the term, the most important of which is phenomenal consciousness. As usually understood, a system is phenomenally conscious when there is something it is like to be that system: so if there is something it is like to be a bat, a bat is phenomenally conscious, and if there is nothing it is like to be a rock, a rock is not phenomenally conscious. One might question the precision of this concept in turn, but it is at least a common and widely defended view (see, e.g., Antony (2006); Simon (2017)) that it picks out a definite and precise property. On this view, phenomenal consciousness comes in a number of varieties, but it is either definitely present or definitely absent in a given system at a given time.

In recent years, theories that give precise mathematically-defined conditions for the presence or absence of consciousness have begun to be developed. The most well-known of these theories is Tononi's integrated information theory (Tononi 2008), which specifies a mathematical structure for conscious states and quantifies them with a mathematical measure of integrated information. Of course it is early days in the science of consciousness, and current theories are unlikely to be final theories. Nevertheless, it is possible to envisage precise theories of consciousness, and to reason about how they might be combined with a consciousness-collapse view to yield precise interpretations of quantum mechanics.

Crucially, when different precise theories of consciousness are combined with the consciousness-collapse view, these yield subtly different experimental predictions. As a result, we have a further motivation for taking consciousness-collapse interpretations seriously: they can be tested experimentally. As we discuss in section 7, there is a long-term research program of experimentally testing consciousness-collapse interpretations and eventually supporting a precise consciousness-collapse interpretation. The required experiments are difficult, but advances in quantum computing may already exclude certain simple consciousness-collapse interpretations. Because of these considerations, the underdetermination of conditions for consciousness does not reflect any fundamental imprecision in consciousness-collapse views. It simply reflects an experimentally testable degree of freedom.

The second common objection to the consciousness-collapse view is that it is committed to dualism: the view that the mental and the physical are fundamentally distinct. The consciousness-collapse view treats consciousness in a special way that seems to exempt it from the standard quantum-mechanical laws governing physical systems. This remark by Peter Lewis (this volume) reflects a common attitude:

> Wigner postulates a strong form of interactive dualism in order to justify a duality in the physical laws. Few will want to follow Wigner down this path: non-physical minds, especially causally active ones, are mysterious at best.

Again, we think the force of this objection is limited.

First: the consciousness-collapse thesis need not lead to dualism. It is compatible with materialist views on which consciousness is a complex physical property. For example, let us suppose a materialist version of integrated information theory on which consciousness is identical to Φ^*, the property of having integrated information above a certain threshold. Then the consciousness-collapse theory will say that Φ^* causes collapse. This interpretation of quantum mechanics will involve a fundamental physical law saying that under the conditions specified by Φ^*, collapse is brought about according to the Born rule. A fundamental law involving a complex physical property may be unlike familiar physical laws, but it involves nothing nonphysical.

Second: where consciousness is concerned, there are reasons to take dualism seriously. There are familiar reasons to question whether any purely physical theory can explain consciousness. One common reason (Chalmers 2003) is that physical theories explain only structure and dynamics (the so-called 'easy problems of behavior and the like), and explaining consciousness (the so-called 'hard problem') requires explaining more than structure and dynamics. These reasons need not lead to substance dualism, on which consciousness involves a separate nonphysical entity akin to an ego or soul, but they have led many theorists to adopt a form of property dualism where consciousness is accepted as a fundamental property akin to spacetime, mass, and charge.

Where physical theories give fundamental physical laws that connect physical properties to each other, a property dualist theory of consciousness gives fundamental psychophysical laws that connect physical properties to consciousness. For example, on a property dualist construal of integrated information theory, there might be a fundamental physics-to-consciousness law saying that when a system has Φ above a certain threshold, the system will have a corresponding state of consciousness. Such a law has a structure akin to the Newtonian mass-to-gravitational-field law, saying that when a system has a certain mass, the system will have a corresponding gravitational field. On a consciousness-causes-collapse theory, there will be an additional consciousness-to-physics law saying that states of consciousness bring about wave function collapse in a certain way. Putting these theories together might yield a mathematically precise version of property dualism that specifies the conditions under which consciousness arises and the role that it plays.

Interestingly, the most common reason among philosophers for rejecting property dualist theories of consciousness is an argument from physics. This argument runs roughly as follows: (1) every physical effect has only physical causes, (2) consciousness causes physical effects, so (3) consciousness is physical. The key first premise is a causal closure thesis, supported by the observation that there are no causal gaps in standard physics that a nonphysical consciousness might fill. But wave function collapse in quantum mechanics appears to be precisely such a gap, and consciousness-collapse models are at least not obviously ruled out by known physics. The situation is that many physicists rule out consciousness-collapse models for philosophical reasons (they are dualistic), while philosophers rule out property dualist models for physics-based reasons (they violate causal closure).

The upshot is that a central reason to reject the consciousness-collapse thesis (it leads to dualism) and a central reason to reject interactionist property dualism (it violates the causal closure of physics) provide no reason to reject the two views when taken together. Perhaps there are other reasons to reject the consciousness-collapse thesis or to reject dualism, but these reasons must be found elsewhere.

A third common objection to the consciousness-collapse thesis is that it is not *necessary* to invoke consciousness in an interpretation of quantum mechanics, as there are alternative interpretations that give it no special role. Even if we retain the measurement-collapse framework, it is possible to understand measurement independently of consciousness, so that nonconscious systems such as ordinary measuring devices can collapse the wave function. Going beyond this framework, a number of alternative interpretations have been developed that give no role to the notion of measurement. These include spontaneous-collapse interpretations (e.g., Pearle (1976); Ghirardi, Rimini, and Weber (1986)) which retain a collapse process but dispense with the need for measurement as a trigger, and hidden-variable interpretations (Bohm 1952) and many-worlds interpretations (Everett 1957), which eliminate collapse entirely.

We agree that one is not forced to accept a role for consciousness in quantum mechanics. At the same time, the mere existence of alternative interpretations is not itself good reason to reject the consciousness-collapse thesis. If it were, we would have good reason to reject all interpretations. Perhaps the underlying thought is that the consciousness-collapse thesis is extravagant and has certain costs, such as dualism. For there to be a serious objection here, an opponent needs to articulate the costs as objections in their own right. As with every other interpretation of quantum mechanics, the consciousness-collapse interpretation has both serious costs (dualism) and serious benefits (taking the standard dynamics at face value, solving the consciousness-causation problem). To assess any interpretation, we need to weigh its costs against its benefits.

In this chapter, we are exploring consciousness-collapse models rather than endorsing them. In particular, we are not asserting that these interpretations are superior to other interpretations of quantum mechanics. Both of us have considerable sympathy with other interpretations and especially with many-worlds interpretations (see Chalmers (1996, ch.10) and McQueen and Vaidman (2019)). But we think that consciousness-collapse interpretations deserve close attention. If it turns out that these interpretations have fatal flaws, they can be set aside. But if there are consciousness-collapse interpretations without clear fatal flaws, then these interpretations should be taken seriously as possible descriptions of quantum-mechanical reality.

In our view, by far the most important challenge to consciousness-collapse models is not the issue of imprecision or of dualism, but the question of *dynamic principles*. Can we find a simple, coherent, and empirically viable set of dynamic principles governing how consciousness collapses the wave function? If we can find such principles, consciousness-collapse models should be placed alongside other dynamic models (including Bohmian hidden-variable models, Everettian many-worlds models, and Pearle-GRW style spontaneous collapse models) as serious contenders to be the correct interpretation of quantum mechanics. If we cannot, then consciousness-collapse models may remain an important speculative class of models, but they will stay on the second tier of interpretations until they are cashed out with dynamic principles.

In what follows, we will explore the prospects for consciousness-collapse interpretations of quantum mechanics. We will do this mainly by exploring and evaluating potential dynamic principles. We focus especially on what we call *super-resistance models*, according to which there are special properties that resist superposition and trigger collapse. When these models are combined with the consciousness-collapse thesis, we obtain models in which consciousness or its physical correlates resist superposition and trigger collapse. We think super-resistance consciousness-collapse models are worth investigating, and in this chapter we investigate some of them.

In this chapter we are not trying to solve the hard problem of how physical processes give rise to consciousness. We are giving an account of the causal role of consciousness that can be combined with many different approaches to the hard problem. Our approach is consistent with both materialist views, on which consciousness is identified with a complex physical property, and dualist views, on which consciousness is a primitive property that correlates with physical properties. Our approach is also consistent with many different theories of consciousness that correlate consciousness with underlying physical processes. For concreteness we will often assume a Tononi-style theory of consciousness on which consciousness is identical to or correlated with integrated information, but much of what we say should translate straightforwardly to other theories of consciousness.

We will not be addressing problems that come up for collapse models of quantum mechanics quite generally. For example, collapse models face important challenges stemming from the theory of relativity (collapse seems to require a privileged reference frame (Maudlin 2011)), and the tails problem (collapse leaves wave functions with tails (McQueen 2015)). The collapse models we consider certainly face these challenges. These are important challenges, but for present purposes we will be happy if consciousness-collapse interpretations can be shown to be about as viable as widely discussed spontaneous-collapse interpretations. Interpretations in both classes will still face the general problems. A number of ideas about how to deal with them have been put forward, but this is a topic for another day.

Our aim is to set out the best consciousness-collapse model that we can and to assess it. Our discussion is speculative and our conclusions are mixed. We articulate both positive models and serious limitations. We first articulate a simple consciousness-collapse model on which consciousness is entirely superposition-resistant. This model is subject to a conclusive objection (distinct from those outlined above) arising from the quantum Zeno effect. We then articulate a model that is not subject to this objection, combining integrated information theory with Pearle's continuous-collapse theory. We explore the prospects of empirically testing these models, and discuss some objections. The model is still subject to both empirical and philosophical objections, but there are some potential ways forward. The upshot is not that consciousness-collapse interpretations are clearly correct, but that there is a research program here worth exploring.

2. Consciousness as Super-Resistant

One can clarify the options for a consciousness-collapse theory by asking a crucial question for any collapse model of quantum mechanics: What is the locus of collapse? That is, which observable determines the definite states that the collapse process projects superposed states onto? Here there are two options: there can be a *variable locus* (different observables serve as the locus on different occasions of collapse) or a *fixed locus* (the same observable always serves as the locus of collapse).

A variable-locus model is closest to standard formulations of quantum mechanics. On a standard understanding, many different observable quantities (e.g. position, momentum, mass, and spin) can be measured and thereby serve as the locus of collapse. Every observable is associated with an operator. Upon measurement, the wave function collapses probabilistically into an eigenstate of that operator, and the measurement reveals the corresponding eigenvalue for the observable (such as a specific position for the particle), with probabilities determined by the prior quantum state according to the Born rule.

Henry Stapp's consciousness-collapse model (Stapp 1993) is a variable-locus model, on which consciousness collapses whatever observable is being consciously observed at a given time. The variable-locus approach has some attractions, but it also faces some hard questions. Not least is the question: what determines which observable is being measured? This question is hard enough that Stapp's model postulates an entirely separate process that determines the locus of collapse. Stapp calls this process "asking a question of nature," which is supposed to be something that takes place in the mind of an observer. Stapp takes this to be a third process distinct from von Neumann's standard dual processes of collapse itself and Schrödinger evolution. Stapp takes this third process as primitive. There are options for analyzing it (perhaps via a precisely specified observation relation between observers and observables, e.g., or by building awareness of observables into the structure of consciousness), but it is clear that such a theory will be complex.

One option for a variable-locus consciousness-collapse theory invokes the idea that consciousness represents certain objects and properties in its environment. For example, visual experiences typically represent the color, shape, and location of observed objects, while auditory experiences represent locations, pitches, and the like. A consciousness-collapse view may hold that when consciousness represents observable properties of an observed object, the object collapses into a definite state of those observables. For example, perceiving the location of a ball that was previously in a superposition will collapse the ball into a definite location. One trouble here is that on standard representationalist views, the represented properties are built into a state of consciousness but the represented objects are not. In some cases an experience as of a single object may be caused by no object or by multiple objects in reality, so there is still a difficult question about which object if any undergoes collapse. This approach may work better with relationist views where consciousness involves direct awareness of specific objects and properties, but there will still be many complications.[2]

Fixed-locus models are simpler in a number of respects, and we will focus on them. In a fixed-locus measurement-collapse model, there are special properties that serve as the locus of collapse. In a fixed-locus consciousness-collapse model, consciousness itself (or perhaps its physical correlate) serves as the locus of collapse. It is this idea that we will develop in what follows.

One natural way to develop a fixed-locus collapse model is through the idea of superposition-resistance, which we will sometimes abbreviate as super-resistance. The idea is that there are special superposition-resistant observables, which as a matter of fundamental law resist superposition and cause the system to collapse onto eigenstates of these observables (with probabilities given by the Born rule). The corresponding class of models are *super-resistance models* of quantum mechanics.[3] There are a number of different ways to make the dynamics of super-resistance precise, some of which we will explore in the following sections. A strong version of super-resistance invokes fundamental *superselection rules* (Wick, Wightman, and Wigner 1952), according to which certain observables are entirely forbidden from entering superpositions. A weaker version invokes principles according to which these superpositions are unstable and tend to collapse.

There are super-resistance models of collapse that give no special role to consciousness or measurement. One well-known super-resistance model is Penrose's model (Penrose 2014) of quantum mechanics on which spacetime structure is superposition-resistant: when the structure of spacetime evolves into superpositions over a certain threshold, these superpositions collapse onto a definite structure. One can also see the GRW interpretation of quantum mechanics as an interpretation on which position is mildly superposition-resistant: superpositions of position tend to collapse, though with low probability for isolated particles.

[2] For representationalist views, see Tye (1995). For relationist views, see Byrne and Logue (2009). These views may face a version of the Zeno problem in the next section, arising from whether the states of consciousness themselves can enter superpositions.

[3] In earlier versions of this chapter, we called superposition-resistant observables "m-properties" (short for "measurement properties") and super-resistance models "m-property models."

Super-resistance models work well with measurement-collapse interpretations of quantum mechanics. In the context of these interpretations, we can think of a super-resistant property not as a *measured* property (e.g., particle position) but as a *measurement* property (e.g., a pointer position or a conscious experience). To sketch the idea intuitively: suppose there is a special class of measurement devices (e.g., oscilloscopes) which have special measurement properties (e.g., meter readings or pointer locations) that (as a matter of fundamental law) resist superposition and tend to collapse. When a measurement takes place, a measured property affects a measurement property. Suppose that we have a quantum system (e.g., a particle) in a superposition of locations a and b, which we represent (simplifying by omitting amplitudes) as the quantum state $|a\rangle + |b\rangle$. The particle interacts with a measurement system such that if not for this principle, it would yield an entangled superposition $|a\rangle\,|M(a)\rangle + |b\rangle\,|M(b)\rangle$, where $M(a)$ and $M(b)$ are the states of the measurement system. Because M is superposition-resistant, the particle and measurement system will instead evolve into a collapsed state $|a\rangle\,|M(a)\rangle$ or $|b\rangle\,|M(b)\rangle$, with probabilities given by the Born rule. The effect will be much the same as if the measured property collapsed directly, but now the measurement properties serve as a single locus of collapse.

Superposition-resistance is an especially natural idea in the context of consciousness-collapse models of quantum mechanics. The idea that consciousness resists superposition is suggested in a brief passage in Wigner (1961), and is later developed by Albert (1992), and Chalmers (2003). Wigner writes:

> If the atom is replaced by a conscious being, the wave function $\alpha(\phi_1 \times \chi_1) + \beta(\phi_2 \times \chi_2)$ (which also follows from the linearity of the equations) appears absurd because it implies that my friend was in a state of suspended animation before he answered my question. It follows that the being with a consciousness must have a different role in quantum mechanics than the inanimate measuring device: the atom considered above. In particular, the quantum mechanical equations of motion cannot be linear.
>
> (Wigner 1961, p. 180)

Wigner's suggestion seems to be that a state of consciousness cannot be superposed because it would require being in a "state of suspended animation." Wigner does not suggest a dynamic process for collapse here, but potential processes are fleshed out a little by Albert and Chalmers.

Albert suggests a picture on which the physical correlates of consciousness immediately collapse once superposed:

> All physical objects almost always evolve in strict accordance with the dynamical equations of motion. But every now and then, in the course of some such dynamical evolutions (in the course of measurements, for example), the brain of a sentient being may enter a state wherein (as we've seen) states connected with various different conscious experiences are superposed; and at such moments, the mind connected with that brain (as it were) opens its inner eye, and gazes on that brain, and that causes the entire system (brain, measuring instrument, measured system, everything) to collapse, with the usual quantum-mechanical probabilities, onto one or another of those states; and then the eye closes, and everything proceeds again in accordance with the dynamical equations of motion until the next such superposition arises, and then that mind's eye opens up again, and so on. (Albert 1992, pp. 81–82)

Albert is entertaining the view mainly for the sake of argument, and he almost immediately rejects it in the passage quoted earlier about the imprecision of consciousness. Chalmers writes more sympathetically:

> Upon observation of a superposed system, Schrödinger evolution at the moment of observation would cause the observed system to become correlated with the brain, yielding a resulting superposition of brain states and so (by psychophysical correlation) a superposition of conscious states. But such a superposition cannot occur, so one of the potential resulting conscious states is somehow selected (presumably by a nondeterministic dynamic principle at the phenomenal level). The result is that (by psychophysical correlation) a definite brain state and state of the observed object are also selected. The same might apply to the connection between consciousness and non-conscious processes in the brain: when superposed non-conscious processes threaten to affect consciousness, there will be some sort of selection. In this way, there is a causal role for consciousness in the physical world. (Chalmers 2003, pp. 262–263)

Chalmers in effect combines Wigner's suggestion that consciousness cannot superpose with Albert's suggestion that consciousness collapses its physical correlates. The key idea here is that consciousness is a superposition-resistant property and that its physical correlates therefore resist superposition too. That is, it is difficult or impossible for a subject to be in a superposition of two different states of consciousness, and this results in the collapse of physical processes that interact with consciousness.[4]

Here the relevant states are total conscious states of a subject at a time. The total conscious state of a subject is what it is like to be that subject: if what it is like to be subject A is the same as what it is like to be subject B, then A and B are in the same total conscious state. A subject's total conscious state at a time may include many aspects: visual experience, auditory experience, the experience of thought, and so on. Like position or mass or color or shape, consciousness in this form can take on many specific values. Its specific values are the vast range of possible total conscious states of a subject at a time.

This view assumes that there is a physical correlate of consciousness (PCC): a set of physical states that correlate perfectly with a system's conscious states. For simplicity, we can start by assuming a materialist view where the total conscious state and its physical correlate are identical. Things work best if we also assume that the physical correlate of consciousness (PCC) can itself be represented as a quantum observable with an associated operator. This assumption is nontrivial, as not every physical property is an observable; we return to it later. A PCC observable will have many different eigenstates corresponding to distinct total states of consciousness. This makes it straightforward to treat consciousness as a super-resistant property.

To illustrate how this works, we can again suppose an electron in a superposition of locations (again omitting amplitudes for simplicity) $|a\rangle + |b\rangle$. The electron registers on a measurement device and then the result is perceived by a human subject. Assuming the measurement device is not conscious, then at the first stage the electron and the device will go into an entangled state $|a\rangle\,|M(a)\rangle + |b\rangle\,|M(b)\rangle$. When the human looks, this result will affect the eye (E), early areas of the nervous system and brain (B), and eventually the physical correlates of consciousness (PCC). Under Schrödinger evolution, we would expect the electron, device, and subject to go into an entangled state $|a\rangle\,|M(a)\rangle\,|E(a)\rangle\,|B(a)\rangle\,|PCC(a)\rangle + |b\rangle\,|M(b)\rangle\,|E(b)\rangle\,|B(b)\rangle\,|PCC(b)\rangle$. However, this superposed state would yield a superposition of states

[4] Halvorson (2011) also argues for a picture on which mental states cannot be superposed and therefore bring about collapse in the physical world.

of consciousness. So at the point where the PCC is affected, the system will collapse. It collapses into $|a\rangle\,|M(a)\rangle\,|E(a)\rangle\,|B(a)\rangle\,|PCC(a)\rangle$ or $|b\rangle\,|M(b)\rangle\,|E(b)\rangle\,|B(b)\rangle\,|PCC(b)\rangle$, with Born rule probabilities. In effect, at the point where the measurement reaches consciousness, the electron, the measurement device, and the brain will collapse into a definite state.

On a dualist view on which consciousness merely correlates with physical properties, things are a little more complicated. We focus on forms of dualism where there are psychophysical laws correlating physical states of a system with states of consciousness. There will be a set of physical correlates of consciousness (which may be disjunctive if necessary) that are in one-to-one correspondence with total states of consciousness. A subject will be in a given state of consciousness if and only if it is in the corresponding PCC state. We can assume as before that the PCC is a quantum observable. Psychophysical laws connect unsuperposed PCC eigenstates to unsuperposed states of consciousness. They also connect superpositions of PCC states to the corresponding superpositions of states of consciousness. A given subject's PCC is in a superposition of PCC eigenstates with certain amplitudes if and only if the subject's conscious experience is in a superposition of the corresponding total states of consciousness with the same distribution of amplitudes.

On a dualist view, a fundamental principle will say that consciousness resists superposition. Whenever Schrödinger evolution plus the psychophysical laws entail that a system enters or is about to enter a superposition of total states of consciousness, the system will collapse into a definite total state of consciousness. As a result, the PCC will also collapse into an eigenstate, and other physical entities that are entangled with the PCC will collapse as described above.

One motivation for the super-resistance consciousness-collapse model is given by Wigner's suggestion that superpositions of consciousness are "absurd." That is, something about the very nature of consciousness or the concept of consciousness rules out total states where consciousness is superposed. It is certainly at least very hard to imagine subjects who are in superposed states of consciousness (at least without these states becoming total states of consciousness in their own right). If something about the nature of consciousness explains why it cannot be superposed, then this might provide a possible explanation of why collapse comes about. This explanatory motivation might be seen as a further motivation for understanding consciousness as the trigger of collapse.

Taking Wigner's motivation seriously leads to the idea that consciousness is *absolutely superposition-resistant*: that is, that it can never enter superpositions, even brief and unstable ones. Invoking absolute superposition-resistance leads to a clean and simple dynamic model for collapse involving superselection rules. Unfortunately this model leads to a fatal problem for absolute super-resistance, which we explore in the next section.

3. Superselection and the Zeno Problem

To develop super-resistance models in more detail, we can start by thinking of them independently of consciousness. In principle any observable could serve as a super-resistant observable, with distinct models of quantum mechanics arising from taking different observables to resist superposition. Later we can consider the special case where consciousness or its physical correlates serve as super-resistant observables.

The simplest (albeit fatally flawed) super-resistance model invokes superselection: the strong form of super-resistance where certain superpositions are ruled out entirely. In particular, it invokes the familiar concept of a superselection rule: a rule postulating that superpositions of a specified observable are forbidden.

Superselection rules are invoked for a number of purposes in quantum mechanics.[5] Sometimes they are postulated to analyze quantum-mechanical properties that are never found in superpositions, such as the difference in charge between a proton and a neutron. Sometimes they are used to help analyze quantum-mechanical symmetries. Sometimes they are used to help address measurement in quantum mechanics, most often through the idea that superselection can emerge through interaction with the environment by Schrödinger evolution alone.

[5] Superselection rules were introduced by Wick, Wightman, and Wigner (1952). There are many somewhat different definitions of superselection rules, analyzed thoroughly by Earman (2008). Here we use a common informal definition. Superselection rules are invoked in analyses of the measurement process by Bub (1988), Hepp (1972), Machida and Namiki (1980), and others. Thalos (1998) gives an excellent review. The most common strategy is to argue that superselection rules can emerge from the Schrödinger dynamics governing the interaction of a system with its environment. It is unclear to us whether anyone has explicitly proposed a superselection collapse interpretation, but we are open to pointers.

Here we are exploring a somewhat different idea: the idea of a *superselection collapse model*, with a fundamental superselection rule governing the collapse process. Such a model will specify a superselection observable, such that physical systems must always be in eigenstates of the operator corresponding to the observable. The associated collapse postulate says that whenever a system would otherwise enter a superposition of eigenstates (with distinct eigenvalues) of this operator (given Schrödinger dynamics alone), it instead enters a definite eigenstate, with probabilities given by the Born rule. In the special case where consciousness (or its physical correlate) is a superselection observable, then whenever consciousness would otherwise be about to enter a superposition, it must collapse to a definite state according to the Born probabilities.

To specify the dynamics better, we can first suppose that the collapse takes place at a time interval of Δt, so that if the system has evolved (according to the Schrödinger equation) in the preceding Δt into a non-eigenstate of the superselection observable, it collapses probabilistically into an eigenstate of that operator, with probabilities given by the Born rule. This yields a well-defined stochastic process. For the absolute super-resistance model, the dynamics is the limiting case of this process as Δt approaches zero.

The superselection collapse model has a dynamics that is already familiar in quantum mechanics: it is precisely the dynamics that would obtain (on a traditional measurement interpretation) if the resistant observable were being continuously measured by an outside observer. The current approach does not require that there are any outside observers, or that resistant properties themselves are ever measured, or that continuous measurement ever takes place (though to aid the imagination, one could metaphorically suppose that God is continuously measuring the resistant properties of the entire universe). All that it requires is the mathematical dynamics associated with continuous measurement of resistant properties, which is fairly straightforward.

Unfortunately, the dynamics of continuous measurement leads to a well known effect, the quantum Zeno effect, which renders any superselection collapse model empirically inadequate. The quantum Zeno effect is the effect whereby the more often one measures a quantum observable, the harder it is for the system to enter different states of that observable. In the extreme case where an observable is measured continuously, it cannot change at all.

The source of the quantum Zeno effect lies in the mathematical fact that for a system to evolve under Schrödinger evolution from some initial eigenstate of an operator to some other eigenstate of that operator, it must evolve through superpositions of eigenstates.[6] Eigenstates are orthogonal to each other, so the continuous process of Schrödinger evolution cannot evolve directly from eigenstate to eigenstate. If a system governed by this process cannot pass through superpositions of these eigenstates, then the system cannot change from one eigenstate to another. Another way to put things is that if small superpositions are permitted, an initial superposition will assign probability 1-ϵ (where ϵ is negligible) to the initial eigenstate. So if there is a measurement of this observable in the first moment, the superposition will collapse to the initial eigenstate with probability 1-ϵ. Continuous measurement will therefore force the system to remain in that initial eigenstate.

This leads to the Zeno problem for superselection collapse interpretations. If there is a superselection observable (one that can never enter superpositions), every system will remain forever in a single eigenstate of that observable. This consequence may be acceptable for standard superselection observables in physics (such as the charge difference between a proton and a neutron), but it is clearly unacceptable for observables tied to measurement that serve as triggers of the collapse process.[7] For example, if a superselection observable corresponds to the position of the pointer on a measurement device, then that pointer will be forever stuck in one location and unable to give useful measurement results.

We can illustrate the Zeno problem by taking the superselection observable to be consciousness (or its physical correlate). We know that systems have different conscious states at different times, and sometimes evolve from being unconscious to being conscious. If consciousness or its physical correlate was a superselection observable, it would obey the dynamics of continuous measurement so it could not change at all. If we started in an unconscious state, we could never become conscious. The unfortunate consequence would be that we could never wake up from a nap. Furthermore, if there is no consciousness in the early universe, then consciousness could never emerge later.[8]

[6] One could argue that this mathematical fact is the common explanation both of the Zeno effect and of the problem for superselection collapse models, rather than the Zeno effect explaining the problem. Still, the problem is still aptly called a Zeno problem, tied to the impossibility of motion.

[7] Mariam Thalos (1998, p. 538) raises a version of this problem for superselection-based accounts of measurement, arguing that if a classical quantity is governed by a superselection rule, it can never change its magnitude in evolution over time.

[8] Barry Loewer (2002) raises a different early-universe problem for consciousness-collapse theories: if the first collapse requires the universe to be in a non-null eigenstate of consciousness, then this will never happen, while if collapse is triggered by any superposition of consciousness, then the first collapse will happen too early. The absolute super-resistance model takes the second horn.

The Zeno problem is not just a problem for superselection collapse interpretations. In "Zeno Goes to Copenhagen," we argue that the Zeno problem is a serious problem for almost any measurement-collapse interpretation of quantum mechanics. Any such interpretation faces the question of whether measurement itself can enter quantum superpositions. If measurement can enter superpositions, the standard dynamics of collapse upon measurement is ill-defined, and new dynamics is required. If measurement cannot enter superpositions, the quantum Zeno effect suggests that measurements can never start or finish, at least if measurement is an observable. One way out is to deny that measurement is an observable, but this option leads to further commitments (embracing a strong form of dualism or construing measurement as a special wave-function property) that themselves require a highly revisionary approach.

In this chapter, however, we are focusing on the Zeno problem as a problem for super-resistance interpretations. To handle the Zeno problem in this framework, the obvious move is to abandon superselection (on which superpositions of the relevant observable are entirely forbidden) for a weaker version of super-resistance. An *approximately super-resistant* observable is one that can enter superpositions but nevertheless resists superposition, at least in some circumstances. On a simple version of this view, superpositions of the observable in question are unstable and they probabilistically tend to collapse over time.

To make the idea of approximate super-resistance precise, we require nonstandard physics. Fortunately, there is a wealth of resources for developing such physics in the literature on modern dynamical collapse theories (Bassi et al. 2013). In section 6, we show how these theories can be adapted to yield a model on which consciousness is approximately super-resistant. The rough idea is that as a total state of consciousness (and/or its physical correlate) enters increasingly large superpositions (where a large superposition is roughly one that gives significant amplitude to distant states), this yields higher probabilities of collapse of consciousness onto a more definite state. Admittedly it is far from clear what a superposition of states of consciousness would amount to. We return to this matter in the final section.

On this view, Loewer's "collapse too early" problem can be minimized by having conditions for consciousness that are not satisfied in the early universe (so that in its early stages, the universe will be in a null eigenstate of consciousness), and also by noting that most initial collapses when they occur will be onto a null state of consciousness. The Zeno problem as it arises for the early universe is the distinct but related problem that all collapses will be onto a null state of consciousness.

4. Integrated Information Theory

There are many ways to spell out the details of a consciousness-collapse super-resistance model. We can combine the view with many different theories of consciousness, and with various different accounts of the collapse dynamics. In what follows we spell out one way of working out some details, by combining the theory with a specific theory of consciousness (integrated information theory, or IIT) and a specific model of approximate super-resistance dynamics (inspired by Pearle's continuous spontaneous localization interpretation of quantum mechanics).

We focus on IIT for several reasons. First, it is one of the few mathematically precise theories of consciousness. Second, unlike many competitors it purports to be a fundamental theory of consciousness that offers basic and universal principles connecting consciousness to physical processes. Third, it offers a specific physical correlate for total states of consciousness, using its notion of a Q-shape (qualia shape). Fourth, it has a distance metric between total states of consciousness, which plays an important role in our framework. None of this means that we are endorsing IIT. Many objections have been made to IIT (e.g., Aaronson (2014), Bayne (2018), Barrett and Mediano (2019), Doerig et al. (2019), Hopkins and McQueen (2022)) and they raise important issues. Our approach could in principle be combined with any theory that has the four properties just listed.

IIT is a theory that associates systems with both quantitative amounts of consciousness and qualitative states of consciousness. Its systems are classical Markovian networks made up of interconnected units that interact with each other according to deterministic or probabilistic rules. Each unit can take on a number of states, and the state of the system is made up of the states of each of the units in the system.

One limitation of IIT as it stands (Barrett and Mediano (2019)) is that its assigns amounts and states of consciousness to discrete Markovian network systems but not to real physical systems. To apply it to real physical systems, we need to combine it with a mapping from physical systems to network structures. In what follows we will assume such a mapping (or some other generalization of IIT) so that IIT applies to physical systems.

IIT is derived from phenomenological axioms rather than from experimental evidence. Experimental support for it is somewhat limited to date, especially because it is impractical to measure and calculate its measures of consciousness in biological systems. However, some measurable approximations of its quantitative measures have been shown to correlate with level of consciousness, (see Massimini et al. (2005), Casarotto et al. (2016), Leung et al. (2021), and Afrasiabi et al. (2021)). Additionally, spatiotemporal patterns of integrated information (approximating IIT's qualitative measures) have been derived from brain areas and correlated with the contents of conscious perceptions of faces and other objects (Haun et al. (2017)). In any case, we will treat IIT as a potential empirical theory of consciousness. Much of our discussion should generalize to other theories.

IIT is built around the notions of information and integration. The information in a system is a measure of the extent to which the present state of a system constrains its potential past and future states. One centerpiece of IIT is its measure of integration, which it labels Φ. Φ is a measure of the extent to which the information in a system is irreducible to the information of its components. It quantifies how much the causal powers of a system fail to be accounted for by any partitioned version of it.

The simplest system with nonzero Φ is a dyad: a network AB with two interacting nodes A and B that swap their states. If A is on or off, B turns on or off at the next time step, and vice versa. In this case, AB has causal powers that are not reducible to those of A and B taken alone, and $\Phi(AB) = 1$. (We spell out the mathematics in an appendix.) By contrast, if A and B are not interacting, then the causal powers of AB are reducible to those of A and B taken alone, so $\Phi(AB) = 0$.

IIT says that a system is conscious if and only if it is a maximum of Φ: that is, if the system has higher Φ than any system nested within it and higher Φ than any system it is nested within. The amount of consciousness in a system is Φ^{max}, which is equivalent to Φ if the system is a maximum and 0 if the system is not. In what follows we drop the superscript for simplicity.

One way to combine IIT with a super-resistance model is to say that Φ is super-resistant. That is, Φ resists superposition and superpositions of Φ trigger collapse. Unfortunately, this view faces a fatal problem. It fails to suppress superpositions of qualitatively distinct conscious states with the same value of Φ. Consider a conscious subject and a screen in a dark isolated room. The screen can display green or blue. If it is put into a superposition of displaying both, then the subject will be put into a superposition of experiencing green and experiencing blue. There is no reason to assume that these experiences differ in their Φ value. But then there is no Φ superposition, and so no collapse. The subject remains in a superposition of qualitatively distinct total states of consciousness. Such a theory therefore will not yield determinate experiences for many crucial observations. The underlying problem is that Φ is not a genuine physical correlate of consciousness—that is, it is not a physical correlate of a total state of consciousness. It is merely a physical correlate of a scalar degree of consciousness, where the same degree can be present in many different conscious states.[9]

Fortunately, IIT also postulates a physical correlate of total states of consciousness. The Q-shape (Tononi 2008 calls this a "shape in Q", where Q is "qualia space"; Oizumi et al. 2014 call it a "maximal irreducible conceptual structure") of a system is an entity that serves as an abstract representation of the structure of the integrated information in a system. IIT specifies a mathematical mapping from network structures to Q-shapes. If we assume (as above) that the total physical state of a system determines a network structure, then IIT will derivatively specify a mapping from total physical states to Q-shapes.

The Q-shape of S is a set of weighted points, one for each mechanism in S. A mechanism is a subsystem m of S—that is, a nonempty set of elements of S—with $\varphi(m) > 0$ (as defined in Appendix A). If S has n elements, then it has up to $N = 2^n - 1$ mechanisms. For example, in the dyad system AB, which has two elements A and B, the subsystems are A, B, and AB, and the mechanisms are A and B. The weight associated with a mechanism m is $\varphi(m)$, a non-negative real number representing the integrated information associated with m. The point associated with m is given by two probability distributions over the 2^n states of S, the so-called maximally irreducible cause repertoire and maximally irreducible effect repertoire associated with m.

According to IIT, a system's Q-shape determines (at least nomologically) the total state of consciousness associated with that system. A Q-shape is itself a mathematical entity, and it is not obvious just how a Q-shape determines a state of consciousness. What matters most for our purposes is that according to IIT, (i) having a given Q-shape is a physically definable

9 We canvassed the idea of using Φ as an absolutely super-resistant property in an early version of this article that raised the Zeno problem for absolute super-resistance and suggested approximate super-resistance via continuous localization as a possible solution. In an article responding to our early presentation and building on the ideas there, Okon and Sebastián (2020) develop the idea that Φ could be an approximately super-resistant property using continuous localization. Okon and Sebastian respond to our current objection by saying that decoherence makes it extremely unlikely that there will be superposed conscious states with the same value of Φ. The blue/green case seems a clear case of this sort of superposition, however, as does any ensuing state resulting from interactions with their environment that makes no difference to their total state of consciousness. The dyad system discussed in the main text and the appendix gives a simple illustration of a superposition of states with different Q-shapes but with the same value of Φ. In addition, the Q-shape collapse model is much better suited for giving all aspects of consciousness a causal role, whereas the Φ-collapse model gives degree of consciousness a causal role and leaves everything else epiphenomenal.

property (we might call it physical Q-shape), (ii) Q-shape is a physical correlate of consciousness, in that any two physical systems with the same associated physical Q-shape will have the same state of consciousness. It will also be helpful to assume the stronger theses that (iii) the mathematical structure of a conscious state is given by a Q-shape (call this a system's phenomenal Q-shape) and (iv) as a matter of psychophysical law, a system has a given phenomenal Q-shape (that is, it has a conscious experience with a given structure) if and only if has the isomorphic physical Q-shape (that is, it has a physical state with the same structure as defined by IIT). These claims are far from obviously correct, but something like them seems to be intended by IIT.

As before, it does not matter too much for our purposes whether these claims of IIT are correct. It is plausible that a final mathematical theory of consciousness will specify *some* mathematical structure for consciousness (though there may be more to consciousness than its mathematical structure, as inverted qualia cases suggest). And it is plausible that this mathematical structure should be realized in some way in the physical correlates of consciousness. If necessary, we can replace Q-shape by that mathematical structure. What matters most is that there is some precise theory of consciousness for which psychophysical isomorphism principles like this are correct.

Different states of the dyad system AB discussed earlier can be associated with different Q-shapes. Consider state 10, where A is on and B is off, and state 00, where both A and B are off. As we show in the appendix, both states have $\Phi = 1$, but they are associated with distinct Q-shapes. In principle one can prepare a dyad system in a superposition of these two states 10 and 00: we might call this *Schrödinger's dyad*. If Q-shape is super-resistant, Schrödinger's dyad will be unstable and will collapse into a state with a definite Q-shape. We discuss a framework for combining IIT with quantum mechanics along these lines in the next two sections. In section 7, we discuss possible experimental tests, which are likely to rule out the simple Q-shape collapse interpretation but which suggest a program for empirically refining collapse interpretations.

5. Combining IIT with Quantum Mechanics[10]

The standard IIT framework (Oizumi, Albantakis, and Tononi (2014)) maps classical network states to Q-shapes. We have assumed a derivative mapping

[10] This section is co-authored with Johannes Kleiner (Munich Center for Mathematical Philosophy, Ludwig Maximilian University of Munich).

from classical physical states to Q-shapes. To combine IIT with quantum mechanics, we need to extend the IIT mapping so that it maps quantum physical states to Q-shapes or to superpositions of Q-shapes. The core idea of a Q-shape collapse model is that systems in superpositions of Q-shape always collapse toward having a determinate Q-shape.

To extend the IIT mapping to quantum physical states, the obvious way to proceed is to use IIT's physical definition of Q-shape to define a set of Q-shape collapse operators, one for each dimension of Q-shape. The joint eigenstates of these operators will be physical states with determinate Q-shapes.

A challenge to defining these Q-shape operators is that in the classical IIT framework, Φ and Q-shape depend on probabilities of state-transitions in a network, which may depend on the position and momentum of the system's parts. Position and momentum are noncommuting operators, so physical systems cannot be in joint eigenstates of them. High-mass systems may have precise enough position and momentum to determine Φ and Q-shape, but these quantities may not be defined for low-mass entities such as electrons in quantum systems (McQueen (2019b, p. 97)).

There are various options for addressing this challenge. We could redefine Φ and Q-shape so they depend only on positions or mass densities of elements of the system. We could also give special treatment for low-mass systems, for example modifying Φ to stipulate that $\Phi = 0$ for systems with mass below a certain threshold, or we could invoke a coarse-grained or "smeared" version of Φ and Q-shape observables, with significant smearing mainly required for systems with very low mass.

Alternatively, we can invoke newer versions of IIT that are defined over quantum states. One framework for an IIT-driven collapse model has been developed by Kremnizer and Ranchin (2015), who define a new measure of quantum integrated information QII for quantum systems. On their model, a system's QII determines the probability of collapses onto a position basis, so that systems with higher QII are more likely to collapse on to the position basis. However, Kremnizer and Ranchin's interesting model is a super-resistance theory only in a weak sense: the properties that trigger collapse (QII) are quite distinct from the collapse basis (position), and position resists superposition only in certain contexts with high QII. Also, while Kremnizer and Ranchin speculate that their quantity QII may be a measure of consciousness, this will yield at best a limited causal role for consciousness, on which the scalar amount of consciousness determines probability of collapse but the specific conscious state of a subject plays no role.

Zanardi, Tomka, and Venuti (2018) have developed a more thoroughgoing quantum-mechanical version of IIT, defining quantum mechanical operators for each IIT notion (including Q-shape as well as Φ) across a broad class of quantum-mechanical networks. (These are networks of finite-dimensional non-relativistic qudits, interacting via Markovian trace preserving completely positive maps.) Further generalizations have been given by Kleiner and Tull (2021). These models do not yet give a complete mapping from physical states to Q-shapes, but they come closer to doing this than standard IIT. In what follows, we will assume a fully developed model along these lines with a complete mapping from physical states to Q-shapes.

Quantum IIT specifies a mapping $\mathbb{E}$ from states of quantum systems to *quantum Q-shapes*. Quantum Q-shapes are are quantum analogs of classical Q-shapes, the Q-shapes invoked in standard IIT. Classical Q-shapes for an n-element system S can be represented as $N = 2^n - 1$ weighted points, one for each subsystem of S, where points are pairs of probability distributions and weights are non-negative real numbers (for a subsystem that is not a mechanism, the weight will be zero). Quantum Q-shapes likewise involve N weighted points, where points are now pairs of density operators associated with the Hilbert space of S and weights are non-negative reals. Where the space of classical Q-shapes is the Cartesian product of N copies (one for each subsystem) of $Pr(S) \times Pr(S) \times R_0^+$, the space of quantum Q-shapes is the Cartesian product of N copies of $D(S) \times D(S) \times R_0^+$. Here $Pr(S)$ is the space of probability distributions over S, whose quantum analog $D(S)$ is the space of density operators over S. R_0^+ is the set of non-negative real numbers.[11]

If there is a quasi-classical basis $|s_i\rangle$ of the Hilbert space of S, then there is a natural mapping from classical Q-shapes to a subclass of quantum Q-shapes, deriving from a mapping from $Pr(S)$ to $D(S)$, defined as follows: $(p(s_i)) \mapsto \sum_i p(s_i)\,|s_i\rangle\langle s_i|$. We can call this distinguished subclass of quantum Q-shapes the *quasi-classical* Q-shapes. Any quantum Q-shape can be seen as a superposition of quasi-classical Q-shapes.

Quantum IIT as it stands does not say much about how quantum Q-shapes correspond to states of consciousness. For our purposes we can add the further claims that (i) quasi-classical Q-shapes correspond to determinate states of consciousness, exactly as the corresponding classical Q-shapes

[11] If S is a network of elements with binary states, each weighted point will have $2^{n+1} + 1$ dimensions (two 2^n-dimensional probability spaces plus a real number), so classical Q-space has $(2^n - 1)(2^{n+1} + 1)$ dimensions. In quantum IIT, the 2^n dimensional probability-spaces are replaced by 2^{2n}-dimensional density spaces, so quantum Q-space has $(2^n - 1)(2^{2n+1} + 1)$ dimensions.

do in classical IIT, and (ii) other quantum Q-shapes are superpositions of quasi-classical Q-shapes and correspond to superpositions of the corresponding states of consciousness.

We can define the quasi-classical *states* of a quantum system as those quantum states that quantum IIT associates (via the mapping $\mathbb{E}$) with a quasi-classical Q-shape. If $\mathcal{C}$ is the class of quasi-classical Q-shapes, the class of quasi-classical quantum states is $\mathbb{E}^{-1}(\mathcal{C})$, the preimage of $\mathcal{C}$ under $\mathbb{E}$. Every state of a quantum system can then be represented as a superposition of quasi-classical states, and its associated Q-shape will be a superposition of the corresponding quasi-classical Q-shapes. We can then set up collapse operators so that quantum systems always collapse toward these quasi-classical states with quasi-classical Q-shapes.

One limitation of quantum IIT as it currently stands is that these quasi-classical states (picked out as those that quantum IIT associates with quasi-classical Q-shapes) may not closely correspond to what we usually think of as quasi-classical quantum states such as mass density eigenstates. As a result, the Q-shape collapse dynamics need not lead to collapse toward standard "classical" states such as mass density eigenstates and may result in a superposition of these states (along with a relatively determinate state of consciousness). If we want to avoid these quantum superpositions as physical correlates of determinate consciousness, there is at least a research program of developing a version of quantum IIT on which quasi-classical Q-shapes and determinate states of consciousness are associated with more "classical" quantum states. In what follows it may be helpful to assume such a version of the framework.

We can now define Q-shape collapse operators. Recall that a Q-shape is a weighted point in the direct product of 2N copies of the density operator space $D(S)$. Any density operator in $D(S)$ can be represented (in the quasi-classical basis $|s_i\rangle$) as

$$\rho = \sum_{i,j} c_{ij} |s_i\rangle \langle s_j| \,. \tag{1}$$

The Q-shape for any given quantum state ψ consists of $2N$ density operators of this kind and N non-negative real numbers. The Q-shape can therefore be represented by $2N$ sets of coefficients c^k_{ij} which we denote as $c^k_{ij}(\psi)$ (for $k = 1, \ldots, 2N$), and N non-negative real numbers which we denote $\varphi^k(\psi)$ (for $k = 1, \ldots, N$). For notational simplicity, we duplicate each of the latter, so that for each $k = 1, \ldots, 2N$, we have a $c^k_{ij}(\psi)$ which describes the first or second factor in $D(S) \times D(S) \times R_0^+$ and a $\varphi^k(\psi)$ which describes the third factor.

We can then define an ensemble of orthogonal self-adjoint collapse operators as follows:

$$\hat{Q}^k_{ij} := \sum_{\psi \in \mathbb{E}^{-1}(\mathcal{C})} \varphi^k(\psi)((c^k_{ij}(\psi) + (c^k_{ji}(\psi)) |\psi\rangle \langle\psi| . \qquad (2)$$

The sum has been restricted so that it runs over the class $\mathbb{E}^{-1}(\mathcal{C})$ of quasi-classical quantum states, that is, those whose Q-shapes are quasi-classical. The index k ranges from 1 to $2N$ (where $N = 2^n - 1$) and i and j range over the dimension of the Hilbert space of S. If S consists of n components, each of which has a two-dimensional Hilbert space, the system will be associated with $2^{2n+1}(2^n - 1)$ collapse operators.[12]

6. Continuous Collapse Dynamics

To complete our picture of super-resistant consciousness-based collapse, we need an account of the dynamics of super-resistant collapse. Fortunately, there exist models of dynamic collapse (due to Philip Pearle and Lajos Diósi, among others) that can be generalized to model the continuous collapse of any observable. It is not difficult to adapt these models to model the continuous collapse of consciousness and its physical correlates such as Q-shape. We start by informally reviewing these models and the adaptation to consciousness-collapse models, before providing formal details.[13]

We start with the continuous spontaneous localization (CSL) model due to Pearle (1976, 1999, 2021), with related contributions by Gisin (1984, 1989). Pearle's model is a continuous relative of the well known GRW model, on which the position of isolated particles undergo spontaneous localization of position with low probability at any given time. On CSL, wave functions undergo a gradual stochastic collapse process at all times. The

[12] For reasons tied to the role of weights within IIT, we have combined the real weights φ^k and the density operator coefficients c^k_{ij} in defining the Q-shape collapse operators. As a result there are N fewer collapse operators than dimensions of Q-space. Alternatively one can define separate operators for the coefficients and real weights as follows: $\hat{Q}^k_{ij} := \sum_{\psi \in \mathbb{E}^{-1}(\mathcal{C})}(c^k_{ij}(\psi) + c^k_{ji}(\psi)) |\psi\rangle \langle\psi|$ and $\hat{B}^l := \sum_{\psi \in \mathbb{E}^{-1}(\mathcal{C})} \varphi^l(\psi) |\psi\rangle \langle\psi|$.

[13] Thanks to Maaneli Derakhshani, Philip Pearle, and Johannes Kleiner for their extensive help with the material in this section.

model provides continuous collapse onto mass density: the amount of mass present at various locations. It provides a dynamics by which superpositions of mass density gradually collapse toward definite states of mass density, with faster collapse in high-mass systems. In effect, CSL is a model on which mass density is super-resistant.

Pearle's model can be informally motivated by an analogy between gradual collapse and the gambler's ruin game in classical probability theory (Pearle 1982). In the gambler's ruin, a number of gamblers play against each other until all but one of them is "wiped out." Consider two gamblers, G_1 and G_2, who have \$100 between them such that G_1 has \$60 and G_2 has \$40. They toss a coin: if heads G_1 gives a dollar to G_2, if tails G_2 gives a dollar to G_1. As they keep playing, their respective amounts fluctuate, but the total remains the same. Eventually, the game ends, as one player acquires \$100. It turns out that G_1 wins 60% of the time while G_2 wins 40% of the time. That is, the probability that a given gambler wins is determined by the initial stakes.

In CSL, the squared amplitudes in a superposition (in the preferred basis) play a continuous stochastic gambler's ruin game against each other, fluctuating up and down until one "wins," thereby completing the collapse. The probability that a given state vector "wins" a collapse in the long run is determined by its initial squared amplitude according to the Born rule. Crucially, we may control the speed at which the games are played in terms of certain (experimentally bounded) parameters. This allows large superpositions to collapse quickly and small superpositions to collapse at a negligible rate.

Like the GRW theory, Pearle's theory involves a weak sort of super-resistance. Mass density resists superposition weakly, in that an isolated particle will only gradually collapse toward a definite position and so a definite mass density. At the fundamental level, superpositions of mass density will be ubiquitous. However, when many particles are entangled in a macroscopic system, the mass density of the system as a whole will collapse extremely fast, so that we will never encounter macroscopic systems in large superpositions of mass density.

Continuous collapse models can be adapted to work with super-resistant properties other than position and mass density. Given any observable, we can postulate a continuous collapse process with a version of the Pearle dynamics applied to this observable. Squared amplitudes for eigenstates of the observable engage in a stochastic gambler's ruin process, so that systems in superpositions of the observable collapse quickly or slowly toward their eigenstates via a gamblers-ruin process.

A related collapse process is postulated in the Penrose (2014) model of gravitational collapse, where spacetime curvature is super-resistant. Superpositions of spacetime curvature collapse onto definite states. Unlike Pearle, Penrose does not give a fully defined dynamics for collapse. He defines a superposition lifetime, $\hbar/\Delta E_G$, where $\hbar$ is Planck's constant and ΔE_G is the gravitational self-energy of the difference between the mass distributions belonging to the two states in the superposition. But the dynamics of collapse during this lifetime are not specified.[14]

An account of the dynamics of gravitational collapse has been independently provided by Lajos Diósi (1987). Diósi sets out a stochastic version of the Schrödinger equation on which there is a continuous collapse process onto spacetime structure. Diósi's dynamic collapse process is closely related to Pearle's continuous spontaneous localization process, with some differences arising from the use of a collapse onto gravitational structure as opposed to just mass density, and from what the collapse rates depend on.

It turns out that the Diósi and Pearle dynamics are both instances of a general formulation of continuous collapse dynamics which can be applied to any collapse operator. Such a formulation has been presented by Angelo Bassi and coauthors (2017).[15] We will adapt this formulation to set out a dynamics for continuous collapse onto consciousness.

In the context of IIT, we can use this general dynamics to develop a view on which Q-shape is super-resistant. Informally: Suppose a system is in a superposition of two Q-shapes, each with an associated amplitude. We can stipulate a "localization" dynamics for this superposition that works much like Pearle's except that collapse is toward eigenstates of Q-shape. The amplitudes trade off probabilistically with each other over time, in effect playing gambler's ruin at a rate proportional to the distance between the two Q-shapes. In the long run, the system will collapse onto a specific Q-shape with probability given by its initial squared amplitude.

[14] The Hameroff and Penrose (2014) "Orch OR" model extends Penrose's model of collapse into a model of consciousness. The Penrose-Hameroff model is not a consciousness-collapse model either: Penrose and Hameroff hold that collapse is triggered by superpositions of spacetime curvature rather than by consciousness or measurement, and that collapse causes consciousness rather than vice versa. Our approach might be considered a distant cousin of the Penrose-Hameroff model, with the main differences on our approach being: (i) consciousness causes collapse rather than vice versa, (ii) collapse is onto Q-shape rather than onto spacetime curvature, (iii) the collapse dynamics corresponds somewhat more closely to Pearle's model rather than Diósi-Penrose's, and (iv) as discussed later, we make no claims about quantum coherence and quantum computation in the brain.

[15] See also Pearle (1999, eqn.10) and Bassi et al. (2013, eqn.14).

We can spell out the mathematical details as follows. The general framework for continuous collapse rests on using a modified version of the Schrödinger equation that includes a nonlinear and stochastic term for collapse as well as the standard linear deterministic evolution. To be consistent and compatible with constraints such as no superluminal signalling, nonlinear modifications to the Schrödinger equation must take a highly constrained stochastic form. This yields the following general form for continuous collapse models (Bassi et al. 2017, p. 27):

$$d\psi_t = [-i\hat{H}_0 dt + \sqrt{\lambda}(\hat{A} - \langle\hat{A}\rangle_t)dW_t - \frac{\lambda}{2}(\hat{A} - \langle\hat{A}\rangle_t)^2 dt]\psi_t. \tag{3}$$

Here ψ_t is the wave function state at t, $\hat{H}_0$ is the Hamiltonian, λ is a real-valued parameter governing collapse rate, $\hat{A}$ is a collapse operator, $\langle\hat{A}\rangle_t$ is its expected value at t, and W_t is a noise function allowing for stochastic behavior. The equation allows continuous stochastic collapse toward an eigenstate of the operator $\hat{A}$ at a rate governed by λ and W, with probabilities given by the Born rule.

It is straightforward to generalize this equation to multiple collapse operators.[16] Using our Q-shape collapse operators defined in (2), we can propose the following dynamics:

$$\begin{aligned} d\psi_t = [-\frac{i}{\hbar}\hat{H}dt + \sqrt{\lambda}\sum_{\alpha}(\hat{Q}_\alpha - \langle\hat{Q}_\alpha\rangle_t)dW_{\alpha,t} \\ -\frac{\lambda}{2}\sum_{\alpha}(\hat{Q}_\alpha - \langle\hat{Q}_\alpha\rangle_t)^2 dt]\psi_t. \end{aligned} \tag{4}$$

Here $\alpha = i, j, k$ is a multi-index that comprises the indices in (2). If there is little difference in the superposed Q-shapes, then the first term on the right hand side (representing Schrödinger evolution) dominates. Otherwise, the system collapses toward a joint eigenstate of the collapse operators, at a rate proportional to the sum of the difference between their eigenvalues.

The noise function $W_{\alpha,t}$ is responsible for the stochastic "gambler's ruin" collapse behavior described earlier.[17] In CSL and other mass density collapse models, the collapse operators correspond to local mass densities $\hat{m}(x)$ (the amount of mass at location x). The CSL noise function is given by Wiener

[16] Bassi et al. (2013, eqn. 36).

[17] For a simple illustration of how this works, see Pearle (1999, sec. 2.2).

processes $W_t(x)$, representing Brownian motion through time at location x. The noise at different spatial locations x and y is correlated by a spatial correlation function $G(x - y)$, which in CSL is a Gaussian function of the distance between x and y. This ensures that collapse rate depends on the distance between mass density distributions.

In our IIT-based collapse model, we can define the collapse rate so that it depends on the extended Earth movers distance EMD* between Q-shapes (see the appendix). To ensure this, we can stipulate that the spatial correlation function involved in the noise functions $W_t(x)$ is defined in terms of Earth movers' distance: specifically, $G(x, y) = 1/EMD^*(x, y)$ (with an appropriate cut-off for when $EMD^*(x, y)$ is small or zero; we omit the details).

In CSL it is also standard to "smear" the mass density operator with the same Gaussian $G(x - y)$, so that collapse is onto smeared mass density eigenstates rather than precise mass density eigenstates, thereby avoiding large violations of energy conservation.[18] Our equation is simpler because our collapse operators do not correspond to points (or smeared regions) in a continuous space but instead correspond to a discrete set of mechanisms. We therefore do not need to include a smearing function in our equation. In principle, however, it is straightforward to add such a smearing function as in mass density models.

Our equation (4) assumes that all superposed Q-shapes are Q-shapes of a single system (network of units) with a fixed number of units and a fixed causal structure. It does not address the case where we have a superposition involving Q-shapes of systems with different numbers of units or causal structures. Extending the current framework to handle those cases is a further project.

The overall theory may look complex, but the underlying principles are fairly simple. First, there is an IIT-style quasi-classical psychophysical theory linking physical Q-shape by a structural isomorphism to phenomenal Q-shape in states of consciousness. Second, there is a generalization of this theory to the quantum realm, so that superpositions of physical Q-shape are linked to superpositions of phenomenal Q-shape and so to superpositions of states of consciousness. Third, there is the key claim that consciousness is super-resistant. More specifically, phenomenal Q-shapes

[18] Smearing the mass density operator results in the following equation: $d\psi_t = [-\frac{i}{\hbar}\hat{H}dt + \sqrt{\lambda}\int d^3x(\hat{m}(\mathbf{x}-\langle\hat{m}(\mathbf{x})\rangle_t)dW_t(\mathbf{x}) - \frac{\lambda}{2}\int d^3x\int d^3yG(\mathbf{x}-\mathbf{y})(\hat{m}(\mathbf{x})-\langle\hat{m}(\mathbf{x})\rangle_t)(\hat{m}(\mathbf{y})-\langle\hat{m}(\mathbf{y})\rangle_t)dt]\psi_t$.

resist superposition via a Pearle-style principle of continuous collapse for Q-shapes, so that superpositions of consciousness rapidly become more determinate. Putting these elements together: superpositions in the environment lead to superpositions of Q-shape in the brain, which lead to superpositions of consciousness. These superpositions of consciousness will rapidly collapse, yielding collapse in the correlated Q-shapes and collapse in the brain states and the environmental states that are entangled with Q-shape.

7. Experimental Tests

Different super-resistant collapse models make different predictions. For any proposed super-resistant property, in principle it is possible (though usually extremely difficult) to test whether a system is in a superposition of that property. This means that in principle (although not yet in practice) it is possible to test which systems can collapse quantum wave functions, and in virtue of which of their properties. For example, in principle we can test whether atoms, molecules, cells, worms, mice, dogs, or humans, as well as oscilloscopes, computers, and other devices have the capacity to collapse a wave function.[19]

To test whether a given property supports superpositions, one can use an interferometer for this property, which detects interference between superposed quantities in much the same way that a double-slit experiment detects interference between superposed positions. In practice it is extraordinarily difficult to set up interferometers for complex properties instantiated by complex systems, because of the need to prepare the relevant system in complete isolation from environmental effects. To date, the most complex such measurements have detected interference in large molecules with around 2000 atoms (Fein, Geyer, Zwick, et al. 2019). Current limitations are practical rather than principled, and measurements for more complex properties are certainly possible in principle.

[19] It is occasionally suggested that we know from existing results that ordinary measuring devices collapse the wave function, perhaps because we always find them in definite states, or because their measurements do not lead to quantum interference. However, it is easy to see that these observations are all equally consistent with a view on which only humans (say) collapse wave functions, and measurement devices are observed by humans and entangled with their environment. Sophisticated variants of this objection are made by Koch and Hepp (2006) and Carpenter and Anderson (2006). Okon and Sebastián (2016) explain what goes wrong in these objections.

These tests have clear implications for super-resistance models. In absolute super-resistance models, superpositions of super-resistant observables are impossible. In approximate super-resistance models, these superpositions are unstable. So at least on a first approximation: if we detect widespread superpositions of an observable, that tends to disconfirm models on which that observable is super-resistant.

On a second approximation, all this depends on just how unstable the superpositions are. We can distinguish *fast-collapse* models on which large superpositions of a super-resistant observable are rare, from *slow-collapse* models on which large superpositions are common. Here a large superposition of an observable is a superposition of significantly different eigenstates of the observable with significant amplitudes for significant periods (where "significant" is a placeholder for now). If we frequently detect large superpositions of an observable, this tends to disconfirm at least fast-collapse super-resistance models involving that observable. These results do not disconfirm slow-collapse models as easily. Still, where consciousness-collapse models are concerned, fast-collapse models are arguably preferable to slow-collapse models, as the latter allow that large superpositions of conscious states are common. So for now, we will focus on fast-collapse models, returning to slow-collapse models shortly.

We may already be in a position to test fast-collapse models in which Q-shape is super-resistant. This project is aided by the fact that even quite simple systems (such as a dyad) can have nonzero Φ and nontrivial Q-shapes, as we have seen. To test the hypothesis, we need only prepare a quantum computer to enter superpositions of Q-shape. The simplest example is Schrödinger's dyad (from section 4): two units A and B in a superposition of connected and disconnected states with distinct Q-shapes. If we find the interference effects predicted by standard quantum mechanics (which assumes that simple systems do not perform measurements and evolve according to Schrödinger dynamics), this will falsify the hypothesis that Q-shape is super-resistant, at least on a fast-collapse model. If we do not find these effects, this will suggest that these superpositions are impossible or unstable and will tend to support the hypothesis that Q-shape is super-resistant.

Something along these lines could be done with a quantum version of a Fredkin crossover gate.[20] A classical Fredkin gate involves three bits, a control bit and two other bits A and B. If the control bit is 1, bits A and B are swapped. If the control bit is 0, bits A and B are left as is. In a quantum

[20] Thanks to Scott Aaronson for this suggestion.

version of the Fredkin gate, the control bit can be in superposition, and the AB system will then be in a superposition of bit-swapping and staying constant. As a result, IIT appears to suggest that the AB system will be in a Q-shape superposition. If Q-shape is super-resistant in a fast-collapse model, we should expect this superposition to collapse.

In fact, a quantum Fredkin gate has recently been constructed (Patel et al. 2016), with results indicating a successful superposition. However, in this example, it does not seem that the conditions for $\Phi(AB)=1$ are met, because there is no two-way feedback interaction between gates A and B. In IIT, purely feedforward networks typically have zero Φ. A feedforward network can have nonzero Φ if it has overlapping inputs and overlapping outputs, but this does not appear to be happening in the quantum Fredkin gate.[21]

How might we properly construct feedback systems such as AB using quantum computers? In the quantum computing literature, two primary types of quantum feedback are distinguished. The traditional type is *measurement-based feedback*. Here, a quantum system performs some (usually feedforward) processing and is measured, and the measurement result is then fed back into the quantum system as input. This will not help for our purposes. A more recent development is *coherent quantum feedback* (Lloyd 2000), where feedback connectivity obtains in the quantum system itself. Superpositions of coherent quantum feedback could be used to build our dyad system in a superposition of states.

For example, consider the ion-trap example discussed by Lloyd (2000, p. 4). The initial state of the system is $|\psi\rangle_s|0\rangle_m|\phi\rangle_c$, where $|\psi\rangle_s$ is the unknown state of the "system" ion, $|\phi\rangle_\iota$ is the prepared state of the "controller" ion and $|0\rangle_m$ is the vibrational mode cooled to its ground state. Lloyd explains how certain directed pulses can evolve the system from $|\psi\rangle_s|0\rangle_m|\phi\rangle_c$ to $|\downarrow\rangle_s|\psi'\rangle_m|\phi\rangle_c$, to $|\downarrow\rangle_s|\phi'\rangle_m|\psi\rangle_c$, and finally to $|\phi\rangle_s|0\rangle_m|\psi\rangle_c$. In effect, the initial unknown state of the system ion is swapped with the initial state of the controller ion. Schrödinger's dyad may then be constructed by putting the input pulses into a superposition of implementing this swap and not implementing this swap, yielding: $\alpha|\psi\rangle_s|0\rangle_m|\phi\rangle_c + \beta|\phi\rangle_s|0\rangle_m|\psi\rangle_c$. If the two terms in the superposition yield distinct Q-shapes, then our model

[21] This points to another test case that can be realized by a quantum computer. Perhaps the simplest feedforward system with nonzero Φ is a dyad system CD that forms a layer of a feedforward network, whereby a node from a previous layer gives input to both C and D, and both C and D give input to a node in a subsequent layer. For illustration, see Oizumi, Albantakis, and Tononi (2014, Fig. 7(B)).

predicts that this superposition is unstable and will eventually collapse, even if the system remains isolated.

The issue is not entirely straightforward, as it might be denied that the full conditions for $\Phi(AB)=1$ are met (perhaps because of the role of the vibrational mode or the pulses). Still, it seems likely that some technologically feasible quantum computation involves a superposition of Q-shapes. If found, such a superposition will falsify the combination of standard IIT (on which Q-shape is the physical correlate of consciousness) and the fast-collapse consciousness-collapse thesis.

More generally, most proponents of quantum computing predict that superposed states in larger and larger systems will gradually be demonstrated. It would be foolhardy to bet against these predictions. In the face of these results, one could maintain an IIT-collapse view by modifying IIT somewhat: for example to say that a system is conscious (and has a Q-shape) only when Φ is above a certain threshold, or by adding other constraints to the definition of Φ so that the relevant simple systems have $\Phi = 0$. Alternatively one could adopt a slow-collapse version of the model; one could reject IIT entirely for a different theory of consciousness; or one could reject the consciousness-collapse thesis. Still, this shows how even near-term experimental results from quantum mechanics can have some bearing on theories of consciousness.

All this brings out that the consciousness-collapse thesis in its fast-collapse version is not easy to combine with panpsychist theories of consciousness on which consciousness is found even in very simple systems. A strong panpsychist fast-collapse view on which position or mass or charge quickly collapses the wave function is straightforwardly refuted by standard experimental results showing interference effects. The more recent results of Fein et al. demonstrating superpositions of position in 2000-atom systems tend to suggest that the threshold for collapse lies somewhere beyond that level. There are some quasi-panpsychist collapse views involving slightly more complex properties distinct from position that have not yet been tested, but we should easily enough be able to test them as above, and few would expect them to be supported. The consciousness-collapse thesis (in fast-collapse versions) tends to fit more comfortably with non-panpsychist views on which consciousness arises only in relatively complex systems. These views are consistent with existing and likely near-term-future observations, while still being subject to experimental test eventually.

There remains the possibility of slow-collapse models on which superpositions of consciousness tend to collapse slowly across long periods. If these models allow widespread large superpositions of human states of consciousness, these views are hard to reconcile with introspection, and it also becomes less clear why we should accept the consciousness-collapse view over an Everett-style view where one's consciousness is constantly in large superpositions. Perhaps there could be a CSL-style slow-collapse panpsychist model on which superpositions of consciousness are common but unstable at the microphysical level, in the way that superpositions of mass distribution are common but unstable at the microphysical level in CSL. In CSL, large superpositions of macroscopic mass distributions are nevertheless uncommon. Likewise, a panpsychist slow-collapse view might have the consequence that large superpositions of human consciousness are uncommon, especially on a constitutive panpsychist view on which human consciousness is constituted by patterns of microconsciousness. Such a view will face the notorious combination problem of how this constitution works, and it may also have less of an irreducible causal role for human consciousness than other collapse views. Still, there are various versions of a slow-collapse model worth exploring.

There are also empirical constraints on super-resistance models tied to energy conservation (collapses tend to produce excess energy, so they cannot be too frequent or too dramatic[22]) and to the quantum Zeno effect (a super-resistance model must allow superpositions to persist long enough to avoid Zeno effects, while not persisting so long that measurements do not have definite outcomes). All these phenomena impose constraints that narrow the class of available super-resistance models: super-resistant properties are not too simple and not too complex, while collapses are not too frequent and not too slow.

For a super-resistance model to be empirically supported, we will eventually have to find systems and properties that resist superposition. One key (if currently far-fetched) experiment would use an interferometer on a human isolated from their environment, preparing them to enter a superposition of

[22] The main difficulty in the experimental detection of such effects involves controlling all the possible ways of cooling. Thus, in their discussion of testing GRW and CSL, Feldmann and Tumulka (2012) consider the Kubacher Kristallhöhle, the largest natural cave in Germany, which is 9°C all year around. When surface temperatures are low, heat spontaneously created in the cave cannot be transported away, thereby suggesting a way of obtaining an empirical bound on the rate of spontaneous warming. It is much more difficult to see how we could find empirical bounds on spontaneous warming in conscious systems, but it may not be impossible.

conscious states and seeing if interference effects are observed. If interference effects are not observed, one will have experimental support for the claim that humans can collapse wave functions. As before this would not decisively demonstrate that consciousness is doing the work, but it would give reason to take that view seriously. If interference effects are not observed, one will have experimental support for the claim that humans cannot collapse wave functions. This will also tend to falsify any measurement-collapse formulation of quantum mechanics, and in particular will tend to falsify the view that consciousness collapses the wave function. In this way the framework of this chapter may ultimately be subject to empirical test.

Admittedly, it is not clear that it will ever be possible to isolate and test a conscious human brain in this way. Perhaps somewhat more feasible in the long term could be running a detailed simulation of a human brain on a quantum computer. If interference effects are not observed, one will have experimental support for the claim that the computational structure of the human brain can collapse wave functions. If they are observed, one will have evidence against this claim. However, this result will leave open the hypothesis that other features of the human brain that are not replicated in a simulation, such as biological features, are responsible for wave-function collapse. It may be especially difficult to test biological collapse models, as many standard methods of isolating systems to test for superposition require low temperatures where the biology may break down. Still, these quantum computing experiments might at least give us evidence for or against a consciousness-collapse model where the correlates of consciousness are computational. In the long run, advances in quantum computing are likely to heavily constrain the prospects for consciousness-collapse models.

8. The Causal Role of Consciousness

On the picture we have sketched, superpositions of physical Q-shape drive collapse. How does this yield a causal role for consciousness?

On a materialist view which identifies physical Q-shape (a physical property) with phenomenal Q-shape (a property of consciousness), the causal role is straightforward. Superpositions of consciousness involve superpositions of phenomenal Q-shapes, which trigger collapse onto more definite phenomenal Q-shapes, which are themselves more definite physical Q-shapes, leading to more definite physical consequences.

On a dualist view, physical Q-shape may be ontologically distinct from phenomenal Q-shape, so a causal role for the former is not yet a causal role for consciousness. The simplest way to derive a causal role for phenomenal Q-shape is to assume (i) that consciousness has a quantum structure whereby subjects are in superpositions of phenomenal Q-shapes iff they are in corresponding superpositions of physical Q-shapes, and (ii) a fundamental principle saying that phenomenal Q-shape is super-resistant and obeys the collapse dynamics we have developed. When subjects are in superpositions of phenomenal Q-shapes, these Q-shapes collapse according to the dynamics. Phenomenal Q-shapes are perfectly correlated with physical Q-shapes, so collapse of phenomenal Q-shapes leads to collapse of physical Q-shapes, and the standard ensuing physical effects of collapse.

Someone might object that we do not give a genuine causal role to nonphysical consciousness at all. Instead, all the causal work is done by the physical correlates of consciousness.

One version of this objection notes that on a dualist consciousness-collapse interpretation, there will be PCC states (e.g., physical Q-shapes, on the IIT framework) that correlate perfectly with consciousness. One can then develop a physicalist collapse interpretation on which the primary locus of superposition-resistance is the PCC states. Collapse of the PCC states does all the causal work, and collapse of consciousness is causally irrelevant. There will at least be a possible world (we might think of it as a quantum zombie world) where collapse works this way. In that world, the physical wave function will evolve just as in our world. So even in our world, consciousness may seem redundant.

In response: on the dualist interpretation spelled out above, it is consciousness that directly causes the wave function to collapse. There is a fundamental principle saying that consciousness resists superposition. (In the IIT framework, phenomenal Q-shapes resist superposition.) This leads to probabilistic collapse toward determinate states of consciousness. This collapse of consciousness brings about physical collapse to a more determinate PCC state, because of a psychophysical law ensuring that states of consciousness and their physical correlates (in the IIT framework, phenomenal Q-shapes and physical Q-shapes) are always in alignment. So consciousness is causally responsible for collapse in our world. There may be other models where physical correlates cause collapse directly, but that is not how things work on the dualist interpretation we have specified.

The quantum zombie scenario does suggest that there is a sort of structural/mathematical explanation that might be given for our actions without mentioning consciousness. Still (as is familiar from discussions of panpsychism and Russellian monism), this structural explanation would not provide a complete explanation of our actions, precisely because it leaves out the role of consciousness in grounding that structure. Like many structural explanations, it leaves out the actual causes. In the actual world consciousness is causing the relevant behavior, and consciousness may explain why it is that we behave determinately at all.

A related objection asks: In the actual world, how do we know that it is consciousness that triggers collapse, and not its physical correlates? As we discussed in the last section, if there is a perfect correlation between the two, these hypotheses cannot be distinguished experimentally. Still, insofar as we already have reason to believe that consciousness is a fundamental property, then the hypothesis that consciousness triggers collapse has at least two advantages. First, this way the fundamental law of collapse involves a fundamental property. Second, this way we have a causal role for consciousness, cohering with a strong pretheoretical desideratum. These virtues give reasons to favor the view over the alternative.

One might also object that even if our models give consciousness a causal role, they do not give consciousness the kind of causal role that we pretheoretically would expect it to have. One worry is that collapsing consciousness may affect the objects we perceive, but we want consciousness to affect action, producing intelligent behavior and verbal reports such as 'I am conscious.'

One worry is that the most obvious effects of collapse point the wrong way: collapse of consciousness will collapse perceived objects such as measurement instruments, but what we want is for consciousness to affect action. In response, we can note that a collapse of consciousness will collapse an associated PCC state in the brain, and this brain state will be entangled with action states or will at least cause a corresponding action state, so a collapse of consciousness will help bring about a determinate action. For example, if consciousness probabilistically collapses into an experience of red rather than an experience of blue, this collapse will bring about a PCC state associated with experience of red, which will tend to lead to an utterance of 'I am experiencing red' rather than 'I am experiencing blue.'

Furthermore, consciousness also involves the experience of agency and action: say, the experience of choosing to lift one's left hand rather than one's right hand. Superpositions of these states will collapse into definite states, which will lead to actions such as raising one's left hand.

This picture naturally raises issues about free will. On this view, the experience of choice plays a nondeterministic causal role in bringing about action. On some popular conceptions of "free will," on which what matters for free will is nondeterminism and a role for consciousness, this picture may vindicate free will in the relevant sense. Others may object that the choices are themselves selected probabilistically, and that random choices are no better than deterministic choices when it comes to free will. We think the issues are far from straightforward, so we will set aside issues about free will here, but we note that a causal role for consciousness can be expected to have some bearing on those issues.

Another objection is that if consciousness always collapses via the Born role, then any effect of consciousness on action will at best be a sort of dice-rolling role. It will probabilistically select between different available outcomes, but it will not yield a qualitatively special outcome. Under a hypothesis where PCC states collapse the wave function, purely physical quantum zombies would have behaved the same way. So consciousness will not make outcomes on which humans behave intelligently or on which they say 'I am conscious' any more likely than they would have been if some other property had collapsed the wave function. One might even simulate the dynamics in a classical computer (with a pseudorandom number generator), with no role for consciousness, and the same patterns of behavior would ensue.

Most of what this objector says is correct. The quantum zombie scenario suggests that there is a sort of structural/mathematical explanation that might be given for our actions without mentioning consciousness. Still, this structural explanation would not provide a complete explanation of our actions, precisely because it leaves out the role of consciousness in grounding that structure. (Like many structural explanations, it leaves out the actual causes.) In the actual world consciousness is causing the relevant behavior, and consciousness may explain why it is that we behave determinately at all. One might have liked a stronger, more transformative causal role for consciousness that could not even in principle have been duplicated without consciousness, but it is not clear why such a role is essential.

If one does want a stronger role for consciousness, the most obvious move is to suggest that the role for consciousness in collapse is not entirely constrained by the Born probabilities. Perhaps perceptual consciousness obeys those constraints (thereby explaining our observations in quantum experiments), but agentive experience does not. For example, collapses due to agentive experience might be biased in such a way that more "intelligent" choices that lead to more intelligent behavior tend to be favored than they would be according to the Born rule. This picture sacrifices the great simplicity of the original quantum dynamics, and it could perhaps be disconfirmed through the right sort of experiments and simulations, but it is arguable that our current evidence leaves room open for it. We do not find this picture especially attractive, but it is at least worth putting it onto the table.

9. Philosophical Objections

We have already considered many objections to our account. Some are technical issues specific to the use of IIT: for example, whether IIT applies to real physical states, whether Q-shape operators can be defined, and whether a Q-shape/collapse theory has already been falsified by existing experimental results. These are serious issues that may require modifying IIT or moving to a different theory of the physical correlates of consciousness. Some are versions of objections that arise for many objective collapse theories: for example, consistency with relativity and the tails problem. These are also serious issues that we have set aside for now with the preliminary aim of getting consciousness-collapse models closer to the level of seriousness of existing objective collapse theories. Another technical issue is whether the parameters of a consciousness-collapse theory can be set to avoid the Zeno effect.

In this final section we consider a number of philosophical objections. We have already considered objections concerning the causal role of consciousness. The largest objection remaining concerns superposed states of consciousness.

Objection 1: What is a superposed state of consciousness?
As we saw earlier, Wigner said that it is "absurd" to suppose that a subject could be in a state of "suspended animation," that is, in a superposition of multiple states of consciousness. However, the approximate super-resistance model we have developed requires that subjects can be in such superposed states. Large superpositions of consciousness (those between significantly different states with significant amplitude for significant periods) will be rare, at least on a fast-collapse model, but they will be possible. Small

superpositions of consciousness (those that are like large superpositions except that they are brief, or low-amplitude, or between closely related states) may be ubiquitous. In fact, on these models it may be that most or all conscious subjects are in small superpositions of consciousness most or all of the time. This raises the questions: are superpositions of consciousness possible, and if so how can we understand them?

There are a few different ways of trying to understand superposed states of consciousness (Gao (2019) has some discussion). First, one could try to understand them as familiar states: for example, a superposition of seeing an object at positions A and B might be a state of double vision. However, double vision is an ordinary state of consciousness that can enter superpositions. It leads to reports such as "I see an object at A and at B." The superposed state does not. It leads to reports such as "I see an object at A" (if the introspection and report process triggers collapse), or at worst a superposition of "I see an object at A" and "I see an object at B" (if no collapse is triggered). This brings out that the sort of superpositions we need are not introspectible or reportable and will be quite different from familiar states such as double vision.[23]

A more radical alternative says that superposed states of consciousness involve multiple subjects having distinct total states of conscious experience. We will set aside this option as extravagant (do subjects pop into and out of existence in superposition and collapse?), though it is perhaps worth some attention.

A third option is to say that a superposition of states of consciousness is a state that the subject is in, but it is not itself a total state of consciousness. That is, when a subject is in a superposition of conscious states A and B, there is no subjective experience of being in this superposition. There is something it is like to be in A, and something it is like to be in B, but nothing it is like to be in A and B simultaneously. The subject has the experience of being in A and the experience of being in B, without having any conjoint experience of being in the superposition. This violates the Unity Thesis articulated by Bayne and Chalmers (2003) holding that whenever a subject is in multiple conscious states, they are also in a single conscious state that subsumes and unifies them. Some theorists hold that the Unity Thesis is false, at least for split-brain patients and other fragmented subjects: these subjects do not have a single determinate total conscious state, but instead have multiple conscious states

[23] Shimony (1963) reads London and Bauer (1939) as allowing superpositions of consciousness and critiques the idea in part by arguing that phenomena such as blurred vision and indecision do not really involve superpositions.

as fragments.[24] It is far from obvious what is really going on in these cases, and any analogy with superposed states seems fairly distant. Still, these cases at least bring out that the Unity Thesis and the corresponding assumption that every subject is in a single determinate total state of consciousness is not non-negotiable.

A fourth option is to say that a superposition of total states of consciousness is itself a total state of consciousness—albeit one quite unlike the ordinary total states of consciousness that we are introspectively familiar with. On this view, when a subject is in a superposition of conscious states A and B, there is something it is like to be in this superposition. It presumably involves some combination of the experience of being in A and the experience of being in B, combined by some novel phenomenal mode of combination. This mode of combination is not something we could introspect or report for the reasons discussed above, so it would have to be something that we have no introspective familiarity with. The phenomenological role of amplitudes is also not clear. Perhaps amplitudes give the ordinary states of consciousness relative weights in the combined states. As a result, it is far from clear what the phenomenology of a superposed state would be like. Still, it is far from obvious that a mode of combination like this is impossible.

We think that the fourth option is perhaps the most worthy of consideration, followed by the third. On the fourth option, we can no longer say that total states of consciousness correspond one-to-one with PCC eigenstates. Instead, ordinary non-superposed total states of consciousness will correspond to PCC eigenstates, and superposed total states of consciousness will correspond to superpositions of these eigenstates.

There is precedent to the thought that there are states of consciousness that we cannot introspect or report. Theorists (e.g., Block) who believe in an "overflow" of consciousness outside attention often postulate such aspects: if introspecting and reporting a state always involve attending to it, unattended states cannot be introspected or reported. One can perhaps make unnoticed superpositions more palatable by noting that on a fast-collapse model they will usually be small superpositions, involving very similar states of consciousness, very low amplitudes, and/or very brief periods of time. As

[24] On split-brain cases, see for example Nagel (1971) who argues for indeterminacy here. Bayne and Chalmers (2003) argue that in these cases there is a single subject with a single determinate state of consciousness, while Schechter (2017) argues that there are multiple subjects each with a determinate state of consciousness.

a result, the superpositions may largely fall below the grain of our ordinary introspective access.

Still, the fact that our super-resistance model has to postulate superposed states of consciousness is a significant cost of the view. Is it possible to develop a super-resistance consciousness-collapse model that avoids superpositions of consciousness while also avoiding the Zeno problem? Such a model would need to give up on the tight connection between definite conscious states and PCC eigenstates, in order that never-superposed conscious states do not lead to never-superposed PCC states and so to the Zeno effect. At the same time, it would need to retain enough of a connection between consciousness and physical states that the definiteness of consciousness leads to collapse in its physical basis. It is not easy to meet both demands at once. One path invokes a looser connection between consciousness and PCC eigenstates, whereby superposed PCC states can coexist with definite states of consciousness at least briefly. For example, one might hold that superposed PCC states determine a definite state of consciousness probabilistically according to the Born rule, and that this definite state of consciousness leads to collapse onto a corresponding PCC state but only after a time delay. Perhaps this view and others in the neighborhood are at least worth developing.

In any case: in ordinary quantum mechanics, many theorists say that they cannot really imagine what it is for a physical state to be in a superposition. At the same time, they adopt the idea and run with it, and the idea seems to be theoretically fruitful. Our suggestion is that we do something like this for superpositions of states of consciousness, at least for now. We should simply adopt the idea and see whether it is fruitful. If it is, we can later return to the question of just what superposed states of consciousness involve.

Objection 2: How do quantum effects make a difference to macroscopic brain processes?

Quantum theories of brain processes are sometimes criticized on the grounds that it is hard to see how low-level quantum processes can affect high-level processing in neurons. A more specific version of this objection is that on some accounts (e.g., Hameroff and Penrose), quantum coherence at the neural level is required for distinctively quantum effects in neural processing, but the high temperatures in the brain are likely to lead to decoherence below the neural level. These objections do not apply to our approach, which does not involve any special effects of low-level quantum processes on neural processes and is entirely consistent with decoherence at relatively low levels. In fact, in our central illustrations, we have treated brain states as superpositions of numerous decoherent eigenstates, which themselves may involve relatively classical processing in neurons. The only high-level quantum process that plays an essential role in our framework is the collapse process, which selects one or more of these eigenstates as outlined above. Our picture is consistent with further macroscopic quantum effects, but they are not required.

Objection 3: What about macroscopic superpositions?

One might worry that on a consciousness-collapse view ordinary macroscopic objects such as measurement devices will exist in states of superposition until they are observed. Our view does not necessarily lead to this consequence. For a start, if a correct theory of consciousness associates these devices with some amount of consciousness (as may be the case for IIT), then the devices will collapse wave functions much as humans do. Even if these devices are not conscious, it is likely that typical measuring devices will be entangled with humans and other conscious systems, so that they will typically be in a collapsed state too. Still, in special cases where such a device is entirely isolated from conscious systems and records a quantum interaction, it will enter a macroscopic superposition. Of course we will never observe such a superposition, as our observation will collapse the state of the system. But we might in principle get empirical evidence of this superposition if we can eventually measure associated interference effects. Perhaps the existence of macroscopic superpositions is counterintuitive, but many cosmological theories already allow macroscopic objects to be in superposition in the early universe where there are no observers. It is unclear why allowing this in the current universe is any worse.

Objection 4: What about the first appearance of consciousness in the universe?

As we saw earlier, if consciousness is absolutely super-resistant, the quantum Zeno effect entails that it can never emerge for the first time in the development of the universe. On an approximate super-resistance model, there is less of a problem. For eons, the universe can persist in a wholly unconscious superposed state without any collapses. At some point, a physical correlate of consciousness may emerge in some branch of the wave function, yielding a superposition of consciousness and unconsciousness (or their physical correlates) with low amplitude for consciousness. With high probability the universe will collapse back toward an unconscious state. As this happens repeatedly in many branches of the wave function, there will eventually be a low probability collapse toward a state of consciousness, and consciousness will be in a position to take hold.

10. Conclusion

The results of our analysis are mixed. We have developed a consciousness-collapse model with a reasonably clear and precise dynamics. But it must be admitted that the model we have developed is not as simple and powerful as the original (simple if imprecise) measurement-collapse framework.

Our initial superselection collapse model was simple, but it leads to the Zeno problem. Avoiding the Zeno problem has led to a number of complications. First, we have had to countenance superpositions in states of consciousness, and it is not at all clear that this is possible. Second, we have had to introduce Pearle-style collapse dynamics along with parameters for the rate of collapse, and these parameters have to be constrained carefully in order to yield empirically acceptable results. We have also had to invoke a complex theory of consciousness – though this is less of a cost, since a theory of consciousness is needed even in the absence of the quantum measurement problem.

Is this consciousness-collapse model the best that we can do? We have seen that to avoid countenancing superposed states of consciousness while also avoiding the Zeno problem, a consciousness-collapse model will need to break the strong link between definite states of consciousness and eigenstates of a PCC observable. Perhaps there are alternative models on which the physical correlates of consciousness involve a more complex wave-function property, or on which consciousness can vary independently of any physical properties. There also remain the possibility of variable-locus models, though these may also need to break the strong link between consciousness and its physical correlates to avoid the Zeno problem. In any case, models along these lines are certainly worth exploring.

Overall: the model we have developed is perhaps not as simple or powerful as some of the leading interpretations of quantum mechanics. If it is the best we can do, then the upshot may be that consciousness-collapse models are subject to principled limitations. Nevertheless, it at least serves as an existence proof for a relatively precise consciousness-collapse model. The model is open to empirical test, and it is not out of the question that a more powerful model along these lines could be developed. In the meantime, the research program of consciousness-collapse models deserves attention.

Appendix A: Calculating Q-Shape for a Dyad System in IIT 3.0 and Quantum IIT

In this appendix, we illustrate some mathematical details of standard IIT (IIT3.0) and quantum IIT (QIIT), by showing how Φ and Q-shape are determined in simple dyad systems with two elements. The IIT formalisms are complex, but dyads avoid some complications. We will also define a distance measure between Q-shapes which is important for the collapse dynamics.

We begin with IIT3.0.[25] We assume a dyad system with two elements A and B, each of which can be in one of two states: [1] or [0]. The composite system AB can be in one of four possible states: [11], [00], [10], or [01]. The transition rules are a simple swap: the state of A at one time is determined by copying the state of B at the previous time and vice versa. We can stipulate that in the system under consideration, the current state of AB is [10]. The next state is thereby determined to be [01]. Subsystems of AB are the nonempty sets of elements of the system: {A}, {B}, and {A, B}, which we will abbreviate as A, B, and AB when there is no chance of confusion. Mechanisms are subsystems with nonzero weight.

The Q-shape of a system consists of a location $L(m)$ for each mechanism m in the system, weighted by the measure $\varphi(m)$. $L(m)$ is a point in a 2^{n+1}-dimensional space with two dimensions for each of the 2^n possible states of the system, where n is the number of elements. $L(m)$ is determined by conjoining two probability distributions over the states S of the system: $p_m(S)$ and $p'_m(S)$, where the former is defined in terms of the effects of m and the latter is defined in terms of the causes of m. Each distribution is associated with a φ value. The weight $\varphi(m)$ is the minimum of these two values. The Q-shape of AB lives in an 8-dimensional space, as AB has four possible states. As we will see, of the three subsystems of AB, only A and B yield mechanisms with nonzero weight. Hence, The Q-shape of AB consists in two weighted points located in an 8-dimensional space.

IIT3.0 distinguishes two notions of integrated information: φ (small phi), which applies to individual mechanisms, and Φ (big phi), which applies to the total system. To know $\Phi(AB)$ we must first calculate AB's Q-shape. To know AB's Q-shape we must first calculate φ for AB's mechanisms. To begin with, we illustrate how the probability distribution $p_m(S)$ and $\varphi(m)$ are calculated and then used to define Q-shape and Φ.

The distribution $p_m(S)$ is a distribution over future states S of the system, reflecting their probability of occurrence given that the elements of m are fixed to their current state (while any other elements are allowed to vary). For the candidate mechanism AB, both elements will be fixed to their current value [10]. $p_{AB}(S)$ is the probability that the

[25] Thanks to Nao Tsuchiya and Leo Barbosa. Our calculations follow the supporting information in Mayner et al. (2018) especially S1: Calculating Φ. See also Oizumi, Albantakis, and Tononi (2014) and Tononi et al. (2016). For the earlier, simpler IIT formalism for calculating $\Phi(AB)$, see Tononi (2004, fig. 5), Tsuchiya (2017), and McQueen (2019a). The reader can experiment with calculating Φ for various systems including the dyad AB at http://integratedinformationtheory.org/calculate.html. Details of the underlying software can be found in Mayner et al. (2018).

following state will be S, given the current state [10]. The following state is guaranteed to be [01], so p_{AB} assigns probability 1 to [01] and probability 0 to the other three states.

Recall that $L(m)$ is determined by conjoining two probability distributions over the states S of the system: $p_m(S)$ and $p'_m(S)$. If we consider just $p_m(S)$, then the location $L(AB)$ can be seen as a point in 4-dimensional space corresponding to the distribution $p_{AB}(S)$. Let us say the four dimensions are ordered as [00], [01], [10], [11]. Then L(AB)=[0,1,0,0], which assigns 1 to [01] and 0 to the other states. While $p_m(S)$ is a distribution over possible future states, $p'_m(S)$ is a distribution over possible preceding states (i.e. the probabilities that the preceding state was S, given the current state). For our system AB the two distributions are the same, so the 8-dimensional location will be a repeated version of the 4-dimensional location: that is, L(AB)=[0,1,0,0,0,1,0,0].

For candidate mechanism A, $p_A(S)$ is the probability of the following state being S given that element A is fixed to its current value [1], while the other element B can vary with probability 0.5 for each value [0] or [1]. Under these conditions, the following state may be either [01] or [11], and p_A will assign these two states probability 0.5 each. Likewise, p_B will assign probability 0.5 each to states [00] and [01], the two states that can follow a state where B is fixed to 0. As with AB, $p_m(S) = p'_m(S)$ for A and for B. As a result, $L(A) = [0, 0.5, 0, 0.5, 0, 0.5, 0, 0.5]$ and $L(B) = [0.5, 0.5, 0, 0, 0.5, 0.5, 0, 0]$.

The integrated information [small phi] $\varphi(AB)$ is determined by considering the difference between the probabilistic effects of the subsystem AB with the effects of a partitioned subsystem A-B where we consider only the effects of A and B taken separately on each other. We can define a probability distribution p_{A-B} as the tensor product of two distributions: a distribution $p_{A|B}$ over states of A given that B is fixed to its current value 0 (so A=[1] has probability 1) and a distribution $p_{B|A}$ over states of B given that A is fixed to its current value 1 (so B=[0] has probability 0). The product distribution p_{A-B} assigns 1 to [10] and 0 to every other state.

We can then define $\varphi(AB) = EMD(p_{AB}, p_{A-B})$. For two probability distributions p_1 and p_2 over the same state-space, EMD(p_1, p_2) is the Earth mover's distance between p_1 and p_2. This can be defined as the minimal amount of work required to turn p_1 into p_2 by moving the "Earth" of probability from some points in the 2^n-dimensional space to other points, where work is measured by the amount of probability moved multiplied by the Hamming distance between the points. In the case just described, p_{AB} and p_{A-B} are exactly the same distribution, so the Earth mover's distance between them is 0. So $\varphi(AB) = 0$.

The quantity $\varphi(A)$ can be defined as a related Earth-mover's distance over states of B, comparing the distribution over those states with A fixed to its current value of [1] (resulting in probability 1 to B=[0]) to a distribution that ignores the value of A (resulting in probability 0.5 each to B=[0] or B=[1]). In this case, $\varphi(A) = 0.5$. Likewise, $\varphi(B) = 0.5$.

As a result, we can fully specify the Q-shape Q_{AB} of the system AB. It consists of location $L(AB) = [0, 1, 0, 0, 0, 1, 0, 0]$ with associated weight $\varphi(AB) = 0$, location $L(A) = [0, 0.5, 0, 0.5, 0, 0.5, 0, 0.5]$ with associated weight $\varphi(A) = 0.5$, and location $L(B) = [0.5, 0.5, 0, 0, 0.5, 0.5, 0, 0]$ with associated weight $\varphi(B) = 0.5$.

In the above calculations we took a shortcut that should now be made explicit. For each candidate mechanism, we chose to consider the probability distribution assigned to the future (or past) states *of a particular subsystem*. For candidate mechanism AB we chose subsystem AB. For candidate mechanism A we chose subsystem B. And for B we chose A. These choices are not arbitrary, but are the result of an optimization procedure. For each candidate mechanism, we in fact consider all possible subsystems and choose the subsystem that maximizes φ. For example, when considering candidate mechanism A, it turns out that A has more integrated information about B than about AB. After all, there are three possible ways of disconnecting A from AB: disconnect A to A, A to B, A to AB. Nothing happens by disconnecting A to A (there was no connection there to begin with!). But then that is the minimal information partition, implying that A has zero φ about AB. On the other hand, there is only one way to disconnect A from B, and that disconnection does make a difference, giving nonzero φ. For details see Barbosa et al. (2021).

We can define the distance between two Q-shapes Q_1 and Q_2 (defined over the same states S, with associated probability distributions $p_{m,1}$ and $p_{m,2}$ and weights $\varphi_1(m)$ and $\varphi_2(m)$) as an extended Earth mover's distance $EMD^*(Q_1, Q_2)$:

$$EMD^*(Q_1, Q_2) = \sum_i (|\varphi_1(m_i) - \varphi_2(m_i)| \times (EMD(p_{m_i,1}, p_{m_i,2})) + EMD(p'_{m_i,1}, p'_{m_i,2}))). \tag{5}$$

This distance is the minimal amount of work required to transform the φ_1 distribution over mechanisms m into φ_2 by repeatedly moving the "Earth" of φ from one mechanism m_1 in Q_1 to another mechanism m_2 in Q_2. (A complication is that in some cases (where Q_1 has more total φ than Q_2), we need to send the excess to an unconstrained distribution p_{uc} associated with Q_2.)

We can then define $\Phi(AB)$ as the minimal value of $\text{EMD}^*(Q_{AB}, Q_{AB^*})$, across all partitions AB^* of AB. A partition of a system requires cutting one or more causal connections between its units. For system AB, a partition cuts the connection from A to B or from B to A or both. In this case, either cut reduces φ to zero for both mechanisms A and B, and their probability distributions are flattened. The reason why cutting just one of these two connections destroys both mechanisms is tied to the fact that $\varphi(m)$ is defined as the minimum of two φ values, the one that pertains to the future state and the one that pertains to the past state. Each cut will send one of these φ values to zero.

Recall that Q_{AB} assigns $\varphi(A) = \varphi(B) = 0.5$, where these serve as weights for $L(A) = [0, 0.5, 0, 0.5, 0, 0.5, 0, 0.5]$ and $L(B) = [0.5, 0.5, 0, 0, 0.5, 0.5, 0, 0]$. Q_{AB^*} instead assigns zero weights to both $L(A^*)$ and $L(B^*)$, where $L(A^*) = L(B^*) = [0.25, 0.25, 0.25, 0.25, 0.25, 0.25, 0.25, 0.25]$. We thus have: $EMD^*(Q_{AB}, Q_{AB^*}) = |\varphi(A) - \varphi(A^*)| \times (EMD(p_A, p_{A^*}) + EMD(p'_A, p'_{A^*})) + |\varphi(B) - \varphi(B^*)| \times (EMD(p_B, p_{B^*}) + EMD(p'_B, p'_{B^*})) = (0.5 \times (0.5 + 0.5)) + (0.5 \times (0.5 + 0.5)) = 1$.

The crucial quantity $\Phi^{max}(AB)$ is defined as $\Phi(AB)$ if AB is a maximum of Φ, and 0 otherwise. Here AB is a maximum of Φ if $\Phi(AB) > \Phi(S)$ for all systems S such that S has elements in common with AB. In our case, we can stipulate that AB is isolated from its environment so that no other system containing A or B has higher Φ. In this case, AB is a maximum of Φ, so $\Phi^{max}(AB) = 1$. According to IIT, Φ^{max} is a measure of consciousness, so system AB has one unit of consciousness.

In section 4 we noted that if AB is in a different state (either 01, 00, or 11), than the calculation for Φ is the same, but the Q-shape is different. This can now be seen by the fact that changing the initial state changes the locations but not their weights. Thus, if the initial state is instead 00, then we still have two mechanisms A and B, each with weight 0.5, but their locations become $L(A) = [0.5, 0.5, 0, 0, 0.5, 0.5, 0, 0]$ and $L(B) = [0.5, 0, 0.5, 0, 0.5, 0, 0.5, 0]$. This is not enough to change Φ, but it is enough to change the Q-shape. We can thus define Schrödinger's dyad as AB in a superposition of 10 and 00. A collapse model base only on Φ would fail to collapse this superposition, despite it being a superposition of conscious states.

We now move to quantum IIT (QIIT).[26] To simplify the calculations of the dyad, it is easier to start A and B in the same initial state ($|00\rangle$ or $|11\rangle$) so that they remain stationary. We add the further stipulation that A (B) maintains its own state over time. We may now consider A and B to be AND gates that each take two inputs as depicted:

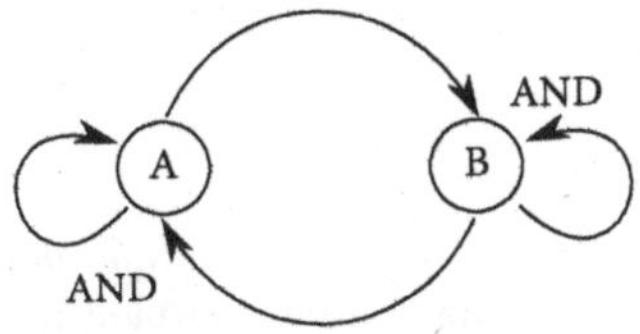

State $|00\rangle$ has zero Φ and Q-shape, that is, $\Phi(|00\rangle) = Q(|00\rangle) = 0$. For if we partition the system by replacing one of the directed edges with random input, the inputs are still only either 00 or 01, whereas the AND gates require an input of 11 to change state. Partitioning does not make a difference.

Partitioning makes a difference if the system is instead in state $|11\rangle$: If any of the edges are removed and replaced by random input, at least in half the cases it will feed a 0 to its target, so that in light of the AND gate the state of the target will change from 1 to 0. This implies that the system in that state has non-zero Φ value, and its Q-shape isn't null.

We can therefore introduce collapse operators for the Q-shapes of these two states, and then use them to define a small consciousness superposition.

Our new dyad still has three subsystems (AB, A, and B). For each we consider the integrated information φ of both future and past states. So for the collapse operators Q^k_{ij}, the k index runs from 1 to 6. Since the Hilbert space of the system in this case is 4 dimensional, the indices i and j run from 1 to 4 each.

The $c^k_{ij}(\psi)$ in (2) are the coefficients of the operator ρ which is the kth component of the Q-shape of ψ. Because $Q(00) = 0$, it follows that $c^k_{ij}(\psi_0) = 0$. Since $|00\rangle$ and $|11\rangle$ are wave functions with classical Q-shapes, they are contained in $\mathbb{E}^{-1}(\mathcal{C})$ and are summed over in (2). It follows that

$$\hat{Q}^k_{ij}\,|00\rangle = \sum_{\psi\in\mathbb{E}^{-1}(\mathcal{C})} \varphi^k(\psi)c^k_{ij}(\psi)\,|\psi\rangle\langle\psi|\,|00\rangle = \varphi^k(00)c^k_{ij}(00)\,|00\rangle = 0\,|00\rangle \tag{6}$$

We have assumed the wave functions with classical Q-shapes are orthogonal. Thus $|00\rangle$ is an eigenvector of every operator $\hat{Q}^k_{ij}$ with eigenvalue 0. We also have

$$\hat{Q}^k_{ij}\,|11\rangle = \sum_{\psi\in\mathbb{E}^{-1}(\mathcal{C})} \varphi^k(\psi)c^k_{ij}(\psi)\,|\psi\rangle\langle\psi|\,|11\rangle = \varphi^k(11)c^k_{ij}(11)\,|11\rangle \tag{7}$$

so that $|11\rangle$ is an eigenvector of Q^k_{ij} with eigenvalue $c^k_{ij}(|11\rangle)$.

Letting $|11\rangle$ be the alive (conscious) state and $|00\rangle$ be the dead (unconscious) state, we can provide (in addition to the section 4 example) another example of Schrödinger's dyad:

[26] Thanks to Johannes Kleiner. Our calculations are intended to follow Zanardi, Tomka, and Venuti (2018) and Kleiner and Tull (2021). Our dyad example is an idealization and would require ancilla qubits if it were to be implemented by a quantum computer.

$$|\Psi\rangle_{AB} = \alpha\,|00\rangle + \beta\,|11\rangle \tag{8}$$

Our dynamics (in section 6) predicts that this state is not completely stable, but continuously collapses towards one of the two Q-shape eigenstates, in accord with the Born rule.

References

Aaronson, S. (2014). *Why I Am Not An Integrated Information Theorist (or, the Unconscious Expander)* http://www.scottaaronson.com/blog/?p=1799.

Afrasiabi, M. et al. (2021). Consciousness depends on integration between parietal cortex, striatum, and thalamus. In: *Cell systems*, 12(4), 363–373.

Albert, D. Z. (1992). *Quantum Mechanics and Experience*. Cambridge University Press.

Antony, M.V. (2006). "Vagueness and the metaphysics of consciousness." In: *Philosophical Studies* 128, pp. 515–538.

Barbosa, L. S. et al. (2021). "Mechanism Integrated Information." In: *Entropy (Basel, Switzerland)* 23 (3), p. 362.

Barrett, A. B. and P. A. M. Mediano (2019). "The Phi measure of integrated information is not well-defined for general physical systems." In: *Journal of Consciousness* 26 (1–2), pp. 11–20.

Bassi, A. et al. (2017). "Gravitational decoherence." In: *Class. Quantum Grav.* 34 (193002).

Bassi, A. et al. (2013). "Models of wave-function collapse, underlying theories, and experimental tests." In: *Rev. Mod. Phys.* 85 (2), pp. 471–527.

Bayne, T. (2018). "On the axiomatic foundations of the integrated information theory of consciousness." In: *Neuroscience of Consciousness* 1.

Bayne, T. and D. J. Chalmers (2003). "What is the unity of consciousness?" In: *The unity of consciousness: binding, integration, and dissociation* Ed. by A. Cleeremans. Oxford University Press.

Bell, J. S. (1990). "Against Measurement." In: *Physics World*, 3(8), 33.

Bohm, D. (1952). "A Suggested Interpretation of the Quantum Theory in Terms of "Hidden" Variables. I." In: *Phys. Rev.* 85 (2), pp. 166–179.

Bub, J. (1988). "How to solve the measurement problem of quantum mechanics." In: *Found Phys* 18, pp. 701–722.

Byrne, A. and H. Logue (2009). *Disjunctivism: Contemporary Readings*. MIT Press.

Capra, F. (1975). *The Tao of Physics*. Shambhala Publications.

Carpenter, R. H. S. and A. J. Anderson (2006). "The death of Schrödinger's cat and of consciousness based quantum wave-function collapse." In: *Annales de la Foundation Loisi de Broglie* 33, pp. 45–52.

Casarotto, S. et al. (2016). "Stratification of unresponsive patients by an independently validated index of brain complexity." In: *Annals of Neurology* 80 (5), pp. 718–729.

Chalmers, D. J. (1996). *The Conscious Mind*. Oxford University Press.

Chalmers, D. J. (2003). "Consciousness and its place in nature." In: *Blackwell Guide to the Philosophy of Mind* Ed. by S. Stich and T. Warfield. Blackwell.

Diósi, L. (1987). "A universal master equation for the gravitational violation of quantum mechanics." In: *Physics Letters A* 120.8, pp. 377–381.

Doerig, A. et al. (2019). "The unfolding argument: Why IIT and other causal structure theories cannot explain consciousness." In: *Consciousness and Cognition* 72, pp. 49–59.

Earman, J. (2008). "Superselection rules for philosophers." In: *Erkenntnis* 69, pp. 377–414.
Everett, H. (1957). " 'Relative State' Formulation of Quantum Mechanics." In: *Rev. Mod. Phys.* 29 (3), pp. 454–462.
Fein, Y. Y., P. Geyer, P. Zwick et al. (2019). "Quantum superposition of molecules beyond 25 kDa." In: *Nat. Phys.* 15, pp. 1242–1245.
Feldmann, W. and R. Tumulka (2012). "Parameter diagrams of the GRW and CSL theories of wavefunction collapse." In: *Journal of Physics A: Mathematical and Theoretical* 45(6).
French, S. (2020). "From a Lost History to a New Future: Is a Phenomenological Approach to Quantum Physics Viable?" In: *Phenomenological Approaches to Physics* Ed. by H. Wiltsche and P. Berghofer. Springer.
Gao, S. (2019). The measurement problem revisited, Synthese 196, 299–311.
Ghirardi, G. C., A. Rimini, and T. Weber (1986). In: "Unified dynamics for micro-scopic and macroscopic systems." In: *Phys. Rev. D* 34 (2), pp. 470–491.
Gisin, N. (1984). Quantum measurements and stochastic processes. In: *Physical Review Letters*, 52(19), 1657.
Gisin, N. (1989). Stochastic quantum dynamics and relativity. In: *Helvetica Physica Acta*, 62(4), 363–371.
Halvorson, H. (2011). "The measure of all things: Quantum mechanics and the soul." In: *The Soul Hypothesis* Ed. by M. Baker and S. Goetz. Continuum Books.
Hameroff, S. and R. Penrose (2014). "Consciousness in the universe: a review of the 'Orch OR' theory." In: *Phys Life Rev.* 11 (1), pp. 39–78.
Haun, A. M. et al. (2017). "Conscious Perception as Integrated Information Patterns in Human Electrocorticography." In: *eNeuro* 4 (5).
Hepp, K. (1972). "Quantum theory of measurement and macroscopic observables." In: *Helvetica Physica Acta* 45, pp. 237–248.
Hopkins, A. R., and McQueen, K. J. (2022). Filled/non-filled pairs: an empirical challenge to the integrated information theory of consciousness. *Consciousness and Cognition*, 97, 103245.
Kleiner, J. and Tull, S. (2021). The mathematical structure of integrated information theory. *Frontiers in Applied Mathematics and Statistics*, 6, 74.
Koch, C. and K. Hepp (2006). "Quantum mechanics in the brain." In: *Nature* 440 (611).
Kremnizer, K. and A. Ranchin (2015). "Integrated Information-Induced Quantum Collapse." In: *Found Phys* 45, pp. 889–899.
Leung, A., Cohen, D., Van Swinderen, B., & Tsuchiya, N. (2021). Integrated information structure collapses with anesthetic loss of conscious arousal in Drosophila melanogaster. In: *PLoS Computational Biology*, 17(2), e1008722.
Lloyd, S. (2000). "Coherent quantum feedback." In: *Phys. Rev. A* 62 (2), p. 022108.
Loewer, B. (2002). "Consciousness and quantum theory: Strange bedfellows." In: *Consciousness: New Philosophical Perspectives* Ed. by Q. Smith and A. Jokic. Oxford University Press.
London, F. and E. Bauer (1939). *La théorie de lóbservation en mécanique quantique.* Paris: Hermann. *English translation in Quantum Theory, Measurement*, edited by J. A. Wheeler and W. H. Zurek (Princeton University, Princeton, NJ, 1983), pp. 217–259.
Machida, S. and M. Namiki (1980). "Theory of measurement in quantum mechanics: Mechanism of reduction of wave packet, part I." In: *Progress of Theoretical Physics* 63, pp. 1457–1473.
Massimini, M. et al. (2005). "Breakdown of cortical effective connectivity during sleep." In: *Science (New York)* 309 (5744), pp. 2228–2232.

Maudlin, T. (2011). *Quantum Non-Locality and Relativity* 3rd Edition. Blackwell Publishing.

Mayner, W. et al. (2018). "PyPhi: A toolbox for integrated information theory." In: *PLOS Computational Biology* 14, pp. 1–21.

McQueen, K. J. (2015). "Four tails problems for dynamical collapse theories." In: *Studies in History and Philosophy of Modern Physics* 49, pp. 10–18.

McQueen, K. J. (2019a). "Illusionist Integrated Information Theory." In: *Journal of Consciousness Studies* 26 (5-6), pp. 141–169.

McQueen, K. J. (2019b). "Interpretation-Neutral Integrated Information Theory." In: *Journal of Consciousness Studies* 26 (1-2), pp. 76–106.

McQueen, K. J. and L. Vaidman (2019). "In defence of the self-location uncertainty account of probability in the many-worlds interpretation." In: *Studies in History and Philosophy of Modern Physics* 66, pp. 14–23.

Nagel, T. (1971). "Brain bisection and the unity of consciousness." In: *Synthese* 22, pp. 396–413.

Oizumi, M., L. Albantakis, and G. Tononi (2014). "From the Phenomenology to the Mechanisms of Consciousness: Integrated Information Theory 3.0." In: *PLOS Computational Biology* 10, pp. 1–25.

Okon, E. and M.Á. Sebastián (2016). "How to Back up or Refute Quantum Theories of Consciousness." In: *Mind and Matter* 14 (1), pp. 25–49.

Okon, E. and Sebastián, M.Á. (2020). A consciousness-based quantum objective collapse model. In: *Synthese*, 197(9), 3947–3967.

Patel, R. B. et al. (2016). "A quantum Fredkin gate." In: *Science Advances* 2 (3).

Pearle, P. (1976). "Reduction of the state vector by a nonlinear Schrödinger equation." In: *Phys. Rev. D* 13 (4), pp. 857–868.

Pearle, P. (1982). "Might god toss coins?" In: *Foundations of Physics* 12 (3), pp. 249–263.

Pearle, P. (1999). "Collapse models." In: *Open Systems and Measurement in Relativistic Quantum Theory* Ed. by H. P. Breuer and F. Petruccione. Springer Berlin Heidelberg, pp. 195–234.

Pearle, P. (2021). "Dynamical collapse." In: *Three Roads to Quantum Reality: Pilot Waves, Dynamical Collapse, Many Worlds* Ed. by S. Saunders A. Valentini P. Pearle. Oxford University Press.

Penrose, R. (2014). "On the Gravitization of Quantum Mechanics 1: Quantum State Reduction." In: *Foundations of Physics* 44, pp. 557–575.

Schechter, E. (2017). *Self-consciousness and "split-brains": The minds' I*. Oxford University Press.

Shimony, A. (1963). "Role of the observer in quantum theory." In: *American Journal of Physics* 31 (1), pp. 755–773.

Simon, J. A. (2017). "Vagueness and zombies: why 'phenomenally conscious' has no borderline cases." In: *Philosophical Studies* 174, pp. 2105–2123.

Stapp, H. (1993). *Mind, Matter, and Quantum Mechanics* Springer.

Thalos, M. (1998). "The trouble with superselection accounts of measurement." In: *Philosophy of Science* 65, pp. 518–544.

Tononi, G. (2004). "An information integration theory of consciousness." In: *BMC Neurosci* 5 (42).

Tononi, G. (2008). "Consciousness as integrated information: a provisional manifesto." In: *Biol Bull* 215 (3), pp. 216–242.

Tononi, G. et al. (2016). "Integrated information theory: from consciousness to its physical substrate." In: *Nat Rev Neurosci* 17, pp. 450–461.
Tsuchiya, N. (2017). " "What is it like to be a bat?"—a pathway to the answer from the integrated information theory." In: *Philosophy Compass* 12 (3), e12407.
Tye, M. (1995). *Ten Problems of Consciousness*. MIT Press.
von Neumann, J. (1955). *Mathematical Foundations of Quantum Mechanics* Princeton University Press. German original: Die mathematischen Grundlagen der Quantenmechanik. Berlin: Springer, 1932.
Wick, G. C., A. S. Wightman, and E. P. Wigner (1952). "The Intrinsic Parity of Elementary Particles." In: *Phys. Rev.* 88 (1), pp. 101–105.
Wigner, E. P. (1961). "Remarks on the mind-body question." In: *The Scientist Speculates* Ed. by I. J. Good. Heineman.
Zanardi, P., M. Tomka, and L. C. Venuti (2018). *Towards Quantum Integrated Information Theory* arXiv: 1806.01421 [quant-ph] https://doi.org/10.48550/arXiv.1806.01421
Zukav, G. (1979). *The Dancing Wu Li Masters*. William Morrow.

2

The Subjective-Objective Collapse Model

Virtues and Challenges

Elias Okon and Miguel Ángel Sebastián

1. Introduction

Quantum mechanics certainly is one of the most successful physical theories ever constructed. However, its standard interpretation suffers from a grave conceptual issue known as the *measurement problem*. Such a problem consists of the fact that the framework crucially depends on notions such as *measurement* or *observer*, but such notions are never formally defined within the theory. A direct consequence of this is an ambiguity regarding where in the causal chain between the interaction of the system under study with a measuring device and our "subjective perception" is the proverbial collapse of the wave function to be placed.

The idea that consciousness causes the collapse of the wave function was first suggested in London and Bauer (1939) and then developed in Wigner (1967). The latter proposed that consciousness, unlike other properties, does not admit superpositions. This, however, seems to suggest that consciousness is not a physical property or, at least, not a standard one. More recently, other works have explored similar ideas—see, e.g., Chalmers and McQueen (MS); Stapp (2007). The problem, though, is that this avenue of research has never been popular among physicists, as it seems to commit us to some form of dualism, in which the interaction between consciousness and the physical remains obscure—if not downright inconsistent with certain basic principles of physics, such as conservation of energy (Averill and Keating, 1981; Larmer, 1986).

In Okon and Sebastián (2018), we presented a physicalist-friendly model in which, in a well-defined sense, consciousness "causes" the collapse of the wave function. The model consists of an objective collapse scheme, where the collapse operator is associated with consciousness as a physical property—

Elias Okon and Miguel Ángel Sebastián, *The Subjective-Objective Collapse Model: Virtues and Challenges*
In: *Consciousness and Quantum Mechanics*. Edited by: Shan Gao, Oxford University Press.
 DOI: 10.1093/oso/9780197501665.003.0003

or with a physical property that perfectly correlates with consciousness. We call it the "Subjective-Objective Collapse" (SOC) model. SOC is such that superpositions of conscious states are dynamically suppressed in a way that is fully compatible with our subjective experience. As such, it opens up new lines of research, both in fundamental physics and consciousness studies.

In this chapter we evaluate the virtues of the model and analyze some possible objections and challenges. For this purpose, the chapter is organized as follows. Section 2 presents the measurement problem and the solution offered by the SOC model. In section 3 we question the compatibility of the model with certain theories of consciousness and, in particular, with the idea that consciousness might be multiply realizable—an idea that is, *prima facie*, in tension with the model. Finally, in section 4 we examine important aspects of the SOC model in light of considerations in the field of quantum foundations.

2. The Measurement Problem and the Subjective-Objective Collapse Model

2.1 The Measurement Problem

The measurement problem can be stated as the fact that standard quantum mechanics crucially depends on notions such as measurement or observer, even though such notions are never formally defined within the theory—see, e.g., Bell (1990). In particular, measurements or observers are employed to decide when the indeterministic collapse process is supposed to interrupt the deterministic Schrödinger evolution. The problem, of course, is that, without a firm grasp of such higher-level notions, one ends up with undesirable vagueness in an otherwise fantastically successful theory.

To explore the issue, let us suppose, for now, that everything evolves according to the Schrödinger equation *at all times*. With this in mind, let us consider an apparatus with a ready state $|R\rangle_M$ that, when fed with a spin-1/2 particle, behaves as follows:

$$|R\rangle_M|+\rangle_p \xrightarrow{\text{Schrödinger}} |+\rangle_M|+\rangle_p \quad \text{and} \quad |R\rangle_M|-\rangle_p \xrightarrow{\text{Schrödinger}} |-\rangle_M|-\rangle_p, \quad (1)$$

where $|+\rangle_M$ and $|-\rangle_M$ are states of the apparatus in which it displays spin-up and spin-down as the result of the experiment. That is, the apparatus correctly measures the spin—say, along z—of the particle. Now, what happens if the apparatus is fed with a particle in a *superposition* of $|+\rangle_p$ and $|-\rangle_p$? Well, the linearity of the Schrödinger equation leads to

$$|R\rangle_M \{\alpha|+\rangle_p + \beta|-\rangle_p\} \xrightarrow{\text{Schrödinger}} \alpha|+\rangle_M|+\rangle_p + \beta|-\rangle_M|-\rangle_p. \qquad (2)$$

We see that the apparatus ends up in a superposition of displaying spin-up and spin-down, which seems odd, to say the least.

To push the matter further, consider the introduction of an observer in a state $|R\rangle_O$, in which she is ready to see the display of the apparatus during the measurement. Once again, the linearity of the Schrodinger equation leads to a superposition, but this time of the observer perceiving the apparatus displaying spin-up and spin-down:

$$|R\rangle_O|R\rangle_M \{\alpha|+\rangle_p + \beta|-\rangle_p\} \xrightarrow{\text{Schrödinger}} \alpha|+\rangle_O|+\rangle_M|+\rangle_p + \beta|-\rangle_O|-\rangle_M|-\rangle_p, \qquad (3)$$

where $|+\rangle_O$ and $|-\rangle_O$ are the states in which the observer perceives that the apparatus displays spin-up and spin-down. The problem, of course, is that the final state above describes a *superposition of perceptions*, which is not what we seem to experience when we perform this type of experiments. Of course, according to standard quantum mechanics, the final state in equation (3) is not the final state of the system because the collapse postulate has been left out of this discussion, but we have already mentioned that the introduction of such a postulate in the standard framework is problematic. So what could be done in order to solve the problem. Well, a satisfactory solution to the measurement problem consists of a formalism which:

1. Is fully formulated in precise, mathematical terms (with notions such as *measurement*, *observer* or *macroscopic* not being part of the fundamental language of the theory).
2. Reproduces the empirical success of standard quantum mechanics at the microscopic level.
3. Explains why certain macroscopic superpositions formally allowed by the theory never seem to occur.[1]

[1] Regarding this last point, it is important to stress an often overlooked distinction between a *superposition of perceptions* and a *perception of a superposition*. That is, between:

At least three strategies have been suggested in order to achieve this. The first one, introduced by Everett (1957), consists of attempting to read the final state in equation (3) not as a state in which the observer does not have a well-defined perception, but one in which the observer simultaneously, but *independently*, has both perceptions. Alternatively, one could avoid having to interpret the final state in equation (3) as one in which the observer does not have a well-defined perception by adding extra elements to the picture—such as Bohmian particles—that determine which of the two terms of the superposition actually obtains in the world—see Goldstein (2013). Finally, one could modify the dynamics in such a way that the actual final state of the system is not a macroscopic superposition, but one of the alternatives; this is the option that SOC explores.

2.2 Objective Collapse Models

The *objective collapse* (or *dynamical reduction*) program aims at constructing a modified quantum dynamics that merges the standard unitary evolution and the collapse mechanism. The idea is to add non-linear, stochastic terms to the Schrödinger equation that suppress unwanted macroscopic superpositions.

The simplest collapse model is know as GRW (Ghirardi et al. (1986)). In it, all elementary particles are postulated to suffer spontaneous localization processes around positions selected according to a probability distribution that approximates the Born rule. Since the frequency of collapses is proportional to the number of particles, macroscopic objects are very likely to undergo them, even if the collapse rate at the level of elementary particles is extremely small. Additionally, due to the fact that the collapse of one of the particles of a macroscopic superposition—such as that in the final state in equation (3)—is sufficient for the whole body to get localized, GRW

- A superposition of incompatible perceptions, such as the final state in equation (3): $\alpha|+\rangle_O|+\rangle_M|+\rangle_p + \beta|-\rangle_O|-\rangle_M|-\rangle_p$.
- A well-defined perception of a macroscopic system in a superposition of different positions: $|S\rangle_O\{\alpha|+\rangle_M|+\rangle_p + \beta|-\rangle_M|-\rangle_p\} = |S\rangle_O|S\rangle_{M+p}$ (where $|S\rangle_{M+p} \equiv \alpha|+\rangle_M|+\rangle_p + \beta|-\rangle_M|-\rangle_p$ and $|S\rangle_O$ corresponds to a state in which the observer experiences a well-defined perception of the measurement apparatus displaying a superposition of spin-up and spin-down).

We have already explained how the first case may arise and why it represents a problem. Regarding the second, from the *quantum* point of view, both $|S\rangle_{M+p}$ and $|S\rangle_O$ are states which are on a par with states such as $|+\rangle_O$ or $|+\rangle_M|+\rangle_p$, and the only reason they seems strange is because we in fact do not experience them. The key question, of course, is why.

guaranties superpositions of well-localized macroscopic states to die quickly, and in accordance with the statistics of standard quantum mechanics.

In the objective collapse model known as Continuous Spontaneous Localization, or CSL Pearle (1989): the discontinuous localization events of GRW are replaced by a continuous, stochastic process. In more detail, in CSL, particular non-linear, stochastic terms, which are designed to drive any initial wave function into one of the eigenstate of a, so-called, collapse operator, are added to the Schrödinger equation. In the simplest case, the CSL equation is such that its solutions are given by

$$|\psi(t)\rangle_B = e^{-\left\{iH+\frac{1}{4\lambda t}[B(t)-2\lambda\hat{A}]^2\right\}}|\psi(0)\rangle, \tag{4}$$

where $\hat{A}$ is the, so-called, collapse operator and $B(t)$ is a classical Brownian function, stochastically chosen with a probability density given by

$$\mathcal{P}_t\{B\} = {}_B\langle\psi(t)|\psi(t)\rangle_B. \tag{5}$$

To see why this achieves what it is supposed to, for simplicity we take $H=0$ and expand $|\psi(0)\rangle$ in terms of eigenstates of $\hat{A}$:

$$|\psi(0)\rangle = \sum_i c_i|a_i\rangle, \tag{6}$$

which leads to

$$|\psi(t)\rangle_B = \sum_i c_i e^{-\frac{1}{4\lambda t}[B(t)-2\lambda t a_i]^2}|a_i\rangle \tag{7}$$

and

$$\mathcal{P}_t\{B\} = \sum_i e^{-\frac{1}{2\lambda}[B(t)-2\lambda t a_i]^2}|c_i|^2. \tag{8}$$

Since the last equation implies that the most probable $B(t)$'s to occur are $B(t) \approx 2\lambda t a_j$, with probabilities $|c_i|^2$, we finally obtain

$$|\psi(t)\rangle_B \approx c_j|a_j\rangle + \sum_{i\neq j} e^{-2\lambda t[a_i-a_j]^2}|a_i\rangle \xrightarrow{t\to\infty} c_j|a_j\rangle, \tag{9}$$

which means that, as $t \to \infty$, the CSL dynamics drives the state of the system into the jth eigenstate of the operator $\hat{A}$, with probability $|c_j|^2$; i.e., the CSL dynamics includes, along with the standard Schrödinger evolution, a "measurement" of the observable represented by the collapse operator.

In standard CSL models, the collapse operator $\hat{A}$ is chosen to be constructed out of the position operator. This is so because such a choice directly induces the desired suppression of superpositions of macroscopic objects at different locations. In fact, it has even been argued that a choice of this sort is the only alternative that could work—see Bassi and Ghirardi (2003). We showed in Okon and Sebastián (2018) that a very different choice for a collapse operator can also lead to a solution of the measurement problem. The point was that, in order to explain why superpositions of macroscopic objects are never actually perceived, (at least) two options present themselves: one can construct models in which such macroscopic superpositions never occur—as in standard collapse theories—or one can maintain that, although such superpositions do occur, we never encounter them because they collapse a soon as we observe them. We exploit this second group of alternatives by developing a CSL model in which the collapse operator is associated with consciousness: the SOC model.

2.3 The Subjective-Objective Collapse Model

The idea that consciousness causes the collapse of the wave function has never been popular, mainly because it appears to be in tension with physicalism. Moreover, if, as Wigner, one assumes that consciousness never superposes, then systems would never get entangled with conscious beings and it would not be clear how a consciousness-based collapse would help explaining why we do not observe macroscopic superpositions.

Chalmers and McQueen (MS) have proposed, as a solution to the measurement problem, the introduction of what they call m-properties or superposition-resistant observables, whose superpositions are postulated to be either forbidden—which we have just seen that is problematic—or to be "unstable" or "more likely to collapse." Moreover, they suggest that some physical correlate of consciousness, such as integrated information, could be a superposition-resistant observable. What is missed is a concrete dynamical model that accommodates these suggestive ideas. For this to work, one needs to make sure that these superpositions quickly evolve into states of well-defined consciousness in such a way that we would fail to notice these transitions in our experience.[2] The SOC model solves this problem. In a few words, it consists of a CSL model in which the collapse operator depends on consciousness. By doing so, we arrive at a model in which, as has been suggested throughout the years, consciousness plays a role in the collapse of the wave function. The advantage of this proposal, of course, is that we incorporate consciousness into quantum theory in a perfectly well-defined way, both mathematically and conceptually, and in a way which is fully compatible with the truth of physicalism if consciousness is physical.

The first thing we need in order to build our model is a physical property upon which consciousness is supposed to depend; let's call this property Φ. What is needed next is the construction of a quantum version of Φ. For this, first we note that, according to quantum theory, to every well-defined property corresponds a Hermitian operator. If Φ is a well-defined property, it follows that there must be a corresponding operator $\hat{\Phi}$. The proposal, then, is that only states with well-defined values of $\hat{\Phi}$ correspond to conscious states, i.e., only eigenstates of $\hat{\Phi}$ are conscious.

With $\hat{\Phi}$ in our hands, we can finally introduce our model. For a given initial state $|\psi(0)\rangle$, the SOC model has as solutions

$$|\psi(t)\rangle_B = e^{-\left\{iH+\frac{1}{4\lambda t}\left[B(t)-2\lambda\hat{\Phi}\right]^2\right\}}|\psi(0)\rangle, \tag{10}$$

with $B(t)$ a classical Brownian motion function selected randomly with probability density

$$\mathcal{P}_t\{B\} = {}_B\langle\psi(t)|\psi(t)\rangle_B. \tag{11}$$

Therefore, according to what we said about the CSL dynamics above, the SOC model is such that it drives any initial state into an eigenstate of the $\hat{\Phi}$ operator. If, as we proposed above, eigenstates of $\hat{\Phi}$ are indeed related to conscious states—by measuring consciousness or a property that perfectly correlates with it in the actual world—then, in the same way that standard CSL quickly destroys superpositions of macroscopic objects localized at

[2] One might be puzzled at first sight by the idea of a conscious state we fail to notice. It is nonetheless easy to make sense of it by means of the conceptual distinction in philosophy of mind between phenomenal consciousness—the experience we have—and access consciousness (Block, 2002)—what we come to notice. There is even empirical evidence that supports the claim that, in cognitive systems like ours, the mechanisms underlying these faculties are in fact segregated (Block, 2011, 2014; Sebastián, 2014).

different places, the above dynamics quickly kills superpositions of incompatible conscious states, leading to states of well-defined consciousness.

Looking back at the list of requirement for a satisfactory solution to the measurement problem presented above in section 2, we see that the SOC model seems promising. To begin with, it is fully formulated in precise, mathematical terms and, as long as the λ parameter in equation (10) is small, it reproduces the empirical success of standard quantum mechanics at the microscopic level—see section 4.1. Regarding an explanation of why certain macroscopic superpositions allowed by the standard theory never seem to occur, the model clearly takes care of the complication by not letting those states last for long.[3]

It is important to note that the standard choice for a collapse operator in CSL, in terms of position, is of course well justified by observations, but lacks an explanation or an independent motivation. In SOC, in contrast, the fact that we never observe superpositions in the position of macroscopic objects is simply a contingent fact, derived from the cognitive architecture that happens to give rise to consciousness in our case.

Despite these virtues, SOC faces its own devils. In the rest of the chapter we deal with some of them. In particular in section 3, we analyze the adequacy of the SOC model in light of different theories of consciousness present in the literature, and in in section 4, we analyze the adequacy of the SOC model as a quantum theory.

3. The SOC Model, Theories of Consciousness and Multiple Realizability

According to the SOC model, superpositions of conscious states are quickly suppressed, which in turn explains why we fail to observe macroscopically superposed objects. The SOC model achieves this by driving any initial state into an eigenstate of an operator that measures consciousness. In the original

[3] As we mentioned in footnote 1, it is very important to highlight an often overlooked distinction between two different scenarios: i) superpositions of incompatible perceptions and ii) well-defined perceptions of a macroscopic system in a superposition of different positions. Our model straightforwardly takes care of the first by not letting those states last for long. However, models in which the collapse of the wave function depends on consciousness, admit the possibility of macroscopic asstates which correspond to well-defined perceptions of a macroscopic system in a superposition of different positions are *not* suppressed by the model. Therefore, a way of restricting access to those states is required. We deal with this case in detail in Okon and Sebastián (2018).

proposal, such an operator ($\hat{\Phi}$) measures integration of information which, according to IIT, is a measurement of consciousness. However, all the SOC model really requires is for there to be a physical property that corresponds to consciousness; that is, for consciousness to be identical with such a property or, at least, for consciousness to perfectly correlate with it in the actual world. The SOC model is then neutral on the fundamental nature of consciousness. If the nature of consciousness is not physical, then SOC can still offer a dynamics that accommodates the idea that "consciousness collapses the wave function" insofar as there is a physical property that perfectly aligns with it in our world. More interestingly, the SOC model can provide such a dynamics even if consciousness is indeed a physical property. Unfortunately, it is not obvious that the SOC model is compatible with most theories of consciousness. In particular, it does not seem to be compatible with theories that endorse the possibility of very different physical set-ups underlying conscious states. To see why, we need to get some clarity regarding the notion of a "physical property."

According to standard quantum mechanics, there is a one-to-one correspondence between *physical properties* and Hermitian operators. This technical use of the notion of a physical property departs from common sense or ordinary use. Although, in the absence of an exhaustive definition, we might disagree on what counts as a physical property in such an ordinary sense—and hence disagree on whether, e.g., the property of *being a human* shall count as one of those—a clear case thereof suffices to illustrate the difference. *Being a table* is definitely a physical property in the ordinary common sense, but not obviously so in the required technical sense because one might reasonably think that there is no Hermitian operator that corresponds to such a property. The reason is that there are many different ways in which tables can be physically realized, and hence no underlying physical property in the required technical sense that all, and only, tables share.[4]

The problem for the SOC model is that it does not seem to be compatible with theories that endorse the multiple realizability of conscious states. Is there any reason to believe that, in the actual world, consciousness can be multiply realized? Yes, there is. Chalmers, e.g., has argued that two systems with a sufficiently fine-grained functional organization—to fix the mechanisms responsible for the production of behavior, and to fix behavioral

[4] From now on, unless otherwise indicated, we restrict our use of of the term "physical property" to the technical sense.

dispositions (Chalmers, 2010)—will be equally conscious and enjoy the same kind of experiences. Accordingly, what matters for consciousness is a certain—sufficiently fine-grained—functional organization, and once this functional organization is satisfied, we can abstract from its particular realization. As Chalmers presents the idea:

> [W]hat matters for the emergence of experience is not the specific physical makeup of a system but the abstract pattern of causal interaction between its components. (*ibid.*, p. 24)

Imagine that the required sufficiently fine-grained functional organization at which behavioral dispositions are fixed is that of neural networks. Neurons in our brain have a certain biochemical composition, but this composition is irrelevant if Chalmers is right. If a silicon chip can satisfy the same pattern of causal interaction as a neuron, then it would be possible to replace our neurons by those silicon chips without a change in the required functional organization and, therefore, in our conscious experience. This is what Chalmers calls the principle of *organizational invariantism*. Although this idea has not gone without controversy, Chalmers (1996, ch. 7) provides two convincing and complementary arguments in its favor: the fading/absent qualia and the dancing qualia arguments. Roughly, the arguments go as follows.

In the fading/absent qualia argument, we are asked to consider someone having, say, an experience of pain and, for the sake of a *reductio*, the possibility of a functional duplicate with a "brain" made out of silicon chips, which does not experience the pain at all. As the two systems have the same functional organization, we can imagine gradually transforming one into the other by replacing neurons by the corresponding silicon chips without changing the functional set-up. Two things might happen during the transformation: either the replacement of a single neuron switches off consciousness or the experience fades slowly along the process with every replacement. None of the alternatives is plausible, or so argues Chalmers. The first one because it requires that "there would be brute discontinuities in the laws of nature unlike those we find anywhere else" (*ibid.*, p. 238). The second one because it would require that a system, whose cognitive processes are not malfunctioning and that is conscious, be systematically wrong about its own experience, complaining about its horrible pain while it is merely having a really mild one.

In the dancing qualia argument, we also consider a transformation process from a system with a neuronal brain to a system with a silicon brain. However, in this case, we assume, again for the sake of a *reductio*, that they have different experiences; e.g., that after the replacement the subject has an experience as of blue while looking at a red apple. If the principle of organizational invariantism were false, when we switch from the neuronal brain to the silicon one, the subject's experience would change from an experience as of red to an experience as of blue, but such a change in experience would go unnoticed for her. What is more, we can imagine flipping the switch back and forth so that "the red and blue experiences 'dance' before [S's] eyes" (Chalmers, 1996, p. 253), but S does not notice any change. This does not seem plausible according to Chalmers. The fading and the dancing qualia arguments provide good support for thinking that consciousness is multiply realizable and hence, as we have suggested, incompatible with the SOC model.

Most (philosophical) theories of consciousness accommodate multiple realizability, and hence, seem to be in tension with the SOC model. Let us briefly discuss the details of the relation between different families of theories of consciousness and multiple realizability in some detail.

Dualist positions—which deny that consciousness is ontologically different from, and does not metaphysically depend on, physical properties—are perfectly compatible with the SOC model insofar as there is a physical property that, in the actual world, perfectly aligns with consciousness. But dualists can also deny that consciousness perfectly correlates with a physical property, endorsing the principle of organizational invariantism (Chalmers, 1996) and in tension with the SOC model.

Panpsychists defend the idea that our conscious experience depends upon the very nature of the most fundamental particles. Multiple realizability seems to be in tension with panpsychism because different systems can satisfy the same sufficiently fine-grained organization and yet differ significantly at the fundamental level (Sebastián, 2015). This makes panpsychism *prima facie* perfectly compatible with the SOC model.

Materialist theories can endorse an identity theory that would be compatible with the SOC model. However, most materialist theories accommodate multiple realizability.

Functionalists commonly hold that conscious states are those that satisfy a certain causal role to be determined either *a priori*, as in Lewis (1978);

Dennett (1991), or *a posteriori*, as in Baars (1988); Prinz (2012).[5] If such a causal role can be multiply realized, then functionalism seems also in tension with SOC as it is reasonable to assume that there is no physical property commonly instantiated by all the possible realizers.

Representationalists claim that conscious states are representational states, i.e., states that have adequacy conditions. Our conscious states represent the world and ourselves as being a certain way and we say that they are adequate or correct depending on whether the world is that way. For example, the conscious experience we have when looking outside the window can be evaluated as correct or incorrect depending on whether it is the result of the interaction with the environment or of the consumption of certain toxic substance.[6] The consistency of representationalism both with materialism and with functionalism depends upon a theory of mental content that makes it explicit what it takes for a system to be in the required representational state. Those that endorse a naturalistic approach typically and very roughly maintain that representational states are those that have the teleological function—one that determines what a trait should do rather than what it actually does—of carrying information about its object. For example, oversimplifying, one might think that the teleological function of the traits of biological entities like us depend on natural selection, and hence that a state to carry information as a state requires two things, i) there being reliable correlations between the state and the object, and ii) there being a certain kind of sender-receiver structure that allows us to exploit such correlations. This does require a common physical property, as the SOC model seems to assume.

Most theories of consciousness are consistent with multiple realizability. If any of those theories is correct, then it seems reasonable to think that there is no unique physical property that conscious states share, and hence that we cannot construct the operator that the SOC model requires. In reply, one could think of consciousness as being identical (or perfectly correlating), rather than with a physical property, with the property that results from the disjunction of all the possible realizers—the property of being this or that physical property. This in turn would require for the model to have as many equations as physical properties related to consciousness the realizers might

[5] Enactivist views can be seen as complex functionalist views (Hutto and Myin, 2013; Noë, 2005).

[6] Representationalists disagree on the representational properties of the corresponding state, and can thereby be taxonomized depending on the alleged content of experience.

have. This would make the model not only unappealing, but also hardly palatable. At any rate, we are going to argue that multiple realizability is not incompatible with the SOC model, as it *prima facie* seems.

According to the dynamic equation that the SOC model proposes, equation (10), superpositions of eigenstates of property Φ, associated with the Hermitian operator $\hat{\Phi}$, are quickly suppressed. If Φ measures consciousness, then we have an explanation of why we fail to observe superpositions of macroscopic objects. However, for the model to be satisfactory, we need to be able, in principle, to define $\hat{\Phi}$. We typically think of the properties associated with an Hermitian operator in a suitable Hilbert space as the fundamental ones that physics postulates. And one might reasonably think that, if consciousness is multiply realizable, there is no fundamental property that distinguishes conscious and unconscious states, in the very same sense as there is no distinctive fundamental property shared by, and only by, all tables. If we can only define Hermitian operators over fundamental properties, then it is not possible to construct the operator $\hat{\Phi}$ that the model requires.

Fortunately for the SOC model, the latter assumption is wrong: we can construct Hermitian operators that correspond to properties that are not fundamental. Conforming to any theory of consciousness that we would want to consider, there are going to be arranged collections of fundamental particles that correspond to conscious states and collections that do not—and maybe, collections such that it is indeterminate whether they correspond to conscious states or not. If there is a fundamental property that individuates the set of all conscious states, then we can uncontroversially construct $\hat{\Phi}$. If, on the other hand, there is none, then consciousness cannot be reduced to a set of fundamental properties—as the property of *being a table* arguably cannot either. This, however, does not mean that we cannot construct $\hat{\Phi}$, where Φ is either the property of being conscious or a perfect correlate. Given the adequate Hilbert space, all that we need to do is to consider all states that are conscious and note that all those states, being microscopically distinguishable, must be represented by vectors that are orthogonal among themselves. Then, since for every set of orthogonal vectors there is a Hermitian operator that has the vectors of the set as eigenstates, there must be a Hermitian operator that has all conscious states as eigenstates. That operator is the $\hat{\Phi}$ we were looking for. This is a perfectly legitimate Hermitian operator, which simply might not correspond to any fundamental property, but that does correspond to the property of consciousness. This is all that is required, so the SOC model, after all, is compatible with theories of consciousness that contemplate multiply realizability—such as functionalism or representationalism.

4. The SOC Model as a Quantum Theory

In this section we examine some aspects of the SOC models in light of considerations in the field of quantum foundations. First we explore the issue of fixing the value of the collapse parameter and then we enquire about the physical interpretation of the formalism.

4.1 Choosing the Value of the Collapse Parameter

The value of the collapse parameter λ in standard collapse models has to satisfy competing constraints. On the one hand, λ cannot be too large because microscopic phenomena, which we know are well-described by a purely unitary evolution, would get disturbed. Moreover, a large λ would lead to a type of quantum Zeno effect (QZE) [see Dagasperis et al. (1974)], in which the collapse terms dominate and eigenstates of the collapse operator suffer recurring collapses and effectively freeze. On the other hand, if λ is too small, these models would not achieve their purpose of suppressing undesirable macroscopic superpositions. Of course, one can allow for these macroscopic superpositions to persist for some time, but one needs to make sure for them to quickly die out before we are able to notice them. The beauty of these collapse models is that there are values for λ that provide empirically successful models of the world around us (see Adler (2007); Feldmann and Tumulka (2012); Bassi *et al.* (2003)).

One could worry that our model could run into trouble with QZE. If present, the effect would mean that conscious states would not be able to change, contradicting our experience. The situation, however, is perfectly analogous with standard, position-based, CSL models which, as we just mentioned, are not affected by QZE. That is, in the same way that standard CSL does not imply that the position of macroscopic objects would not change, our model does not imply eigenstates of Φ to freeze. This is so because the actual CSL evolution is always dictated by the interplay between the Hamiltonian and the collapse terms, so the actual evolution of a system involves a competition between the two. Of course, the result of this struggle is to be decided by the strength of the collapse terms, which is determined by the parameter λ, so the key question is if there is a possible value for λ in our model that avoids these problems and yields empirically successful predictions.

In our model, as in standard CSL, λ is a free parameter, a new constant of nature if you will, that controls the strength of the collapse terms. Its value must then be chosen, if possible, to make sure the model is empirically adequate. In particular, λ must be chosen to avoid, on one hand, the QZE, and, on the other, perceptible Schrödinger cat states. But one might worry, though, that this may not be possible for our model.

To begin with, we notice that in GRW, λ is taken to be a very small number ($\approx 10^{-16}\ s^{-1}$). However, the *effective* collapse rate of a system is given by $N\lambda$ where N is the number of particles of the system which are entangled and in a superposition of different positions. A macroscopic system in such a state of superposition is then extremely likely to undergo a collapse very soon, and if such collapses would continue at such a rate, then it would suffer from a QZE-type issue and practically freeze. This does not happen, however, because an initial collapse destroys the superposition, dramatically lowering the number of particles entangled, and thus the effective collapse rate. It is not clear, though, that a mechanism of this sort is present in our case. The issue described above contains in fact two related worries, one potentially afflicting CSL in general, and one specific to our model:

1. Unlike GRW, our model does not seem to contain a mechanism that suppresses collapses once an eigenstate (or something close to it) is reached; this could lead to a QZE-type problem.
2. Unlike GRW and standard CSL, the collapse rate in our model does not seem to scale with the number of particles.

Regarding 1, in spite of appearances to the contrary, the CSL model we employ does contain a mechanism that suppresses the collapse terms once the state is close to an eigenstate. This is not obvious from the CSL equation described above, which does not preserve the norm of the state (note the c_j multiplying the eigenstate at the end of equation (9)). However, it becomes clear if one writes an equation for a physical state that remains normalized during the evolution process. By doing so, one notices that the collapse terms are a function of the collapse operator *minus* its expectation value (see equaiton 3.13 in Pearle (1990)). Since, acting on an eigenstate, such a subtraction is zero, the collapse mechanism is, as desired, ineffective on such states.

Regarding 2, in GRW one postulates a collapse process at the micro level with fix rate λ_{GRW}, and depends on the state of the whole system having special correlations for the collapse to be significant at the macro level. As a result, one ends up with an effective collapse rate at the macro level that depends on the details of the macro state. In CSL, on the other hand, one postulates a collapse directly at the level of the whole system. So, regardless of the specifics of the initial state, the system will be driven to an eigenstate of the collapse operator. The strength of this collapse is always governed by the parameter λ_{CSL}, again, regardless of the details of the initial state, i.e., of there being or not special correlations in the state.

Moreover, in standard CSL, the collapse rate grows as N does because the collapse operator is constructed as a sum of single particle operators. In our model, this is not the case because Φ of the total system is not the sum of single particle operators. As we said before, this might seem problematic because our model does not appear to be able to distinguish micro and macro states, and to treat them accordingly. In order to deal with this, we can just *postulate* the collapse rate to be a function of the number of particles in the system; in other words, we postulate that the collapse rate has to be *renormalized* when changing the scale of the system under study. This might seem odd, but of course there is no guarantee that the laws of nature are such that the values of their parameters are scale-invariant (and, in fact, this even happens in quantum field theory with the change, or *running*, of the value of the parameters at different scales). Alternatively, instead of renormalizing λ as described above, one could simply add an N to the definition of our collapse operator. In fact, given that Φ is meant to measure consciousness, and thus, expected to capture some sort of complexity within the system, that N might even naturally appear in the definition of Φ.

4.2 Interpreting the Model

The standard interpretation of quantum mechanics subscribes to the so-called Eigenvalue-Eigenvector (EE) rule, which holds that a physical system possesses the value α for a property represented by the operator O if and only if the quantum state assigned to the system is an eigenstate of O with eigenvalue α. Such a rule is essential in order to link the mathematical apparatus of the standard formalism and predictions, i.e., it plays the role of a (partial) physical *interpretation* of the theory.

A legitimate concern regarding collapse models arises from the fact that such formalisms lead systems to states which are very close to eigenstates of the collapse operator, but not exactly to such eigenstates. Therefore, if one continues to subscribe to the EE rule, systems under collapse dynamics never actually possess well-defined values for the property associated with the collapse operator—ostensibly in contrast with what one was hoping for. As a result, one cannot *interpret* collapse theories in terms of EE and a different interpretation—that specifies the relation between the mathematical and the physical objects—is required.

In the context of standard collapse models, one could solve this issue with the use of the, so-called, *fuzzy link* interpretation introduced in Albert and Loewer (1996), in which one allows for some tolerance away from an eigenstate to ascribe the possession of well-defined properties. There are, however, complications with this approach. First off, it is not clear how to define in a non-vague and non-arbitrary way how close a state needs to be to an eigenstate in order for the value of a property to be well-defined. Moreover, this type of interpretation remains as ontologically obscure as the standard interpretation because it only talks about possession of properties but remains silent regarding what are supposed to be the property bearers according to the theory.

A more atractive alternative, again, in the context of standard collapse models, is to construct out of the wave function a, so-called, primitive ontology and to interpret it as the stuff that populates the world (see Allori (2005)). The most popular options in this direction are the flash ontology, in which the centers of the GRW collapses are taken to constitute the primitive ontology, and the mass density ontology, available both for GRW and CSL, in which a mass density in 3D space is constructed out of the wave function as the expectation value of the mass density operator. While promising, these approaches face some open issues (see, e.g., McQueen (2015)).

As with standard collapse models, one may worry about the fact that the SOC model does not really lead systems to eigenstates of Φ, but only to states that are very close to those eigenstates. Also, as with standard collapse models, if one adheres to the EE rule, one gets into trouble because one concludes that SOC leads to a scenario in which conscious states never actually occur. The solution, as with standard collapse models, is to deviate from the EE rule and introduce some alternative. Following what we said above, one option would be to introduce some sort of fuzzy link that ascribes consciousness to states which are close enough to Φ eigenstates. However, as with the standard case, it seems difficult to rigorously define such a "close enough." The other option, of course, is to introduce a primitive ontology, such as mass density, and use it to interpret the SOC model.

There are of course a number of concerns with collapse models, such as the so-called tails problem, that are not specific to our model. In fact, any proposal to solve the measurement problem on the market nowadays contains at least some open issues. We would be satisfied if our model turns out to be no worse than standard collapse models.

Acknowledgments

EO acknowledges partial financial support from DGAPA-UNAM project IN102219.

References

Adler, S. L. (2007). Lower and upper bounds on CSL parameters from latent image formation and igm heating. Journal of Physics A: Mathematical and Theoretical, 40(12):2935.

Albert, D. Z. and Loewer, B. (1996). Tails of schrodinger's cat. In Clifton, R., editor, Perspectives on Quantum Reality: non-relativistic, relativistic, field-theoretic, page 81–92. Kluwer.

Allori, V. (2015). Primitive ontology in a nutshell. International Journal of Quantum Foundations, 1:107–122.

Averill, E. W. and Keating, B. (1981). Does interactionism violate a law of classical physics? *Mind*, 90:102–107.

Baars, B. J. (1988). *A Cognitive Theory of Consciousness*. Cambridge University Press.

Bassi, A. and Ghirardi, G. C. (2003). Dynamical reduction models. Phys. Rep., 379:257–426.

Bassi, A., Lochan, K., Satin, S., Singh, T. P., and Ulbricht, H. (2013). Models of wave-function collapse, underlying theories, and experimental tests. Rev. Mod. Phys., 85:471.

Bell, J. S. (1990). Against measurement. In Miller, A. I., editor, Sixty-two Years of Uncertainty. Plenum Press.

Block, N. (2002). Some concepts of consciousness. In Chalmers, D., editor, *Philosophy of Mind: Classical and Contemporary Readings*. Oxford University Press.

Block, N. (2011). The higher order approach to consciousness is defunct. *Analysis*, 71(3):419–431.

Block, N. (2014). Rich conscious perception outside focal attention. *Trends in Cognitive Sciences*, 18(9):445–447.

Chalmers, D. (2010). *The Character of Consciousness*. Oxford University Press.

Chalmers, D. J. (1996). *The Conscious Mind: In Search of a Fundamental Theory*. Oxford University Press.

Chalmers, D. and McQueen, K. J. (MS). Wave-function collapse theories of consciousness. https://www.youtube.com/watch?v=UL1h-QgeD9c, https://www.youtube.com/watch?v=R-jOfW9UIEA\& feature=youtu.be\&list=PLl_UXfN1hubVda8RyXj1FLgwK4tW_AqEA\&t=2016.

Dagasperis, A., Fonda, L., and Ghirardi, G. (1974). Does the lifetime of an unstable system depend on the measuring apparatus? *Il nuovo cimento A.*, 21(3):471–484.

Dennett, D. C. (1991). *Consciousness Explained*. Back Bay Books.

Everett, H. (1957). 'relative state' formulation of quantum mechanics. Rev. Mod. Phys., 29(3).

Feldmann, W. and Tumulka, R. (2012). Parameter diagrams of the GRW and CSL theories of wavefunction collapse. Journal of Physics A: Mathematical and Theoretical, 45(6):065304.

Ghirardi, G. C., Rimini, A., and Weber, T. (1986). Unified dynamics for microscopic and macroscopic systems. Phys. Rev. D, 34:470–491.

Goldstein, S. (2013). Bohmian mechanics. In Zalta, E. N., editor, The Stanford Encyclopedia of Philosophy

Hutto, D. and Myin, E. (2013). *Radicalizing Enactivism: Basic Minds Without Content*. MIT Press.

Larmer, R. (1986). Mind-body interactionism and the conservation of energy. *International Philosophical Quarterly*, 26:277–285.

Lewis, D. (1978). Mad pain and martian pain. In Block, N., editor, *Readings in the Philosophy of Psychology*, volume 1. Harvard University Press.

London, F. and Bauer, E. (1939). La th orie de l'observation en m canique quantique. Actualit s scientifiques et industrielles, 755.

McQueen, K. J. (2015). Four tails problems for dynamical collapse theories. Studies in History and Philosophy of Modern Physics, 49:10–18.

Noë, A. (2005). *Action in perception*. Bradford Books.

Okon, E. and Sebastián, M. Á. (2018). A consciousness-based quantum objective collapse model. *Synthese*.

Pearle, P. (1989). Combining stochastic dynamical state vector reduction with spontaneous localization. Phys. Rev. A, 39:2277–2289.

Pearle, P. (1990). Toward a relativistic theory of statevector reduction. In Sixty-Two Years of Uncertainty, pages 193–214. Plenum Press.

Prinz, J. (2012). *The Conscious Brain*. Oxford University Press.

Sebastián, M. Á. (2014). Dreams: An empirical way to settle the discussion between cognitive and non-cognitive theories of consciousness. *Synthese*, 191(2):263–285.

Sebastián, M. Á. (2015). What panpsychists should reject: On the incompatibility of panpsychism and organizational invariantism. *Philosophical Studies*, 172(7): 1833–1846.

Stapp, H. (2007). *Mindful Universe*. Springer.

Wigner, E. (1967). Remarks on the mind-body question. In Symmetries and Reflections, page 171–184. Indiana University Press.

3

Quantum Mentality

Panpsychism and Panintentionalism

J. Acacio de Barros and Carlos Montemayor

1. Introduction

Panpsychism is gaining momentum as a theoretical explanation of phenomenal consciousness. Recent volumes and discussions about scientific interpretations of panpsychist views are evidence of this. A powerful combination of arguments and interpretations of quantum mechanics strengthens the case for panpsychism as a plausible view of fundamental reality. In fact, according to some authors, the case for panpsychism is now so strong that it should be considered superior to the, until recently prevailing, physicalist view, particularly in light of the hard problem of consciousness and considerations regarding the nature of intrinsic properties.

The theoretical and empirical arguments in favor of panpsychism are based on epistemic and metaphysical considerations. The main arguments are roughly as follows. If physicalism were true, the relational and functional vocabulary of physics should be capable, at least in principle, of explaining all the properties of fundamental reality. However, it is clear that such sparse vocabulary cannot account for all the properties that are metaphysically fundamental. In particular, scientific models can only capture the relational properties of fundamental entities that are described in terms of mathematical relations. This clearly indicates that such a vocabulary is insufficient since those relational, dynamical, structural, and dispositional properties must depend on their categorical bases. These properties, the categorical bases, must specify what fundamental entities are intrinsic in themselves and not only in relation to other fundamental entities. Therefore, metaphysicians have the task of complementing the physicalist picture with these fundamental, intrinsic properties.

J. Acacio de Barros and Carlos Montemayor, *Quantum Mentality: Panpsychism and Panintentionalism*
In: *Consciousness and Quantum Mechanics*. Edited by: Shan Gao, Oxford University Press.
 DOI: 10.1093/oso/9780197501665.003.0004

This argument about intrinsic properties can be combined with an intuitive argument about the epistemic status of consciousness. The epistemic argument states that the best candidates for intrinsic properties are those associated with phenomenal consciousness. The redness of red, for instance, is fully displayed in our experience of red. It is intrinsic to the experience. We could build mathematical models regarding how to describe red in physics as part of the spectrum of light and our knowledge of light detection by our retina and signal processing by the brain. But what red is, essentially and intrinsically, is fully given in our experience of red. This is not only an aspect of our experience, but the sole epistemic access we have to the redness of objects in our immediate environment. Thus, properties associated with our conscious awareness are epistemically and ontologically fundamental.

There are, then, good reasons to conclude that the categorical bases of the dispositional properties defined in fundamental physics are conscious because qualia are the intrinsic properties we know in an immediate and robust manner (throughout, by qualia, we mean the specific phenomenal properties of conscious experiences). One way of presenting this conclusion is that microphysical properties are conscious, in some relevant sense of "conscious." They either combine into larger conscious structures like us or are related through some kind of emergence relation to robustly phenomenal conscious experiences. These are two very different formulations of panpsychism, but they are both committed to the presence of consciousness at a fundamental level.

A substantial portion of the debate surrounding these two versions of panpsychism focuses on two key difficulties: the combination problem and the problem of micro to macro causation. The main approaches to solve these problems are the constitutive and non-constitutive views. The constitutive view claims that macro phenomenal properties depend on, and are realized by, micro properties that are intrinsically phenomenal (non-relational or dispositional). This view has the advantage that the same relation of composition holds between micro and macro properties, and this helps with the thorny problem of causation. The key difficulty for the constitutive view is the combination problem: How exactly is it that the micro properties organize themselves into macro structures with conscious experiences, such as ours? What kinds of macro structures are conscious like us? How are the experiences of macro conscious beings different from the conscious states of elementary particles?

The non-constitutive view avoids the combination problem because it appeals to an emergence-like relation that does not depend on the same fundamental composition relation at the micro level. The cost of this advantage is that the problem of causation (causal overdeterminism and downward causation) becomes intractable because the micro level causation is not directly responsible for macro level causal interactions, thereby generating the need for new macro to micro and macro to macro causal relations that cannot be explained at the micro level. Thus, even though the micro properties are conscious, it is unclear how exactly they casually determine the macro properties. Moreover, one might suspect that this non-constitutive view is compatible with a dual-aspect monistic view or even with standard physicalist emergentism, because the emergent properties may be radically different than the micro properties and, thus, the macro could be conscious while the micro is semi-unconscious or just "minimally" conscious. We will not explore this difficulty here and interpret this view as somehow requiring the micro to be conscious (for a review, see Bruntrup and Jaskolla, 2017).

If panpsychism is interpreted as the claim that the micro level properties, the categorical bases for physical dispositional and relational properties, are conscious, it would be important to explain how exactly this claim should be scientifically understood. It seems that without some scientific explanation, the panpsychist is merely playing with intuitions about the metaphysics of dispositions, appealing to infinite regresses that might not be relevant for a scientifically rigorous understanding of fundamental reality. This is particularly pressing for panpsychism since phenomenal consciousness is the type of thing we intuitively attribute only to creatures that look very much like us—intuitively, only animals very similar to us seem to be robustly conscious and, certainly, we never think inanimate matter is conscious. Therefore, the argument in favor of panpsychism cannot purely proceed in terms of intuitive appeal because there are strong intuitive reactions against it.

Here is where quantum mechanics comes in, allegedly providing strong support to panpsychism. An explanation of the relation between quanta and qualia can be provided by many of the major interpretations of quantum mechanics (e.g., Copenhagen, many-minds, observer-dependent). The appeal is neither unintuitive nor unnatural because, according to many physicists working on the foundations of quantum mechanics, the explanation of how the theory works must appeal somehow to a (conscious?) observer. This provides the crucial missing piece for the argument in favor of panpsychism: panpsychism might go against our commonsensical views about what kind of living organisms are conscious (only those that strongly resemble us in social skills and biology), but this is no obstacle because any explanation of fundamental reality, such as quantum mechanics, is unintuitive in that exact way.

We will not dispute, for the sake of argument, that micro-physical properties are conscious in *some* sense. Our objection is that phenomenal consciousness is not the best characterization of the type of mentality that is needed to understand fundamental reality because it does not capture essential features of the type of mentality physicists assume in their interpretations of quantum mechanics. We also think there are good reasons to reject panpsychism as a view of fundamental reality (see Montemayor, 2017, for an argument based in terms of the nature of information). Our focus here, however, is on developing an objection to the claim that the type of mentality involved at the micro level is *necessarily* phenomenally conscious. This is a challenge that affects, as far as we can see, the two prominent versions of panpsychism described above. We specifically target the key empirical claim that quantum mechanics might provide the scientific explanation of why panpsychism is the best view of fundamental reality, but we also provide reasons for thinking that, even for the epistemic argument, there might be reasons to distinguish between rational access to information and phenomenal consciousness.

2. Rational Categorical Bases and Indexical Information

A key aspect of our argument is to explain how anything else, other than phenomenal consciousness, could be intrinsic. Intentionality, as Franz Brentano first noticed, is essential to the mind. Mental states have aboutness in an intrinsic way—they are directed towards semantic contents intrinsically, without necessitating the existence of the objects or situations they are about in any specific extrinsic way. It is because of this that Brentano called this property of mental states their "intentional inexistence." This property of aboutness, moreover, provides rational access to semantic contents in the sense that, for a creature with conceptual capacities like us, being directed towards a semantic content provides immediate access to how that content is related to other contents in a coherent way. Any creature with language will likely have a mind that is intrinsically intentional. The question is whether such a mind needs to also have the kind of phenomenology our minds have.

Although there are good reasons to think that only qualia can provide the mind with its intrinsic intentionality, there are at least equally powerful reasons to reject this view. One of them is based on the possibility of unconscious mental representation. The mind eliminates possibilities concerning contents in a way that can only be explained in terms of intentionality, even though in many cases it does so without any specific phenomenology. Think about syntax processing, for instance. We understand the content of words, and their meaning is part of the phenomenology of listening to and thinking in language. But the rules of syntax we use in decoding the meaning of words have no phenomenology. This is intentional aboutness, intrinsic to our mental states concerning language, but without phenomenology.

This kind of content, in spite of its lack of phenomenology, is still intrinsic to our minds. It is qualitative, rather than purely quantitative and mathematically modeled information. It makes possible the drastic qualitative difference between human beings and species without language. But this qualitative informational difference, which is intrinsic to the human mind, need not be defined phenomenally—this is a case of qualitative information without qualia. This does not entail that qualia play no role in determining some intrinsic features of our mind. On the contrary, qualia are indeed obvious candidates for intrinsic properties, and we are not disputing this. Our claim is that there are other aspects of our mind that are also intrinsic to the mind and that might be independent from qualia. The case of unconscious perception and cognition is one example.

2.1 Argument for Panintentionalism

1. There are intrinsic properties of our mind without phenomenal character: they are *purely intentional* properties.
2. Purely intentional properties suffice to explain the mentality involved in quantum mechanics and the micro properties.
3. Qualia are not necessary to explain the mentality involved in quantum mechanics.

Premise (1) is incompatible with the view that all intentionality depends on phenomenology, but it is compatible with many other views of the mind, including functionalist and physicalist views. In fact, this view is compatible with views that give phenomenal consciousness a primary or foundational epistemic role but also make room for unconscious mental representation. We take premise (1) to be uncontroversial enough to accept it without further defense. Defending premise (2) is the purpose of the remainder of this paper.

One could go further than this sufficiency argument and propose that qualia are *inadequate* to explain the kind of mentality involved in quantum mechanics. Although this stronger claim is compatible with some of our views, we will defend the weaker claim since that is enough to show that quantum mechanics is best understood in terms of panintentionalism. More precisely, our main conclusion is that, although both purely intentional properties and qualia are intrinsic, it is easier and more parsimonious to assume that intrinsic properties at the micro level are purely intentional. The central claim is, therefore, that purely intentional properties of the mind related to content and qualitative information specification are better suited to explain the type of mentality involved in quantum mechanics. Panintentionalism is a form of panpsychism to the extent that it assumes some type of mentality at the micro level, but it is incompatible with the prevailing view of panpsychism that assumes these mental properties at the micro level are necessarily phenomenal.

A key aspect of our argument is that semantic relations represented in purely intentional properties are informational in an indexical way, without invoking the qualitative character of experiences. They eliminate possibilities in a way that is intrinsically related to the perspective of a mind or an observer that has at least some rational basis to determine an outcome from her perspective. Information in this sense need not be essentially relational, and in fact, quantum mechanics itself demands a distinction between purely relational or mathematically modeled information and qualitative or indexical information.

As Šafránek et al. (2019) argue, information cannot just be relational in fundamental physics, and a distinction is needed between actual and potential information. But as it is clear from his discussion, this does not entail that intrinsic information is necessarily phenomenal (IIT assumes this, but it need not be an essential assumption of intrinsic indexical information; see De Barros). Šafránek and colleagues say that such an understanding of indexical information must account for the fundamental physical notions of thermodynamics and entropy at the quantum level, a notion Šafránek et al. call "observational entropy" (Šafránek et al. 2019).

Influential interpretations of the central cases that are used to argue for the alleged metaphysical necessity of phenomenal consciousness as independent from any physical descriptions, such as Jackson's Mary, have explained the change Mary undergoes in terms of indexical information and not qualia (de Barros et al., 2017; Stalnaker, 2008; Perry, 1979) or as an observational ability Mary gains (which can also be understood in terms of indexical content; see Kwon, 2017).

One of the reasons offered in favor of qualia as natural candidates for being the categorical bases of physical dispositions is that they are associated with subjectivity and semantic content in an irreducible intrinsic manner. One of the more parsimonious ways to explain this claim is in terms of contextualized forms of agency (what David Lewis' (1979) "two Gods" lack). In the case of Lewis' two Gods, their omniscience of all facts is insufficient to provide them with the indexical belief "This is who I am." This type of knowledge need not be phenomenal, as Perry's messy shopper illustrates. What it certainly requires is a new type of belief with indexical content (a de se belief).

The essential change in Mary is a change in agency, which involves a kind of *personal centering* that can be understood in terms of new agential abilities (Kwon, 2017). Perry characterizes the new type of information Mary learns in terms of a specific kind of content, which he calls "reflexive content" that does not require a new property but a new way of relating to it: a fundamentally contextual and self-referential way of referring to it. Perry writes: "Identities can be informative because there are two ways of knowing, two modes of presentation. If modes of presentation are limited to attributive conditions of reference, the situation cries out for *another property* to explain the informativeness of the identity. My solution has been to explain the twoness, in its various forms, not at the level of what is known about, but at the level of *what is involved in the knowing*: the level of reflexive content" (Perry, 2001, 207; our emphasis).

Contextual information for a specific kind of agency (an observation that is capable of grounding the reflective endorsement of further observation and action) is essential to draw the distinction between actual information centered on an agent and general or potential information concerning possibilities. Such capacities seem to require attention on the part of the agent, but not necessarily phenomenal consciousness (Fairweather and Montemayor, 2017). This centering and indexical relation depends on abilities to rationalize and expand our own contexts of information evaluation. As Stalnaker says:

> We begin . . . in more local contexts. We then develop means for expanding our representational resources, and for incorporating information from different contexts into more inclusive contexts. Doing this will involve representing ourselves and our local contexts within a more robust and inclusive context, and representing, in our conception of the world as it is in itself, *the relation between ourselves and the things we represent* (what John Perry called reflexive content). In the end, we must recognize that even our most stable and robust representations have the content that they have in virtue of our relations to what we represent.
>
> (Stalnaker, 2008, 137; our emphasis)

So far, we have argued that the type of mentality required for certain interpretations of quantum mechanics should comply with a constraint: categorical bases at the micro level must explain the rational and semantically evaluable measurements that determine outcomes in quantum mechanics. If access consciousness or attention (or non-phenomenally conscious cognition) suffices to specify such observational capacities, then the argument for phenomenal consciousness is unjustified. Access consciousness or attention without qualia suffices to explain these capacities. Therefore, there is no need to postulate phenomenal properties at the micro level to explain the kind of mentality required for quantum mechanics.

3. The Notion of Observer in Quantum Mechanics

Let us now discuss the role of the observer in quantum mechanics in order to support our claims concerning access consciousness. In classical physics, the state of a system of particles is determined by their position and momenta, represented as a point in phase space (the space of all possible position and momenta, usually $\mathfrak{R}^{6N}$, where N is the number of particles). This comes from the fact that, in classical mechanics, the dynamics in phase space is given by a set of differential equations that are first order on time, and therefore the position and momenta for each particle determine their position and momenta at any other time.

Quantum mechanics has something similar. The state of a quantum system of particles is not their position and momenta, which cannot be simultaneously measured in the quantum world as accurately as we want, but is described by a vector in a Hilbert space.[1] The dimension of the Hilbert space depends on the number of independent properties that can be measured for this quantum system and can be very large. The vector representing the state of a system of quantum particles is also governed by a first order differential equation, and given this vector at time t_0, its state vector at another time $t \geq t_0$ is also determined. However, there is a crucial caveat about the state at a later time: it is determined if no measurement is made. If a measurement is made, then the dynamics of the quantum system may go from deterministic to probabilistic.

This aspect of a measurement, where it can change the type of dynamics for a quantum system, leads to some important questions. What constitutes a measurement? What happens during a measurement? How do we model measurements? What makes measurements special, i.e., why are their interactions with a system different from other non-measurement interactions? Those issues are still the subject of intense discussion in the physics community, and they are intimately related to what is known as the *measurement problem.* So, given its importance and relationship to the notion of observers in quantum mechanics, we will briefly discuss the measurement problem here.

The measurement problem is a consequence of two important characteristics of quantum theory: its linear and unitary dynamics, and its state representation via vectors in a Hilbert space. Let us unpack each one of those and show how it leads to some difficulties. First, let us start with the representation with vectors.

For simplicity, let us start with a two-dimensional Hilbert space where we have two linearly independent and orthogonal vectors **v** and **w**. Each vector **v** and **w**, in quantum theory, represents a possible property for the quantum system, which we will represent using the projection operators P_v and P_w, defined as $P_v\mathbf{u} = (\mathbf{v} \cdot \mathbf{u})\mathbf{v}$ and $P_w\mathbf{u} = (\mathbf{w} \cdot \mathbf{u})\mathbf{w}$ for any vector u in the Hilbert space.[2] Projectors P_v and P_w, which can also be called observers, have two

[1] A Hilbert space is a vector space that is complete, i.e., all Cauchy sequences in it converge to an element in it, and is endowed with a metric.

[2] Usually, the bracket notation is used in quantum theory, but this needs the concept of a dual space, which we do not assume the reader is familiar with. Instead, we keep our discussion within elementary linear algebra.

possible eigenvalues, 0 or 1, and their eigenvectors are, respectively, **v** and **w**. These eigenvalues are interpreted as the two possible values of the properties associated with the projectors, with 1 being equivalent to "the system has the property" and 0 to "the system does not have the property." If the state is **w**, it follows that it has property p_w but not P_v due to the orthogonality of **v** and **w**, and similarly for **v**.

Of course, there is nothing necessarily special about **v** and **w**, and we could use another set of orthogonal vectors to represent the two properties. However, a consequence that the dynamics of quantum theory is linear and unitary is that it is possible to create an evolution of the quantum system such that it starts in a state, say, **w** and ends up in another state $\mathbf{s} = c_v\mathbf{v} + c_w\mathbf{w}$, with $|c_v|^2 + |c_w|^2 = 1$. The state **s** is said to be in a superposition of **v** and **w**.

What can we say about the properties of **s**? In particular, what are the values of P_v and P_w? A quick computation reveals that $P_v\mathbf{s} = c_v\mathbf{v}$ and $P_w\mathbf{s} = c_w\mathbf{w}$, which shows that **s** is not an eigenvector of the projectors, and therefore does not have a well-defined property before the measurement. However, after a measurement, either P_v or P_w is true, and the system "jumps" to their eigenvector, depending on the observed value. This jump, known as the *collapse of the state vector*, does not happen deterministically, but stochastically, with probabilities given by Born's rule, i.e., $|c_v|^2$ for P_v and $|c_w|^2$ for P_w. So, superposition states of the measurement basis lead to a probabilistic dynamics when interacting with a measurement device.

To understand what was special about superpositions, von Neumann (von Neumann, 1983) attempted to model the interaction of a quantum system with a measurement apparatus. His idea was that, if quantum mechanics was a universal theory, we should be able to explain the outcomes of a measurement using only quantum theory. He then constructed a model where the measurement apparatus, when interacting with a system in the state **v**, would end in a state where its "pointer" indicated that property P_v was true and that P_w was false. Similarly, when interacting with a system in the state **w**, the "pointer" would indicate property P_v as false and P_w as true. However, as von Neumann remarked, because the quantum evolution is unitary and linear, it follows that if the system starts at the superposition $\mathbf{s} = c_v\mathbf{v}+c_w\mathbf{w}$, it will evolve such that the measurement apparatus transforms itself into superposition and not in a quantum state where it has definite properties such as "pointer indicating P_v is true" or "pointer indicating P_v is false." In other words, the interaction between the measurement apparatus and the system does not create the quantum jumps, nor does it explain why only one outcome happens.

But von Neumann went even further. He said that we could model also the interaction between our eyes, themselves considered a type of measurement apparatus, and the system composed of "original system + measurement apparatus." If we do that, we end up with our eyes also in a state of superposition. We can then model the interaction between our eyes and our optic nerves, or our brain, or even all the matter in the universe, and we will find that we still end with a superposition. But, as von Neumann emphasized, we never *see* a superposition. We never see Schroedinger's cat in a state that is neither dead nor alive. So, von Neumann claims, the interaction of the *observer* and the totality of system and measurement apparatuses is what causes the collapse of the wave function.

Von Neumman's idea was pushed even further by London and Bauer (London & Bauer, 1939), who argued that if we exhaust all matter, what is left, the observer, is not matter. Eugene Wigner (Wigner, 1961), and later Henry Stapp (Stapp, 2009) went so far as to say that this implied a dualist view of nature, where matter satisfied the linear and deterministic dynamics of Schroedinger's equation (or unitary evolution) only when it didn't interact with consciousness, itself a non-material entity not describable by quantum theory. So, from this line of reasoning, consciousness enters the quantum realm through the interaction between the conscious observer and the quantum system.

Could we confirm the idea that consciousness causes the collapse (CCC) of the wave function, i.e., what leads to the probabilistic dynamics that ends in a state with a definite property is the interaction of the system/measurement device and a conscious being? It so happens that confirming or falsifying this is very difficult, in fact probably impossible in principle. To understand this, let us examine a simple thought experiment that is, maybe even with today's technology, executable (de Barros & Oas, 2017).

Imagine a box isolated from any external environment. Inside the box, an animal whose eyes sensitive to single photons (e.g., a cockroach or locust) is stimulated with a single photon either on its left or right eye. This animal is conditioned to respond differently to photons on the right eye than from photons on the left eye. The response activates the release of a photon on output 1 of the box if the animal responds to a photon on the right, and output 2 if responding to a photon on the left (see de Barros & Oas (2017) for a detailed description of this thought experiment). If, after the interaction with the photon, the animal is guided back to its original state, we would be left with two possible outcomes. If CCC, then regardless of what type of input we enter in the box, the output would always be either a 1 or 2 photon, but never a superposition. However, if the CCC hypothesis is false, then we would end up with a superposition of output 1 and 2 photons. In principle, with repeated experiments, one could distinguish a superposition from a non-superposition (what is called a *proper mixture*).

The difficulty with the above experiment is the following. In order to preserve the quantum superposition, we need to ensure that the animal and box are brought back to their original state. To see how difficult this is, imagine that we conditioned a cockroach to respond to different photons. This means bringing each atom[3] in the cockroach's box back to the state that they initially were in before the photon arrived. This is a gargantuan task, something that has never been accomplished even for such a small living system such as a tardigrade. But we have an additional issue. On top of bringing the system back to its original state, we also must ensure that the cockroach does not couple with the thermal bath, as this would lead to irreversible loss of quantum information, and make the experiment inconclusive (i.e., we would always see a mixture, regardless of whether CCC is true or not).

The difficulties above can be of fundamental importance to falsify the CCC hypothesis. Let us start with the fact that the experiment described above is with an insect, say, a cockroach. It is not obvious that insects have phenomenal consciousness, and the idea that a locust or a cockroach is conscious is certainly not uncontroversial. However, on top of that, the experiment requires the cockroach to be set at a temperature close to 0 K. First of all, it is hard to imagine behavioral responses happening at this temperature (probably impossible). But if it were possible, it is hard to imagine that any type of (uncontroversial) conscious process would exist. So, this experiment, if found to falsify the CCC hypothesis, is open to many criticisms and would be inconclusive.

But the measurement problem and its relationship to the observer are not the only argument about the importance of the observer in quantum physics, and it is worth discussing the other argument. The starting point is the famous Kochen and Specker's theorem (Kochen & Specker, 1967).

[3] At least the ones coupled with the impinging photons.

We are not going to detail its proof, and the interested reader is referred to their original paper or to the easier proof given by de Barros and Oas (de Barros & Oas, 2017), but we will sketch its main idea. Imagine you have a Hilbert space with a dimension greater than 3. Because the dimension is 3, it is always possible to find a set of three orthogonal vectors representing three distinct and compatible properties, P_{v_a}, P_{v_b}, P_{v_c} such that only one of them is true for the system. This means that there is an experiment we can perform where we can measure each of those properties, and ask which one is true for the quantum system. Now, imagine that we get this first set of three vectors and keep one of them fixed (i.e., the one associated with P_{v_a}), and rotate the three vectors around the fixed one. We now end with two more properties compatible with the first one, namely P_{v_d}, P_{v_e}. It is "reasonable" to assume that if we measure the system, we will find the same answer for P_{v_a}. So, the question we can ask is this: Can we now assign truth values to all properties associated with the set of possible measurements for this system? The answer is no: if we construct a set of observables that assume such truth values can be assigned, independent of the experiment where it is obtained, we reach a contradiction.

One key aspect of the Kochen/Specker result is the assumption that we can assign truth values independent of the experimental conditions (for details, see de Barros, Jorge, & Holik, 2021). In other words, the assumption that P_{v_a} is the same when we measure it with P_{v_b}, P_{v_c} or with P_{v_d}, P_{v_e} leads to a contradiction. This characteristic of quantum systems is called contextuality: the values of a property can change from one context to the other.

Now the observer enters the picture by selecting the context. When Alice, in her lab, decides to measure P_{v_d}, P_{v_e} or P_{v_d}, P_{v_e}, if we assume Alice has free will, she is deciding to create the values of P_{v_a} for one of those contexts, and not for the other. The experimental choice of the observer affects the observed quantities. This is a different issue from the measurement problem above.

Let us go back to the issue of the observer and consciousness in both cases: the measurement problem and quantum contextuality. In the first one, we have what is called the *Consciousness Causes Collapse hypothesis* (CCCH), where the presence of a conscious observer changes the dynamics of the system: deterministic and linear without consciousness, stochastic and non-linear with consciousness. In the second case, even if we do not postulate the CCCH, the outcomes of experiments dictate that the property of a quantum system depends on the choices of a free-willing observer.

In the CCCH, the argument is simple: if we believe that the linear and deterministic dynamics of Schroedinger's equation (or unitary evolution) applies to all matter, except when interacting with a conscious observer, we are almost forced to conclude that the observer, or its mind, is not matter, as if it were, the evolution of its different mental states would also be in a superposition. This mind, with its recollection of the recorded experimental events never being in a superposition, is a conscious mind. It is this conscious realization of an observation that collapses the wave function. As we mentioned above, this is not an argument that can be easily dismissed empirically, as any experiment attempting to falsify this theory needs to isolate a "conscious" observer from its surrounding thermal bath, virtually creating an unsurvivable environment for organisms that one could uncontroversially refer to as conscious. So, criticisms of the CCCH would have to come from metaphysical arguments. In the contextual case, the connection between quantum mechanics and consciousness is a little less clear. Alice, the conscious observer, decides which sets of properties she measures at the same time. It is her (possibly) conscious decision that creates the objective reality of quantum properties. Consciousness comes here from a more indirect path: the observer's decision making and her free will.

Regardless of how we want to approach it, we hit an important issue. In both cases, CCCH or decision making and free will, their proponents use the idea of a conscious being. However, neither case requires phenomenal consciousness. In CCCH, what is required is the irreversibility of memory registrations on the observer's "mind," which can be thought as related to access consciousness and not phenomenal consciousness. In decision-making and free-willing decision making, we also need access consciousness, not phenomenal consciousness.

As we emphasized before, access consciousness is ideally suited to play the mental role that is fundamental to certain interpretations of quantum mechanics because of its relation to rationality and semantic content. Phenomenal consciousness is strictly dependent on appearance and qualitative properties. Access consciousness, by contrast, is strictly dependent on rational access to contents. The observer in quantum mechanics is an access-conscious observer, rather than a phenomenally-conscious observer, because measurements are not entirely determined by merely appearance properties of experiences, but rather by concrete interventions in an environment by a rational agent with specific goals that have unique theoretical meaning.

The observer in quantum mechanics, even in a panintentionalist or panpsychic interpretation, is a rational agent that speaks the language of mathematics and expects certain regularities from her environment. The act of observation is not merely "the universe reflecting on its experiential aspects" but rather, the universe understood under a rationally controlled *mental action*: the act of measurement. The extent to which mentality is involved in quantum mechanics at a fundamental level is not thoroughly undefined, unanalyzable, and beyond analysis. Even if mentality is intrinsic to the universe, the role of mentality must be understood in terms of acts of measurement, and this can be fully captured by access consciousness.

4. Conclusion

We have argued that even if one grants that panpsychism is true, i.e., even if mentality is an intrinsic property of fundamental reality, it is inadequate to conclude that this kind of mentality must be understood in terms of phenomenal consciousness or subjective qualitative character. We propose that access consciousness is the type of mentality that best characterizes panpsychic mentality in quantum mechanics. One may object that this deprives the universe of the intense interest and value we attribute to the universe. This objection would be misplaced because our argument concerns fundamental reality and not the value we humans might attribute to the universe. But even if value is a concern, it is not clear that phenomenal consciousness is the only source of value one should consider—access consciousness is at least a plausible candidate, if not a superior one, as Levy (2014) proposes.

References

Brüntrup, G., & Jaskolla, L. (Eds.) (2017). *Panpsychism: Contemporary Perspectives.* New York: Oxford University Press.

de Barros, J. A., & Oas, G. (2017). Can We Falsify the Consciousness-Causes-Collapse Hypothesis in Quantum Mechanics? *Foundations of Physics, 47*(10), 1294–1308. https://doi.org/10.1007/s10701-017-0110-7

de Barros, J. A., Montemayor, C., & de Assis, L. P. G. (2017). Contextuality in the Integrated Information Theory. In de Barros, J. A., Pothos, E., & Coecke, B. (Eds.), *Quantum Interaction: Lecture Notes in Computer Science (10106)* (pp. 57–70). New York: Springer International.

de Barros, J. A., Jorge, J. P., & Holik, F. (2021). On the assumptions underlying KS-like contradictions. *arXiv preprint arXiv:2103.06830*.

Fairweather, A., & Montemayor, C. (2017). *Knowledge, Dexterity, and Attention: A Theory of Epistemic Agency*. Cambridge, UK: Cambridge University Press.

Kochen, S., & Specker, E. P. (1967). The Problem of Hidden Variables in Quantum Mechanics. *Journal of Mathematics and Mechanics, 17*, 59–87.

Kwon, H. (2017). Mary and the Two Gods: Trying Out and Ability Hypothesis. *Philosophical Review, 126*(2), 191–217.

Levy, N. (2014). The Value of Consciousness. *Journal of Consciousness Studies, 21*(1–2), 127–138.

Lewis, D. (1979). Attitudes de dicto and de se. *Philosophical Review, 88*, 513–543.

London, F., & Bauer, E. (1939). *La théorie de l'observation en mécanique quantique*. Paris: Hermann.

Montemayor, C. (2017). The Problem of the Base and the Nature of Information. *Journal of Consciousness Studies, 24*(9–10), 91–102.

Perry, J. (1979). The Problem of the Essential Indexical, *Nous* 13: 3-21.

Perry, J. (2001). *Knowledge, Possibility, and Consciousness*. Cambridge, MA: MIT Press.

Šafránek, D., Deutsch, J. M., & Aguirre, A. (2019). Quantum coarse-grained entropy and thermalization in closed systems, *Phys. Rev. A, 99*(1), 012103.

Stalnaker, R. C. (2008). *Our Knowledge of the Internal World*. New York, NY: Oxford University Press.

Stapp, H. P. (2009). Mind, Matter, and Quantum Mechanics. In H. P. Stapp (Ed.), *Mind, Matter and Quantum Mechanics* (pp. 81–118). Heidelberg: Springer Berlin.

von Neumann, J. (1983). *Mathematical Foundations of Quantum Mechanics* (translated by Robert T. Beyer from the 1932 German edition). Princeton, NJ: Princeton University Press.

Wigner, E. (1961). Remarks on the Mind-Body Problem. In Good, I.J. (Ed.), *The Scientist Speculates* (pp. 284–302). London: William Heinemann, Ltd.

4

Perception Constraints on Mass-Dependent Spontaneous Localization

Adrian Kent

1. Introduction

Finding a theory that unifies quantum theory and gravity is universally agreed to be a fundamental unsolved problem in physics. Finding a theory that explains the apparent emergence of classicality from quantum theory, resolving the so-called measurement problem or reality problem is thought by many to be another, and there are several well-known lines of thought on possible solutions. Explaining the emergence of consciousness from either classical or quantum physics is also thought by many to be a fundamental problem; those who think this mostly think we do not currently have lines of thought that promise anything like a complete solution.

One popular approach to the measurement problem is to propose explicit laws governing wave function collapse. Wigner [1] considered the possibility that collapses take place when observations are made by conscious observers. Diosi [2] and Penrose [3] suggested that unifying quantum theory and gravity may require that superpositions collapse whenever they would otherwise create superpositions of distinguishable spacetimes. Ghirardi-Rimini-Weber-Pearle [4, 5] developed spontaneous collapse models, in which unitary quantum dynamics are replaced by stochastic differential equations that are proposed as fundamental laws, from which the unitary Hamiltonian evolution of micro-systems and the effective collapse of macroscopic superpositions emerge as special cases. In the currently preferred versions of these models, collapse rates are proportional to mass densities. This avoids the need to treat composite particles such as nucleons as composed of definite numbers of elementary particles, which would be

Adrian Kent, *Perception Constraints on Mass-Dependent Spontaneous Localization* In: *Consciousness and Quantum Mechanics.* Edited by: Shan Gao, Oxford University Press. © Oxford University Press 2022.
DOI: 10.1093/oso/9780197501665.003.0005

difficult to reconcile with current theory. It also maintains consistency with current experiments, which appear to exclude the original GRW [4] model. Moreover, appealingly, it suggests a link with gravity.

Although all of these justifications are certainly questioned, collapse hypotheses thus also risk over-motivation. It is not immediately obvious that a collapse law designed to prevent spacetime superpositions necessarily also explains the appearance of classical outcomes of all measurements, or even that it is possible to find a single law that does both and remains consistent with known experiment. In principle, of course, one could postulate two or even more collapse laws: Wigner and Diosi-Penrose could be pointing to independent fundamental collapse phenomena, for example. For most theorists, though, this seems at least one law too many. We would like any alternatives to unitary quantum dynamics to be as simple and elegant as possible and to explain as much as possible.

To define and analyze the question quantitatively, we need to consider specific dynamical collapse models. I focus here on mass-dependent spontaneous collapse models, and on a pioneering paper [6] by Bassi, Deckert and Ferialdi (BDF) which considered the implications of these models for events associated with visual perception. These are certainly not the only models linking gravity with collapse, and indeed do so less directly than other proposals. However, they are better developed than most, their experimental implications have been carefully analyzed, and they include two parameters that allow the predictions of other models to be compared and fitted in a given experimental regime.

On this complex topic, it is natural that some assumptions may be debatable, and progress is likely to be incremental. Indeed, reanalysing BDF's arguments, I note some problems both with the calculations and the approximations. These make a significant enough difference—a factor of at least ≈ 10 and perhaps significantly more in the lower bound on the collapse rate—that they cast doubt on the conclusion that the relevant collapse models can be consistent both with known experiment and with collapse taking place within human perception times.

That said, a definitive conclusion would require a very complicated analysis, including a detailed understanding of physical chemistry, microscopic cell biology and the correlates of conscious visual perception in the human brain. I am unable to present such an analysis, and indeed not certain that the present state of understanding of these topics will allow precise and reliable estimates of collapse rate bounds from perception. Nonetheless, more progress can surely be made, and I hope that this discussion will stimulate further work.

2. BDF on Continuous Spontaneous Localization

To ensure that we represent BDF accurately, we quote directly from their analysis in this and the next section. BDF begin by presenting the stochastically modified Schrödinger equation that defines the mass proportional version of the Ghirardi-Pearle-Rimini [5] continuous spontaneous localization model:

$$|d\psi_t\rangle = \left(-\frac{i}{\hbar}Hdt + \sqrt{\gamma}\int d^3x(M(\mathrm{x}) - \langle M(\mathrm{x})\rangle_t)dW_t(\mathrm{x}) \right.$$
$$\left. -\frac{\gamma}{2}\int d^3\mathrm{x}(M(\mathrm{x}) - \langle M(\mathrm{x})\rangle_t)^2 dt\right)|\psi_t\rangle. \tag{1}$$

Here H is the Hamiltonian and $M(\mathrm{x})$ is a smeared mass density operator. It takes the form

$$M(\mathrm{x}) = \frac{1}{m_N}\int d^3y g(\mathrm{x}-\mathrm{y})\sum_s m_s a_s^\dagger(\mathrm{y})a_s(\mathrm{y}), \tag{2}$$

where the sum is over particle species s with mass m_s. BDF take m_N to be the mass of a nucleon, in an approximation in which the difference between the proton and neutron masses is negligible. The smearing function is taken to be

$$g(\mathrm{x}) = \frac{1}{(2\pi r_C^2)^{3/2}}\exp(-\mathrm{x}^2/(2r_C^2)). \tag{3}$$

Here the coupling constant γ and the length scale r_C are parameters of the collapse model. These may be varied independently, and a complete analysis would consider all ranges of both. In their analysis BDF set $r_C \approx 10^{-5}$cm and consider the bounds implied for γ, or equivalently for the collapse rate

$$\lambda = \frac{\gamma}{8\pi^{3/2}r_C^3}. \tag{4}$$

BDF then consider a superposition of states of N particles, of the form

$$\alpha'|\bar{\mathbf{x}}'\rangle + \alpha''|\bar{\mathbf{x}}''\rangle, \tag{5}$$

where $\bar{\mathbf{x}}' = \mathbf{x}'_1, \mathbf{x}'_2 \ldots \mathbf{x}'_N$ and $\bar{\mathbf{x}}''$ is similarly defined. (Here BDF implicitly assume that each particle has the nucleon mass m_N: an atom with atomic mass x dalton is effectively treated as a system of x tightly bound nucleons in their discussion.) They set the Hamiltonian to zero, and write the stochastic average density matrix as

$$\rho_t = \mathbb{E}\left[|\psi_t\rangle\langle\psi_t|\right]. \tag{6}$$

They then give the time evolution of the off-diagonal elements:

$$\frac{\partial}{\partial t}\langle \bar{\mathbf{x}}' \,|\, \rho_t \,|\, \bar{\mathbf{x}}'' \rangle = -\Gamma(\bar{\mathbf{x}}', \bar{\mathbf{x}}'')\langle \bar{x}' \,|\, \rho_t \,|\, \bar{\mathbf{x}}'' \rangle. \tag{7}$$

Here

$$\Gamma(\bar{\mathbf{x}}', \bar{\mathbf{x}}'') = \frac{\gamma}{2}\sum_{i,j=1}^{N}\left[G(\mathbf{x}'_i - \mathbf{x}'_j) + G(\mathbf{x}''_i - \mathbf{x}''_j) - 2G(\mathbf{x}'_i - \mathbf{x}''_j)\right] \tag{8}$$

and

$$G(\mathbf{x}) = \frac{1}{(4\pi r_C^2)^{3/2}}\exp(-\mathbf{x}^2/(4r_C^2)). \tag{9}$$

Now if $|\mathbf{x}'_i - \mathbf{x}''_i| \ll r_C$ for all i, then the first two terms in each summand in Eqn. (8) cancel the third, up to negligible contributions, and so the decay rate is negligible. If $|\mathbf{x}'_i - \mathbf{x}''_i| \geq 3r_C$ for all i while $|\mathbf{x}'_i - \mathbf{x}'_j| \ll r_C$ and $|\mathbf{x}''_i - \mathbf{x}''_j| \ll r_C$ for all distinct i, j, then $\Gamma \approx \gamma(4\pi r_C^2)^{-3/2}(N^2 - 2N)$ and so $\Gamma \approx \gamma(4\pi r_C^2)^{-3/2}N^2 = \lambda N^2$ to leading order in N. If first and second (or third) conditions hold, while the third (or second) set of separations are larger than $3r_C$, then $\Gamma \approx \frac{\gamma}{2}(4\pi r_C^2)^{-3/2}N^2 = \frac{\lambda}{2}N^2$, again giving a quadratic leading order dependence. If $|\mathbf{x}'_i - \mathbf{x}'_j| \geq 3r_C$ and $|\mathbf{x}''_i - \mathbf{x}''_j| \geq 3r_C$ for all distinct pairs (i, j), while $|\mathbf{x}'_i - \mathbf{x}''_j| \geq 3r_C$ for all (i, j) then only the terms with $i = j$ in the first two sums contribute, giving $\Gamma \approx \gamma(4\pi r_C^2)^{-3/2}N = \lambda N$, i.e., a linear dependence.

More generally, consider a superposition of two states, in each of which the particles are clustered in groups, with separations $\ll r_C$ within the clusters and $\gg r_C$ between the clusters. Suppose that the separations between the states of each cluster in the two components are $\gg r_C$ and that there are n_i particles in cluster i. Then to leading order the collapse rate is

$$\Gamma = \lambda \sum_i n_i^2. \tag{10}$$

As noted above, an atom of mass x is treated as a cluster of x nucleons. As this suggests, one can extend the result to the general case in which particle type i has mass m_i, giving [7]

$$\Gamma = \frac{\lambda}{m_N^2} \sum_i m_i^2 n_i^2. \tag{11}$$

3. BDF on Visual Perception

BDF consider a human observing a superposition state of a few photons, arranged so that one component causes the photons to impinge on the retina while the other does not. The components of the photon state may be very widely separated: non-relativistic collapse models generally do not assume any spontaneous collapse of photon states, and in any case the collapse rate for a few particles is negligible and effectively independent of the state separation l in the regime $l \gg r_C$.

The goal of collapse models is to explain the appearance of classicality. Humans do indeed perceive definite outcomes—namely, observing photons or not—when observing such states. Hence, BDF argue, a plausible collapse model must imply that a superposition reaching the eye must collapse before it is transformed into a perception in the brain. Human reaction time for weak light perceptions is $\approx$100 ms, so, BDF argue, this requires a collapse within that time. This appears reasonable, though of course there is room for discussion. Three points seem worth elaborating on.

First, our reports and memories of perceptions might not be entirely reliable. Theoretically, one could imagine that collapses take place at a much later point—hours or days after the interaction—leaving us with post-collapse memory states indistinguishable from memories of a (near) real time observation. However, if we are happy to accept theories in which the appearance of classicality is a false post hoc construct, we may struggle to explain why we are not happy with some version of many-worlds quantum theory [8], undercutting entirely the motivation for considering collapse models.

Second, one could imagine that collapse takes place not as a result of events within the brain, but as a result of our physiological responses to these events. Perhaps neither the eye detecting the photons, nor our visual cortices processing the information, are sufficient to cause collapse. Perhaps, instead, collapse only takes place when we blink, or subtly shift position, or report our observation orally or in writing. Though this is not a ridiculous hypothesis, it is not completely evident that it is consistent with our experience. It is tempting, if perhaps naive, to feel one would surely notice the photons even if one's head and body were completely immobilized. Perceptions of conscious events are notoriously tricky and sometimes deceptive, though (e.g., [9]). Perhaps subtle but macroscopic involuntary physiological responses could be crucial to conscious observation.

This possibility has been discussed in the past by some advocates of CSL [10]. At present, my impression is that there is no consensus among advocates of CSL as to how seriously to take it. For example, a recent analysis [11] proposes lower bounds on collapse model parameters without allowing for the possibility that physiological responses induce the relevant collapses. Everyone should agree, at least, that if a purportedly fundamental physical theory such as a CSL model can only be kept alive by invoking the hypothesis, then (a) anyone advocating the theory should be very clear about this, and (b) we should try to test the hypothesis directly as far as possible (difficult though this is). Since my focus here is on BDF's arguments, which do not involve the hypothesis, I will not consider it further here.

Third, one could imagine that collapse takes place as a result of events within the brain, but not necessarily within the eye. As BDF note, some authors have produced bounds for CSL models on this hypothesis. BDF consider it dubious: they argue that it would imply that "animals with a simpler visual apparatus could perceive . . . superpositions which we consider rather unlikely." Here, if I understand BDF correctly, I disagree. Animals with simpler brains would not necessarily perceive superpositions if no collapse took place as a result of events in their brains before their reaction time. They might have no conscious perception at all of these observations, or they might have delayed perceptions. They might not necessarily have conscious memories of these perceptions, and if they do these may not necessarily give them the same impression of time sequencing that our memories give us. So I consider the hypothesis that collapse takes place within the human brain, but not necessarily within the human eye, within ≈ 100 ms, perfectly reasonable. However, my aim here is to discuss BDF's arguments. These assume that collapse takes place within the eye, and this is certainly an interesting and prima facie plausible hypothesis. I will argue that there are problems with those arguments, which make it very hard to produce precise bounds for mass-dependent CSL collapse rates. The same issues arise in considering information processing elsewhere in the brain, and so I will not pursue this hypothesis further here either.

BDF's account of the biochemical processes involved in photodetection in the eye considers the following stages. Each photon is absorbed by a rhodopsin molecule, transforming it. The transformed molecule interacts with ≈ 20 transducin molecules, splitting off α-subunits from each. Each subunit diffuses over the rod disc and binds to a phosphodiesterase (PDE) molecule, activating it. Each active PDE converts a cyclic guanosine monophosphate (cGMP) molecule to guanosine monophosphate (GMP). The reduction in cGMP causes the closure of ≈ 300 ionic channels on the rod cell membrane, each preventing ≈ 10 sodium ions (Na^+) from entering the rod. This generates an electric signal which is transmitted to the optic nerve.

Using the approximations described in the previous section, BDF argue that there are three relevant components in the superposition state of detecting and not detecting a photon. First, the ≈ 20 α-subunits either remain attached to the transducins or diffuse over the rod disc surface, in which case they become separated from one another by $> r_C$. They then bind to PDE. Second, in the absence of photons cGMP molecules bind to the ion channels, while converted GMP molecules diffuse in the cytoplasm. Third, $\approx 10^3$ Na^+ ions either enter or fail to enter the rod membrane through ion channels.

BDF argue that Eqn. (10) can be applied to obtain contributions to the collapse rate from each of these three components. They take the first component as effectively giving a contribution of $n_1^2 N_1$, where $n_1 = 3.9 \times 10^4$ is the molecular weight of the α-subunits in daltons, and $N_1 = 20$ is the number of subunits separated by $> r_c$. The second component is taken to give a contribution $n_2^2 N_2$, where $n_2 = 363$ is the molecular weight of GMP and $N_2 = 2000$ the number of molecules.

The third component is taken to give a contribution $n_3^2 N_3$, with two different hypotheses assigning different values. One (BDF's "most likely case") takes $n_3 = 5 \times 3 \times 23$, corresponding to 5 channels within distance r_C, clusters of 3 ions separated by $< r_C$, each with molecular weight 23. In this case there are ≈ 60 groups of 5 channels, and ≈ 333 clusters of ions, and

BDF take $N_3 = 60 \times 333$. The second (BDF's "extreme case") assumes all ions passing through a channel are separated by $<r_C$, giving them 10^3 ions for each of 5 channels in a cluster and $n_3 = 5 \times 10^3 \times 23$, and 60 groups of 5 channels, giving $N_3 = 60$.

Accepting these values for the moment, this gives

$$n_1^2 N_1 \approx 3 \times 10^{10}, \qquad n_2^2 N_2 \approx 3 \times 10^8 \tag{12}$$

and two estimates defining a range for the third contribution

$$n_3^2 N_3 \approx 2 \times 10^9 - 8 \times 10^{11}. \tag{13}$$

Summing these, BDF argue, gives the effects of one photon, and multiplying by 6 gives the effects of 6 photons, which they take to be the fewest detectable by the human eye. Thus their final estimate is

$$6 \times \sum_{i=1}^{3} n_i^2 N_i, \tag{14}$$

with the n_i and N_i given above.

4. Problems in the BDF Analysis

4.1 Problems in BDF's Calculations

The second term in BDF's sum is dominated by the first and third, and so may be neglected. The first term lies within the range of estimates for the third, so that the sum lies in a compressed range

$$6 \times \sum_{i=1}^{3} n_i^2 N_i \approx 2 \times 10^{11} - 5 \times 10^{12}. \tag{15}$$

BDF's estimates for λ appear, however, to be based only on their estimates for the range of $n_3^2 N_3$, neglecting the contribution of $n_1^2 N_1$. This gives them a larger range than should follow from their assumptions and estimates.

BDF adopt the criterion that a superposition is taken to have collapsed when $\Gamma t \approx 10^2$, meaning that one term is $\approx e^{100}$ times smaller than the other. As they note, this is reasonable but arbitrary, and a factor of 10 either way could reasonably be included. The equation $\Gamma t = 10^2$, with a time $t = 100$ ms, implies $\Gamma = 10^3 \mathrm{s}^{-1}$.

Using the corrected range, we find from BDF's estimates and Eqn. (10) a range for the collapse rate given by

$$\lambda \approx 5 \times 10^{-9} - 2 \times 10^{-10} \tag{16}$$

rather than BDF's estimate of

$$\lambda \approx 5 \times 10^{-9} - 2 \times 10^{-11}. \tag{17}$$

As BDF note, both ranges could reasonably be multiplied by $10^{\pm 1}$ given the arbitariness noted in the previous paragraph.

4.2 Allowing for the Cytoplasm

BDF's calculations effectively model visual perception as though the only relevant massive particles are the specific particles they discuss: the α-subunits, the GMP molecules, and the Na^+ ions. Their estimates of the collapse rate are thus derived from Eqns. (8) and (10), where the sums include these particles and no others.

This would be a valid approximation if the interactions between incoming photons and these three types of particles took place in otherwise empty space. In fact, of course, they take place within rod cells, which have membranes and other structures filled with cytoplasm, a gel-like substance containing many proteins and ions.

To see immediately that this is likely to affect the calculations significantly, note that Eqn. (1) depends on $M(\mathbf{x})$ through $(M(\mathbf{x}) - \langle M(\mathbf{x}) \rangle_t)$, and that $M(\mathbf{x})$ itself is a smeared mass density, with the smearing function (3) having characteristic scale r_C.

4.2.1 Considering the cytoplasm as homogeneous

We thus cannot apply Eqn. (11) directly, taking m_i as the actual masses for the relevant particles, for superpositions arising in an otherwise homogeneous fluid. A more relevant approximation would be to take

$$m_i' = m_i - \rho V, \tag{18}$$

where m_i is the actual particle mass, ρ the average smeared density of the fluid, and V the volume of fluid notionally displaced by the particle. More precisely, we could take $\rho V = k_i m$, where k_i is the (not necessarily integer) average number of fluid particles absent in a volume of r_C^3 when that volume contains a particle of type i and m is the mass of each fluid particle.

This is significant because the densities of the relevant particles in BDF's analysis and of the cytosol and other components of the cytoplasm are likely not dissimilar. It is hard to be precise, because the details depend on the properties of the relevant particles when suspended in the cytosol environment, which itself is complex. I have found it hard to locate data even for aqueous suspensions. The best I can offer are very crude estimates, which nonetheless illustrate the problem and the need for closer analysis.

For example, the density of metallic sodium, 968 kgm^{-3}, is very close to that of water, 997 kgm^{-3}. While data on the effective density of Na^+ ions in water solution are harder to find, one crude estimate is given by comparing the estimated effective radius of Na^+ in water [12] (218 pm), by that of Na atoms (227 pm). If (which is admittedly not clearly justified by the cited data) we could approximate the effective density of Na^+ in water by $(227/219)^3\, 968 \approx 1078\ \text{kgm}^{-3}$, we would get an effective m'_i for sodium ions in water of approximately 0.08 m_i, thus multiplying the estimated collapse rate by $<10^{-2}$.

To get a crude estimate for the α-subunits and GMP molecules, we could compare the typical density of proteins, $\approx 1200 - 1400\ \text{kgm}^{-3}$, with either the density of water or, presumably better, the density of the rod cytosol or cytoplasm (perhaps 1100 kgm^{-3}). This gives an effective $m'_i \approx 0.3 m_i$, multiplying the estimated collapse rate in the rod by $\approx 10^{-1}$ in this case.

Allowing for these factors gives a collapse rate estimate in the rod of

$$6 \times \sum_{i=1}^{3} (m'_i)^2 n_i^2 N_i \approx 2 \times 10^{10} - 5 \times 10^{10}. \tag{19}$$

This would imply bounds in the range

$$\lambda \approx 2 \times 10^{-8} - 5 \times 10^{-8}. \tag{20}$$

Figures 4.1 and 4.2 give schematic illustrations of superposition states illustrating the relevance of relative densities. Here the darker gray dots represent idealized ions and the lighter gray dots idealized fluid molecules.

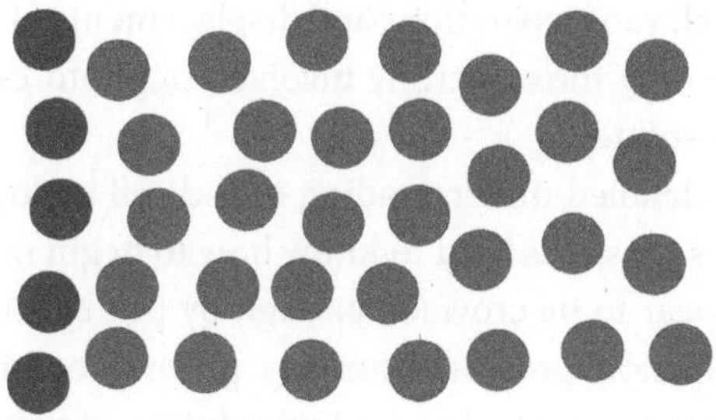

Figure 4.1 Ions concentrated on left.

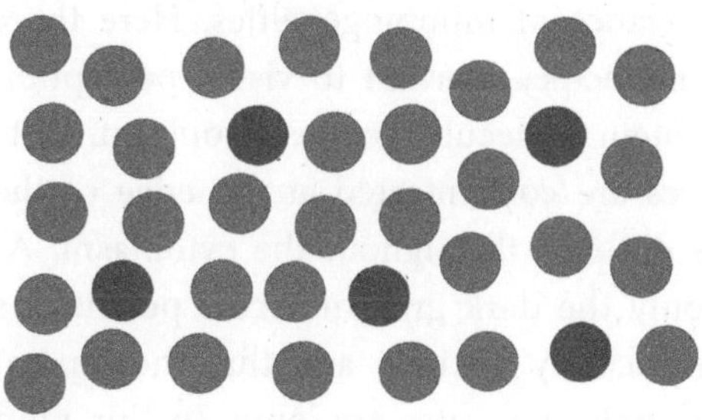

Figure 4.2 Ions diffused.

In the first state, the ions are concentrated at the edge of the volume; in the second, they have diffused throughout the fluid. In our simplified model, the ions have the same mass and volume as the fluid molecules and diffuse so that the molecule positions are identical (although different molecule types occupy some positions) in the two components. An approximation which considers only the ion positions would suggest that the two states are significantly distinct, and hence that the mass-dependent CSL model should predict the superposition will collapse. However, when all the particles are taken into account, the two states have identical mass distributions. Hence the mass-dependent CSL model predicts no collapse.

4.2.2 Allowing for cytoplasmic inhomogeneity

Even these last estimates, however, are based on an invalid model. Cytosol and cytoplasm are not at all homogeneous on the relevant scales. To calculate the difference in smeared mass density distributions between a superposition component in which some number of proteins have or have not diffused around the cell, for example, one thus has to consider all the proteins and other components that may have been relocated in the course of the diffusion. To then apply (11), one needs to know—or at least plausibly estimate—all the relevant separations and displacements of all these proteins (including but not only those actively involved in photo-detection), and all the ions and other solutes.

Without a very detailed understanding of rod cell biology and biochemistry at very small scales, it is hard to know how to begin making a plausible estimate. Cells appear to be crowded enough by proteins of various shapes and sizes that diffusion processes for any given protein cannot be well modelled by treating the cytosol as a dilute solution of that protein [13].

Figures 4.3 and 4.4 give schematic illustrations of superposition states illustrating the relevance of inhomogeneities. Here the darker gray dots represent protein molecules relevant to visual perception and the lighter gray dots other protein molecules in the cytoplasm. In the first state, the dark gray molecules are concentrated at the edge of the volume; in the second, they have diffused throughout the cytoplasm. An approximation which considers only the dark gray molecule positions suggests that the two states are significantly distinct, and that the separations relevant to the two dark gray molecule states are large. In this model, the molecule positions are different in the two components. Thus, when all the particles

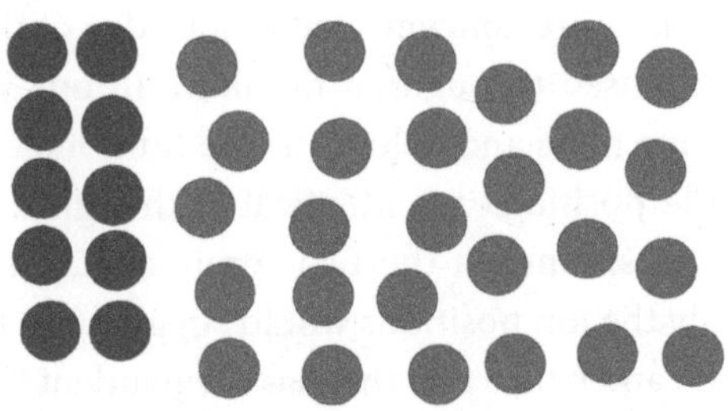

Figure 4.3 Dark gray protein molecules concentrated on left.

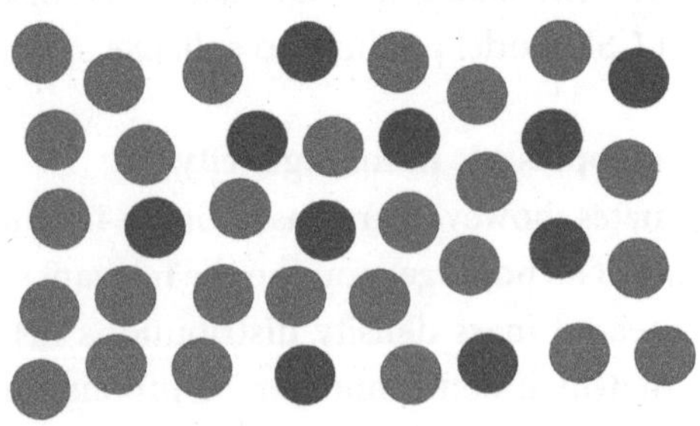

Figure 4.4 Dark gray protein molecules diffused.

are taken into account, the two states still have distinct mass distributions. However, if the dark and light gray protein molecules have identical masses and densities, the relevant separations are those between dots of either shade in the two component states. These are much smaller than the typical differences between dark gray molecule positions in the two states or the typical separations between dark gray molecule positions in the second state.

More realistic illustrations can be found in Figure 4.1 and in the supplementary information of Ref. [13]. These suggest a very complex picture of intra-cellular protein diffusion. The interaction of any given diffusion process with collapse dynamics may not be easily captured without very precise information about the relevant cell environment. It is thus impossible to say for sure, but to me the most plausible guess is that an accurate estimate would produce significantly higher collapse rate bounds than those derived from Eqn. (12).

Similar comments apply to the collapse rate bounds derived from the diffusion of sodium ions. Modelling the sodium ions as inhomogeneities in an otherwise homogeneous aqueous solution, as above, is likely misleading. One needs to consider the precise environment within the solution, including all ions and other solutes, and allowing for the smearing defined by Eqn. (3). Again, I am not sure of the likely result, but find it plausible that the result would be significantly higher collapse rate bounds than those given in Eqn. (20).

4.3 Limits of Human Perception

Since BDF's work, evidence has been presented [14] suggesting that humans are able to detect single photons. The evidence is not as yet compelling: results are reported for three individuals, and their responses were statistically significant but not perfectly reliable.

If it could be shown that humans can reliably detect single photons, BDF's collapse rate bounds, and others similarly derived, would be increased by a further factor of 6. Given the uncertainty in interpreting the evidence, I do not include this additional factor here. It is worth keeping in mind, though, given that it would increase the bounds by close to a further order of magnitude.

5. Conclusions

Dynamical collapse models in general, and mass-dependent continuous spontaneous localization models in particular, are well motivated and experimentally testable alternatives to quantum mechanics. It is an intriguing question whether these models can be excluded with forseeable technology, or even are already excluded by existing experimental and observational data. Lower bounds on the model collapse rates can only ultimately be justified by assuming that collapses take place within human perception times, so that the models predict that humans should perceive one component of a superposition.

BDF's pioneering work gives a basis for deriving such bounds. However, their assumptions and approximations are questionable enough that it seems unwise to rely on the bounds they suggest. Further detailed work is needed to decide whether mass-dependent continuous spontaneous localization models remain viable (for some parameter choices) or are already effectively excluded.

Acknowledgments

This work was partially supported by Perimeter Institute for Theoretical Physics. Research at Perimeter Institute is supported by the Government of Canada through Industry Canada and by the Province of Ontario through the Ministry of Research and Innovation. I thank Angelo Bassi and Philip Pearle for very helpful discussions.

References

[1] Eugene P. Wigner. Remarks on the mind-body question. In *Philosophical Reflections and Syntheses*, pages 247–260. Springer, 1995.

[2] Lajos Diosi. A universal master equation for the gravitational violation of quantum mechanics. *Physics Letters A*, 120(8):377–381, 1987.

[3] Roger Penrose. On gravity's role in quantum state reduction. *General Relativity and Gravitation*, 28(5):581–600, 1996.

[4] Gian Carlo Ghirardi, Alberto Rimini, and Tullio Weber. Unified dynamics for microscopic and macroscopic systems. *Physical Review D*, 34(2):470, 1986.

[5] Gian Carlo Ghirardi, Philip Pearle, and Alberto Rimini. Markov processes in Hilbert space and continuous spontaneous localization of systems of identical particles. *Physical Review A*, 42(1):78, 1990.

[6] Angelo Bassi, D.-A. Deckert, and Luca Ferialdi. Breaking quantum linearity: Constraints from human perception and cosmological implications. *EPL (Europhysics Letters)*, 92(5):50006, 2010.

[7] Stephen L. Adler. Lower and upper bounds on CSL parameters from latent image formation and igm heating. *Journal of Physics A: Mathematical and Theoretical*, 40(12):2935, 2007.

[8] Simon Saunders, Jonathan Barrett, Adrian Kent, and David Wallace. *Many Worlds?: Everett, Quantum Theory, & Reality*. Oxford University Press, 2010.

[9] Benjamin Libet. Unconscious cerebral initiative and the role of conscious will in voluntary action. Behavioral and brain sciences, 8(4):529–539, 1985.

[10] Franca Aicardi, Antonio Borsellino, Gian Carlo Ghirardi, and Renata Grassi. Dynamical models for state-vector reduction: do they ensure that measurements have outcomes? Foundations of Physics Letters, 4(2):109–128, 1991.

[11] Marko Toros and Angelo Bassi. Bounds on quantum collapse models from matter-wave interferometry: calculational details. Journal of Physics A: Mathematical and Theoretical, 51(11):115302, 2018.

[12] Zhong-Hua Yang. The size and structure of selected hydrated ions and implications for ion channel selectivity. *RSC Advances*, 5(2):1213–1219, 2015.

[13] Tadashi Ando and Jeffrey Skolnick. Crowding and hydrodynamic interactions likely dominate in vivo macromolecular motion. *Proceedings of the National Academy of Sciences*, 107(43):18457–18462, 2010.

[14] Jonathan N. Tinsley, Maxim I. Molodtsov, Robert Prevedel, David Wartmann, Jofre Espigulé-Pons, Mattias Lauwers, and Alipasha Vaziri. Direct detection of a single photon by humans. *Nature Communications*, 7:12172, 2016.

PART II

CONSCIOUSNESS IN QUANTUM THEORIES

5

Quantum Mechanics and the Consciousness Constraint

Philip Goff

Quantum mechanics is one of the best predictive machines humankind has ever produced. Much of our modern technology, from computers to smartphones to GPS, is reliant on its predictive power. The trouble is nobody knows what quantum mechanics is telling us about reality. There are numerous proposals but no consensus on which is most probable. As things stand, the empirical data seem to underdetermine the theory.

In this kind of situation, philosophy has an important role to play, helping us to evaluate the evidential situation with respect to the various hypotheses. But it is generally assumed in this context that philosophy is not able to offer us *new data*, over and above the scientific data of observation and experiment. The usual expectation is that the philosopher of physics will contribute conceptual clarity and perhaps a cost-benefit analysis of the various interpretations of quantum mechanics in terms of theoretical virtues, such as simplicity, parsimony, non ad-hocness, etc.

In contrast to this standard assumption, I'm inclined to think that philosophy *does* have new data to offer, and that this data might have bearing on the ontology of quantum mechanics. What I have in mind is data pertaining to the reality of consciousness. Consciousness is not something that we know about through observation and experiment. If we were just going off the data of third-person observation and experiment, we would have no need to postulate subjective experiences, as Daniel Dennett (2007) has argued very effectively. Nonetheless, contra Dennett, we do know that consciousness is real: we know that it's real in virtue of the immediate awareness of each of us of our own feelings and experiences. Any theory of reality unable to account for the reality of consciousness is at best incomplete. In this sense, the reality of consciousness is a datum in its own right. I call the theoretical obligation to account for this datum "the consciousness constraint."

Philip Goff, *Quantum Mechanics and the Consciousness Constraint* In: Consciousness and Quantum Mechanics. Edited by: Shan Gao, Oxford University Press. © Oxford University Press 2022.
DOI: 10.1093/oso/9780197501665.003.0006

I believe that scientists and philosophers of the future will be baffled by the fact that their late-20th-century/early-201st-century ancestors did not make more use of the consciousness constraint. There is a certain phenomenon known to be real with something close to certainty, and yet the ontological implications of that phenomenon are completely ignored by most theoretical scientists, and even most metaphysicians. It is true that the problem of consciousness, broadly understood as the challenge of understanding "how brains produce consciousness," is now taken to be a serious scientific problem. However, this is generally assumed to be a problem that will go away with a bit more neuroscience. But the problem of consciousness is radically unlike any other scientific problem, not least because the fundamental datum that needs to be accounted for does not come from observation or experiment. Consciousness is something we know about *independently* of third-person empirical science; as such, it is a valuable source of information that needs to be *added* to the data of observation and experiment.[1]

The bearing of consciousness on quantum mechanics has been very little explored. Of course, a small number of heterodox thinkers have tried to make sense of the old idea that consciousness might have a role at the heart of quantum mechanics (see Chalmers and McQueen, this volume). But this has never been articulated as part of a general approach of working out how the reality of consciousness constrains theory choice in this area. This paper will take a first step in rectifying this, by tentatively exploring the question of whether *wave function monism*—a popular interpretation of the ontology of quantum mechanics—is able to satisfy the consciousness constraint.

1. The Consciousness Constraint in More Detail

I will assume that the reality of consciousness is at least as certain as anything else we know about the contingent world. But I think we can go further than this. Not only do we know that consciousness exists, but we know that it exists *as we ordinarily take it to be*. I don't mean by this that all of

[1] It may turn out that the postulations we make to account for the data of observation and experiment can also account for the reality of consciousness, but this cannot be assumed from the start. Indeed, when you think about it, it would be strange if postulations that have been tailored to fulfill one theoretical task (accounting for observation and experiment) just happened to be suited to account for a distinct, and not obviously related, theoretical task (accounting for subjective experience).

our reflective theoretical judgements about the nature of consciousness are correct. Rather, my claim is that we know that certain of our paradigmatic *concepts* of consciousness are an accurate representation of consciousness itself. The concepts I have in mind are those that have become known in the literature as "phenomenal concepts." This is just the philosophical jargon for the concepts we deploy when we think about our experience in terms of how it feels, or more broadly in terms of what it's like to have it. You're in pain; you attend to your pain, and think about it in terms of how it feels; in doing so, you form a phenomenal concept of your pain. I want to remain open to the possibility that some phenomenal concepts misrepresent; someone might, for example, mistakenly think they're having an orange experience, whereas when they attend to it in more detail, they realize it was actually a red experience. It may be, as David Chalmers (2003) has defended, that there are a subset of beliefs involving phenomenal concepts that are in a certain sense infallible. However, it will be sufficient for my purposes to make the more modest claim that at least *some* phenomenal concepts—some instances of that general category—are satisfied.

Hence, we can define the consciousness constraint as follows:

> *The Consciousness Constraint.* Any adequate theory of reality must entail that at least some phenomenal concepts are satisfied (where a concept is satisfied just in case it corresponds to reality). (Goff 2017, Chapter 1)

This gives consciousness a unique status in metaphysics. With respect to most of our concepts—time, free will, color—we are happy to have them molded and revised by scientific developments. Perhaps our initial concept of time was such that time is absolute and the present has a privileged status. After reflecting on relativity, many philosophers have become persuaded that that pre-theoretical understanding of time cannot be quite accurate. This doesn't force us to a position of saying that time does not exist, but it does mean revising our concept of time to fit better with the picture of reality science has forced upon us. Similarly, many believe that our pre-theoretical understanding of free will as involving choices that lack prior causes is incompatible with our scientific knowledge of the brain, and hence must be replaced with a softer notion of freedom, one compatible with our choices being causally determined.

In contrast, in the case of consciousness, we have a class of concepts that we know accurately depict reality. While some individual classifications of one's subjective experience may slightly misfire—I think I'm experiencing red when, in fact, I'm experiencing orange—the general understanding that each of us has of subjective experience is correct. When I entertain the proposition <there is something that it's like to be me>, I know that *that proposition*, and not some revised form of it, is true. In this sense, the knowledge that subjective experience is real is insulated from empirical refutation.

This is a powerful and much neglected tool for metaphysical enquiry. In the public mind, the only task that for a scientific theory to fulfill is to account for experimental data. But there's something we know about reality independently of experiments, and a theory of reality must be capable of accounting for that, too. Could explicit recognition of this theoretical constraint help us make progress on the ontology of quantum mechanics? It is to this matter I now turn.

2. Can the Wave Function Ground Ordinary Objects?

Wave function monism, the view that fundamental reality consists of and only of the wave function, is thought by many philosophers of physics to be the most simple and straightforward theory as to what quantum mechanics implies about reality (Albert 2013: 53). There are a few different ontological interpretations of what the wave function is. I suspect the challenges I am raising here would generalize to all of them. For the sake of space, however, I will focus on one prominent view (Albert 1996, Loewer 1996, Ney 2012, 2012) according to which the wave function consists of a complex-valued field in high dimensional space.[2] The space of the wave function is not the three-dimensional physical space of which we are familiar, but is rather "configuration space," so called because there is a correspondence between each location in configuration space and a configuration of particles in ordinary space. Such a configuration space is thought essential in order to capture the ways in which particles are constrained by facts about entanglement. Suppose, for example, we have two entangled particles in a superposition such that:

[2] I found Ney (2013) an extremely helpful resource for understanding wave function monism. See Chen 2019 for an excellent overview of different ontological interpretations of the wave function.

- The particles are at either location L1 or location L2, and are equally likely to be found at one as the other.
- If either is measured to be at L1, the other will be measured to be at L2, and vice versa.

In the corresponding configuration space, there will be:

- A location corresponding to both particles being at L1, and a location corresponding to both particles being at L2.
- Locations corresponding to one particle being at L1 and one particle being at L2.

The wave function will have zero amplitude at the former locations and non-zero amplitude at the latter locations, corresponding to the fact that the particles could not be measured to be in the same locations but could be measured such that one is at L1 and the other at L2.

In the above description, we are considering just two particles. For there to be facts about the wave function corresponding to all possible ways in which particles might be arranged in three-dimensional space, the configuration space in which it is housed must have $3 \times N$ dimensions, where N is the total number of particles in the universe.

It might be tempting to think of configuration space as a way of *representing* what's going on in physical space. But for the wave function monist this gets things the wrong way around. The wave function in configuration space is the fundamental reality; the particles in three-dimensional space—if they exist at all (more below)—are grounded in facts about configuration space.

The wave function is clearly a peculiar entity, and the idea that reality consists of such a thing seems in sharp contrast to the evidence of our senses. The claim of wave function monists, of course, is that we should look to science, not everyday experience, to find out what fundamental reality is really like. If wave function monism is the simplest interpretation of our best empirical theory, then wave function monism is what we ought to believe.

However, other philosophers of physics have questioned whether wave function monism could possibly be supported empirically, given its clash with experience. After all, what is empirical support other than what is known on the basis of experience? To be sure, scientific support involves experience in highly specific, controlled, repeatable experimental circumstances. Nonetheless, at the end of the day, we only know the results of experiments through using our senses, and whenever we use our senses, we seem to experience a world of objects in three-dimensional space. This has led Tim Maudlin (2007) to conclude that wave function monism is "empirically incoherent": the theory cannot account for the very evidence upon which it supposedly relies.

The obvious response for the wave function monist is to dispute the charge that they cannot account for the three-dimensional world we perceive with our senses. Just because three-dimensional objects do not exist *at the fundamental level*, it does not follow that they do not exist *at all*. If the wave function theorist can claim that three-dimensional objects are *grounded in* facts about the wave function, then they can thereby account for the existence of the evidence on which their theory is based. Unfortunately, this is easier said than done.

What is required in general for there to be a grounding relationship? Or to put it another way: What grounds a grounding relationship? In my book *Consciousness and Fundamental Reality*, I argue that the crucial feature of a grounding relationship is that the less fundamental entities are *nothing over and above* fundamental entities. I like party examples. Suppose Rod, Jane, and Freddy are dancing and drinking one night. You've thereby got a party. But it's not as though the party is this wholly new thing that the reveling brings into being, in the way that the dances of witches may bring into being a demonic spirit. There's a clear sense in which the party is *nothing extra* to be the people having a good time.

The "nothing over and above" relationship is prima facie paradoxical. How can *X* be *distinct* from *Y*, while nonetheless be *nothing more than Y*. I suggest that an account of grounding needs to resolve this paradox; I call this the "free lunch constraint," after David Armstrong's (1997: 12) famous term "ontological free lunch" for an entity that is nothing over and above already postulated facts. Building on the work of others, I have argued that the "nothing over and above" relationship should be accounted for in terms of an *analysis* of the grounded entities. For a given entity *e*, an analysis is a description of *what the reality of* e *consists in*, or, equivalently, *what is essentially required for* e *to be real*. In the case of a party, we can give the following analysis:

> *Party Analysis*. What is essentially required for there to be a party is for there to be people reveling.

For almost all the entities we talk about, it's impossible to give complete necessary and sufficient conditions for their existence; any proposal for the definition of "reveling" will inevitably lead to a long-winded game of *counterexample leading to refining of the analysis, leading to counterexample, leading to refining of the analysis*, and so on ad infinitum. Despite the fact that we can't give a precise definition of what is required for reveling, the very fact that we can engage in the game of "spot the counterexample" entails that we have an implicit grasp of what is required.

Why is the party at Rod's house nothing over and above the fact that Rod, Jane, and Freddy are dancing? Because all that is essentially required for there to be a party is for there to be people reveling, and the fact that Rod, Jane, and Freddy are reveling logically entails that there are people reveling. The revelers are not strictly speaking identical with the party (they will go on existing when the party ends), but they provide all that is essentially required for the party to exist, and this gives us a clear sense in which the party is "nothing over and above" the fact that Rod, Jane, and Freddy are reveling; in other words, we have satisfied the free lunch constraint.

I call this way of accounting for grounding relationships "grounding by analysis," defined as follows:

Fact *X* is grounded by analysis in fact *Y* iff:

- *X* is grounded in *Y*, and
- *Y* logically entails what is essentially required for the entities contained in *X* (including property and kind instances) to be part of reality.

Let's return to the putative grounding of three-dimensional objects in facts about the wave function. David Wallace (2012) has proposed that we should think of classical objects as *patterns* in the wave function. But what we want to know is why certain patterns in the wave function are sufficient to ground facts about three-dimensional objects. If we are thinking in terms of grounding by analysis, we need to make a case, via an analysis of, let's say, a table, that what is essentially required for that table to exist is logically entailed by facts about patterns in the wave function. I argue in *Consciousness and Fundamental Reality* that we can analyze facts about macro-level objects, such as tables, in terms of *patterns of penetration resistance among regions of space*. For any given three-dimensional object, it is impossible to precisely define the relevant pattern, but this is consistent with our having an implicit grasp of it (compare to the case of "reveling" discussed above). We can say that the fact that there is a table in front of me consists in the fact that there is a table-ish region of space in front of me that resists penetration, for example, in the sense that if I put a cup down on it, it won't fall to the floor. This offers us a kind of rough and ready functionalist analysis of what it is for a table to exist.

If we were in a world of classical physics, it is plausible that arrangements of particles in three-dimensional space could entail that there is the right kind of pattern of penetration resistance for there to be a table. The trouble is that a description of the wave function does not logically entail that anything has the causal property of *resisting penetration*. There is no inconsistency in giving a complete description of the wave function and conjoining it with the denial that anything resists penetration.[3] Of course, we can point to the aspects of the wave function that *correspond* to a given pattern of penetration resistance, and we can argue that those aspects of the wave function are physically salient, due to the process of decoherence (Wallace 2010), but this does not in itself show us that those aspects of the wave function *logically entail* that there really is a pattern of penetration resistance. As already noted, we need more than a correspondence relation to secure a genuine grounding relationship.

Of course, there may be other ways of analyzing what it is for a table to exist, or other ways of accounting for the grounding relationship. But then proponents of wave function monism are obliged to come up with the goods. David Albert (2013, 2015) has offered perhaps the most detailed attempt to do this. Albert effectively gives an analysis of what it is for a certain three-dimensional system of particles to exist by specifying the Hamiltonian of the system. Without going too much into the details, we can think of a Hamiltonian as capturing how the system's behavior across time is dependent on the masses and velocities (along three dimensions) of each of the particles involved in the system and the distances between them. Albert then argues that facts about the evolution of the wave function over time realize that functional profile.

However, as Alyssa Ney (2015) has pointed out, the problem with Albert's account is that while the Hamiltonian of the system of particles in three-

[3] Note that what we are talking about here is strict logical inconsistency. Likewise, in a particle-ontology, a complete description of the particle facts does not logically entail any facts about composite objects. The difference in the latter case, however, is that facts about particles (in three-dimensional space) *can* logically entail the existence of the pattern of penetration essentially required for there to be composite objects.

dimensional space characterizes it in terms of *velocities and distances*, the corresponding facts about the wave function involve *velocities and locations but not distances*. As a result, the relevant wave function facts do not meet the requirement specified by Albert's own account of the essential nature of a system of particles in three-dimensional space. As in Wallace's account, all we really have are *correspondences* between particle facts and wave function facts without an adequate account of how the latter ground the former. This is a quite general challenge: it's hard to see how we could capture what is essentially required for there to be a system of particles in three-dimensional space without mentioning features like *distance* or *three-dimensionality*, features whose existence does not logically follow from facts about the wave function.

Despite her skepticism about Albert's account, Ney is herself a wave function monist. Rather than offering a functionalist account of the grounding of three-dimensional objects in the wave function, Ney offers a *priority monist* account. Priority monism is the ancient position pioneered in recent times by Jonathan Schaffer (2010), according to which there is one fundamental entity, usually taken to be the universe as a whole. Philosophers tend to assume that facts about big things are grounded in facts about little things, with all facts ultimately grounding in facts concerning arrangements of particles. Schaffer turns this on its head: facts about little things are grounded in facts about big things, with all facts ultimately grounded in facts about the universe.

Thus, Ney aspires to ground particles as *parts* of the wave function. One obvious difficulty with this ambition is that particles on the one hand and the wave function on the other do not share a common space. Particles exist in familiar three-dimensional space (or perhaps four-dimensional spacetime); the wave function exists in a high dimensional configuration space. Ney responds by suggesting that sharing a common space is not a general requirement for a part-whole relationship, on the grounds that some entities that stand in such a relationship do not exist in space at all: "For example, 'egalité' is part of the national motto of France" (Ney 2015).

While it is true that non-spatial entities can stand in a part-whole relationship, we have no examples of *spatial* entities that stand in a part-whole relationship despite the fact that they fail to share a common space. In the absence of such examples, Ney's claim that there is a part-whole relationship here seems quite obscure. She suggests analyzing mereological relationships in terms of their formal features, such as being a partial ordering relation and principles such as Supplementation: If x is a proper part of y, then $\exists z$ (z is a part of y and it's not the case that z overlaps x). However, one may be skeptical that one can give a complete analysis of the part-whole relationship in terms of these purely formal features, or one may (as seems plausible to me) want to stipulate that it's a necessary condition for X to be a part of Y that if X and Y are spatial entities, then X and Y share a common space. Without this stipulation, we leave open the possibility that entities in spatio-temporally distinct parallel universes could stand in part-whole relationships, which, to my ears, sounds incoherent.

I think there is a deeper challenge for Ney's view. Note that her account does not involve grounding by analysis; she does not account for the grounding relationship in question in terms of *what is essentially required* for three-dimensional objects to exist. And in the absence of such an analysis, Ney has not demonstrated that she can account for a *nothing over and above* relationship between wave function facts and particle facts.

One way to see why this matters is to note that there are radical emergentist forms of priority monism, that is, views according to which the parts of the universe depend on the universe but are nonetheless genuine additions in being relative to the universe.[4] These views are just priority monist version of more familiar forms of radical emergence, such as were defended by the British emergentists of the 19th and early 20th centuries.[5] According to the British emergentists, chemical, biological, and mental properties are fundamental entities in their own right that arise from physical properties without being in any sense reducible to them. If the British emergentists had been priority monists, they would have held that fundamental chemical, biological, and mental properties arise from the physical nature of the universe without being in any sense reducible to facts about the universe.

It is standardly assumed that materialists must give some way of distinguishing their view from radical emergentism. For the materialist, all facts are nothing over and above the fundamental physical facts. Likewise, Ney is obliged to distinguish her view from a radically emergentist form of priority monism. Indeed, a view in which particles radically emerge from the wave function would *not* be a form of wave function monism, but would rather be an interpretation of quantum mechanics according to which there is *both* a

[4] Shani (2015).

[5] For examples of British emergentism, see Mill (1843), Broad (1925), and Alexander (1920). For a good discussion of British emergentism, see McLaughlin (1992).

wave function and matter (with the latter arising from the former). To make sense of wave function monism, we need to show that all facts are *reducible* to facts about the wave function. The most obvious way to do this would be to offer an analysis of what is essentially required for systems of particles in three-dimensional space to exist, and then to show that facts about the wave function meet that requirement. But, as Ney herself has argued, no extant account has managed to do this.

I hope to have shown that at the very least there are serious difficulties for a wave function monist wanting to account for three-dimensional objects. John Hawthorne (2010) has suggested that this is a serious "explanatory gap" analogous to that which seems to hold between the physical facts and the facts about consciousness. In one sense, the wave function/three-dimensional objects explanatory gap looks even more challenging: in the case of the physical/experiential gap, we at least know what a functionalist account would look like, whether or not we find it plausible. On the other hand, in another sense, the gap is less serious, as we lack the certainty of the existence of three-dimensional objects that we enjoy with respect to consciousness. If the wave function/three-dimensional objects gap can't be closed, it might be an option for the wave function monist to simply deny the existence of three-dimensional objects, provided she can give some response to Maudlin's charge of empirical incoherence.

While accepting that I have not here made anything like a conclusive case that that gap can't be closed, in what follows I would like to explore what follows for wave function monism if this gap cannot be closed.

3. The Nature of Evidence

In the last section, we critiqued Alyssa Ney's attempt to close the wave function/three-dimensional objects explanatory gap and thereby to respond to Maudlin's charge that wave function monism is empirically incoherent. In an earlier paper (Ney 2015), presumably before she had formulated the account critiqued above, Ney suggests an alternative way in which a wave function monist might respond to Maudlin's challenge. After conceding, at least for the sake of discussion, that the wave function monist cannot ground objects in three-dimensional space, Ney suggests that an option for the wave function monist is simply to revise our understanding of what "evidence" and "confirmation" consist in, on the grounds that "[w]e should reason from what it is reasonable to believe our evidence is like given our best scientific theories, not from how our evidence pre-theoretically seems" (Ney 2015: 3120):

> On this view, confirmation is not going to involve causal relationships between spatially localized bits of a three-dimensional world. Instead confirmation is going to involve the entire state of the world moving from one that is properly described (nonexhaustively) as "Theorists have constructed quantum theory T," to one that is properly described (nonexhaustively) as "Theorists have acquired evidence for theory T." For this to be viable, we should be able to connect such descriptions with elements of a wave function ontology (this is where decoherence can help), but this does not require solving the macro-object problem. Wave function realists should be open to the epistemic possibility that we may have to substantially revise how we think (in ontological terms) of things like "theorists" and processes like "constructing theories." (Ney 2015: 3122)

This seems to put the cart before the horse. Surely, we need grounds for thinking the theory is true *before* we are entitled to reinterpret our evidence in terms of the theory. And to have grounds for thinking the theory is true, we need *already* to have evidence that supports it. If it were permissible to reinterpret evidence in terms of that theory, then a fundamentalist Christian could interpret the Bible as the inerrant word of God, and thus find plenty of confirmation.

This raises the tricky question of what our basic evidential support for scientific theories consists in. Does our evidence for the standard model of particle physics consist of the experimental results we perceive with our senses? These sensory experiences (of the results of experiments) seem to represent objects in three-dimensional space and time, which would seem to imply that what they give us grounds for believing (if anything) is the existence of objects in three-dimensional space and time. And it would seem to follow that our scientific theories cannot be inconsistent with the existence of such objects, on pain of undermining the evidential support we have for those theories.

However, I agree with Ney's concern that this seems to put undue limitations on the potential of science to revise our pre-theoretical views about the underlying structure of the physical world. As she puts it, "We should be highly skeptical that we should be able to legislate as an a priori matter what science can and cannot reveal about the spatial structure of our world" (Ney 2015: 3123). Indeed, as Ney points out, for most of the history of philosophy, the dominant view about the nature of reality was idealism: the view that reality is fundamentally mental. The idealist need not—contrary to what Ney claims in this paper—think that the three-dimensional space and its contents are illusory: Berkeley believed in physical objects but thought they were constructed of ideas. However, some idealists do contend that the physical world is an illusion, and it seems wrong to think that we can rule out this view, as Samuel Johnson famously tried to do, merely by kicking a stone.

One obvious way around this is to define our evidence in terms of consciousness, and there are two ways in which we might do this. We could take our evidence simply to be our conscious states, collectively, and judge a theory in terms of its how well it explains the fact that we have the experiences we do. Obviously, this approach leaves open the possibility of idealism (which is not to say that idealism is plausible). Alternately, we can define the content of our experiences in terms of causal connections between experience and the external word. Thus, the pointer on a dial can be defined as "whatever causes our pointer experiences."[6] On the former approach, our evidence is the conscious experiences themselves; on the latter approach, our evidence consists of entities that are (or at least might be) mind-independent but that are latched onto in virtue of their relationships to experience. In terms of their evidential import, there might end up being not too much difference between the two approaches. The crucial point for our purposes here is that either approach allows a great deal more flexibility in what science can end up telling us about the physical world. We are not tied, just by virtue of our use of empirical evidence, to a universe of three-dimensional objects in space and time. That which accounts for our conscious experience, or that which our conscious experiences latch onto, may turn out to be very different to how we ordinarily conceive of it.

However, this way of conceiving of our evidence does not leave our metaphysics entirely unconstrained. For it commits us to the reality of human conscious states, either explicitly (if our evidence just is our conscious states) or implicitly (where our connection to our evidence is fixed by virtue of our conscious states). Hence, even if Ney is right that the wave function monist is

[6] Putting this terms of Chalmers' (2004) two-dimensional semantic framework, we would say that the concept's primary intention picks out whatever is causally related in the right way to consciousness. The primary intention corresponds to our epistemic situation, and hence in each epistemically possible world in which the concept refers, consciousness exists.

not obliged to account for a three-dimensional world, they do end up being obliged to account for consciousness.

Can we construe evidence in a way that avoids this commitment to consciousness? Perhaps one might try to adopt a purely externalist theory of perceptual content. Thus, we might say that in perception we latch on to *something*, but that reference is fixed wholly by facts outside of the content of the perception. In this way, appeal to perceptual evidence would not presuppose a commitment to the three-dimensional world, nor would it presuppose a commitment to consciousness, as neither (on this view) enter into the basic content of perceptual experience.[7]

However, this approach would not leave the wave function monist unconstrained; it would constrain her to commit to *reference*.[8] And if we can't ground three-dimensional objects in the wave function, it's hard to see how we could ground perceptual reference in the wave function. In general, naturalistic theories of reference ground reference in causal relationships between perceiver and environment.[9] If the wave function realist doesn't have either perceiver or environment, then they're going to have to give an account of reference purely in terms of the wave function. At the very least, nobody (as far as I know) has shown how this can be done.

In any case, even if there is a way to make sense of evidence without a commitment to consciousness (in a way that avoids being tied to a three-dimensional world), the fact remains that we know that consciousness is real. Even if it's possible to construe empirical data in a way that makes no reference to subjective experience, nonetheless we have at least as much justification for believing in subjective experiences as we do in trusting the empirical data. Regardless of its connection to evidence, the wave function

[7] On Chalmers' two-dimensional framework, such concepts aren't possible as they would lack primary intentions. However, many philosophers think such concepts are possible. See Goff (2017: 4.3) and Goff and Papineau (2014) for more discussion. Also, one might assume that any appeal to perception necessarily commits us to consciousness, given that perceptual states are conscious states. However, I am imagining here a view according to which perceptual evidence is nothing more than receipt of information from the external world.

[8] I'm not claiming here that a commitment to reference is implied by the content of a perceptual state (e.g., by its primary intention). My point rather is that the theory that is being signed up to here, that is, "in perception we latch on to *something*, but that reference is fixed wholly by facts outside of the content of the perception," involves a commitment to reference, and hence that someone signing up to that theory is obliged to account for that commitment.

[9] There are related concerns in Braddon-Mitchell and Miller (2019), which discusses whether reference could be grounded in a world in which time is not fundamental. There are also phenomenal intentionality theories (Kriegel 2013, Mendelovici 2018) that ground facts about reference in facts about consciousness. I will not discuss these here, as the next section concerns whether the wave function monist can account for consciousness.

monist is obliged to account for consciousness simply because we know that consciousness exists.

4. Can the Wave Function Monist Account for Consciousness?

As we have seen, the crucial question for wave function monist is whether she can account for the reality of consciousness. If she can, then she can construe our evidence in terms of our conscious experiences. If she can't, then wave function monism cannot be true in any case, as there's something real that the theory can't account for. Is accounting for consciousness harder for wave function monism than for other theories?

The answer to this question may depend on whether we aspire to give a materialist or a non-materialist account of consciousness. Let us begin by focusing on materialist accounts of consciousness. There are, broadly speaking, two kinds of materialist theory of consciousness, which David Chalmers (2002) labeled "type A" and "type B." Type A materialists aspire give an a priori functionalist reduction of consciousness, that is to say, to analyze conscious states in terms of their causal role, a causal role that is realized by physical states.[10] Thus, for example, the type A physicalist might hold that, by definition of the word "pain," all that is essentially required for someone to be a pain is for there to be an inner state that negotiates between bodily damage and avoidance behavior in the distinctive way that pain does. Could facts about the wave function realize these functional facts? There is no bodily damage or avoidance behavior in the wave function, and so the best option for the wave function monist would be to ground the existence of particles in three-dimensional space and then to suggest that systems of particles realize the functional states constitutive of mentality. But as we saw above, the wave function monist struggles to give a functionalist account of systems of particles in three-dimensional space, because such an analysis would at the very least need to involve reference to distance relations between particles, and distance relations are not present in the wave function. Without particles, it's hard to see how the wave function monist can give a functionalist reduction of consciousness.

[10] For some examples of type A physicalism, see Putnam (1967); Lewis (1970); Armstrong (1968).

Type B materialists, in contrast, do not rely on the *meaning* of mental terms in order to account for the grounding of consciousness in the physical.[11] This doesn't entail rejecting the grounding by analysis model outlined above; it just requires holding that an analysis of consciousness is to be given *a posteriori* rather than a priori. In other words, the type B analysis holds that empirical work, rather than introspective reflection, reveals the essential nature of consciousness. Typically, the type B materialist proposes that our phenomenal concepts refer to physical states, but that it is not a priori knowable that this is the case (because the reference is fixed outside of what we have a priori access to). In the toy example that has become standard in the literature, "pain" refers to c-fibers firing in the brain, but one cannot know just through reflection on pain experientially conceived that this is so.

The wave function monist adopting this approach will hold that our mental terms latch on to features of the wave function (although, of course, this cannot be known a priori). The difficulty here is that some account must be given of how precisely our mental terms latch on to features of the wave function. And as discussed in the last section, accounting for reference without three-dimensional objects in space and time is not going to be easy.

In summary, the challenges faced by the wave function monist in accounting for three-dimensional objects translate into challenges for the wave function monist desiring to give a materialist account of consciousness. What about non-materialist accounts of consciousness? Naturalistic dualists postulate basic psychophysical laws that ensure that certain facts about consciousness arise from certain physical facts. Given that there is a correspondence between facts about the wave function and the putative facts concerning familiar physical entities in three-dimensional space, I can see no reason in principle that psycho-physical laws formulated in terms of the latter could not be "translated" into laws concerning the former. The result would be psycho-physical laws according to which facts about consciousness arise from facts about the wave function. In the final chapter of The Conscious Mind (1997), David Chalmers defends an Everettian version of this view; on the Everettian view the wave function does not collapse. Alternately, one might also combine this kind of dualist approach with the view, explored in this volume by Chalmers and Kelvin McQueen, that consciousness collapses the wave function.

[11] For examples of type B physicalism, see Loar (1990/1997); Balog (1999); Papineau (2002); Diaz-Leon (2010); Howell (2013).

However, one big challenge for this combination of wave function monism and mind-body dualism is that it's hard to see how it can give an adequate account of what's going on in perception, at least if we're still working with the assumption that wave function monism cannot ground ordinary objects in three-dimensional space. On wave function monism, facts about my environment correspond to facts about the wave function, and hence presumably in perceptual experience I am tracking certain facts about the wave function. However, in the absence of three-dimensional objects, we can't explain this in the familiar way, in terms of objects in the environment causally impacting on my consciousness. It looks like the correspondence between my perceptual consciousness and certain facts about the wave function must be left as brute fact, which seems like a very unattractive feature of this view.

One currently popular alternative to both physicalism and dualism is Russellian monism.[12] Russellian monism starts from the thesis that physical science fails to give us the complete story of the nature of physical reality. This is put in various slightly different ways, but the basic idea is that physical science is confined to telling us about the *structure* of physical reality—those features that can be captured in a mathematico-causal vocabulary—leaving us in the dark about the underlying intrinsic natures of the entities that realize that structure. This is sometimes called "the problem of intrinsic natures." The Russellian monist proposes locating consciousness, or perhaps proto-consciousness, in the intrinsic nature of physical entities.

The attraction of this approach is that it promises to avoid the difficulties both of the materialist and of the dualist. The challenge for materialism is that there are good arguments to the conclusion that facts about subjective qualities cannot be fully accounted for in terms of kind of structural facts conveyed by physical science. The challenge for dualism starts from the putative evidence that physical reality forms a causally closed system, which seems to leave non-physical consciousness with no role to play in the generation of human or animal behavior. By taking consciousness (or at least proto-consciousness) as basic, the Russellian monist hopes to avoid the problems with physicalism; by placing proto-consciousness within the intrinsic nature of the physical, the Russellian monist hopes to avoid the problems with dualism.

[12] For a good collection of essays on Russellian monism, see Alter and Nagasawa (2015). See also Goff (2017).

A wave function monist taking this approach would locate (proto) consciousness in the intrinsic nature of the wave function. Are there problems with this form of Russellian monism, not shared with other forms? According to Russellian monism, physical science tells us nothing of the nature of physical reality—whether it consists of a wave function or particles in three-dimensional space—and hence difficulties won't arise in connection with the intrinsic nature of the wave function. Or at least, given that we are equally ignorant of both the intrinsic nature of the wave function and the intrinsic nature of particles in three-dimensional space, we can't possibly have grounds for thinking one is more inhospitable to mind than the other. If it is harder to squeeze consciousness into the wave function than it is to squeeze it into other putative forms of physicality, this must be down to the only thing physical science does reveal to us about the wave function: its structure.

The wave function certainly does have a bizarre structure. To make things clear, we can take a simple scenario that we don't, in fact, find in the real world: one in which the amplitude of the wave function is entirely focused at one location. This corresponds to a situation in which all particles have completely determinate locations. Now imagine a panpsychist Russellian monist wanting to ground facts about conscious particles in terms of this wave function fact. On such a view:

F1: a large number of conscious particles with determinate locations is grounded in:

F2: a high dimensional space with a property (i.e., amplitude) instantiated at one location.

I can't see any way of definitively ruling out such a grounding relationship. Perhaps if we knew the intrinsic nature of configuration space and the intrinsic nature of amplitude, it would just be obvious that F2 grounds F1. Nonetheless, the very different structure of the two facts makes it, to say the least, puzzling how they could stand in a grounding relationship. Certainly, we don't have the more familiar panpsychist picture of little conscious things combining to produce bigger conscious things, or even the cosmopsychist picture of a conscious universe fragmenting into conscious parts (at least not parts in the same space).[13] It is true that there are also puzzles about

[13] Of course, if we could ground conscious particles, we could then try a more familiar panpsychist grounding story to get to macro-level consciousness, but this brings us back to the problems already discussed.

structural mismatch in these more familiar forms of Russellian monism (Chalmers 2016; Goff 2017: Chapter 8)—the structure of consciousness seems different to the structure of the physical brain—but the structural mismatch faced by the Russellian wave function monist seems to be of a wholly different order.

Notice that F1 involves reference to locations in three-dimensional space, which entails the existence of distances in three-dimensional space. Does the Russellian wave function monist face the same difficulties grounding three-dimensional distances that we discussed in Section 2? The advantage for the Russellian monist is that she is committed to locations and/or distances having an *intrinsic nature*, which opens up the possibility that an analysis of the intrinsic nature of locations/distances may yield a condition (specifying what is essentially required for there to be locations/distances) that can be satisfied by facts about the intrinsic nature of the wave function. Of course, we have no grasp of a possible intrinsic nature of location/distance that could do this, but Russellian monists tend to be comfortable with admitting ignorance about the intrinsic nature of physical reality on the grounds that we wouldn't expect as evolved creatures to have ways of accessing the underlying intrinsic nature of matter.

However, we do have access to the intrinsic nature of consciousness. And therefore, if the Russellian monist wants to ground F1 in F2, then she must offer an analysis of consciousness that "opens it up" to this kind of reduction. To make clear what I have in mind, return to the analysis of a party. For there to be a party is for there to be people reveling; this analysis opens "there is a party" up to being reduced to facts about revelers. I don't think that something analogous can be done with respect to consciousness. The only analysis we can give to the proposition <there are *N* subjects of experience> is the trivial one that what is essentially required for this proposition to be true is for there to be *N* subjects of experience. And it's hard to see how F2 could satisfy this condition. The only possibility is to identify the many dimensions with the *N* subjects; but, in this case, the fact that the amplitude is focused in one location would seem to have no role in the grounding of F2.[14]

I argue in Chapter 9 of *Consciousness and Fundamental Reality* that given how "thin" the analysis of consciousness is, the only hope for grounding

[14] Perhaps there could be a causal relationship between the fact that the amplitude is in one location and the fact that each of the dimensions is a subject, but this wouldn't be a grounding relationship.

consciousness is to do so via what I call "grounding by sumsumption." In cases of grounding by subsumption, a less fundamental entity is irreducibly subsumed in a more fundamental entity. To give an analogy, some Christians think about the trinity in these terms: the three persons are irreducible conscious subjects but are nonetheless subsumed in a single substance. I suggest the Russellian monist may hold that conscious minds are grounded by subsumption in the universe: although irreducible, they are subsumed in the more expansive reality of the universe. For this model to be compatible with wave function monism, human and animal conscious minds would have to be irreducibly present in the wave function (compare: the three persons are irreducibly present in the Godhead, on this model of the trinity). This would be the case with respect to F2 only if each dimension was identical with a conscious subject; but, as already stated, this would make the fact that the amplitude is focused in one location redundant in the grounding of F1.

In summary, there doesn't seem to me to be a way for the wave function monist to account for the facts of consciousness, at least on the assumption that they can't account for the facts about three-dimensional objects. And if wave function monism cannot account for consciousness, then wave function monism cannot be true.

Could these challenges be avoided if one gives up wave function monism and adopts an interpretation of quantum mechanics that postulates matter or particles in addition to the wave function, such as the Bohmian view? This would be a solution for the dualist. Once we have matter in the picture, the dualist can proceed to give the familiar account of visual perception in terms of the interactions between material entities and our conscious minds. One concern for materialists or Russellian monists is that these interpretations can be understood as rendering matter epiphenomenal: something that is guided by the wave function rather than having any causal efficacy in its own right. If consciousness is grounded in matter, then consciousness would also lack causal power. One option for a materialist or a Russellian monist is to adopt realism about matter accompanied by the thesis that the wave function has the status of a law, specifying the causal powers of matter, rather than being a physical entity in its own right.[15] Another option for a panpsychist Russellian monist is to adopt the 'pan-agentialist' view I have explored (Goff 2020, Goff 2019: ch. 5), according to which particles have free will but behave predictably because, lacking rational deliberation, they

[15] For an example of this kind of view, see Allori et al. (2007).

inevitably follow their conscious inclinations. A Bohmian pan-agentialist could hold that the wave function determines the conscious inclination of particles, which particles then freely choose to follow.

5. Much Work to Be Done!

I have tentatively argued that wave function monism is false, on the grounds that it cannot account for the reality of consciousness. I don't take these arguments to be conclusive; rather, they present challenges that future work may overcome. The crucial point is that there is much work to be done here properly assessing whether or not wave function monism, and other interpretations of the ontology of quantum mechanics, can account for the reality of consciousness. There is a desperate need for a new generation of philosophers who know the physics and who are serious about consciousness. How to interpret quantum mechanics has been one of the biggest scientific/philosophical challenges for nearly a hundred years. At the same time, the potential for our knowledge of consciousness to assist in metaphysical progress has been woefully neglected for at least as long. It's just possible that proper attention to consciousness—the one phenomenon we know with certainty to exist—might allow us to make progress on this issue.

References

Allori, Valia, Goldstein, Sheldon, Tumulka, Roderich, & Zanghi, Nino (2007). "On the common structure of Bohmian mechanics and the Ghirardi-Rimini-Weber theory." *British Journal for the Philosophy of Science* 59: 353–389.

Albert, D. Z. (2013). "Wave function realism," in Ney & Albert (2013).

Albert, David Z. (2015). *After Physics*. Cambridge, MA: Harvard University Press.

Alexander, Samuel (1920) *Space, Time and Deity*. New York: Macmillan.

Alter, Torin, & Nagasawa, Yujin (2015). *Consciousness in the Physical World*. Oxford: Oxford University Press.

Armstrong, David (1968). *A Materialist Theory of Mind*. London: Routledge and Kegan Paul.

Armstrong, David M. (1997). *A World of States of Affairs*. Cambridge, UK: Cambridge University Press.

Balog, Katalin (1999). "Conceivability, possibility, and the mind-body problem." *Philosophical Review* 109: 4, 497–528.

Braddon-Mitchell, David, & Miller, Kristie (2019). "Quantum gravity, timelessness, and the contents of thought." *Philosophical Studies* 176: 7, 1807–1829.

Broad, C. D. (1925). *The Mind and its Place in Nature.* London: Routledge and Kegan Paul.

Chalmers, David J. (1997). *The Conscious Mind*, New York: Oxford University Press.

Chalmers, David J. (2002). "Consciousness and its place in nature," in Chalmers, David J. (Eds.), *Philosophy of Mind: Classical and Contemporary Readings*. Oxford: Oxford University Press.

Chalmers, David J. (2003). "The Content and Epistemology of Phenomenal Belief," in Quentin Smith & Aleksandar Jokic (Eds.), *Consciousness: New Philosophical Perspectives*, pp. 220–272. Oxford: Oxford University Press.

Chalmers, David J. (2004). "Epistemic two-dimensional semantics." *Philosophical Studies* 118: 153–226.

Chalmers, David J. (2016). "The combination problem for panpsychism," in Godehard Brüntrup & L. Jaskolla (Eds.), *Panpsychism*. Oxford: Oxford University Press.

Chalmers, David J., & McQueen, Kelvin J. (this volume). "Consciousness and the collapse of the wave function."

Chen, Eddy, K. (2019). "Realism about the wave function." *Philosophy Compass* 14(7).

Dennett, Daniel C. (2007). "Heterophenomenology reconsidered." *Phenomenology and the Cognitive Sciences* 6: 1–2, 247–270.

Diaz-Leon, Esa (2010). "Can phenomenal concepts explain the explanatory gap?" *Mind* 119: 476, 933–951.

Goff, Philip (2017). *Consciousness and Fundamental Reality*. Oxford: Oxford University Press.

Goff, Philip (2019) Galileo's Error: Foundations for a New Science of Consciousness, New York: Pantheon.

Goff, Philip (2020). "Panpsychism and free will: A case study in liberal naturalism." *Proceedings of the Aristotelian Society* 120: 2, 123–144.

Goff, Philip, & Papineau, David (2014). "What's wrong with strong necessities?" *Philosophical Studies* 167: 3, 749–762.

Hawthorne, John (2010). "A metaphysician looks at the Everett interpretation," in Saunders et al. (2010).

Howell, Robert (2013). *Consciousness and the Limits of Objectivity: The Case for Subjective Physicalism*. Oxford: Oxford University Press.

Kriegel, Uriah (Ed.) (2013). *Phenomenal Intentionality*. Oxford: Oxford University Press.

Loar, Brian (1990/1997) "Phenomenal states," originally published in James Tomberlin (Ed.), *Philosophical Perspectives 4: Action Theory and Philosophy of Mind*, Ridgeview; reprinted in substantially revised form in N. Block, O. Flanagan, & Guüzeldere (Eds.) (1997), *The Nature of Consciousness: Philosophical Debates*, Cambridge, MA: MIT Press.

Loewer, Barry (1996). "Humean supervenience," *Philosophical Topics*, 24: 4, 609–620.

Lewis, David (1970). "An argument for the identity theory." *Journal of Philosophy* 63: 17–25.

Maudlin, Tim W. E. (2007). "Completeness, supervenience and ontology." *Journal of Physics A: Mathematical and Theoretical* 40: 12.

Mendelovici, Angela (2018). *The Phenomenal Basis of Intentionality*. Oxford: Oxford University Press.

Mill, John Stuart (1843). *System of Logic*. London: Longmans, Green, Reader and Dyer.

McLaughlin, Brian (1992). "The rise and fall of British emergentism," in Ansgar Beckerman, Hans Flohr, and Jaegwon Kim (Eds.) (1992) *Emergence or Reductionism*? Berlin: De Gruyter.

Ney, Alyssa (2012). "The status of our ordinary three dimensions in a quantum universe. *Nous* 46: 3, 525–560.

Ney, Alyssa (2013). "Introduction," in Ney, Alyssa, & Albert, David Z. (Eds.) (2013). *The Wave Function: Essays on the Metaphysics of Quantum Mechanics*. Oxford: Oxford University Press.

Ney, Alyssa (2015). "Fundamental physical ontologies and the constraint of empirical coherence: A defense of wave function realism." *Synthese* 192: 10, 3105–3124.

Ney, Alyssa (2020). "Finding the world in the wave function: Some strategies for solving the macro-object problem." *Synthese.*

Papineau, David (2002). *Thinking About Consciousness*. Oxford: Clarendon Press.

Putnam, Hilary (1967). "The nature of mental states," reprinted in his *Mind, Language and Reality*, Cambridge, UK: Cambridge University Press, 1975.

Schaffer, Jonathan (2010). "Monism: The priority of the whole." *Philosophical Review* 119: 1, 31–76; reprinted in Philip Goff (2012). *Spinoza on Monism*. London: Palgrave Macmillan, 9–50.

Shani, Itay (2015). "Cosmopsychism: A Holistic Approach to the Metaphysics of Experience," *Philosophical Papers* 44: 3, 389–437.

Wallace, David (2010). "Decoherence and ontology," in Saunders, Simon, Barrett, Jonathan, Kent, Adrian, & Wallace, David (Eds.) (2010). *Many Worlds? Everett, Quantum Theory, and Reality*. Oxford: Oxford University Press.

Wallace, David (2012). *The Emergent Multiverse*. Oxford: Oxford University Press.

6

Against "Experience"

Peter J. Lewis

1. Against "Measurement"

Thirty years ago, J. S. Bell gave a talk called "Against 'measurement'" (Bell 1990).[1] In it, he laments a particular kind of imprecision in the standard formulation of quantum mechanics. He locates the source of this imprecision in the use of certain "bad words" in textbook discussions: "system, apparatus, environment, microscopic, macroscopic, reversible, irreversible, observable, information, measurement" (1990, 34). He singles out the word "measurement" as *particularly* bad: "[T]he word has had such a damaging effect on the discussion, that I think it should now be banned altogether in quantum mechanics" (1990, 34). I argue here that Bell's list of bad words needs to be extended to include *experience* and its cognates like *awareness*, *perception*, *observation*, and *consciousness*. The word "experience," too should be banned altogether in quantum mechanics.

Of course, Bell's suggestion is hyperbolic: you couldn't, and shouldn't, excise the word "measurement" from physics altogether. His point is rather about the *role* the word should play: it's fine for the word "measurement" to appear in an *application* of a physical theory, but it's not OK for it to appear in the *formulation* of the theory. Similarly, my suggestion is hyperbolic: there is no need to completely excise the word "experience" from physics—although in this case, it seems like you *could.* It's fine to discuss how the experiences of human observers arise in various *applications* of quantum mechanics—and indeed, as we shall see, such discussions play an important role in the philosophical literature concerning the interpretation of quantum mechanics. But it's not OK for the word "experience" to appear as a primitive in the theory itself.

[1] The conference was 62 Years of Uncertainty, Erice, Sicily, August 5–14, 1989.

Peter J. Lewis, *Against "Experience"* In: *Consciousness and Quantum Mechanics.* Edited by: Shan Gao, Oxford University Press. © Oxford University Press 2022. DOI: 10.1093/oso/9780197501665.003.0007

The basic reasons for the elimination of "experience" are parallel to those Bell gives for "measurement." One of Bell's concerns is that there is no clear answer to the question "What exactly qualifies some physical systems to play the role of 'measurer'?" (1990, 34). Many kinds of physical systems can constitute measuring devices, and there is at best a vague line between those physical interactions that count as measurements and those that do not. Vague terms have no place in the foundations of a precise physical theory. Another of his concerns is that any measuring device "can be seen as being made out of atoms" (1990, 36). In that case, since atoms are subject to the laws of quantum mechanics, an appeal to measurement in the foundations of quantum mechanics becomes viciously circular, "like a snake trying to swallow itself by the tail" (1990, 36).

The same can be said of experience. What exactly qualifies some physical systems to play the role of "experiencer"? Many kinds of physical systems can constitute experiencers, including vertebrates with centralized nervous systems and cephalopods with more distributed ones (Godfrey-Smith 2016). The limits of experience are vague within a species (Brown, Lydic, and Schiff 2010) and across species (Schwitzgebel 2020). And experiencing systems are, of course, made of atoms, so an appeal to experience in the foundations of quantum mechanics would be viciously circular.

Thus, one might take my thesis here to be a simple extension of Bell's. Indeed, some of Bell's arguments seem to be directly *about* experiencing systems rather than measuring instruments: "Was the wavefunction of the world waiting to jump for thousands of millions of years until a single-celled living creature appeared? Or did it have to wait a little longer, for some better qualified system . . . with a PhD?" (1990, 34). But, in fact, the character of our arguments is somewhat different.

Bell is primarily taking to task the standard theory of quantum mechanics and the textbook writers who present that theory. He gives substantial evidence that talk of "measurement" is pervasive and problematic in such textbooks. And his message has largely been taken on board, at least by the foundations of physics community. Bell can be taken as giving a particularly forceful expression of the *measurement problem*, and the main business of work in the foundations of quantum mechanics in recent years has been to solve, dissolve, or explain away the measurement problem. One can no longer blithely appeal to measurement in the formulation of a version of quantum mechanics; any such appeal would at least require a *lot* of footnotes.

Quantum mechanics textbooks do not typically appeal to "experience." They do appeal to "observation," but this can typically be taken as a synonym for "measurement" rather than as a synonym for "experience." So my complaint is not against textbook formulations. Rather, it largely concerns *responses* to the measurement problem—attempts to formulate quantum mechanics without appealing to measurement as a primitive term. A few of these directly appeal to experience as a primitive term—e.g., Wigner (1961). But more often, "experience" and its cognates show up in the surrounding discussion.

It is trickier, I think, to say what's wrong with these appeals to "experience" than with appeals to "measurement." In part, this is because it is rather obvious, when it is pointed out, that measurements don't constitute a primitive, unanalyzable kind. This is not so obvious when it comes to *experience*—it seems to some people that experience or consciousness *is* primitive or unanalyzable, and this makes it harder to say what's wrong with *treating* it as such. A second difficulty with saying what's wrong with appeals to "experience" is that there are legitimate and important discussions of experience in the literature on responses to the measurement problem, so this can't be a *blanket* condemnation. The best way to proceed, I think, is to by way of some illustrative examples—good, bad, and ugly.

2. The Good

If, as Bell insists, you can't use "measurement" in the formulation of quantum mechanics, the theory needs to be reformulated to avoid any such use. There are, of course, various ways of doing this. One way is to replace the textbook "collapse on measurement" with a *spontaneous* collapse process, as in the GRW theory (Ghirardi, Rimini, and Weber 1986). The spontaneous collapse approach raises various interpretational questions, and these questions can't be answered without talking about measurement, and about experience. These uses of "measurement" and "experience" are perfectly good.

The interpretational questions arise because of the way the GRW theory works. For a single particle, or a small collection of particles, the probability of a spontaneous collapse to a well-localized position during a reasonable period of time is very small. But for a macroscopic collection of particles, the probability of such a collapse is high, even during a small fraction of a second. So as long as all measuring devices record their outputs using the position of some macroscopic object, the GRW collapse process ensures that measurements have determinate outcomes.

This raises a legitimate question about the nature of measurement: Is it safe to assume that measuring devices always display their outputs using the position of a macroscopic object? Albert and Vaidman (1989) consider this question and answer it in the negative. In particular, they note that a cathode-ray TV screen can be used to display the output of a quantum measurement, and the production of a visible dot on such a screen involves too few particles to reliably produce a GRW collapse.

Note that nothing in Albert and Vaidman's critique of the spontaneous collapse program treats "measurement" as an unanalyzed primitive; indeed, the whole point is to *analyze* our actual measuring devices to see if they fulfill the assumptions required for the GRW approach to deliver determinate measurement outcomes. Hence, their use of "measurement" does not fall foul of Bell's prohibition.

Albert and Vaidman go on to ask a similar question about *experience*. If, as they argue, a TV screen-measuring device won't precipitate a GRW collapse, will the observation of such a device by a human being precipitate a collapse? The question is whether the human brain represents distinct visual experiences in terms of distinct positions of a macroscopic number of particles. This question is answered affirmatively by Aicardo et al. (1991): even if we use a measuring device that doesn't reliably precipitate a GRW collapse, because of the way human brains work, such a collapse will be reliably produced in the brain of a human observer.

Again, nothing in this investigation treats "experience" as an unanalyzed primitive; the whole point is to *analyze* the way visual experiences are actually grounded in human neurophysiology to see if they fulfill the assumptions required for the GRW approach to deliver determinate visual experiences. This use of "experience" is entirely appropriate.[2]

There are further questions that arise in this context. For example, even if actual human brains reliably precipitate GRW collapses, is it the case that *any possible* brain would do so? Albert (1992, 107) uses thought-experiments involving people with brain implants to argue that the answer to this question is "no." In particular, he envisions a person for whom an experience is grounded in the position of a single particle. For such a person,

[2] See Gao (2017, 134) for a similar "good" example: Gao considers the details of neurophysiology to argue that an observer will have determinate experiences under his preferred collapse model.

a measurement using a cathode-ray TV screen probably won't precipitate a GRW collapse even in the person's brain, and so probably won't produce a determinate experience. This thought-experiment raises tricky issues; I return to it in Section 5.

3. The Bad

Let us now turn to a case where I think that the word "experience" is used badly. The case concerns Bohm's (1952) hidden variable theory, another way of reformulating quantum mechanics to avoid "collapse on measurement." Rather than reconceiving collapse as a spontaneous process, Bohm's theory stipulates that the quantum state never undergoes a collapse at all. Instead, Bohm's theory supplements the quantum state with a set of particles with determinate positions, which move such that the probability distribution for the particle locations is always given by the Born rule.

Measurement has a natural interpretation in Bohm's theory: the quantum state gives us a probability distribution over all the *possible* particle locations, and measurement reveals where the particles *actually* are. But there is a potential issue here that needs attention: after a measurement, we *know* where the particles are, and yet because there is no collapse, the quantum state hasn't changed. This threatens the Born rule, which says that the probability distribution for any observable quantity, including particle locations, is *always* given by $|\psi|^2$ (the squared amplitude of the quantum state, written in the relevant basis).

Dürr, Goldstein, and Zanghì (1992) address this issue, seeking to show that the Born rule can be given a consistent interpretation within Bohm's theory. They ask whether it is possible to learn more about the locations of the Bohmian particles than is given by $|\psi|^2$. They derive a result they call "absolute uncertainty" that answers this question in the negative: "[N]o devices whatsoever, based on any present or future technology, will provide us with the corresponding knowledge. *In a Bohmian universe such knowledge is absolutely unattainable!*" (1992, 884).

This result secures the consistency of the Born rule, but at an apparently devastating cost: since the particle locations in Bohm's theory are supposed to tell us the actual outcome of our measurement, if we can't find out the locations of the Bohmian particles with greater precision than given by $|\psi|^2$, and there is no collapse, then we can't find out the outcome of our measurement! That is, "absolute uncertainty" might be taken to entail that Bohm's theory doesn't solve the measurement problem after all (Stone 1994).

But this conclusion is based on a misreading of Dürr et al.'s result: the $|\psi|^2$ appearing in "absolute uncertainty" is the *effective* state, not the full quantum state of the system (Maudlin 1995). If the state is written in an appropriate basis, it can typically be written as a sum of terms, only one of which is relevant (for practical purposes) to the future motion of the Bohmian particles; this term is the effective state. When the position of a particle is measured, the effective state depends on the accuracy of the measurement. That is, if your position measurement has an accuracy of 1 mm, the post-measurement effective state has a width of 1 mm, and you learn the actual position of the particle to within 1 mm. After the measurement, you don't know the position of the particle with greater accuracy than is given by (post-measurement) $|\psi|^2$, but you *do* know the position of the particle with greater accuracy than is given by *pre*-measurement $|\psi|^2$. Hence, the particle positions do, after all, tell you the actual outcome of the measurement.

The possibility of misreading "absolute uncertainty" is not my concern here.[3] My concern is with an *exception* to "absolute uncertainty" suggested by Dürr, Goldstein, and Zanghì: "There is one situation where we may, in fact, know more about configurations than what is conveyed by the quantum equilibrium hypothesis $\rho = |\psi|^2$: when we ourselves are part of the system!" (1992, 903). This exception is where things go bad.

Why should there be such an exception? Stone (1994), because he (mistakenly) thinks that "absolute uncertainty" precludes us from learning *anything* about the locations of the particles, suggests that Dürr et al. postulate the exception in a vain attempt to secure determinate outcomes to our measurements. That is, Stone suggests that their idea is that direct awareness of the locations of certain particles in your brains tells you the outcome of your measurement. But this attempt is in vain, Stone argues, because Dürr et al.'s "absolute uncertainty" means that the locations of these particles in your brain must be inaccessible to the *rest* of your brain, and hence you couldn't, after all, be aware of them.

As discussed above, there is no need for an exception for this purpose: you *can* learn about the locations of Bohmian particles in your measuring device, and by the same token, the locations of the Bohmian particles in one part of

[3] See Lewis (2019) for further discussion of this misreading, and of the related question of whether knowledge of measurement outcomes would allow superluminal signaling.

your brain are accessible by the rest of your brain without any exception to "absolute uncertainty."[4] So again: Why the exception?

I suspect the reason goes deeper than Stone suspects: it is not that Dürr et al. postulate the exception as an ad hoc fix to a problem in their account, but rather that they think that *any* physical theory must contain some such exception. It is not hard to construct a tempting argument for this claim. Consider a spin measurement on a particle whose state is a symmetric superposition of spin-up and spin-down eigenstates. If the particle is passed through an inhomogeneous magnetic field, the result is a superposition of two wave packets, one deflected upwards and the other deflected downwards. The Bohmian particle follows one component or the other depending on its initial position. But how do we know which wave packet contains the particle?

Well, to find out, we can run the particle into a fluorescent screen, resulting in a superposition of two terms, one describing a flash at the top of the screen and one describing a flash at the bottom. Again, the Bohmian particles are associated with one term or the other. But how do we know which term contains the particles? Well, to find out, we can consider the interaction of the flash with the human eye. We get a superposition of two terms describing the retina, where the Bohmian particles are associated with one of them. But how do we know which term contains the particles? We can proceed into the human brain: we get a superposition of two terms describing the state of the visual cortex, where the Bohmian particles are associated with one of them. But how do we know which one? We could consider the interaction of the visual cortex with some other region of the brain, but this is going nowhere. Rather, we just have to assume that at some stage we are *directly aware* where the particles are—we are directly aware of the configuration of Bohmian particles in parts of our brains. Otherwise, we could never know which wave packet contains the particles.[5]

[4] Gao (2017, 100–103) argues that Bohm's theory is problematic under *any* assumption about how experience supervenes on the ontology of the theory. In particular, he argues that the *instantaneous* Bohmian particle configuration cannot ground determinate experience. However, I think this overlooks the possibility that the evolution of the particle configuration over time grounds experience. Indeed, Gao later notes the dynamical nature of neurochemical explanations of experience (2017, 134).

[5] There is nothing distinctly Bohmian about this argument: we could replace the Bohmian particles with the underlying ontology of any other physical theory, and the question of how we know the state of that ontology can be asked at any stage of the measurement process. I consider the historical roots of this kind of "tempting argument" in Section 5.

There is a construal of this argument according to which it is unobjectionable: unless *awareness* arises at some stage in the above process, we couldn't have *knowledge* of the outcome of the measurement. But the argument is surely fallacious if it is taken to imply that something needs to be *posited* to accomplish this awareness, something in addition to the process otherwise described. You need something beyond a flash on a screen in order for there to be awareness of the flash: what you need is a correlated brain state. But you don't need to posit something in addition to the brain state in question in order for there to be awareness *of the flash*. Maybe you need something outside that particular brain state in order for there to be experience or knowledge of *that brain state* as an object of scientific inquiry, but that's a separate issue.

Insistence on experience as *exceptional* in this way gives critics of Bohm's theory some extra ammunition. For example, Brown and Wallace suggest that in order to defend Bohm's theory, Dürr et al. (1992) and Maudlin (1995) have to assume that "our conscious perceptions supervene directly *and exclusively* on the configuration of (some subset) of the corpuscles associated with our brain" (2005, 534). But this assumption, they suggest, rests on the further assumption "that consciousness is some sort of bare physical property (like, say, charge) whose connection with physical matter can simply be posited rather as we posit other basic physical laws," which in turn "makes consciousness completely divorced from any assumptions rooted in the study of the brain" (2005, 536). You can't just *posit* a connection between experience and the physical brain; you have to *discover* the connection via neuroscientific study.

My claim here is that the assumptions about experience that Brown and Wallace object to are both entirely unnecessary and counter-productive for the Bohmian. There is no need for Bohmians to say anything special about experience: a measurement correlates some property of the measured system with the configuration of Bohmian particles in the measuring device, and eventually with the configuration of Bohmian particles in the brain of the human observer. As Maudlin (1995) rightly observes, that's all that needs to be said by the interpreter of quantum mechanics. Dürr et al.'s "absolute uncertainty" result, which is exceptionless, ensures that measurements are always governed by the Born rule. There is no need for a further postulate about the connection between Bohmian particles and experience.

Furthermore, a postulate that says that we are directly aware of the positions of the Bohmian particles in our brains is counter-productive, serving only as a lightning rod for criticism. By itself, such a postulate doesn't work: direct "awareness" of the position of an isolated particle would not have the functional connections to other cognitive powers to truly constitute *awareness*. And once you add the functional connections, it is unclear what the *direct* awareness was adding in the first place. Maybe we don't know the full story about how the physical state of a human brain produces experience, but this is not a matter for interpreters of quantum mechanics to worry about. Whatever the full story is, it can surely be expressed in terms of the configuration of Bohmian particles.[6]

4. The Ugly

So far, I have presented two examples of the discussion of experience in the literature on the foundations of quantum mechanics; one I take to be clearly good and the other I take to be clearly bad. But sometimes it is not so easy to tell whether a discussion of quantum mechanics is good or bad. I consider two such cases here, both of which concern the spontaneous collapse approach again.[7]

One of the earliest philosophical discussions of the spontaneous collapse approach is Bell (1987). Bell worries that the GRW theory represents the world using a wave function alone; this is a potential problem because the wave function is defined over a configuration space, with $3N$ dimensions for an N-particle system, not over ordinary three-dimensional space. Here is Bell's proposed solution:

> However, the GRW jumps (which are part of the wavefunction, not something else) are well localized in ordinary space. Indeed, each is centered on a particular spacetime point ($\mathbf{x}$, t). So we can propose these events as the basis of the 'local beables' of the theory. These are the mathematical counterparts in the theory to real events at definite places and times in the real world.... A piece of matter then is a galaxy of such events. As a schematic psychophysical parallelism we can suppose that our personal

[6] It is plausible that whatever the full story about experience is, it can also be expressed in terms of the wave function configuration. This is part of Brown and Wallace's (2005) argument for the redundancy of the Bohmian particles, which I return to in Section 5.

[7] There is, of course, nothing intrinsically ugly about these cases; they are only "ugly" insofar as they make life difficult for me in distinguishing good from bad uses of "experience."

> experience is more or less directly of events in particular pieces of matter, our brains, which events are in turn correlated with events in our bodies as a whole, and they in turn with events in the outer world. (1987, 45)

What is Bell up to here? It looks like he is postulating a direct connection between the physical world and experience as part of his exposition of the GRW theory. Isn't that *bad*?

Something similar can be found in the work of Hameroff and Penrose (1996). They adopt a spontaneous collapse theory incorporating general relativity, according to which superpositions of distinct arrangements of matter produce superpositions of distinct space-time geometries. The latter superpositions are hypothesized to be unstable, collapsing to one matter arrangement or the other. They combine this spontaneous collapse theory with a particular account of conscious experience, according to which a conscious event is associated with a spontaneous collapse occurring in a specific neuronal structure (a coherent superposition involving a number of tubulin molecules). Again, Hameroff and Penrose seem to be postulating a direct connection between physics and experience as part of the exposition of their preferred quantum theory.

In the previous section, I characterized postulating a connection between physics and experience as *bad*. But in each of the cases just described, there are mitigating factors. In Bell's case, I think he's not really postulating such a connection at all; rather, his concern is to show that we can find our ordinary three-dimensional ontology in the $3N$-dimensional wave function of the GRW theory. If we can't, then a peculiar form of empirical inadequacy threatens: the theory says nothing about the three-dimensional world, and hence fails to predict experimental outcomes expressed in three-dimensional terms. But if we *can* associate three-dimensional objects with sets of GRW collapse events, then we *also* find brains and their contents in the GRW ontology, since brains are three-dimensional objects. The "psychophysical parallelism" Bell outlines isn't really an additional postulate, but simply an expression of the nature of physical stuff according to GRW, consistent with more or less *any* neurological theory of experience.[8]

[8] This is not to say that the GRW theory as Bell envisions it will *automatically* succeed in accounting for determinate measurement results or determinate experiences. Parallel concerns to those of Albert and Vaidman (1989) can be posed using Bell's event-based GRW: perhaps there will be no collapse events associated with a particular measurement outcome, or with a particular mental state, because they involve too few particles. These concerns about experience are of the "good" kind explored in Section 2.

Hameroff and Penrose cannot be defended in this way: they make it clear that they are postulating a *particular* connection between quantum collapse events and conscious experience. Rather, their approach here can be taken as exemplifying an *objection* to my claim that postulating a connection between physics and experience is a bad thing in discussions of the foundations of quantum mechanics. We don't know the details of how the brain gives rise to conscious experience. Hameroff and Penrose are offering a *hypothesis* concerning some of those details, a hypothesis that is open to empirical test. Thus, a prohibition on postulating a connection between quantum physics and experience looks like an unwarranted restriction on scientific exploration.

I think I have to grant this point. One can object to Hameroff and Penrose's hypothesis on scientific grounds (e.g., Grush and Churchland 1995), but surely not on general methodological grounds. The unification of apparently disparate phenomena is a feature of several notable episodes of scientific progress. Hameroff and Penrose may or may not succeed at their chosen unification, but you can't fault them for trying.

Note, though, that their proposal about conscious experience is quite cleanly separable from their proposal for solving the measurement problem in quantum mechanics. Penrose (1996) formulates the latter without postulating any particular connection to consciousness—and indeed, the spontaneous collapse theory he develops seems consistent with more or less any such connection.[9] Maybe he always had his eye on a connection to consciousness (Penrose 1994), but the interpretation of quantum mechanics *itself* does not appeal to any such connection. So Hameroff and Penrose (1996) does not constitute an objection to my thesis, construed as the claim that postulates about experience have no place *within* the interpretation of quantum mechanics.

5. The Extent of the Problem

The examples considered in the previous sections show that "experience" and its cognates appear regularly in discussions of the foundations of quantum mechanics, and that such discussions are not always bad. It's fine (and sometimes necessary) to consider whether some interpretation of quantum

[9] Again, subject to caveats of the kind raised by Albert and Vaidman (1989).

mechanics can account for the determinacy of experience. What's not OK is to incorporate a posit about experience within the interpretation itself.

But is this really a significant problem? The example of Section 3 could, for all I have said, be an isolated case—an offhand comment that doesn't reflect anyone's considered opinion about how to approach the foundations of quantum mechanics. However, a look at the history of appeals to experience suggests otherwise.

Consider, e.g., the classic discussion of "psycho-physical parallelism" in von Neumann (1955). Von Neumann describes a simple temperature measurement, and notes that we can choose whether or not to include the thermometer in the physical analysis, and similarly for the eye and the brain of the human observer:

> But in any case, no matter how far we calculate—to the mercury vessel, to the scale of the thermometer, to the retina, or into the brain, at some time we must say: and this is perceived by the observer. That is, we must always divide the world into two parts, the one being the observed system, the other the observer. In the former, we can follow up all physical processes (in principle at least) arbitrarily precisely. In the latter, this is meaningless. (1955, 419)

There is something very like the "tempting argument" of Section 3 in this passage; indeed, I take von Neumann to be the canonical source of such arguments. However, one might read von Neumann as talking about *measurement* rather than *experience* here: the discussion occurs at the beginning of a chapter called "The Measuring Process" in which he elaborates his proposal that measurements obey a different dynamical law from non-measurements. The folly of this approach to quantum mechanics is precisely Bell's point in "Against 'measurement.'"

But this would not be the best reading of this passage, I think. Von Neumann's goal in this chapter isn't just to *describe* his collapse-on-measurement proposal; the discussion above occurs in the context of an attempt to *justify* it:

> Indeed experience only makes statements of this type: an observer has made a certain (subjective) observation; and never any like this: a physical quantity has a certain value. Now quantum mechanics describes the events which occur in the observed world, so long as they do not intersect with

> the observing portion, with the aid of the process 2 [unitary dynamics], but as soon as such an interaction occurs, i.e., a measurement, it requires the application of process 1 [collapse dynamics]. The dual form is therefore justified. (1955, 420)[10]

That is, the duality in the dynamical laws is justified because it reflects a fundamental duality in nature. The distinction between measurements and non-measurements is clearly not fundamental, as Bell forcefully points out: measurements are just a rather loosely defined set of ordinary physical processes. But the distinction between experience and that which is experienced—between subjective and objective—looks at first glance like a reasonable candidate for such a fundamental distinction. Plausibly, it is that distinction that von Neumann is appealing to here.

However, the subjective-objective distinction as von Neumann understands it is unable to deliver the justification he needs for the duality in the laws. Von Neumann thinks that where we place the boundary between the experiencing subject and the experienced object "is arbitrary to a very large extent" (1955, 420). But the point at which the collapse dynamics takes over from the unitary dynamics is *not* a matter of arbitrary stipulation; it is empirically decidable, at least in principle (Albert 1992, 84).

What von Neumann needs is a precise boundary between those entities that precipitate collapse and those that do not. Such a sharp line can be had, at a cost: Wigner (1961) proposes that there are fundamental facts about which systems are *conscious*, and that consciousness acts on physical systems to precipitate collapse. Note that these fundamental facts have to float free of the physical facts about a system if Wigner's proposal is to avoid the problem that experiencing systems are just systems of atoms (see Section 1). That is, Wigner postulates a strong form of interactive dualism in order to justify a duality in the physical laws.

Few will want to follow Wigner down this path: non-physical minds, especially causally active ones, are mysterious at best. But the point is that unless you are willing to go to such extremes, von Neumann's appeal to a fundamental subjective-objective distinction is idle: it doesn't do the work he requires of it. Furthermore, by assuming that the physicist can simply *posit* a connection between the physical world and experience at her convenience, von Neumann opens himself to the objection that he is

[10] Von Neumann cites Bohr (1929) as the source of this line of justification.

treating consciousness as "some sort of bare physical property," rather than paying serious attention to the results of neuroscientific research (Brown and Wallace 2005, 536).

There are, of course, perennial (but controversial) arguments for the irreducibility of the mental to the physical, e.g., based on *qualia* (Chalmers 1996). But the existence of a "hard problem" of consciousness is distinct from the methodological question at issue here: whether or not qualia are irreducible, you can't posit whatever psychophysical connection you like. Neuroscience constrains the connections between brain structures and experience, irrespective of whether what it tells us constitutes a reduction of qualia.

Nevertheless, von Neumann's approach is tempting. It is tempting, I think, because coming up with a consistent interpretation of quantum mechanics is hard, and the freedom to simply *posit* a psychophysical connection appears to give you an additional free variable to play with in constructing an adequate theory. But this appearance is deceptive: the additional variable is only *free* because it is disconnected from anything that might do any genuine explanatory work, and because we are ignoring any *constraints* on the psychophysical connection coming from scientific research on the brain.

Dürr et al.'s exception to "absolute uncertainty" provides one example of succumbing to this temptation. But it is not hard to find others; I only need to look as far as my own CV. In "How Bohm's Theory Solves the Measurement Problem" (Lewis 2007), I attempt to resist Brown and Wallace's (2005) redundancy argument against Bohm's theory. Brown and Wallace argue that since experiences could supervene on the wave function configuration just as easily as on the Bohmian particle configuration, the Bohmian particles are redundant, and Bohm's theory reduces to the many-worlds theory. I respond, in part, that the Bohmian can *posit* that we are directly aware of particle configurations in our brains. I now think this is a mistake: such a posit does no work and ignores neuroscience. This is not to say that the redundancy argument necessarily succeeds: the argument can be resisted (see, e.g., Ney 2013 and Callender 2015), but notably *not* by postulating a special connection between particles and experience.

Even Albert's (1992, 107) thought experiment concerning determinate experience in the GRW theory, which I earlier characterized as "good," raises some concerns. There is an assumption in Albert's argument, namely that an "enhanced human" could, in principle, have an experience that is grounded in the position of a single particle. Is this a (bad) example of simply *positing* a psychophysical connection? Or is it just a conjecture for the sake of argument, one that might be overruled by research on the kinds of physical states that could possibly underlie experience? It is not easy to tell.

This is why my polemic against "experience" is not as straightforward as Bell's polemic against "measurement." It is relatively easy to tell whether a use of "measurement" in a discussion of quantum mechanics is good or bad, and a case for the *badness* of the bad uses can be made in relatively uncontroversial terms. But it is not so easy to tell whether a use of "experience" is good or bad, and even making the case that *some* uses are bad gets us into the contested territory of the nature of the mental.

Nevertheless, I hope to have shown that it is worth trying to disentangle the good from the bad, because there *are* bad uses of "experience," and they have the capacity to sow confusion in the foundations of quantum mechanics just as much as bad uses of "measurement." Whatever the philosophical status of experience, the mental-physical connection is not something that philosophers or physicists can posit at their convenience.

References

Aicardi, F., A. Borsellino, G. C. Ghirardi, and R. Grassi (1991). "Dynamical models for state-vector reduction: Do they ensure that measurements have outcomes?" *Foundations of Physics Letters* 4: 109–128.

Albert, D. Z. (1992). *Quantum Mechanics and Experience*. Cambridge, MA: Harvard University Press.

Albert, D. Z., and L. Vaidman (1989). "On a proposed postulate of state-reduction." *Physics Letters A* 139: 1–4.

Bell, J. S. (1987). "Are there quantum jumps?" In C. W. Kilmister (ed.), *Schrödinger: Centenary of a Polymath*. New York: Cambridge University Press, 41–52. Reprinted in Bell (2004), 201–212.

Bell, J. S. (1990). "Against 'measurement.'" *Physics World* 3(8): 33–40. Reprinted in Bell (2004), 213–231.

Bell, J. S. (2004). *Speakable and Unspeakable in Quantum Mechanics*, 2nd ed. New York: Cambridge University Press.

Bohm, D. (1952). "A suggested interpretation of the quantum theory in terms of "hidden" variables I & II." *Physical Review* 85: 166–193.

Bohr, N. (1929). "Wirkungsquantum und Naturbeschreibung." *Naturwissenschaften* 17: 483–486.

Brown, E. N., R. Lydic, and N. D. Schiff (2010). "General anesthesia, sleep, and coma." *New England Journal of Medicine*, 363: 2638–2650.

Brown, H. R., and D. Wallace (2005). "Solving the measurement problem: De Broglie-Bohm loses out to Everett." *Foundations of Physics* 35: 517–540.

Callender, C. (2015). "One world, one beable." *Synthese* 192: 3153–3177.

Chalmers, D. J. (1996). *The Conscious Mind: In Search of a Fundamental Theory*. New York: Oxford University Press.
Dürr, D., S. Goldstein, and N. Zanghì (1992). "Quantum equilibrium and the origin of absolute uncertainty." *Journal of Statistical Physics* 67: 843–907.
Gao, S. (2017). *The Meaning of the Wave Function*. New York: Cambridge University Press.
Ghirardi, G. C., A. Rimini, and T. Weber (1986). "Unified dynamics for microscopic and macroscopic systems." *Physical Review D* 34: 470–491.
Godfrey-Smith, P. (2016). *Other Minds: The Octopus, the Sea, and the Deep Origins of Consciousness*. New York: Farrar, Straus and Giroux.
Grush, R., and P. S. Churchland (1995). "Gaps in Penrose's toilings." *Journal of Consciousness Studies* 2: 10–29.
Hameroff, S., and R. Penrose (1996). "Conscious events as orchestrated space-time selections." *Journal of Consciousness Studies* 3: 36–53.
Lewis, P. J. (2007). How Bohm's theory solves the measurement problem. *Philosophy of Science* 74: 749–760.
Lewis, P. J. (2019). "Bohmian philosophy of mind?" In José Acacio de Barros and Carlos Montemayor (eds.), *Quanta and Mind*. New York: Springer, 91–102.
Maudlin, T. (1995). "Why Bohm's theory solves the measurement problem." *Philosophy of Science* 62: 479–483.
Ney, A. (2013). "Ontological reduction and the wave function ontology." In A. Ney and D. Z. Albert (eds.), *The Wave Function*. New York: Oxford University Press, 168–183.
Penrose, R. (1994). *Shadows of the Mind*. New York: Oxford University Press.
Penrose, R. (1996). "On gravity's role in quantum state reduction." *General Relativity and Gravitation* 28: 581–600.
Schwitzgebel, E. (2020). "Is there something it's like to be a garden snail?" Philosophical Topics, 48: 39–64.
Stone, A. (1994). "Does Bohm's theory solve the measurement problem?" *Philosophy of Science* 61: 250–266.
von Neumann, J. (1955), *Mathematical Foundations of Quantum Mechanics*. Princeton, NJ: Princeton University Press. Originally published (1932) as *Mathematische Grundlagen der Quantenmechanik*, Springer.
Wigner, E. P. (1961). "Remarks on the mind-body question." In I. J. Good (ed.), *The Scientist Speculates*. Portsmouth, NH: Heinemann, 284–302.

7

Why Physics Should Care about the Mind, and How to Think about it Without Worrying about the Mind-Body Problem

Jenann Ismael

Proposals that connect consciousness to quantum mechanics take place against the background of a larger discussion of the mind and its role in physics. They cast some light on the larger discussion in the form of specific proposals for a role that consciousness might play in solving the measurement problem, but I will argue that they should be treated as dynamical proposals without worrying about the aspects of our mental lives that give rise to the mind-body problem. It is uncontroversial to say that physics does not have a very finely etched understanding how to fit the mind into its account of the natural world.[1] It is not that physics has any particular problem with the brain and body. These are made of the same stuff as, and obey the same laws as, trees and planets. And it is not that we don't know how to talk about the mind when we can describe it in the vocabulary recognizable from our own experience: i.e., as a flow of perception and thought and action we know from a first-person perspective. It is that we don't know how to describe the mind in the terms that physical theory itself provides.[2]

Physics has all kinds of workarounds to avoid focusing on experience. Physicists often talk of "observation," but by "observation," they usually mean something more akin to measurement, so the mind doesn't enter in a substantive way. Physicists talk about evidence, but evidence is typically stated in the language of physics; the positions of pointers on the front of measuring instruments or marks on a photographic plate. Because the evidence for our

[1] I'd like to thank Anthony Aguirre and Max Tegmark for the invitation that prompted this paper, and the audience at the 2016 FQXi International Conference in Banff. I am grateful to Shan Gao for including it here, and David Chalmers for so many years of warm friendship and gracious tolerance.

[2] Although sometimes used in a narrow way to refer to sensations or perceptual offerings, "experience" is used here to mean the full introspectively accessible mental life.

Jenann Ismael, *Why Physics Should Care about the Mind, and How to Think about it Without Worrying about the Mind-Body Problem* In: *Consciousness and Quantum Mechanics.* Edited by: Shan Gao, Oxford University Press.
 DOI: 10.1093/oso/9780197501665.003.0008

theories ultimately comes from experience, however, eventually we have to be able to bring the mind itself firmly under the scope of our physical theories and understand how human experience fits into the picture.

1. The Intrusion of the Observer

Talk of experience has begun to creep into physics in a variety of ways most specifically in the discussion of quantum mechanics. The contrast between what the deterministic dynamics yields and what the observer *sees* has been at the forefront of the theory almost from the beginning. Proposals like those in this volume have offered detailed and specific accounts of a role for consciousness in resolving the conflict.

In the controversies surrounding the status of time, straightforwardly physical questions like whether there is a global present have gotten intertwined with questions about the subjective experience of time. Questions like whether we really have experience of the passage of time, and whether flow is a property of the world or internal to the mind, are used to question the relativistic conception of time.[3]

In discussions of quantum gravity where it is sometimes said that "space disappears" at the fundamental level, there is a need to understand what to make of the spatial character of our experience.[4] In what sense do we see space and what sort of requirement does that place on physical theory?[5]

2. The Pessimistic Reaction

The pessimistic reaction to the intrusion of experience into physics is a kind of horror. The thought is that there are reasons to steer a wide berth from talking about the mind: it's a morass of endless and fruitless debate. Physics

[3] L. Smolin, *Time Reborn* (Boston: Mariner Books, 2013); M. Roubach and Y. Dolev (eds), *Cosmological and Psychological Time*, Boston Studies in the Philosophy and History of Science 285 (New York: Springer, 2015).

[4] T. Maudlin, "Completeness, Supervenience, and Ontology," *J. Phys. A: Math. Theor.* 40 (2007): 3151; J. Ismael, "Do You *See* Space? How to Recover the Visible and Tangible Reality of Space (Without Space)," *Philosophy Beyond Space-Time 2*, ed. Christian Wuthrich and Nick Huggett (Oxford: Oxford University Press, forthcoming).

[5] In cosmology questions about the relations between what our theories say about the world and the experience of the observer are coming under pressure as well because of a disparity between the scope of the theory and the information that is even in principle available from within space-time. Except where anthropic reasoning is involved, the observer is treated as a generic embedded system and the mind doesn't enter in a specific way.

is about the movements of material things. If progress of physics depended on resolution of the mind-body problem, it would be a terrible thing. Physics has gotten as far as it has precisely because it has left the messy business of human experience alone.[6] And it does not help that the people who have been willing to talk about consciousness have sometimes approached from a fringe perspective.[7] A desire to keep physics *physics*, and to avoid the squishiness of philosophical debate is more than enough reason (one might think) to stay away from talk of experience.

I think the pessimistic reaction is too pessimistic. Physics doesn't stop at the surface of the skin. Some understanding of observation as a physical process is always implicit in bringing evidence to bear on theory. The fact that questions about experience are beginning to infect physics at quite fundamental level suggests it is time to bring them into focus. And an increasing amount is known about the mind in purely scientific terms. Because the explanatory emphases and points of departure for physics are very different than from philosophy, problems and issues that are deeply contested and tend to form focal points for debate in those discussions can be set aside. I'm going to give a quick, opinionated account of what matters and what can be sidelined and then a sketch of how to fill in the outlines of the parts that matter with application to the discussions of quantum mechanics and cosciousness.

3. What We Can Ignore: the Mind-body Problem and Physics

First: what can be sidelined. The mind-body problem is one of the oldest and most intractable problems of philosophy. It concerns the relationship between the mind and the body—between the realm of experience and the realm of matter. The question is whether the progression of thoughts, feelings, perceptions, sensations, that make up our mental lives are things that happen in addition to the physical processes in the brain or are themselves just some of those physical processes?

The last forty or fifty years has seen an explosion of scientific progress understanding the human mind, in no small part as the result of the

6 Maudlin 2007.

7 https://www.youtube.com/watch?v=WJoWIEYDzuk; *Cosmology of Consciousness: Quantum Physics and Neuroscience of Mind*, ed. D. Chopra et al. (New York: Science Publishers, 2017). E. H. Walker, *The Physics of Consciousness: The Quantum Mind and the Meaning of Life* (New York: Perseus Books, 2000).

emergence of cognitive science. Cognitive science is an interdisciplinary study of mind and intelligence, embracing philosophy, psychology, artificial intelligence, neuroscience, linguistics, and anthropology. Its intellectual origins are in the mid-1950s, as behaviorism finally fell out of fashion and people began developing theories of mind based on complex representations and computational procedures. The field employs a fruitful mixture of methods. Cognitive abilities are typically functionalized and studied. Mechanisms and neural implementations are sought for things like the ability to discriminate, categorize, and react to environmental stimuli; the ability to integrate information coming through different perceptual pathways; to take its own states as objects of representation; to organize them and report them; to impose coherence and consistency constraints; to regulate attention and control behavior.

In light of all of the progress made understanding the mind in scientific terms, there was a general attitude of optimism that cognitive science might be resolving this age-old philosophical problem. In 1996, David Chalmers threw cold water on that optimism with an article followed by a seminal book, which has organized much of the discussion in the decades since. In the book, he argued that none of this evident progress touches the heart of the mind-body problem.[8] Chalmers separated problems into two classes: Easy and Hard. Easy Problems concern cognitive abilities like those above that can be characterized in functional terms. Once an ability is characterized in functional terms, computational and neural mechanisms that give rise to the ability can be sought. The ultimate goal is to uncover how the brain supports the ability in question. These kinds of problems, though not actually easy, are at least amenable to scientific understanding. The Hard Problem, according to Chalmers, is the problem of accounting for subjective experience. This one, he argues, is not amenable to scientific understanding because it concerns a notion of the qualitative character of mental states that is *not* functionally definable. The notion of qualitative character he has in mind is something you are supposed to recognize from your own case:

> When we see . . . we experience visual sensations: the felt quality of redness, the experience of dark and light, the quality of depth in a visual field. Other experiences go along with perception in different modalities: the sound of

[8] D. Chalmers, "Facing Up to the Problem of Consciousness," *Journal of Consciousness Studies* 2 (1995): 200–19. D. Chalmers, *The Conscious Mind* (Oxford: Oxford University Press, 1996).

a clarinet, the smell of mothballs. Then there are bodily sensations, from pains to orgasms; mental images that are conjured up internally; the felt quality of emotion, and the experience of a stream of conscious thought. What unites all of these states is that there is something it is like to be in them. All of them are states of experience.

The term "phenomenal consciousness" was coined to refer to the property of there being something it is like for a system to be in a given state. Chalmers argues that no matter how detailed an account we give of the cognitive and behavioral capacities that a system possesses or of the physical mechanisms that underwrite those capacities, that will leave undetermined whether it is phenomenally conscious. Whether it is phenomenally conscious is, he argued, a *further fact*, distinct from any collection of outwardly observable abilities, and one that (moreover) no amount of scientific investigation will settle.

Chalmers collected and organized the best of Descartes' arguments, combined them with others that had been floating around in the literature, and provided some of his own, in a powerful case meant to bring the Hard Problem into relief and establish its scientific intractability. Here (briefly) is how the arguments go.[9] Suppose we give some functional specification of what it was for a system to have introspectively accessible states and we offer introspective accessibility as an account of what it is for a state to be conscious. We will be met with arguments like this: we can imagine a being (a robot, for example) who had states that were introspectively accessible, but who was not conscious, so consciousness can't just be introspective accessibility. The same will go for global broadcast, informational integration, and any purely functional specification we can give. It will always seem possible to imagine a creature that satisfied that specification but that isn't conscious and that is supposed to show that consciousness can't simply be a matter of satisfying one of these functional descriptions. Whatever it is to have

[9] There are six canonical arguments. Some of them deal with epistemic possibility (the Open Question argument, Frank Jackson's Black and White Mary argument), some with metaphysical possibility (the zombie argument, the modal argument). The literature surrounding them is now enormous. See also Chalmers, "Consciousness and Its Place in Nature," in *Blackwell Guide to the Philosophy of Mind*, ed. S. Stich and F. Warfield (Blackwell, 2003); and for an overview of the wider literature (with bibliography) see Robert Van Gulick, "Consciousness", *The Stanford Encyclopedia of Philosophy* (Spring 2018 Edition), Edward N. Zalta (ed.), URL = <https://plato.stanford.edu/archives/spr2018/entries/consciousness/>. I have also written about Chalmers's arguments: *The Situated Self* (Oxford: Oxford University Press, 2006/7), particularly Part II.

conscious experience, the argument goes, it is not a matter of having a certain functional organization or cognitive or behavioral capacities because any such organization, and any collection of such capacities, could be reproduced in a system that wasn't conscious.

At this point in the argument, we have a staring match that ensues between those think that consciousness has to be some sophisticated, functionally definable notion (maybe a kind of informational integration and introspective accessibility) and those who think it the most obvious thing in the world that whatever functional characterization you give can be satisfied by an unconscious robot or a zombie. If these arguments are correct, phenomenal consciousness by its nature falls through any attempt to capture it by linking it to something that can be empirically investigated. That's what makes the Hard Problem *hard*. Whether or not you agree with the arguments, there is no question that they have a powerful intuitive force. They have prompted debate that has produced a mountain of baroque argumentation without showing any signs of resolution.

The nice part about all of this for the purposes of physics is that although Chalmers's own purpose was to reestablish the heart of the mind-body problem as impenetrable to scientific resolution, he managed to isolate it almost surgically from making a difference to physics. If there is such a thing as a kind of consciousness that by its nature falls through the net of physical description because it has no functional or causal role of its own—the physicist interested in the role of mind in nature doesn't need to worry about it.[10] The physicist interested in representing the role that minds play in the causal fabric of the world can be serenely unconcerned about whether phenomenal consciousness really is a kind of magic fairy dust that when sprinkled on certain processes, lights them up from the inside. She is concerned only with the shadow those processes cast in the physical world.[11] As soon as consciousness matters to physics—i.e., as soon as it makes an

[10] Our own conscious states can be *associated* with brain states identified in virtue of the fact that they play a certain functional role, and thereby *associated* with states that are uncontroversially physical and integrated into the causal web. The arguments are supposed to establish that the association cannot be one of identity since the functional organization can be reproduced in a non-conscious system. So there is wide agreement that our conscious states can be picked out by extensional definition that lets us associate them with brain states. The issue for proponents of the Hard Problem specifically concerns claims of identity between consciousness and any proposed physical basis.

[11] And to say that (again) is not to agree that that is what phenomenal consciousness is. It is just to say that we can ignore the philosophical debate, because proponents of the Hard Problem succeed in establishing that consciousness is not understandable in physical terms only at the expense of making it irrelevant to physics.

observable impact on the motions of material things—it becomes detectable by that impact and integrated into the causal fabric of the world. And then it becomes (as far as physics is concerned) physical.[12] What this means is that functional interpretations give us everything we need in order to address questions about observation and action as they appear in the problem space of physics.

The second debate that can be sidelined for purposes of the physicist is what we might think of as an analogous Hard Problem of intentionality; again, it is taken as the hallmark of mental states that they represent features of the world and it has become a focal point of debate in the philosophy of mind to say what it is for one to represent something distinct from itself. Some of the central arguments that make this an object of philosophical dispute purport to show that no functional account of what it takes for a state to have representational (or "intentional") content could be right. If anyone offers such an account, they will be met with an argument that the functional description can be satisfied, and there be nothing like full-blooded content present. Searle's Chinese Room argument is the locus classicus of this kind of argument.[13]

Again, here, there is the scientific question: what functional role do representational states play in whatever happens between sensory impact and movement in a human being? And then there is the philosophical question of whether their having a content is purely a matter of playing that role. The Hard Problem of Intentionality is about bridging the gap between these functionally specifiable notions and some more full-blooded notion we are supposed to know in a first-personal way.

And again, it doesn't matter for physics. In this case, there is less consensus about whether the arguments establish that intentionality is by its nature non-physical. The reason is that there is more room for things like embedding the human in a social environment, enhancing it with memory and

[12] At least in the early articles, Chalmers agrees. Defending his own view that there should be a devoted science of consciousness that looks for psychophysical laws he writes:

> Certain features of the world need to be taken as fundamental by any scientific theory. A theory of matter can still explain all sorts of facts about matter, by showing how they are consequences of the basic laws. The same goes for a theory of experience. *This position qualifies as a variety of dualism, as it postulates basic properties over and above the properties invoked by physics.* (Emphasis mine)

[13] Discussion occurs in the disputes surrounding the naturalization of content. See Fred Adams and Ken Aizawa, "Causal Theories of Mental Content," *The Stanford Encyclopedia of Philosophy* (Summer 2017 Edition), Edward N. Zalta (ed.), URL = https://plato.stanford.edu/archives/sum2017/entries/content-causal/, for an overview and bibliography.

reflective processing, and in general imposing more structure on the setting (cognitive or environmental) in which mental states are used which makes a difference to the persuasiveness of the arguments. On a wide conception of physical that includes the social environment, it is not nearly as clear that the features of meaning that make a direct reduction to physics difficult can't be ultimately emergent from social interactions. In this sense, it is harder to make the intuitive case that once all of the Easy problems are solved, there won't be any Hard residue left. What is making the arguments for the Hard Problem of Consciousness so powerfully convincing is the intuition that we have immediate awareness of the phenomenal properties of our mental lives, and nothing that we can learn from a third person perspective seems to cross the divide. There is nothing analogous to that in the intentionality case.

The good thing about all of this from the point of view of someone interested in the role of the mind in the physical world is that the arguments that are supposed to establish that some aspect of mind (e.g., Consciousness, Intentionality) is irreducible to physics do so only at the expense of making it irrelevant to physics. That's bad news if you want to solve the mind body problem. It's good news if you want to do physics without worrying about it. All that you need to care about for physical purposes is those aspects of mind that make a difference to the movements of physical things.

In saying this, I am not saying anything proponents of the Hard Problem in the vein of Chalmers should disagree with. In isolating the Hard Problem and quarantining it from the Easy ones, Chalmers also establishes that the very features of the Hard Problem that make it scientifically intractable also make it irrelevant to physics. Physics can focus on the Easy Problems, which include finding a physical basis for consciousness. It doesn't need to worry about whether the relationship between the physical basis and the first-person accessible phenomenon is analytic entailment, metaphysical necessity, entailment by special psychophysical laws, or something else altogether. The point is just that the features of experience that escape functional characterization in terms of their causal relations to something in the physical world, don't appear in the problem space of physics, and so nothing in the problem space of physics is going to depend on their resolution.[14] As soon as Consciousness and Intentionality make a difference

[14] Note that this isn't behaviorism. Behaviorists held that behavior could be explained by reflex and conditioning. This view recognizes a rich internal life of perception, thought, feelings and volitions. It is simply agnostic about any purported feature of mental life that makes no detectable impact on behavior.

in these ways, they become something that matters to physics. But then they also become something that is characterizable in terms of their physical role. There is a closure in the problem space.

4. Does Quantum Mechanics Make Consciousness Relevant to Physics?

The role that consciousness has played in the discussions of the foundations of quantum mechanics illustrates all of this nicely. Any theory has to predict something about the observer's experience if it is going to make testable predictions. In classical contexts, observation stayed mostly out of view. There was a presumption that observation gives us information about the values of local macroscopic variables and theories were tested by deriving implications for the values of such variables. In quantum mechanics, observation becomes problematic because of a conflict between the linear evolution of the wave function and what an observer sees at the end of a measurement. Linearity entails that an observer coupled to an apparatus carrying out a measurement on a system in a superposition of the measured observable should end up in a superposition of seeing different results but the observer invariably sees a definite result. The difficulty has brought the coupled interaction between observer and measuring apparatus under careful scrutiny and discussion of the observer's experience is made explicit in careful presentations since this is ultimately where the conflict occurs.

Although many of those working in quantum foundations think that solving the problem is a matter of producing definite pointer states for the measuring apparatus, Shan Gao advocates a mentalistic formulation of the problem that makes reference to the observer's experience explicit, since the dynamics on its own is perfectly consistent. As he says, one can see the influential responses to the measurement problem as advocating different kinds of psychophysical linkages:

> The mentalistic formulation of the measurement problem highlights the important role of psychophysical connection in causing the measurement problem. By this new formulation, we can look at the solutions of the problem from a new angle. In particular, Bohm's theory, Everett's theory

and collapse theories correspond to three different forms of psychophysical connection (as well as three different result assumptions).[15]

Each of these three theories assumes some form of psychophysical supervenience.[16] If one assumes psychophysical supervenience, whatever form it takes, there will be a necessary connection between a conscious state and its physical basis and the dynamical role of conscious states won't be any different from that of the brain states on which they supervene. This means that the physicist can safely bracket the phenomenal properties without worrying that he's missing something that makes a difference to the dynamics. If psychophysical supervenience is assumed, physics is not going to know the difference between the conscious state and its physical basis and consciousness is—*for that very reason*—not going to make a difference to the physics.

There is, however, a small and well-established tradition stemming from Wigner that treats consciousness itself as a physical agent. The thought is a natural one given the predicament in quantum mechanics. It is that consciousness comes from outside the known physics and isn't supervenient on anything that falls under the linear dynamics. If the states in which observations terminate come from outside the known physics, they don't need to be governed by Schrodinger's equation and one is free to expand the theory by thinking of consciousness as inducing collapse.

There are two ways of developing the suggestion. One is that there is some heretofore unknown material substance or quantity on which consciousness supervenes which induces collapse. If this is the suggestion, then there is an extension of the physical ontology, but psychophysical supervenience still holds and the situation can be assimilated to the one above. The more interesting way to develop the suggestion - and the one that Wigner seems to have intended - is to hold that the physical interaction between measured

[15] "Why Mind Matters in Quantum Mechanics," http://philsci-archive.pitt.edu/15910/. See also S. Gao, "The Measurement Problem Revisited," *Synthese* (2017), doi:10.1007/s11229-017-1476-y; and A. Oldofredi, "Some Remarks on the Mentalistic Reformulation of the Measurement Problem: A Reply to S. Gao," *Synthese* (2019), doi:10.1007/s11229-019-02101-3.

[16] There are different brands of supervenience, corresponding to different strengths of modal connection between a brain state and its phenomenal properties. There is analytic entailment, metaphysical necessity, even the dualist's special brand of psychophysical necessity. Philosophical distinctions between these different brands of necessity can be more or less ignored by the physicist who cares about dynamical roles. Any form of necessary link between the conscious state and its physical basis will make the conscious state and physical basis interchangeable from the physicist's point of view.

system and human brain is described by Schrodinger evolution, but consciousness itself (not some hitherto unknown not some hitherto unknown material substance or quantity which forms the supervenience base for consciousness) induces collapse. This suggestion denies the supervenience of the mental on the physical and treats *consciousness* as a physical agent in its own right.

As a resolution of the measurement problem, the proposal has been less popular than the alternatives.[17] But it is interesting as a test case for the claim that physics can ignore the Hard Problem of Consciousness by showing us what happens as soon as consciousness becomes relevant to physics. It amounts to a solution to the Hard Problem.

The Hard Problem was supposed to be that there is a kind of consciousness—phenomenal consciousness—that is by its nature undetectable and physically elusive. Whether a system is phenomenally conscious was supposed to be detectable from the inside, only by the system itself. As Chalmers said:

> We know that a theory of consciousness requires the addition of something fundamental to our ontology, as everything in physical theory is compatible with the absence of consciousness.

The proposal that is being entertained here—i.e., that consciousness is a causally active, physical stuff that induces wave-function collapse—amounts to a the denial that everything in physical theory is compatible with the absence of consciousness. If consciousness *itself* induces collapse, consciousness is an ineliminable part of the causal fabric of the physical world, one that can be defined by its causal role, and appears alongside other physical quantities in a unified theory that explains the observable movements of material bodies.[18] Even if difficult to detect experimentally, we could treat

[17] But see Chalmers and Mcqueen, this volume. In its original formulation, the proposal was vague. Which systems exactly are associated with consciousness, and exactly when does it enter the dynamics and induce collapse? See the discussion in J. S. Bell, "Against 'Measurement'," *Physics World* 3.8 (1990): 33–40. Reprinted in Bell (2004), 213–231, 34. There is increasing but not conclusive, evidence that no collapse occurs. See, for example, K. G. Johnson, J. D. Wong-Campos, B. Neyenhuis, J. Mizrahi, and C. Monroe, "Ultrafast Creation of Large Schrödinger Cat States of an Atom," *Nature Communications* 8 (2017), article number 697, doi:10.1038/s41467-017-00682-6, for recent developments.

[18] And there's the question of whether it is even properly thought of as dualism. Chalmers (see Chalmers and Mcqeen, ibid.) continues to regard it as a dualist view, though I'm puzzled why, in light of his given reasons for calling his original position dualist. He writes, "Certain features of the world need to be taken as fundamental by any scientific theory. A theory of matter can still explain

inducing wave function collapse as a test for the presence of consciousness. And for the physicist, consciousness would now be brought firmly into the realm of physics.

So, again, we have the closure of the problem space. As soon as consciousness makes a difference to the dynamics of material things, it becomes something that matters to physics. But then it also becomes something that is characterizable in terms of its physical role. The physicist needs to think about the human mind insofar as internal processes are part of the causal fabric of the world. The Hard Problem can be ignored because if consciousness enters the problem space of physics, it does so by making a difference to the behavior of physical objects.

5. Easy (and Interesting) Problems: is the Mind Properly Characterized in Computational or Dynamical Terms?

Saying that we can put aside the Hard Problems doesn't mean that the remaining questions about how to fit our mental lives into the general machinery of nature are easy, but there have been some fruitful disputes in cognitive science whose resolution has provided the outlines of a way of thinking about the mind that is helpful from the point of view of someone who is trying to understand how it fits into physics.

One of the disputes is a question about the vocabulary we use to describe the mind. It is the question of whether the mind is properly characterized in computational or dynamical terms.

The answer to this is: both.

There was a time when people thought of the mind just in terms of its conscious part; i.e., in terms of the progression of perceptions, thoughts and feelings of which we are consciously aware, and because those are identified and individuated in representational terms from a first-person perspective, that is the vocabulary we use to describe them. Cognitive science has taught us that there's a lot going on in the mind that falls below the threshold of first-person accessibility, but it continues to describe that activity in terms that come naturally when we are describing our conscious lives: representation

all sorts of facts about matter, by showing how they are consequences of the basic laws. The same goes for a theory of experience. *This position qualifies as a variety of dualism, as it postulates basic properties over and above the properties invoked by physics*" (emphasis mine, *op.cit.* 1995).

and computation. A spate of challenges in the last couple of decades (starting around 1995) has argued that this whole vocabulary is misplaced and that we should use the same straightforwardly dynamical vocabulary to describe the mind that we use to describe planets and pendula.[19] Their reasons have to do with specific issues about the role of time and the complexity of certain kinds of causal interactions. The reason that it is an interesting dispute for our purposes is that it forces us to get clear on what the vocabulary of representation and computation is doing and how it relates to the dynamical vocabulary. To illustrate the difference between the representation- and computation-based description, on the one hand, and a straightforward dynamical description, on the other, van Gelder (who is one of the primary figures advocating the dynamical vocabulary) describes a device whose job is to keep constant the speed of a flywheel to which some machinery is connected. The device is called a Watt Governor, because it was invented by James Watt in 1788. There is a tendency for the speed to fluctuate (because of varying steam pressures and workloads). To smooth things out the amount of steam entering the pistons is controlled by a throttle valve. How might such control be achieved?

(1) One way would be to program a device to measure the speed of the flywheel, compare this to some desired speed, measure the steam pressure, calculate any change in pressure needed to maintain the desired speed, adjust the throttle valve accordingly, then start again.
(2) Watt's solution was to instead geared a vertical spindle into the flywheel and attach two hinged arms to the spindle. To the end of each arm, attach a metal ball. Link the arms to the throttle valve so that the higher the arms swing out, the less steam is allowed through. As the spindle turns, centrifugal force causes the arms to fly out. The faster

[19] See T. van Gelder, "What Might Cognition Be, if Not Computation?" *Journal of Philosophy* XCII.7 (1995): 345–81; T. van Gelder & R. Port, "It's about Time: An Overview of the Dynamical Approach to Cognition," in *Mind as Motion: Explorations in the Dynamics of Cognition*, ed. R. Port & T. van Gelder (Cambridge, MA: MIT Press, 1995), 1–44. Related arguments are found in E. Thelen & L. Smith, *A Dynamic Systems Approach to the Development of Cognition and Action* (Cambridge, MA: MIT Press, 1994); S. Kelso, *Dynamic Patterns* (Cambridge, MA: MIT Press, 1995); F. Varela, E. Thompson, & E. Rosch, *The Embodied Mind* (Cambridge, MA: MIT Press, 1991); and in M. Wheeler, "From Activation to Activity. Artificial Intelligence and the Simulation of Behavior" *(AISB) Quarterly* 87 (1994): 36–42. R. Beer & J. C. Gallagher, "Evolving Dynamical Neural Networks for Adaptive Behavior," *Adaptive Behavior* 1 (1992): 91–122, M. Wheeler, *op. cit.* pp. 36–42. For discussion see F. Keijzer and S. Bem, "Behavioral Systems Interpreted as Autonomous Agents and as Coupled Dynamical Systems: A Criticism," *Philosophical Psychology* 9 (1996): 323–46; Clark & Toribio *op. cit.*; and A. Clark & R. Grush, *Towards a Cognitive Robotics. Adaptive Behavior* (forthcoming).

it turns, the higher the arms fly out. But this now reduces steam flow, causing the engine to slow down and the arms to fall. This, of course, opens the valve and allows more steam to flow. Properly calibrated, it maintains engine speed smoothly despite wide variations in pressure, workload, and so on.[20] Here's what it looks like:

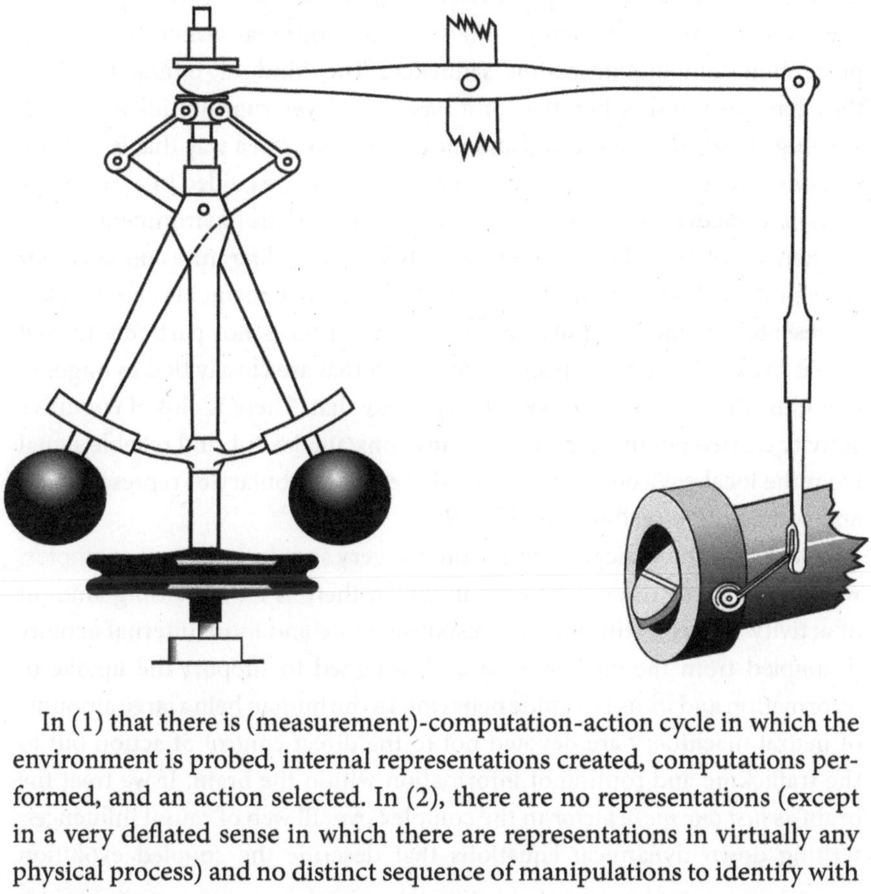

In (1) that there is (measurement)-computation-action cycle in which the environment is probed, internal representations created, computations performed, and an action selected. In (2), there are no representations (except in a very deflated sense in which there are representations in virtually any physical process) and no distinct sequence of manipulations to identify with the steps in a computational process. There's just an ongoing process of continuous reciprocal causation in which the Governor (here, the agent) is coupled to the rest of the system (the engine).

The way that we model a system like that is that we write down a state-space for the engine as a whole and look for a set of differential equations

[20] Van Gelder, *op. cit.*, pp. 347–50.

that describe trajectories through that space.[21] The space may have some interesting structure (attractors and so on) that we use to understand systemic behavior, but what we don't do is try to pull out one component (the Governor) and describe its exchanges with the rest of the system in terms of representation and computation.[22]

The claim of those who advocate this sort of description for the mind is that this sort of process is much closer to the true profile of agent-environment interactions than is the traditional vision of a simple perception-computation-action sequence. The kind of interaction that they have in mind is like that of a baseball player running for a fly ball, keeping visual track of the ball and moving his body in a way that is directly responsive to perceived position so that his actions are guided in a particular kind of unmediated way by an ongoing signal from the environment.

There's a lot to say here and the cognitive scientific literature contains a lot of helpful and often fascinating details that makes a convincing case that the representation-and computation-style explanation is not particularly well suited to explaining the aspects of cognition that are closely tied to ongoing stimulus. But not all cognitive activity is like that. There is alot of cognitive activity carried out in the absence of any constant, lawful, and reliable signal from the local environment. This is where the vocabulary of representation and computation really comes into play.

If we look at biological systems from the very simple to the more complex, we can see a line of development in which there is an increasing amount of activity between stimulus and response: more and more internal activity decoupled from the environment and designed to support the uptake of information and its use to guide behavior. In the human being large amounts of neural machinery are devoted not to the direct control of action but to the trafficking and routing of information within the brain. If we treat the brain as *just one more factor* in the complex overall web of causal influences, writing down dynamical equations that describe the coupled evolution

[21] We could use representational vocabulary to describe this only if "representation" meant something like "information-bearing state," in which case the representational description would add nothing to the dynamical description.

[22] Total state explanation for any dynamical model emphasizes the fact that all aspects of a system are changing simultaneously and invites us to understand the behavior of the system in terms of the possible sequences of changes in total state over time. Trajectories through state spaces populated by attractors, repellers, and so on reflect motion in a space of total states, i.e., states that assign values to all systemic variables and parameters.

of agent and environment, we have a single vocabulary that integrates it smoothly into the rest of physics but we obscure something important.

The key to understanding what we might think of as the specifically *intelligence*-based route to evolutionary success has to do with our ability to exploit information. In systems like human beings, behavior is not keyed directly to an environmental stimulus but depends on the maintenance of many bodies of information and complex goal structures. Systems in which complex information flow plays a key role tend to exhibit a kind of complex articulation in which behavioral flexibility comes from being able to quickly and cheaply alter the inner flow of information in a wide variety of ways. That articulation is revealed in the computational description, but is (for reasons that I'll talk about below) camouflaged in the purely dynamical description. It would not be wrong to say that the computational description highlights a level of functional organization that explains the point of the low-level activity in the brain and that is crucial to understanding the distinctive kinds of flexibility and control characteristic of truly mindful engagements with the world.[23]

So the lesson of this dispute is that to treat the brain as the principal seat of information-processing activity is to recognize that nature discovered the utility of information long before Silicon Valley, and used that insight to build machines capable of highly complex, flexible behavior. In representing the low-level dynamics of brain and body the dynamical vocabulary is still applicable, but we get special insight by seeing how the high-level dynamics guides the flow of information through the mind. The right approach to understanding how the human being fits into physics will be a pragmatic pluralism in which charting the flow of information is *as important* as the low-level dynamics, and in which some *high-level* dynamical features lead a double life as elements in an information-processing economy. It's the information-processing economy that explains the distinctive kinds of behavioral flexibility that human beings exhibit. Actions are no longer spontaneous reactions to stimuli, but temporally extended plans that respond to complex stores of information, with flexible goal structures. The way that human experience fits into this is that we're able to identify at least functional analogs of our conscious mental lives in a way that explains why

[23] Where we confront especially complex interactive causal webs, however, it does indeed become hard to isolate the syntactic vehicles required by the computational approach. Whether the high-level functional organization screens off information from low-level processes is a matter of degree and depends on what aspects of behavior one is trying to explain.

observation is a source of information about the world and how human agency (in the guise of action guided by decision procedures that draw on internally stored information) is possible.

Now, I turn to why the "information-processing economy" is concealed by the dynamical description and that leads us to a second dispute that has led to some helpful insight.

6. Is the Mind a Computer?

The answer to the now hackneyed question of whether the mind is a computer is: in one sense no, and in one sense yes.

The early uses of the computer analogy in trying to understand the mind emphasized computation in a particularly narrow sense and turned out to be too limited to provide a good model of the mind. The formal theory of computation as exemplified in Turing Machine Computationalism is defined only for discrete state machines and digitality is crucial to many classical results in the theory of computability. Neither of these is a general feature of the kind of information-processing that the brain performs. But there is a less formal notion of computation which is tied to the much more general idea of automated information processing and semantically sensible transitions between representational elements. In that much looser sense, the mind *is* a computer but the notion of computation doesn't do much work. Almost any physical process can be thought of as a computation in that sense.[24]

There is a very different lesson that we can learn from computers, which has much more to do with how the high-level functional organization of a computer relates to the bit-level description of the hardware. A typical modern computer can be thought of as having a state represented by a vector giving the bit-values of all the locations in its memory and in its registers and all processes in the computer can be modeled as trajectories through the machine's state space. In practice, software engineers don't think that way at all. They find it more useful to think of various persisting sub-components (strings, arrays, trees, networks, databases, stored programs) as having their

[24] A common worry about the proposal to recognize analog computation is that everything then turns out to be *some* kind of computer and hence the thesis that the brain computes is empty. The response is that we can narrow the focus by looking at the special class of computations whose implementations support flexible behavior and reason-guided action. And we can draw on the intuitive notion of computation as semantically sensible state-transitions without insisting that the notion of computation needs to do anything more than heuristic work.

own changing states which interact with one another. In a standard computer, for example, we find multiple databases, procedures, and operations, and the information-processing power of the device lies in the fact that these can be rapidly and cheaply reconfigured—much more rapidly and cheaply reconfigured than its mechanical components. This way of parsing the activity isn't imposed; it emerges very naturally from the high-level functional organization that captures the contours of processes in a way that highlights what is crucial to what we use computers for: to process information.

If you asked about the relationship between the high-level functional organization and the bit-level description of the hardware, you might have thought that there must be a discernible correspondence between components and operations at the level of functional organization and processes defined over bit-level components.

But it doesn't work like that. Computers give us insight into the complex ways in which higher-level entities (and processes defined over them) can be realized in lower-level hardware by giving us concrete examples of (running) *virtual machines*. A virtual machine is a generic world for a functional duplicate of a real or hypothetical machine made not of mechanical parts, but of virtual components. Examples of virtual machines include a word processing system, a simulation of an earthquake or a growing population, a video game, a calculator, or a chess player. These are all specialized virtual machines that perform specific functions. There are also platform virtual machines (like operating systems) that are capable of supporting many specialized virtual machines. The cool thing about virtual machines is that they can exhibit dynamical behavior (as this variety attests) very different from the physical hardware in which they are implemented. In so doing they show us how layers of structure support high-level functionality in a manner that defies the expectation of reduction or anything like a simple, visualizable supervenience relationship.

In computers, a combination of hardware and software technology produces a complex web of causal (or, if you like, virtuo-causal) relationships between elements displayed on your screen when the machine is running. The support for that network of relationships is all of the layers of accreted structure that developed in stages, sometimes over decades. These include a plethora of interacting software or hybrid hardware-software sub-systems including: schedulers, device drivers, file management systems, memory management systems, compilers, interpreters, interrupt handlers, caches, programmable firmware stores, error-correcting memory, wired and wireless network interfaces, network protocol handlers, email systems, web browsers, and many more. People added to these structures and built on top of platforms once they were in place, without knowing how to relate the structures they built to what was below. The relationship between a running virtual machine and the physical machine on which it is implemented may be no more transparent than all of these levels of accreted structure. And there won't typically be a discernible structural correspondence between the two. There won't be a fixed or readily ascertainable correspondence between components of the virtual machine and components of the physical machine (files stored by a word-program, for instance, won't have a fixed location in computer memory). There won't be processes available on inspection of the hardware that look like the processes executed by the machine.[25] The structure of the virtual machine, moreover, can change significantly without structural changes occurring at the physical level though the physical states of millions of switches may need to change to alter conditional connections. Indeed, that is what explains the power of these devices to quickly and cheaply alter the flow of information.

Evolution has had far more time than we have to discover these cheap and flexible ways of routing information. And it has also had more time to build up layers of virtual machines running on virtual machines. If one is thinking in these terms, it is natural to think of our own experience as a kind of high-level control interface with no direct structural correspondence to anything that goes on in the brain.

The suggestion here—which I take from Aaron Sloman, and which is implicit in the very familiar idea that the mind is the software of the brain—is that the right kind of functionalism is virtual machine functionalism.[26] And the right way to think of the relationship between mental activity and brain activity is implementation: not reduction, not correspondence, none of the much simpler, visualizable relationships that philosophers have

[25] Where there are no fixed physical correlates for virtual machine entities, the processes that operate on them will not be readily discernible in the low-level dynamics. Detection will be possible where there are more or less reliable physical correlates.

[26] A. Sloman, "Phenomenal and Access Consciousness and the 'Hard' Problem: A View from the Designer Stance," *Int. J. Of Machine Consciousness* 2.1 (2010): 117–169; A. Sloman, "Architecture-Based Conceptions of Mind," in *In the Scope of Logic, Methodology, and Philosophy of Science* (Vol. II), ed. P. Gardenfors, K. Kijania-Placek & J. Wolenski, *Synthese* Library Vol. 316 (Dordrecht: Kluwer, 2002), 403–427; A. Sloman and R. Chrisley, "Virtual Machines and Consciousness," *Journal of Consciousness Studies* 10.4–5 (2003): 133–172. The idea was also developed by John Pollock, "What Am I? Virtual Machines and the Mind/Body Problem," *Philosophy and Phenomenological Research* 76.2 (2008): 237–309.

sometimes thought have to hold if the mind is part of the physical world. It's a relationship of vastly more complexity than the imagination by itself can penetrate.

7. Details of Implementation

Once we have introduced the notion of a virtual machine, we can speak freely of mental processes in the vocabulary that comes naturally, treating the mind as a machine that processes and stores information, bringing it to bear on the determination of behavior. This allows us to integrate minds into dynamics in a rather smooth fashion, because it speaks the language of mechanisms and machinery.[27] So conceived, the mind becomes both an object of empirical investigation and one that is integrated into physics, so that we can understand how observation and reason-guided action fit into the larger dynamical framework. But the difficult details of implementation can be left aside to be sorted out by others. Those details are bound to be enormously complex and tend to obscure the high-level functional relationships that matter. It was the information-processing economy that was selected for, because it is the architecture that determines the causal-informational exchanges with the environment.

8. In Sum

Here, then, is an opinionated list of the progress that has been made about resolving some of the disputes in the literature:

- The dispute between dynamical and computational theories of mind. This is resolved by separating aspects of cognition that are closely tied to real-time perceptual input from the environment (e.g., the baseball player running for the flyball) and the more cognitive internal processes that are less tightly coupled to an event stream and are better modeled

[27] And the hope is that natural functional interpretations of properties such as rationality and intentionality will then emerge if we get our designs right. The pre-theoretic notion of consciousness might separate into a collection of functionally definable notions (e.g. introspective accessibility, global broadcast, informational integration).

by talking about information, representation, and computation (e.g., thought and deliberation).

- The dispute about whether the mind is a computer. This is resolved by distinguishing different notions of computation, saying in some senses it is not and in some senses it is.
- The question of how we map talk of representations and computation into a causal map of the brain. Here computers can shed some light by defusing simple ideas about how the information-processing economy of the mind is related to the low-level dynamical description.

Here is a list of what has not been resolved, but can be isolated (I think) from the ways in which human experience should matter to physics:

- The so-called Hard Problem of Consciousness.
- The analogous Hard Problem of Intentionality.
- Details about implementation that don't matter for architecture.

All of this shows that it is quite possible to talk about the aspects of the mind that matter to physics without getting embroiled in the mind-body problem. How should we think about proposals that connect consciousness to the collapse of the wave function in this light? Insofar as they provide a precise, mathematically defined condition for the presence or absence of consciousness that gives consciousness (so conceived) a role in inducing physical collapse, these are straightforwardly dynamical proposals and should be assessed on those grounds. (This includes proposals that treat consciousness as a dynamical primitive; consciousness in the guise of collapse-inducing property of systems is just a new kind of dynamically relevant hidden variable). They should be just as welcome as accounts that provide a precise mathematically defined condition for the presence of consciousness that gives it a role in the production of human behavior. The Hard Problem targets the question of whether these are merely *extensional definitions or constitutive accounts of what it is to be conscious*. It targets the question of whether any objective condition for consciousness can provide a constitutive account of subjective experience. Physics can be sanguine about that question; progress doesn't hinge on it. It cares about Easy problems. It can leave the Hard problems alone.

8

Why Mind Matters in Quantum Mechanics

Shan Gao

Quantum mechanics is a very successful physical theory due to its accurate empirical predictions. The core of the theory is the Schrödinger equation and the Born rule. The Schrödinger equation is linear, and it governs the time evolution of the wave function assigned to a physical system. The Born rule says that the result of a measurement on a physical system is definite but generally random, and the probability is given by the modulus squared of the wave function of the system. However, when assuming the wave function of a physical system is a complete description of the system, the linear Schrödinger equation is incompatible with definite measurement results. This leads to the measurement problem. Maudlin (1995) gave a precise formulation of the problem independent of standard quantum mechanics and the minds of observers. According to this formulation, the measurement problem originates from the incompatibility of the following three claims:

(C1) The wave function of a physical system is a complete description of the system;

(C2) the wave function always evolves in accord with a linear dynamical equation, such as the Schrödinger equation;

(C3) a measurement yields a single definite result.

Recently, I revised Maudlin's formulation and proposed a new mentalistic formulation of the measurement problem (Gao, 2019), which states the incompatibility of the following three assumptions:

Shan Gao, *Why Mind Matters in Quantum Mechanics* In: *Consciousness and Quantum Mechanics.*
Edited by: Shan Gao, Oxford University Press. © Oxford University Press 2022.
DOI: 10.1093/oso/9780197501665.003.0009

(A1) The mental state of an observer supervenes on her wave function[1];

(A2) the wave function always evolves in accord with a linear dynamical equation, such as the Schrödinger equation;

(A3) a measurement by an observer yields a single mental state with a definite record.

Oldofredi (2019) objected to my new formulation and defended Maudlin's original formulation. He argued that the measurement problem and its solutions are independent of the minds of observers. It seems that this view is common and well accepted in the literature. In this chapter, I will defend my mentalistic formulation of the measurement problem. In particular, I will explain more clearly why the new formulation is necessary and why it is more appropriate than Maudlin's original formulation. Moreover, I will argue that the solutions to the measurement problem also need to care about the minds of observers, such as assuming a certain form of psychophysical connection, and their validity depends on our understanding of minds.

Let us first compare the two formulations of the measurement problem. The key is to understand the claim (C3) in Maudlin's formulation of the measurement problem, namely "a measurement yields a single definite result." Where does this claim come from? Is it really true? Here, we only consider the one-world case, since both (C3) and (A3) are not true in the many-worlds case. As Oldofredi (2019) also admitted, the claim (C3) comes from the *empirical* evidence. He said, "[M]acroscopic objects can be in superposition, contradicting empirical evidence, i.e. uniqueness and definiteness of measurement outcomes. This is in essence of the famous measurement problem of quantum theory" (Oldofredi, 2019). If the claim (C3) indeed comes from the empirical evidence, then it actually means that an observer observes that a measuring device yields a single definite result after a measurement. In other words, an observation or a measurement by an

[1] In this chapter, when I say the "wave function of an observer," I mean the "wave function of the relevant part of the brain of an observer that relates to her mental state," and when the observer is entangled with another system, it denotes the entangled state of the composite system. Certainly, it is extremely difficult to know what the wave function of an observer is, let alone its connection with the mental state of the observer. But, as we will see, what I will consider below is a simple situation, in which the connection between the physical state and the mental state is assumed for each result branch of a post-measurement superposition, and the question I want to analyze is only what the psychophysical connection is for the whole superposition.

observer yields a single definite record, such as a pointer being in a definite position, in her mental state.

In fact, what we know with certainty by experience is only that we as observers obtain a definite record after a measurement by having a definite mental state with the record. While we don't know whether a measuring device really obtains a definite result after a measurement. For example, if the mental state is determined randomly by one branch of the post-measurement superposition as in the single-mind theory (Albert and Loewer, 1988), then the pointer of a device does not indicate a definite position after a measurement, but an observer will obtain a definite record after an observation of the position of the pointer.

Therefore, what the claim (C3) in Maudlin's formulation of the measurement problem really means is (A3), namely that a measurement by an observer yields a single definite record in her mental state. Moreover, the three claims in Maudlin's formulation of the measurement problem are not necessarily incompatible (when the measurement is made by an observer). When the three claims are compatible, the formulation is invalid; there is no contradiction and thus no problem in this case. This strongly suggests that a more appropriate formulation of the measurement problem based on (A3) is needed. What is the formulation then? Since (A3) concerns the mental state of an observer, we need an assumption about the psychophysical connection (i.e., the connection between the mental state and the wave function) in order to derive a contradiction. This means that (C1), which says that the wave function of a physical system is a complete description of the system, should be replaced by another assumption about the psychophysical connection. It can be seen that this assumption may be (A1), namely that the mental state of an observer supervenes on her wave function. In fact, (C1) implies (A1) by the principle of psychophysical supervenience.

Thus, we will obtain the mentalistic formulation of the measurement problem as given above. This new formulation of the measurement problem is valid in the sense that the three assumptions in the formulation are always incompatible, and thus it is more appropriate than Maudlin's original formulation of the problem.

The question now is: Can we have a physical formulation of the measurement problem that does not refer to the mental state? It seems that the answer is negative. The essential reason is that we don't know what physical state a mental state corresponds to in quantum mechanics. When removing the reference to the mental state, (A1) can be replaced by (C1) or another assumption, such as that the wave functions represent measurement results (and other things in the world that can be perceived by us). But (A3) cannot be replaced by another assumption that refers only to the physical state, such as "a measurement by a device yields a single definite result," since we don't know to what physical state a definite mental state corresponds. As noted before, although we see the pointer of a device being in a definite position, this does not necessarily imply that the pointer of the device is in a definite position as a matter of fact. Thus, we cannot derive a contradiction between quantum mechanics and experience when removing the reference to experience or the mental state. In other words, there is no physical formulation of the measurement problem. In this sense, the measurement problem of quantum mechanics is essentially the determinate-experience problem indeed (Barrett, 1999).

It should be pointed out that the above conclusion is true only in a strict sense. We may also have a physical formulation of the measurement problem in a restricted sense, namely when assuming what we see reflects what it is truly (in the one-world case). In this case, when we see the pointer of a device being in a definite position, the pointer of the device is indeed in a definite position. Then, the mentalistic formulation of the measurement problem may reduce to a physical formulation of the problem, which states the incompatibility of the following three assumptions:

(B1) The wave function represents the measurement result;

(B2) the wave function always evolves in accord with a linear dynamical equation, such as the Schrödinger equation;

(B3) a measurement by a device yields a single definite result.

The result assumption (B1) seems better than the completeness assumption (C1) in Maudlin's formulation, since in order to lead to a contradiction, the wave function of a physical system is not necessarily a complete description of the system, and it is only required that the post-measurement state of the measuring device represents the measurement result.

The mentalistic formulation of the measurement problem highlights the important role of psychophysical connection in causing the measurement problem. By this new formulation, we can look at the solutions of the problem from a new angle. In particular, Bohm's theory, Everett's theory, and collapse theories correspond to three different forms of psychophysical connection (as well as three different result assumptions). In fact, there are only three types of physical states that may determine the mental state of an observer, which are (1) the wave function in collapse theories, (2) certain branches of the wave function in Everett's theory, and (3) other hidden variables such as particle configuration in Bohm's theory.

Oldofredi (2019) gives a good introduction on how Bohm's theory and collapse theories solve the measurement problem. His line of reasoning seems well accepted. A quantum theory such as Bohm's theory is constructed as follows. We first have an ontology and its dynamics that can explain definite localization of macroscopic objects and definite measurement results consistent with the Born rule. Then, as a result of the ontology and dynamics, the theory ensures that observers' mental states will supervene on well-localized physical states, and thus it can also explain the definite perceptions of observers. Therefore, the measurement problem will be solved by these theories without caring about the minds of observers. He said, "Bohmian mechanics and GRW theories provide clear explanations of the physical processes responsible for the definite localization of macroscopic objects and, *consequently*, for well-defined perceptions of measurement outcomes by conscious observers. . . . [T]hese theories *guarantee*, in virtue of their Primitive Ontology (PO) and dynamical laws, that observers' mental states will supervene on well localized physical states representing measurement outcomes" (Oldofredi, 2019, italics added).

However, this common line of reasoning is problematic. Why assume such ontology and dynamics in the first place? It is arguable that such assumptions are made in order to account for our definite experience. Thus, the actual line of reasoning should be that these theories first assume *implicitly* that observers' mental states supervene on the well-localized physical states representing measurement outcomes, and then they assume certain ontology and dynamics so that the measurement outcomes predicted by them can be consistent with the Born rule. This means that in order to solve the measurement problem, these theories still need to care about the minds of observers by assuming a certain form of psychophysical connection.

Furthermore, the form of psychophysical connection in a theory cannot be simply posited, and it is restricted by our understanding of minds (see also Lewis, 2022). For example, as argued by Brown and Wallace (2005), the form of psychophysical connection assumed by Bohm's theory is inconsistent with the popular functionalism in the philosophy of mind when including the wave function in the ontology of the theory.[2] If functionalism is correct, for the mental state to supervene on the Bohmian particles but not on the wave function, the Bohmian particles must have some functional property that the wave functions do not share. However, in a human brain, where the wave functions are decohered due to the hot, wet, and noisy environment and thus they are effective wave functions in Bohm's theory, the functional behavior of the Bohmian particles is arguably identical to that of the effective wave function in which they reside (when coarse-grained at the level relevant to the mental functions of the brain).

Finally, even if the form of psychophysical connection assumed by a theory is valid, there are still unsolved issues closely related to the minds of observers. Take collapse theories as an example. Since these theories can explain definite localization of macroscopic objects and definite measurement results, and observers' mental states also supervene on these well-localized physical states, it seems that they can readily explain the definite perceptions of observers. However, due to the imperfectness of wave function collapse, the post-measurement state of an observer is an entangled superposition of brain states with different records, although the modulus squared of the amplitude of one state is close to one in general. This leads to the well-known tails problem (Pearle, 1989; Albert and Loewer, 1990, 1996; Lewis, 1995; McQueen, 2015). Moreover, even though the tails problem can be solved, we still need to analyze how the mental state of an observer supervenes on her wave function. Since the collapse time of a single superposed state is an essentially random variable, whose value can range between zero and infinity, there always exist certain measurements with a tiny probability, for which the collapse time is longer than the normal conscious time and the observer after the measurements is in an almost even entangled superposition of brain states with different records. Then, an intriguing question arises: What is it like to be such a quantum observer? A similar question has been asked in the bare theory (Albert, 1992, p. 124;

[2] Note that "functionalism in one version or another remains the dominant view of mind among philosopers" (McLaughlin et al., 2009, p. 149). In particular, "virtually all current major theories of mental content are in one way or another functionalist" (McLaughlin et al., 2009, p. 8). Moreover, in philosophy of mind, it is widely thought that although qualitative properties of consciousness, or qualia, may be not reducible, other mental phenomena are physically reducible, i.e., reducible to how physical matter moves and interacts (Kim, 2005). Thus, if this kind of physicalism is true, then the form of psychophysical connection assumed by Bohm's theory will be problematic (when including the wave function in the ontology of the theory).

Barrett, 1999). This issue is certainly related to the minds of observers (for an initial analysis, see Gao, 2019).

To sum up, I have defended my mentalistic formulation of the measurement problem and argued that it is more appropriate than Maudlin's original formulation. Moreover, I have argued that the solutions to the measurement problem also need to care about the minds of observers, such as assuming a certain form of psychophysical connection, and their validity depends on our scientific and philosophical understandings of the conscious mind.

References

Albert, D. Z. (1992). *Quantum Mechanics and Experience*. Cambridge, MA: Harvard University Press.

Albert, D. Z. and B. Loewer. (1988). Interpreting the Many Worlds Interpretation. *Synthese, 77*, 195–213.

Albert, D. Z. and B. Loewer. (1990). Wanted Dead or Alive: Two Attempts to Solve Schrödinger's Paradox. *Proceedings of the Biennial Meeting of the Philosophy of Science Association, 1*, 277–285.

Albert, D. Z. and B. Loewer. (1996). Tails of Schrödinger's Cat. In R. Clifton, ed., *Perspectives on Quantum Reality* (pp. 81–92). Dordrecht: Kluwer Academic.

Barrett, J. A. (1999). *The Quantum Mechanics of Minds and Worlds*. Oxford: Oxford University Press.

Brown, H. R. and D. Wallace. (2005). Solving the Measurement Problem: de Broglie–Bohm Loses Out to Everett. *Foundations of Physics, 35*, 517–540.

Gao, S. (2019). The Measurement Problem Revisited. *Synthese, 196*(1), 299–311.

Kim, J. (2005). *Physicalism, or Something Near Enough*. Princeton, NJ: Princeton University Press.

Lewis, P. J. (1995). GRW and the Tails Problem. *Topoi, 14*, 23–33.

Lewis, P. J. (2022). Against "Experience." this volume.

Maudlin, T. (1995). Three Measurement Problems. *Topoi, 14*, 7–15.

McLaughlin, B. P., A. Beckermann, and S. Walter, eds. (2009). *The Oxford Handbook of Philosophy of Mind*. Oxford: Oxford University Press.

McQueen, K. J. (2015). Four Tails Problems for Dynamical Collapse Theories. *Studies in History and Philosophy of Modern Physics, 49*, 10–18.

Oldofredi, A. (2019). Some Remarks on the Mentalistic Reformulation of the Measurement Problem: A Reply to S. Gao. *Synthese*. https://doi.org/10.1007/s11229-019-02101-3.

Pearle, P. (1989). Toward a Relativistic Theory of Statevector Reduction. In A. I. Miller, ed., *Sixty-Two Years of Uncertainty* (pp. 193–214). New York: Plenum Press.

9

The Nature of Belief in No-Collapse Everett Interpretations

Paul Skokowski

1. Introduction

In Chapter 4 of *The Quantum Mechanics of Minds and Worlds*, Jeffrey Barrett considers *the bare theory*, which takes the standard formulation of quantum mechanics (what he calls the von Neumann-Dirac formulation), and strips it of the collapse postulate. This yields an Everettian no-collapse interpretation of quantum mechanics in which "the usual deterministic linear dynamics always correctly describes the time-evolution of the quantum-mechanical state" (Barrett 1999, p. 94). There is much to agree with in Barrett's discussion of the bare theory, and I concur with most every one of his conclusions about its shortcomings in the final sections as he closes out the chapter. The bare theory, as Barrett correctly points out, is at its most puzzling when it is applied to human mental states—in particular, human beliefs including introspective beliefs and experiences. And it is here where I think Barrett's excellent analysis falls just a little short, not by any misapplication of quantum mechanical principles, but by not properly accounting for the complex nature of *human* belief states, and by this I mean in particular taking into account the neuroscience of belief (paying attention to its neural vehicles and causal roles), and paying due diligence to the contents and intentional properties of belief. By carefully accounting for these properties of belief within the quantum mechanical analysis, we will come to question two claims arising out of the bare theory about an observer of a superposition. First, when asked in a certain way about the result of the experiment, that the observer actually believes what he reports, and second that the observer would end up with a disjunctive belief about the experiment. In both cases we shall see that no beliefs actually have the content reported, and so the agent has no such beliefs. And the reasons for this have to do with paying close

Paul Skokowski, *The Nature of Belief in No-Collapse Everett Interpretations* In: *Consciousness and Quantum Mechanics*. Edited by: Shan Gao, Oxford University Press. © Oxford University Press 2022.
DOI: 10.1093/oso/9780197501665.003.0010

attention to the vehicles, contents, and causal roles of the beliefs in question. We will find that this is a side-benefit to taking into account details of the neuroscience and content of belief.

2. The Bare Theory, Introspection, and Superposition

As an example of what one is confronted with in the bare theory, Barrett considers an observer M of a Stern Gerlach measurement that will measure the x-spin of a spin 1/2 system S.[1] Before interacting with the device, the system S is in an eigenstate of z-spin, and the observer M is in an eigenstate of being ready to measure the x-spin of the system S. After the measurement, the composite system of the observer M and the spin 1/2 system S, according to the bare theory, will be given by Barrett's equation as follows:

$$|\psi\rangle = 1/\sqrt{2}(|x\text{-spin up}\rangle_M|\uparrow_x\rangle_S + |x\text{-spin down}\rangle_M|\downarrow_x\rangle_S).$$

Or, more compactly,

$$|\psi\rangle = 1/\sqrt{2}(|\uparrow\rangle_M|\uparrow_x\rangle_S + |\downarrow\rangle_M|\downarrow_x\rangle_S).$$

Since we are talking about belief and experience, it will be important to recognize that the observer M, who is a competent observer of experiments, is a human observer. Barrett does initially develop a model of an automaton that represents spin measurements, where "this model requires a close correspondence between physical memory configurations and mental states . . . " (Barrett 1999, p. 95). But ultimately, it is the human experience of a superposition that needs explaining. For example, in the first sentence of the section entitled *The Account of Experience*, Barrett asks, "So just how far can the bare theory go in explaining our experience?" (Barrett 1999, p. 110). We will concentrate then on the nature of human belief when it is involved in measuring a superposition.

Now let us consider Barrett's original example of an observer of the outcome of an experiment that results in a superposition. Barrett proposes asking the observer, who is in the superpositional state, "Did you get a determinate result of either x-spin up or x-spin down?" (Barrett 1999, p. 98).

[1] We can consider the particles to be electrons for the purposes of this paper.

Note that when a person is asked to report about the content of a belief they hold, they need to introspect in order to access that belief and so be able to report the content of that belief. Barrett agrees, saying that *M* "would believe that he knows what the result is" (Barrett 1999, p. 98). Such a *belief* of *M*'s about what he *knows* is an introspective belief, and in this case *M* is being asked to introspect his perceptual belief/knowledge about the result of the experiment. This introspection is a belief about a belief.[2] For example, suppose *M* perceives a yellow daffodil in a green field. Then we would say that he has the occurrent perceptual belief that the flower is yellow. Such an occurrent belief would involve *M*'s visual cortex (Zeki 1993; Seymour et al. 2016). In this situation, upon being asked what color the flower is in front of him, *M* would presumably report, "The flower is yellow." Now let's ask *M* to report if he has a determinate result for his belief about the color of a flower in front of him. *M* might ask in this case, "Are you asking me what the color of the flower is?" To this we would answer, "No, we are asking you if you now have a determinate belief about the color of the flower in front of you. In particular, we are asking you to introspect your belief about the color of the flower in order to verify that you have a determinate belief about the flower's color." In this case, *M* will form an introspective belief—a belief about a belief—because he will need to check his current beliefs to verify that he has a belief about a yellow flower in front of him. This introspective belief is a belief that will have another belief as a content; and in particular, the content of this introspective belief will be *M*'s occurrent perceptual belief.

Note that the fine-grainedness of mental contents ensures that *M*'s belief, or introspective state, or experience that the flower is yellow will always be different from his belief, or introspective state, or experience that the flower is red (Tye 1995; Dretske 1995; Perry 1977; Putnam 1975; Frege 1892). And the same applies when *M* is observing the detector in the experiment: the fine-grainedness of mental contents ensures that *M*'s belief, or introspective state, or sensation that the electron is spin up will be different from his belief, or introspective state, or sensation that the electron is spin down.[3] This fine-grainedness is important: the intentional content of a belief about yellowness is different from the intentional content of a belief about redness.

[2] As knowledge states are typically taken to be some form of true belief, then *M*'s belief about this knowledge state is a belief about a belief; hence, an introspective belief. From here on for consistency, I will not refer to M's perceptual belief of the experimental outcome as knowledge; rather, I will refer to any such occurrent perceptual state as a belief.

[3] Where the content in these cases includes, say, the position of a pointer on the measuring device; for example. pointing to one of the marks "+" or "-."

And the intentional content of a belief that the dial points to "+" is different from the intentional content of a belief that the dial points to "–." Such beliefs then will always differ. So, any sort of mental state with a content about spin up will differ from any mental state about spin down.[4] And, of course, it is the content of the state that allows us to call such states mental representations in the first place (Brentano 1874; Dretske 1988; Tye 1995). The content of a mental state also helps distinguish that type of state from other states, and in any causal theory of mental content, such states will be distinguished because of their differences in content, vehicle, and causal role (Skokowski 1999, 2018).

In order to have an introspective belief that can be used to report the occurrent belief being introspected, this introspective belief must be causally/physically connected with the occurrent perceptual belief through neural connections and neural firings. So, *M*'s introspective belief about a given perceptual belief will involve specific neural connections—as part of the introspective belief vehicle—to the specific portion of visual cortex that is involved with the exemplification of his perceptual belief, which itself involves a different set of neurons. In addition, when *M* has this introspective belief, then specific neural firings of the sort associated with the particular connections to this region of visual cortex will occur. That is, the action potentials that occur within the introspective belief in virtue of its connection with the color belief will be specific to that particular connection between the two sets of neurons. This introspective belief vehicle will therefore be made up of a specific set of neurons, which exemplify a certain pattern of neural firing (action potentials), when it occurs in *M*'s brain. Indeed, imaging studies appear to show that introspective beliefs occur in the pre-frontal cortex (Fleming et al., 2010).

And this analysis will apply for any of *M*'s mental states that are tasked with representing the content of his occurrent perceptual belief. Any such mental state will need to be causally connected to the occurrent perceptual belief, and the intentional content of the analyzing mental state will be fine-grained and hence *about* that occurrent perceptual belief and its content. That is, if *M* has a mental state *B* that is about his occurrent perceptual state and its content—where *B* is an introspective or some other self-analyzing state—then this mental state *B* will have a fine-grained content that is about

[4] Note that this difference in content holds whether the content is the position of a pointer towards either "+" or "-" or whether the content is actual color content "yellow" or "red." The intentional content will be fine-grained in either case.

that occurrent perceptual belief and its content. Hence, any query to *M* about an occurrent perceptual belief state of his will depend for its answer on a mental state of his that represents that occurrent perceptual belief, and the intentional content of that representing mental state will have a fine-grainedness that tracks the fine-grainedness of the occurrent perceptual belief it represents.[5]

So, for example, the eigenstate corresponding to *M*'s introspection of his belief that the electron is x-spin up will contain his introspective state in the pre-frontal cortex as well as his perceptual belief in the visual cortex, and would therefore be written:

$$|\text{Introspect } x\text{-spin up}\rangle_{M-PF}\ |\uparrow\rangle_{M-VC}|\uparrow_x\rangle_S.$$

Or, more compactly,

$$|\text{Introspect}\uparrow\rangle_{M-PF}\ |\uparrow\rangle_{M-VC}|\uparrow_x\rangle_S,$$

where the subscripts "*M-PF*" stand for the introspective state in *M*'s pre-frontal cortex and the subscripts "*M -VC*" stand for the perceptual belief involving *M*'s visual cortex.

Let's apply this analysis to how the initial state of *M*'s brain, and the electron state he is observing, evolve from being "ready" to how things end up after the measurement. Before the measurement, we have:

$$|PF\ \text{ready}\rangle_{M-PF}|VC\ \text{ready}\rangle_{M-VC}(|\uparrow_x\rangle_S + |\downarrow_x\rangle_S),$$

where the first state is the "ready" state of the pre-frontal cortex, the second state is the "ready" state of the visual cortex, and the final state is the superposition of the x-spin directions up and down (that is, an electron initially in an eigenstate of z-spin expanded in the x-spin basis) for the electron that is about to be passed through the detector.

After the measurement, this state evolves into a superposition, call it $|\psi\rangle$, of the form:

$$|\psi\rangle = 1/\sqrt{2}(|\text{Introspect}\uparrow\rangle_{M-PF}|\uparrow\rangle_{M-VC}|\uparrow_x\rangle_S \\ + |\text{Introspect}\downarrow\rangle_{M-PF}|\downarrow\rangle_{M-VC}|\downarrow_x\rangle_S).$$

[5] If the representing state did not have this fine-grainedness, then *M* would not be capable of answering queries about the content of the perceptual state in question.

Note that the first two eigenstates of both terms in the superposition represent different mental/belief states of *M*'s. As such, they will have different neural vehicles and contents from one another. In addition, were we to measure any of their neural properties (for example, which neurons, their action potentials, their connections, etc.), they would yield different eigenvalues. In the first term of the superposition, the first (introspective) state would give a measurement of a state of an introspection of a belief that the electron is spin up, and the second state would give a measurement of a perceptual belief that the electron is spin up. In the second term of the superposition, the first (introspective) state would give a measurement of a state of an introspection of a belief that the electron is spin down, and the second state would give a measurement of a perceptual belief that the electron is spin down. All these states differ in vehicle (the particular state in cortex) and content, which must be the case given the fine-grainedness of mental states.

Since each component of this superposition yields a different eigenvalue for its corresponding operator, then there will *not* be a common eigenvalue between the two components that would serve as an eigenvalue for an operator on the superposition taken as a whole. This is because of the fine-grainedness of mental states. Since each mental state in the superposition has its own vehicle and fine-grained intentional content, then each one will differ: each one will have a different eigenvalue from the other. This has to be the case for mental states, if they are to be individuated *qua* mental states. An introspective state about a belief that an electron is spin up is different from an introspective state about a belief that an electron is spin down. A perceptual belief that an electron is spin up is different from a perceptual belief that an electron is spin down. The vehicles differ: each state—introspective or perceptual—is composed of different neurons and action potentials, so the vehicles are exemplified as different physical states. And as Frege and others have taught us, the contents will also differ, as beliefs are known to be referentially opaque. For we know that even when $m = n$, a belief that m is P is not the same as a belief that n is Q, even when P and Q are co-extensional, and refer to the same things (Frege 1892; Putnam 1975; Dretske 1988; Tye 1995).

There are two important things to note here. First, the state of the system corresponding to *M*'s introspection of his belief that the electron is spin up will not be given by equation. Instead, the state of the system will contain *M*'s introspective state in the pre-frontal cortex as well as his perceptual belief involving the visual cortex, as spelled out above with state $|\psi\rangle$. Second, the superpositional state $|\psi\rangle$ as it stands does not yield the result that Barrett has claimed, which is that there is a measurable eigenvalue that is common to both components of the superposition. If there were such an eigenvalue, then there would be an observable property of *M* for the superposition of brain states he finds himself in according to the linear equations of motion.

Barrett's solution to this problem is to have *M* answer "Yes" to a question about his mental state. The question is for *M*, "Did you get some determinate result to your *x*-spin measurement, either *x*-spin up or *x*-spin down?" (Barrett 1999, p. 97). And so, "*M* would report that he got a determinate *x*-spin result when he did not determinately get up and did not determinately get down" (p. 97).

Note that in order to answer this question, *M* will, in both components of the superposition, need to introspect his perceptual belief about the result of the experiment in order to evaluate the disjunction "*x*-spin up or *x*-spin down." Since the perceptual beliefs in either component of the superposition are different, then the contents of these introspections in either component will also be different from each other. In the first component, *M*'s introspection of the perceptual belief that the *x*-spin is up would reveal the answer to the disjunction to be "up," whereas the second component, *M*'s introspection of the perceptual belief that the *x*–spin is down, would reveal the answer to the disjunction to be "down." Let's refer to the contents of these two introspective states as (UP ∨ DOWN: UP) and (UP ∨ DOWN: DOWN). *M* would need intentional contents with values like these in order to answer Barrett's question correctly on either component of the superposition. And let's say that, upon formulating these introspective beliefs, *M* can now answer the question as posed, "Did you get some determinate result to your *x*-spin measurement, either *x*-spin up or *x*-spin down?"

But note that there is a peculiarity about answering such a question. Answers to questions are formulated in a different part of the brain: Broca's area. And Broca's area is an area of the brain that is tasked with linguistic output—not with introspection. These linguistic outputs include unconscious grammatical processing and the signals required to form the mouth and tongue in a particular configuration, exhaling breath in a certain manner, opening and closing the nasal passages, and so forth.[6] Introspection

[6] Because introspection is conscious (Dretske 1995; Moore 1903), and linguistic processing is unconscious (Pinker 1994), the latter is not a candidate for introspective beliefs.

is tasked with producing beliefs about beliefs (Dretske 1995), whereas Broca's area is not: it is tasked with producing linguistic outputs (Pinker 1994, 1997).

The outputs of intentional mental states are not to be confused with mental states themselves. Consider the simple example of drinking a beer. I believe there's six-pack in the fridge and I desire a cold one. These mental states cause me to open the fridge door, grab a bottle, twist off the cap, and take a drink. The belief and desire are mental states with intentional contents, but the reaching, twisting, and drinking are outputs that are caused by these representational states. These outputs are not themselves representational states: that is, states with intentional contents from a function to represent properties in the environment, and executive capacities to cause action (Dretske 1988; Papineau 1987). They are instead *outputs* of representational states: causal consequences of mental states that *do* have the requisite executive capacities and intentional contents.

Barrett's question therefore is designed to detect a common measurable property in the $M + S$ system by detecting a common output, and not by detecting a common property of the introspective representational eigenstates themselves. This is peculiar, because what is presumably at issue is the content of *M*'s own introspective and occurrent perceptual beliefs.

Note that Barrett is not asking M to introspect what spin result he perceived. The reason for this prescription is clear: The eigenvalue for asking M, "Do you introspect you are perceiving spin up?" will be different on both sides of the superposition, as it will if we decide instead to ask M, "Do you introspect you are perceiving spin down?," and so there will be no common eigenvalue for the superposition $|\psi\rangle$ if either of these questions is asked. So a different question—a disjunctive one—must be asked. But that means that, rather than measuring *M*'s introspective states and their contents with an operator operating on them directly, we are instead being asked to measure a common output of those introspections. The focus has been shifted from introspection itself—and so a question about *M*'s beliefs—to a common *output* of introspective beliefs.

3. Linearity and Deception

The problem of shifting the focus in this way can be illustrated by an example. Note that if we can detect a common eigenvalue in the M-spin-detector superposition by means of a common output, rather than by the representational state itself, then we should be able to detect a common eigenvalue in a *measuring device* by a common output as well. After all, what is important about an electron-spin measuring device is the representational state it ends up in: pointing to "+," or pointing to "–," for example. These pointer states are representational states with a content: they are *about* something, and crucially, they are *about* whether the measured electron is spin up or spin down. That's what makes them representational in a way appropriate for measuring the *actual* spin of a particular electron.

Suppose, for example, that the needle on our electron-spin measuring device slides on the x-axis. It slides left for a spin-up electron and right for a spin-down electron. Now suppose we notice that whenever the measuring device registers the spin of an electron, there is a slight pressure wave of air in the y-axis normal to the x-axis. Due to turbulence effects, this wave is the same whether the electron is spin up or spin down. So we place a pressure detector on the y-axis that detects when such a pressure wave occurs.

Before we had this pressure detector attached, our superposition looked like:

$$|\Phi\rangle = 1/\sqrt{2}(|\uparrow\rangle_M|\uparrow_x\rangle_S + |\downarrow\rangle_M|\downarrow_x\rangle_S).$$

And after we attached the pressure detector, we have (where the subscript p refers to this detector):

$$|\Phi'\rangle = 1/\sqrt{2}(|\text{"y pressure"}\rangle_p|\uparrow\rangle_M|\uparrow_x\rangle_S + |\text{"}y\text{-pressure"}\rangle_p|\downarrow\rangle_M|\downarrow_x\rangle_S).$$

Call the magnitude of this pressure wave λ. Then we can see, by linearity, that a measurement of this pressure wave by an operator O is an observable property of the superpositional state as well as of each component of the superposition. So, by putting an electron in an eigenstate of z-spin through an x—spin detector, we can measure the observable property λ:

$$O|\Phi'\rangle = \lambda|\Phi'\rangle.$$

But notice that measuring a pressure wave like this—that is, an output of a representational/detecting state—does not mean that the detector is confused or deceived about what it's detecting. It just means that the detector produces a measurable pressure wave regardless of whether it has collapsed to one component of the superposition or the other, or that, if Everett is correct, this observable will be measurable even in a superposition (by virtue of linearity). A measurement of a y-pressure wave makes no claim about the final position of the pointer in the x-direction and hence the spin of the electron. This is explicitly spelled out in the state vector $|\Phi'\rangle$, where the two properties occupy different eigenstates, for example, $|$"y-pressure"$>_p$ and $|$"x-spin up"$>_m$, that yield different physical properties for the device. Measurement of one property of the device does not mean deception about another property, as these are separate eigenstates with their own associated operators and eigenvalues. So the detector is not confused or deceived. There is just a common measurable output that occurs regardless of the QM interpretation.

And now we can say the same about the human observer M. When we consider the linguistic output from Broca's area, then before the measurement, we have:

$$|B \text{ ready}\rangle_{M-B}|PF \text{ ready}\rangle_{M-PF}|VC \text{ ready}\rangle_{M-VC}(|\uparrow_x\rangle_S + |\downarrow_x\rangle_S),$$

where the first state is the "ready" state of M's Broca's area, and the rest of the "ready" states are defined as before: M's pre-frontal cortex, his visual cortex, the "ready" state of the measuring device, and the final state is the superposition of up and down spin for the electron that is about to be passed through the detector. And again, M's eigenstates are designated by subscripts $M-B$, M-PF, and M-VC.

After the measurement, this state evolves into a superposition of the form $|\psi'\rangle$, which is different from the superposition $|\psi\rangle$ we considered earlier:

$$|\psi'\rangle = 1/\sqrt{2}(|\text{“Yes”}\rangle_{M-B}|\text{Introspect}\uparrow\rangle_{M-PF}|\uparrow\rangle_{M-VC}|\uparrow_x\rangle_S + |\text{“Yes”}\rangle_{M-B}|\text{Introspect}\downarrow\rangle_{M-PF}|\downarrow\rangle_{M-VC}|\downarrow_x\rangle_S).$$

Then we can see, by linearity, that a measurement of M's linguistic output will be an observable property of the superpositional state as well as of each component of the superposition. That is, when the superposition $|\psi'\rangle$, which includes M's states, is asked whether M has some definite belief in the way prescribed by Barrett, where this question is the operator O, then he will answer "Yes":

$$O|\psi'\rangle = \text{“Yes”}|\psi'\rangle.$$

And this result, of course, is by virtue of this operator O operating on M's state $|$"Yes"$>_{M-B}$.

But this is like the detector example given immediately above. "Measuring" an answer like this—that is, an output of an introspective/representational state—does not mean that *M* is deceived about what he's introspecting. It just means that *M* produces the answer "Yes" regardless of whether he has collapsed to one component of the superposition or the other, or that, if Everett is correct, this answer will be measurable even in a superposition (by virtue of linearity). A measurement of a spoken output "Yes" makes no claim about *M*'s occurrent introspective state and his occurrent perceptual belief about the spin of the electron. This is explicitly spelled out in the state vector $|\psi'\rangle$, where the three properties are given by three different eigenstates $|\text{“Yes”}>_{M-B}$, $|\text{Introspect}\uparrow\rangle_{M-PF}$, and $|\uparrow\rangle_{M-VC}$ of *M*'s Measurement of one property of *M* does not mean deception about another property, as these are separate eigenstates with their own associated operators and eigenvalues. So when Barrett claims that *M* "would believe that he knows what the result is" (Barrett 1999, p. 98) (one of spin up or spin down) based on this spoken output, and that this belief is "false" (p. 98), we see that there is no basis for this claim, as *M* does not actually have the belief in question. That is, there is no single belief state that emerges from the superposition with the singular content being one of spin up or spin down. So *M* is not deceived. There is just a common linguistic output that occurs regardless of the QM interpretation.

In addition, it is important that deception is having a false belief: for *X* to be deceived about *G* is for *X* to believe *G* when *G* is not the case. We see that on either side of the superpositions involving *M*, there are only two beliefs: an introspective belief and a perceptual belief. As deception is a type of belief, these are the only candidates. However, none of these beliefs are instances of deception, because Barrett has stipulated beforehand that "*M* is a *good x*-spin observer in Everett's sense (indeed, one might call him a perfect observer).... " (Barrett 1999, p. 96), so his beliefs within each component of the superposition—perceptual and introspective—about pointer position are accurate. This means any deception about the outcome must be at the level of speech output; that is, at the level of *M* saying, "Yes" about introspecting a definite belief about the spin of the electron. But speech output, as we have shown, is not an introspective state, and indeed, being an output rather than an intentional state, it is not a belief at all about the spin of the electron.

Barrett goes on to point out another peculiar aspect of the bare theory—that is, the nature of what he calls disjunctive experiences. In observing the experiment mentioned at the beginning of this paper, a proponent of this theory "would not say that *M* would determinately believe that he had recorded *x*-spin up, nor would he say that he would believe that he had recorded *x*-spin down; rather, he would say that *M* would determinately believe that he had recorded *x*-spin up *or* *x*-spin down. One might call the experience leading to this disjunctive belief a disjunctive experience" (Barrett 1999, pp. 110, 111).

The problems with a claim like this from a bare theorist are twofold: first, in analyzing the superpositional states of the $M + S$ system in any of the permutations we have considered, the belief states (either introspective or perceptual) are always on both sides of the superposition, with different vehicles, contents, and causal roles within their respective component of the superposition. That is, there is no single meta-belief formed with a disjunctive content like that proposed. The belief-chains, if you will, are formed on either side of the superposition, and are related to one another through their contents, neural connections, action potentials, etc. And neither of these chains can lead to such a disjunctive content, as each chain begins with a different spin for the particle, which is then perceived as that spin, then introspected as that spin, and so forth. The transparency of beliefs on either side of the superposition ensure that the contents of any introspective beliefs are of the singular contents within the respective component. Indeed, even the introspective "disjunctive" contents considered earlier contain components, "UP" on one side of the superposition and "DOWN" on the other side [(UP ∨ DOWN: UP) and (UP ∨ DOWN: DOWN)], that confirm which side of the superposition they reside in, so they are not pure disjunctions like the one given in the quote above. Which leads immediately into the second problem. As G. E. Moore painstakingly showed over a century ago, introspective states are *transparent*. That is, when one introspects another of one's own mental states, one's awareness is of the content of the mental state being introspected, not the mental state itself. Thus, any introspective states are drawn to the contents of the state being introspected: introspection therefore reveals no further phenomenal character or intentional content than the content of the sensation or belief being introspected (Moore 1903; Dretske 1995; Tye 1995). And the beliefs on both sides of the superposition contain contents specific to that side of the superposition ("+" or "–"), but never a content of a disjunctive belief with a content of the form "*x*-spin up *or* *x*-spin down." And no mechanism has been provided by the bare theorist to show that such a belief has been formed. In order to do this, the bare theorist would need to explicate the vehicle, content, and causal role (including its origin) for such a belief state. And that is a tall order that I don't believe the bare theorist has yet provided, because neither component of the superposition contains such a disjunctive content.

4. Conclusion

We have seen that two claims of the bare theory about the observer *M* of a superposition fall short. First, when the observer *M* answers "Yes" to a question about his perceptual beliefs of the measurement, this does not imply that *M* would therefore falsely believe that he *knows* what the result of the measurement is. The reason, we have seen, is that *M* does not actually have the belief in question: there is no single belief state of *M*'s that emerges from the superposition with the singular content being one of spin up or spin down. Instead, as we have seen, there is a common linguistic output that occurs on both sides of the superposition, and such linguistic outputs of intentional states are not themselves intentional, and so are not themselves beliefs. We have also seen that *M* does not end up with a disjunctive belief about the experiment. The reason for this is that there is no mechanism for producing a meta-belief with such a content from the beliefs available to *M*. In both cases, these results were arrived at by carefully considering the vehicles, contents, and causal roles of *M*'s introspective and perceptual beliefs.

Acknowledgments

Thanks to John Perry, Reed Guy and Harvey Brown for very helpful discussions on the issues discussed here. All mistakes and misunderstandings are entirely my own.

References

Barrett, J. (1999), *The Quantum Mechanics of Minds and Worlds*. Oxford: Oxford University Press.

Brentano, F. (1874), *Psychologie vom Empirischen Standpunkt*. Leipzig.

Dretske, F. (1988), *Explaining Behavior*. Cambridge, MA: MIT Press.

Dretske, F. (1995), *Naturalizing the Mind*. Cambridge, MA: MIT Press.

Fleming, S., et al. (2010), "Relating introspective accuracy to individual differences in brain structure." *Science*, 329(5998): 1541–1543.

Frege, G. (1892), "On sense and reference" [Über Sinn und Bedeutung]. *Zeitschrift für Philosophie und philosophische Kritik*, 100: 25–50.
Lee, T. S., et al. (1998), "The role of the primary visual cortex in higher level vision." *Vision Research*, 38: 2429–2454.
Moore, G. E. (1903), "The refutation of idealism." *Mind*, 12: 433–453.
Papineau, D. (1987), *Reality and Representation*. Oxford: Blackwell.
Perry, J. (1977), "Frege on demonstratives." *Philosophical Review*, 86: 474–497.
Pinker, S. (1994), *The Language Instinct*. New York: HarperCollins.
Pinker, S. (1997), *How the Mind Works*. New York: W. W. Norton.
Putnam, H. (1975), "The meaning of 'meaning.'" *Minnesota Studies in the Philosophy of Science*, 7: 131–193.
Seymour, K. J., et al. (2016), "The representation of color across the human visual cortex: Distinguishing chromatic signals contributing to object form versus surface color." *Cerebral Cortex*, 26: 1997–2005.
Skokowski, P. (1999), "Information, belief, and causal role." In Moss et al. (ed.), *Logic, Language and Computation*. Stanford, CA: CSLI Press.
Skokowski, P. (2018), "Temperature, color and the brain: An externalist response to the knowledge argument." *Review of Philosophy and Psychology*, 9(2): 287–299.
Tye, M. (1995), *Ten Problems of Consciousness*. Cambridge, MA: MIT Press.
Zeki, S. (1993), *A Vision of the Brain*. Oxford: Blackwell.

10

The Completeness of Quantum Mechanics and the Determinateness and Consistency of Intersubjective Experience

Wigner's Friend and Delayed Choice

Michael Silberstein and W. M. Stuckey

1. Introduction

For a variety of reasons, there is renewed interest in trying to relate conscious experience to quantum mechanics (QM). Usually someone tries to explain one in terms of the other, e.g., they claim that conscious experience collapses the wavefunction [1]. There are also many attempts to explain conscious experience or some feature of it in terms of some quantum process or property such as wavefunction collapse, entanglement, etc. Such accounts are metaphysically diverse, ranging from strong emergence, to panpsychism, to dual-aspect theories and beyond [2]. Obviously, attempts at explaining some feature of QM in terms of some feature of conscious experience or vice versa require that one be very specific about the interpretation of QM at issue and the particular feature/conception of conscious experience in question. Often when this game is played one assumes that wavefunction realism is true and that qualia is the best way to think about conscious experience; we will jettison both those assumptions in this chapter. There are of course other ways to relate QM and conscious experience, such as simply using the Hamiltonian formalism of QM to model conscious decision making, e.g., thinking thru a decision is like being in a superposition state and making a decision is like wavefunction collapse [2].

More generally, Gao argues that the very measurement problem itself should be characterized as a problem about determinate-experience [3, p. 4]:

Michael Silberstein and W. M. Stuckey, *The Completeness of Quantum Mechanics and the Determinateness and Consistency of Intersubjective Experience: Wigner's Friend and Delayed Choice* In: *Consciousness and Quantum Mechanics*. Edited by: Shan Gao, Oxford University Press. © Oxford University Press 2022.
DOI: 10.1093/oso/9780197501665.003.0011

> The problem is not only to explain how the linear dynamics can be compatible with the appearance of definite measurement results obtained by physical devices, but also, and more importantly, to explain how the linear dynamics can be compatible with the existence of definite experiences of conscious observers.

Of course one could certainly push back on Gao's characterization of the measurement problem. Many would argue that the measurement problem would exist even in a world with no conscious observers or that there are interpretations of QM that yield definite classical outcomes and thus determinate experiences. However, if one assumes wavefunction realism as Gao does and one adopts some no collapse interpretation of QM, then it is certainly reasonable to raise the question of why and how we have determinate experiences. Even though one may be convinced in general that no-collapse interpretations such as Many-Worlds and Bohmian mechanics have no such concern, it is this question of determinate experience and intersubjectively consistent determinate experience that we shall take up herein. This is because, as we are all well aware, there are cases (experiments, gedanken or otherwise) where, at least on certain interpretations, there is a problem about determinate and intersubjectively consistent experience, such as the appearance of definite measurement outcomes that all observers can experience as determinate and will agree upon. Herein we will look at two such cases: Wigner's friend and delayed choice quantum eraser with a twist (see also Lucien Hardy's chapter in this volume for more on the latter). We will discuss why these cases raise concerns for the determinateness and intersubjective consistency of conscious experience at length later. Let us note for now that Wigner's friend has recently raised its ugly head again. There are new Wigner's friend related experiments, alleging to show that QM is either incomplete or incompatible with determinate and intersubjectively consistent conscious experience, such as experiences of measurement outcomes [4, 5]. Recently Hardy has suggested putting conscious decision makers into an EPR delayed-choice type experiment to see if they could violate the predictions of QM [6]. Such an experiment pits the completeness of QM against our everyday experience of free will. Hardy discusses this possibility further in this volume and we will consider a much more idealized version of such an experiment herein.

The specific question we want to address is as follows: is there a take on QM that explains why there is and must always be determinate and intersubjectively consistent experience about all experimental outcomes? A take that accepts the completeness of the theory (i.e., no modification to the formalism is made such as the addition of an objective collapse mechanism), and furthermore requires no invocation of relative states (e.g., outcomes being relative to branches, conscious observers, etc.). And finally, a take that requires no allegedly hybrid models such as claims about "subjective collapse." We want a take on QM that provides a single world wherein all the observers (conscious or otherwise) agree about determinate and definite outcomes, because those outcomes are in fact determinate and definite. We also want a take on QM that explains why conscious human choices cannot violate the predictions of QM, that is nonetheless consistent with our experience of having free will. Herein, we will provide such a take on QM. Our take on QM is a realist psi-epistemic one that is based on principle explanation as opposed to constructive, dynamical or causal mechanical explanation. As we shall see, the principles in question constitute adynamical global constraints ranging over spacetime, not unlike those constraints underlying special relativity (SR). The idea of a principle-based take on QM is not new to us, the quantum information theory (QIT) community has sought such an account for a long time [7]. Our principles constrain what observers (conscious or otherwise) can experience, measure, observe, etc. One such constraint is that no observer occupies a preferred frame of reference. This obtains in QM per Brukner and Zeilinger's principle of Information Invariance & Continuity [81], i.e., "The total information of one bit is invariant under a continuous change between different complete sets of mutually complementary measurements," at the foundation of QIT reconstructions of QM [82], as first explained by Hardy in 2001 [83]. Our take on QM has many additional advantages such as a deflation of the measurement problem, a local explanation of EPR-Bell correlations in terms of adynamical global constraints, and an explanation of the Born rule (for details see [8, 9, 10]). The key to getting all these advantages is to give up naive realism about the wavefunction, which is motivated in part by physicalism and the little questioned assumption that fundamental explanation must always be constructive, causal or dynamical. Let us note again that while our take on QM is psi-epistemic, unlike QBism or pragmatic accounts, our take is fully realist and resides fully in spacetime.

There is also another advantage of our particular principle take on QM and relativity as well. We are now in a position to jettison the second odious assumption that forces an eternal return of the dialogue about how consciousness might explain the collapse of the wavefunction or vice versa. The first offending assumption as we noted, is realism about the wavefunction coupled with collapse. The second troublemaking assumption is dualism about conscious experience, i.e., the idea that conscious experience is best conceived as some sort of qualia that is either somehow produced by brains via strong emergence or must be present in some form in fundamental physics a la panpsychism. Both strong emergence and panpsychism are notoriously problematic and yet, for want of a better alternative, we return to them over and over again [9]. Our view provides a better alternative called neutral monism. The basic idea of neutral monism is that the mental and physical are nondual (not essentially different), they are neutral in virtue of being neither essentially mental nor physical, and in virtue of the fact that they are grounded in something neutral [9]. This ubiquitous "something" does not "pervade" spacetime, it does not "generate" spacetime, it is co-extensional with spacetime. According to neutral monism, our conscious experience of an external world is not some virtual model or construction of the world trapped in the mind and generated by the brain, while the actual external physical world lies outside us forever, as I-know-not-what noumena. Neutral monism is a form of direct realism, wherein there is only the spatiotemporally extended world of experience, of which we are inextricably a part; the discipline of physics is all about the world so described.

It will become clear as we go along what all this entails about the relationship between physics and psychology. For now however, note that given neutral monism, physics doesn't start with experience merely for pragmatic reasons. Physics is inherently all about the possibility of and rules of experience, but not because the world is mind dependent. Just as there is no metaphysical dualism of the "inner" world of experience and the "outer" physical world, there is no inherent dualism of psychology and physics. As Bertrand Russell puts it, "The whole duality of mind and matter . . . is a mistake; there is only one kind of stuff out of which the world is made, and this stuff is called mental in one arrangement, physical in the other" [11, p. 15]. Compare this to the words of William James, "Things and thoughts are not fundamentally heterogeneous; they are made of one and the same stuff, stuff which cannot be defined as such but only experienced; and which one can call, if one wishes, the stuff of experience in general. . . . 'Subjects' knowing 'things' known are 'roles' played, not 'ontological' facts" [12, p. 63]. The point is that when Galileo and others insisted on the primary/secondary quality distinction or any other kind of dualism of inner mental experience and external physical world, they doomed us to the mind/body problem and the hard problem. Such dualism is not empirical data we experience directly, it is a cognitive illusion, an inductive projection of theorizing minds. The way to undo this mistake is not to merely move the location of the mysterious dualism from brains (strong emergence) to fundamental physics (panpsychism), but to reject completely and thoroughly the primary/secondary property distinction altogether with neutral monism. Given neutral monism, what we call physical phenomena are just one mode of description of the neutral base.

We hope the reader will see that if one is willing to give up the assumptions of wavefunction realism and dualism about conscious experience, the very assumptions that have led us into this morass that includes the hard problem, the mysteries of QM and recurring idea that they are somehow deeply related, there is a way out to an entirely different picture of the world and our place in it. Given neutral monism, conscious experience and physics are in fact deeply related, but not in the occult way most people think. Rather, they are deeply related because conscious experience and the physical world are nondual, which means that physics is inherently already all about experience and constraints upon it. It is not hard to see that one of the things motivating these problematic assumptions is physicalism (the idea that everything is physical or at least that physical phenomena are basic) and fundamentalism (the idea that all axiomatic facts must reside in basic physics). Given these assumptions, it isn't surprising that we continue to chase our tail seeking deep connections between the hard problem and the measurement problem, that we keep gravitating toward panpsychism, etc. Neutral monism is a hard rejection of both these assumptions. Sometimes, when a problem remains intractable over long periods of time, the best move is to deflate the problem by rejecting the axiomatic assumptions that got you there. This chapter is an attempt to deflate both the hard problem and the measurement problem in one fell swoop.

In section 2 we will use a recent, much discussed Wigner's friend article that alleges to show we must either give up the completeness of QM, commonsense intuitions about free will, locality, or the determinate and objective nature of reality. This is a very useful set-up for our chapter since we allege to save all these assumptions, with the possible exception of commonsense intuitions about free will—the reader will decide. In section 3 we will introduce the reader to the concept of principle explanation and briefly explain our particular local, principle explanation of EPR-Bell correlations. This will allow the reader to appreciate our analysis of various Wigner's friends experiments and the quantum eraser delayed choice experiment with a twist, both to be discussed in subsequent sections. In section 7 we will summarize and recontextualize everything that we have said in light of neutral monism, in order to show the reader what a different and better world it is.

2. Wigner's Friend Revisited

Let us recall the Wigner's friend thought experiment. We start with the Schrödinger's cat thought experiment (for example) and now we imagine a human observer comes up to the box and opens the lid. If we assume that QM is complete and thus applies to all macroscopic measuring devices including conscious observers, then the Schrödinger equation tells us we will now have an entangled state that includes the human observer and the cat in a box, and so on if we keep adding observers. This is of course just the measurement problem. While the Wigner's friend thought experiment specifically raises the question as to whether a conscious observer could ever be in a QM superposition or entangled state, note that the measurement problem as such does not essentially involve consciousness in any way, e.g., we could have a sent a robot to look in the box and made the same point. However, some people have speculated that if conscious states cannot be placed in a superposition state, perhaps that is evidence that consciousness itself collapses the wavefunction. This is the question Bong et al. hope to explore in a new experiment, the results of which were recently published in Nature [5]. The experiment is a toy one in the sense that in the place of conscious observers or macroscopic measuring devices, they use photons. They believe however to have a "proof-of-principle" that hopefully someday can be scaled up to macroscopic measuring devices and conscious observers. What is it they allege to have proved?

Bong et al. allege to show that we must give up at least one of the following assumptions:

- Assumption 1 (Absoluteness of Observed Events (AOE)): An observed event is a real single event, and not relative to anything or anyone.
- Assumption 2 (No-Superdeterminism (NSD)): Any set of events on a space-like hypersurface is uncorrelated with any set of freely chosen actions subsequent to that space-like hypersurface.
- Assumption 3 (Locality (L)): The probability of an observable event e is unchanged by conditioning on a space-like-separated free choice z,

even if it is already conditioned on other events not in the future light-cone of z.

- Assumption 4 (The completeness of QM): QM unmodified applies to any and all macroscopic measuring devices including human observers.

We will talk about the details of this experiment in section 5, but for now we merely want to use it to frame our chapter. In order to appreciate why some people might think this result is a big deal, we first need to remind ourselves what the Kochen-Spekker (KS) and Bell theorems are supposed to show. KS is supposed to show that measurement outcomes (possessed values) in QM must be contextual. That is, KS is generally taken to show that measurement outcomes (possessed values) in QM are never independent of the measurement context–how the value is eventually measured. KS then is generally taken to be a knock on a hidden variables account of QM. In addition to that, Bell's theorem is supposed to establish that even if one is prepared to concede contextuality, any such explanation must be non-local, i.e., there can be no local hidden variable account for QM. It is well known that both proofs have loopholes and here's the point. Both proofs take as tacit truths some or all of assumptions 1–4 above. The point being that you can get around one or another of the two theorems if you are willing to give up one or more of assumptions 1–4. But obviously there is a price, for each assumption has its strong supporters.

Here is why you might not think this result is a big deal: we already knew all this. Indeed, most every existing interpretation of QM is defined by giving up one or more of these four assumptions. Defenders of the result's importance will note that the result is based on weaker constraints than those assumed by either KS or Bell. What Bong et al. allege to have shown is that given assumption 4, given the claim that QM applies to macroscopic measuring devices and conscious observers, then we must give up one or more of assumptions 1–3. Detractors will reply that since photons are neither macroscopic nor conscious, the new experiment proves absolutely nothing new until at the very least, they scale it up. Regardless of one's opinion, we can say that the new experiment does at least put our choices in stark relief. However, no one has of yet established that this list of assumptions is exhaustive, that is, there may be other choices we could make, other things we could give up. For example, notice that built into assumption 4 is the suppressed assumption that we are realist's about QM. QBists and others reject this assumption and thus reject the idea that QM can tell us what other observers will measure, or they will say QM is only about subjective belief states, not physical states or brain states.

Other people make stronger claims about what the new experiment shows. For example, Renato Renner claims the new theorem is telling us that QM needs to be replaced [13]. We will show that such claims are patently false. QM doesn't need to be replaced, it needs to be reconceived. We will analyze Bong et al. and other related new Wigner's friend type experiments with our reconception of QM in order to show that we can keep every assumption above except possibly number 2. However, we need to make clear that it is wrong to call this assumption "superdeterminism," rather it should be called "measurement independence." Bell in his proof assumes that the properties of a particle are independent of the (future) measurements to be performed on them, this is measurement independence (MI). Often dubbed "retrocausal," there are accounts of QM that reject this assumption in order to provide a local take on EPR-Bell correlations that is consistent with SR. The soul of this idea is as follows [14]:

> In putting future and past on an equal footing, this kind of approach is different in spirit from (and quite possibly formally incompatible with) a more familiar style of physics: one in which the past continually generates the future, like a computer running through the steps in an algorithm. However, our usual preference for the computer-like model may simply reflect an anthropocentric bias. It is a good model for creatures like us, who acquire knowledge sequentially, past to future, and hence find it useful to update their predictions in the same way. But there is no guarantee that the principles on which the universe is constructed are of the sort that happens to be useful to creatures in our particular situation. Physics has certainly overcome such biases before—the Earth isn't the center of the universe, our sun is just one of many, there is no preferred frame of reference. Now, perhaps there's one further anthropocentric attitude that needs to go: the idea that the universe is as "in the dark" about the future as we are ourselves.

While one can find many different instantiations of this idea in the literature, notice that nothing in this description entails any particular mechanism such as literal causal influences or signals coming from the future. Nor does this idea entail superdeterminism. Technically speaking, a superdeterministic world is one in which independence is violated via a past common cause—a common cause of one's choice of measurements and say the particle spin properties, in the case of Bell correlations. In short, superdeterminism is a conspiratorial theory with only past-to-future causation. So while superdeterminism does entail that experimenters are not free to choose what to measure without being influenced by events in the distant past, i.e., it does give up MI, it does so in a particularly spooky way, forcing us to accept some very special conditions at the big bang as a brute fact or seek some sort of physically acceptable explanation for those initial conditions that is presumably not some sort of supernatural conspiracy. In the last few years we have argued that there are really two ideas buried in retrocausal accounts. One is the idea that we should seek out some dynamical or constructive explanation that involves determination relations from the future. The other idea, our idea, is that we should jettison dynamical, causal and constructive explanation altogether when it comes to EPR-Bell correlations and instead seek an explanation in terms of adynamical global constraints [15]. This would be an adynamical and acausal explanation that seeks a constraint-based explanation for the pattern of EPR-Bell correlations in spacetime as given by QM. The idea here isn't merely the retrocausal idea that maybe the future determines the past as much as the other way around. The idea is to stop worrying about determination being time-like versus space-like, it can be both or neither, because the nature of explanation in the case of EPR-Bell correlations is not in any way causal or dynamical. Therefore, even the focus on measurement independence is somewhat misleading, because it immediately makes people think about future measurements somehow retrocausally bringing about the properties of QM systems in the past, when the point is to view the entire experiment in terms of spatiotemporal constraints that don't care about time-like versus space-like.

Before we can explain how the Wigner's friend and delayed choice quantum eraser gedanken experiments of QM bear on conscious experience per our constraint-based approach to physics, we must provide some background. Our perceptions are formed dynamically, i.e., in a causal, temporally sequential fashion, as in a game of chess. For example, move number 25 cannot be made until all moves 1–24 have been made. Therefore, move number 25 can be said to be explained by move number 24 which is explained by move number 23, etc. In contrast, the answers to a crossword puzzle explain each other, no word has any explanatory priority over the other words. A crossword puzzle is analogous to what we mean by "adynamical" or "constraint-based" explanation. The goal in a crossword puzzle is a self-consistent collection of intersecting words in accord with the clues (constraints) given. We propose interpreting modern physics in analogous fashion where the constraints in physics are spatiotemporal and can be understood as constraints on experience per neutral monism. Do such explanations also place a constraint on free will? Yes, of course, as we will show explicitly in our analysis of the delayed choice QM eraser experiment with a twist. But so what? Who thought physics was unquestionably compatible with libertarian free will anyway? Who believed that the laws of physics placed no constraints on the degrees of freedom of human action? However, let us note again, this is not superdeterminism, fatalism or any other kind of conspiracy theory. It is simply the acknowledgment that not all determination relations are dynamical or causal.

We have provided such an adynamical global constraint explanation for EPR-Bell correlations, QM superposition such as seen in the twin-slit experiment, and other QM phenomena in a variety of different publications over the years [8, 15, 9, 16, 10]. As we will illustrate in the next section, most recently we have simplified our adynamical global constraint-based explanation and think it is best appreciated as a principle explanation of the sort we find in SR.

Our principled adynamical global constraints are as follows:

- Axiom 1: Interacting "bodily objects" coextensive with space and time form the context of all self-consistent, shared perceptual information between POs (point of observational origin; need not be conscious) and these perceptions constitute different reference frames in that spacetime model of the Physical (the "real external world").
- Axiom 2: For all of physics, the "perceptions" of any particular PO do not provide a privileged perspective of the Physical. This is known as "no preferred reference frame" (NPRF).

We will discuss these axioms at greater length going forward, but Axiom 1 is basically saying that what QM is really telling us is that to exist (to be a diachronic entity in space and time) is to interact with the rest of the universe creating a consistent, shared set of classical information constituting spacetime. It is in effect a generalization of "the boundary of a boundary principle" $\partial\partial = 0$ [17, 18, 19]. Axiom 2 is just a generalization of the relativity principle.

Putting this all together we can say the following. We understand that each subject is just a conscious point of origin (PO) in spacetime. The "perceptions" of each PO form a context of interacting trans-temporal objects (TTOs) for that PO. Since TTOs are "bodily objects" with world-lines in spacetime, TTOs are coextensive with space and time. When POs exchange information about their "perceptions," they realize that some of their disparate "perceptions" fit self-consistently into a single spacetime model with different reference frames for each PO. Thus, physicists' space-time model of the "Physical" represents the self-consistent collection of shared perceptual information between POs, e.g., "perceptions" upon which Galilean or Lorentz transformations can be performed. That is, spacetime is the self-consistent collection of shared classical information regarding diachronic entities (classical objects), which interact per QM. The consistency of shared classical information of spacetime is guaranteed by the divergence-free (gauge invariant) nature of the adynamical global constraints for classical and quantum physics. We understand that the reader will no doubt be puzzled by all this now. As we unpack and apply these ideas we believe it will all come into focus. We will begin with our principle explanation of EPR-Bell correlations in the next section.

Finally however, it is important that the reader appreciate our realist, psi-epistemic account of QM in a little more detail [15]. Recall that the measurement problem and the worry about Wigner's friend are driven by wavefunction realism. We need to make it clear what our alternative looks like. If one constructs the differential equation (Schrödinger equation) corresponding to the Feynman path integral, the time-dependent foliation of spacetime gives the wavefunction $\Psi(x, t)$ in concert with our time-evolved perceptions and the fact that we do not know when the outcome is going to occur. Once one has an outcome, both the configuration x_o, that is the specific spatial locations of the experimental outcomes, and time t_o of the outcomes are fixed, so the wavefunction $\Psi(x, t)$ of configuration space becomes a probability amplitude $\Psi(x_o, t_o)$ in spacetime, i.e., a probability amplitude for a specific outcome in spacetime. Again, the evolution of the wavefunction in configuration space before it becomes a probability amplitude in spacetime is governed by the Schrödinger equation.

However, the abrupt change from wavefunction in configuration space to probability amplitude in spacetime is not governed by the Schrödinger equation. In fact, if the Schrödinger equation is universally valid, it would simply say that the process of measurement should entangle the measurement device with the particle being measured, leaving them both to evolve according to the Schrödinger equation in a more complex configuration space (as in the relative-states formalism shown below). The Many-Worlds interpretation notwithstanding, we do not seem to experience such entangled existence in configuration space, which would contain all possible experimental outcomes. Instead, we experience a single experimental outcome in spacetime.

This contradiction between theory and experience is called the "measurement problem." However, the time-evolved story in configuration space is not an issue with the path integral formalism as we interpret it, because we compute $\Psi(x_o, t_o)$ directly. That is, in asking about a specific outcome we must specify the future boundary conditions that already contain definite and unique outcomes. Thus, the measurement problem is a non-starter for us. When a QM interpretation assumes the wavefunction is an epistemological tool rather than an ontological entity, that interpretation is called "psi-epistemic." In our path integral, constraint-based account the wavefunction in configuration space is not even used, so our account is trivially psi-epistemic. But, one must also fully understand the classical-quantum contextual implications of our view as expressed in Axiom 1 above. The so-called quantum system is in fact the totality of the entire experimental set-up [20, p. 738] such that different set-ups or configurations are not probing some autonomous quantum realm, but actually constitute different "systems." QM then, just as the KS theorem and more recent related theorems suggest, is a theory that has quantum-classical contextuality at its heart, that is among its deepest lessons about the nature of physical reality. As Ball puts it, "The quantum experiment is not probing the phenomenon but is the phenomenon" [21, p. 90]. Given our Lagrangian approach, the entire experimental set-up includes future boundary conditions: the experiment from initiation to termination in spacetime, no time-evolved configuration space required. For a detailed treatment of the measurement problem, the Born rule, and environmental decoherence see [8].

Finally, we should clarify possible sources of confusion with this quantum-classical contextuality. First, we are not saying that there exist quantum entities with inaccessible properties that are traversing the space between emitters and receivers. We are denying the existence of any worldlines for quantum momentum exchanges, no matter how large the energy or momentum being exchanged. On pain of infinite regress, a quantum exchange of momentum is not a TTO. A quantum of momentum or energy does not have any "intrinsic properties" or intrinsic existence as we discussed above as regards Axiom 1. Second, there is no Bohrian quantum-classical "cut" and they do not exist independently of each other. As we are all aware, there doesn't seem to be any limit on the size of the quantum momentum exchange.

Ironically perhaps, the adynamical global constraints in question actually guarantee a sensible world of classical/dynamical objects that evolve in space and time. That is, classical/dynamical objects (things that persist), time, and space are all interdependent just as general relativity suggests. In short, we will explain why, on our view, this is and must be a world with determinate experience and universal intersubjective agreement.

3. Principle Explanation of EPR-Bell Correlations

Many physicists in quantum information theory (QIT) are calling for "clear physical principles" [7] to account for QM. As Hardy points out, "The standard axioms of [quantum theory] are rather ad hoc. Where does this structure come from?" [22] Fuchs points to the postulates of SR as an example of what QIT seeks for QM [7] and SR is a principle theory [23]. That is, the postulates of SR are constraints offered without a corresponding constructive explanation. In what follows, Einstein explains the difference between the two [24]:

> We can distinguish various kinds of theories in physics. Most of them are constructive. They attempt to build up a picture of the more complex phenomena out of the materials of a relatively simple formal scheme from which they start out. . . .
>
> Along with this most important class of theories there exists a second, which I will call "principle-theories." These employ the analytic, not the synthetic, method. The elements which form their basis and starting point are not hypothetically constructed but empirically discovered ones, general characteristics of natural processes, principles that give rise to mathematically formulated criteria which the separate processes or the theoretical representations of them have to satisfy. . . .
>
> The advantages of the constructive theory are completeness, adaptability, and clearness, those of the principle theory are logical perfection and security of the foundations. The theory of relativity belongs to the latter class.

For those who believe the fundamental explanation for QM phenomena must be constructive, at least in the sense envisioned by Einstein above, none of the mainstream interpretations neatly fit the bill. Not only do most interpretations entail some form of QM holism, contextuality, and/or non-locality, the remainder invoke priority monism and/or multiple branches or outcomes. A perceived with attempting a constructive account of QM is, as articulated by Van Camp, "Constructive interpretations are attempted, but they are not unequivocally constructive in any traditional sense" [25]. Thus, he states [25]:

> The interpretive work that must be done is less in coming up with a constructive theory and thereby explaining puzzling quantum phenomena, but more in explaining why the interpretation counts as explanatory at all given that it must sacrifice some key aspect of the traditional understanding of causal-mechanical explanation.

It seems clear all of this would be anathema to Einstein and odious with respect to constructive explanation, especially if say, statistical mechanics is the paradigm example of constructive explanation. Thus, for many it seems wise to at least attempt a principle explanation of QM, as sought by QIT. A perceived problem with QIT's attempts is noted by Van Camp [25]:

> However, nothing additional has been shown to be incorporated into an information-theoretic reformulation of QM beyond what is contained in QM itself. It is hard to see how it could offer more unification of the phenomena than QM already does since they are equivalent, and so it is not offering any explanatory value on this front.

However, one of QIT's foundational principles in their axiomatic reconstructions of QM is Information Invariance & Continuity [81] and, as we showed [82], this information-theoretic principle entails the invariant measurement of Planck's constant h per NPRF in total analogy with SR.

The term "reference frame" has many meanings in physics related to microscopic and macroscopic phenomena, Galilean versus Lorentz transformations, relatively moving observers, etc. The difference between Galilean and Lorentz transformations resides in the fact that the speed of light is finite, so NPRF entails the light postulate of SR, i.e., that everyone measure the same speed of light c, regardless of their motion relative to the source.

If there was only one reference frame for a source in which the speed of light equaled the prediction from Maxwell's equations ($c = \frac{1}{\sqrt{\mu_o \epsilon_o}}$), then that would certainly constitute a preferred reference frame. Herein, we extend NPRF to include the measurement of another fundamental constant of nature, Planck's constant h ($= 2\pi\hbar$).

As Steven Weinberg points out, measuring an electron's spin via Stern-Gerlach (SG) magnets constitutes the measurement of "a universal constant of nature, Planck's constant" [26, p. 3] (Figure 10.2). So if NPRF applies equally here, everyone must measure the same value for Planck's constant h regardless of their SG magnet orientations relative to the source, which like the light postulate is an empirical fact. By "relative to the source" of a pair of spin-entangled particles, we mean relative "to the vertical in the plane perpendicular to the line of flight of the particles" [27, p. 943] (Figure 10.3). Here the possible spin outcomes $\pm\frac{\hbar}{2}$ represent a fundamental (indivisible) unit of information per Dakic and Brukner's first axiom in their reconstruction of quantum theory, "An elementary system has the information carrying capacity of at most one bit" [28]. Thus, different SG magnet orientations relative to the source constitute different "reference frames" in QM just as different velocities relative to the source constitute different "reference frames" in SR. Per Brukner and Zeilinger [81], a reference frame in this context is established by a set of mutually complementary spin measurements. And, as shown by Höhn and Müller [84], Alice and Bob must interact to establish common coordinate directions for these complementary spin measurements. Borrowing from Einstein, NPRF might be stated [29]:

> No one's "sense experiences," to include measurement outcomes, can provide a privileged perspective on the "real external world."

This is consistent with the notion of symmetries per Hicks [30]:

> There are not two worlds in one of which I am here and in the other I am three feet to the left, with everything else similarly shifted. Instead, there is just this world and two mathematical descriptions of it. The fact that those descriptions put the origin at different places does not indicate any difference between the worlds, as the origin in our mathematical description did not correspond to anything in the world anyway. The symmetries tell us what structure the world does not have.

Why the quantum? = Why the Tsirelson bound?

	CHSH Quantity	
$-2 \leftrightarrow 2$	$-2\sqrt{2} \leftrightarrow 2\sqrt{2}$	PR correlations $\rightarrow 4$
Satisfy Bell inequality	Tsirelson bound	No-signaling max
Classical Correlations	Quantum Correlations	Superquantum Correlations
Violate Constraint	Satisfy Constraint	Violate Constraint

Figure 10.1 Answer to Bub's question, "Why the Tsirelson bound?" The "constraint" is conservation per no preferred reference frame.

That is, there is just one "real external world" harboring many, but always equal perspectives as far as the physics is concerned [9].

We have shown elsewhere that the quantum correlations and quantum states corresponding to the Bell states, which uniquely produce the Tsirelson bound for the Clauser–Horne–Shimony–Holt (CHSH) quantity, can be derived from conservation per NPRF [31]. Thus, Bell state entanglement is ultimately grounded in NPRF just as SR [10]. And again, we showed that the principle of Information Invariance & Continuity at the basis of QIT's reconstructions of QM reveals the relativity principle at the foundation of QM exactly as it exists at the foundation of SR, i.e., in demanding the invariant measurement of a fundamental constant of Nature. For SR, that constant is the speed of light c and for QM, that constant is Planck's constant h [82]. As summarized in Figure 10.1, the quantum correlations responsible for the Tsirelson bound satisfy conservation per NPRF while both classical and superquantum correlations can violate this constraint. Therefore a principle explanation of Bell state entanglement and the Tsirelson bound that be stated in "one clear, simple sentence" [7, p. 302] is "conservation per no preferred reference frame" (Figure 10.1, or in information-theoretic terms, "conservation per Information Invariance & Continuity.").

What qualifies as a principle explanation versus constructive turns out to be a fraught and nuanced question [23] and we do not want to be sidetracked on that issue as such. Let us therefore state explicitly that what makes our explanation a principle one is that it is grounded directly in phenomenology, it is an adynamical and acausal explanation that involves adynamical global constraints as opposed to dynamical laws or causal mechanisms, and it is unifying with respect to QM and SR.

Let us also note that while contrary to certain others [32, 33, 34, 35], we are arguing that conservation per NPRF need not ever be discharged by a constructive explanation or interpretation. This is at least partially distinct from the question in SR for example, of whether facts about physical geometry are grounded in facts about dynamical fields or vice versa. Furthermore, this principle explanation is consistent with any number of "constructive interpretations" of QM. For example, this principle explanation avoids the complaints about Bub's proposed principle explanation of QM leveled by Felline [36]. That is, the principle being posited herein does not require a solution to the measurement problem nor again does it necessarily beg for a constructive counterpart.

The Bell states are

$$\begin{aligned} |\psi_-\rangle &= \frac{|ud\rangle - |du\rangle}{\sqrt{2}} \\ |\psi_+\rangle &= \frac{|ud\rangle + |du\rangle}{\sqrt{2}} \\ |\phi_-\rangle &= \frac{|uu\rangle - |dd\rangle}{\sqrt{2}} \\ |\phi_+\rangle &= \frac{|uu\rangle + |dd\rangle}{\sqrt{2}} \end{aligned} \tag{1}$$

in the eigenbasis of σ_z. The first state $|\psi_-\rangle$ is called the "spin singlet state" and it represents a total conserved spin angular momentum of zero ($S = 0$) for the two particles involved. The other three states are called the (entangled) "spin triplet states" and they each represent a total conserved spin angular momentum of one ($S=1$, in units of $\hbar=1$ for spin-$\frac{1}{2}$ particles). In all four cases, the entanglement represents the conservation of spin angular momentum for the process creating the state.

If Alice is making her spin measurement σ_1 in the $\hat{a}$ direction and Bob is making his spin measurement σ_2 in the $\hat{b}$ direction, we have

$$\begin{aligned} \sigma_1 &= \hat{a} \cdot \vec{\sigma} = a_x\sigma_x + a_y\sigma_y + a_z\sigma_z \\ \sigma_2 &= \hat{b} \cdot \vec{\sigma} = b_x\sigma_x + b_y\sigma_y + b_z\sigma_z. \end{aligned} \tag{2}$$

The correlation functions are given by [10]

$$\begin{aligned} \langle\psi_-|\sigma_1\sigma_2|\psi_-\rangle &= -a_xb_x - a_yb_y - a_zb_z \\ \langle\psi_+|\sigma_1\sigma_2|\psi_+\rangle &= a_xb_x + a_yb_y - a_zb_z \\ \langle\phi_-|\sigma_1\sigma_2|\phi_-\rangle &= -a_xb_x + a_yb_y + a_zb_z \\ \langle\phi_+|\sigma_1\sigma_2|\phi_+\rangle &= a_xb_x - a_yb_y + a_zb_z \end{aligned} \tag{3}$$

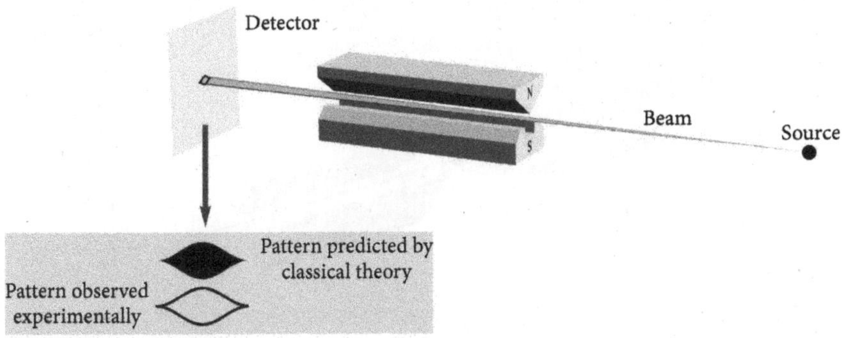

Figure 10.2 A Stern-Gerlach (SG) spin measurement showing the two possible outcomes, up $(+\frac{\hbar}{2})$ and down $(-\frac{\hbar}{2})$ or +1 and −1, for short. The important point to note here is that the classical analysis predicts all possible deflections, not just the two that are observed. This binary (quantum) outcome reflects Dakic and Brukner's first axiom in their reconstruction of quantum theory, "An elementary system has the information carrying capacity of at most one bit" [28]. As the SG magnets are rotated, these binary outcomes continue to obtain per Brukner and Zeilinger's principle of Information Invariance & Continuity, "The total information of one bit is invariant under a continuous change between different complete sets of mutually complementary measurements" [81]. The difference between the classical prediction and the quantum reality uniquely distinguishes the quantum joint distribution from the classical joint distribution for the Bell spin states [37].

The spin singlet state is invariant under all three SU(2) transformations meaning we obtain opposite outcomes ($\frac{1}{2}$ *ud* and $\frac{1}{2}$ *du*) for SG magnets at any $\hat{a} = \hat{b}$ (Figures 10.2 and 10.3) and a correlation function of $-\cos(\theta)$ in any plane of physical space, where θ is the angle between $\hat{a}$ and $\hat{b}$ (Eq. (3)). We see that the conserved spin angular momentum ($S = 0$), being directionless, is conserved in any plane of physical space. Again, $\hat{a} = \hat{b}$ means Alice and Bob are in the same reference frame.

The invariance of each of the spin triplet states under its respective SU(2) transformation in Hilbert space represents the SO(3) invariant conservation of spin angular momentum $S = 1$ for each of the planes xz ($|\phi_+\rangle$), yz ($|\phi_-\rangle$), and xy ($|\psi_+\rangle$) in physical space. Specifically, when the SG magnets are aligned (the measurements are being made in the same reference frame) anywhere in the respective plane of symmetry the outcomes are always the same ($\frac{1}{2}$ *uu* and $\frac{1}{2}$ *dd*). It is a planar conservation and our experiment would determine which plane.

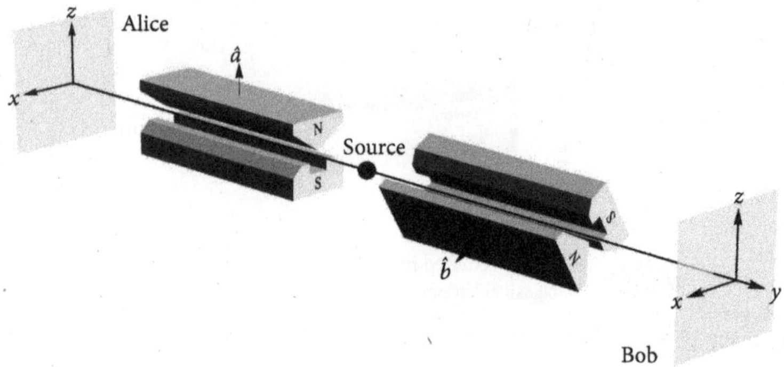

Figure 10.3 Alice and Bob making spin measurements on a pair of spin-entangled particles with their Stern-Gerlach (SG) magnets and detectors in the *xz*-plane. Here Alice and Bob's SG magnets are not aligned so these measurements represent different reference frames. Since their outcomes satisfy Information Invariance & Continuity in all reference frames and satisfy explicit conservation of spin angular momentum in the same reference frame, they can only satisfy conservation of spin angular momentum on *average* in different reference frames.

We will explain the spin singlet state correlation function, since the spin triplet state correlation function is analogous. That we have opposite outcomes when Alice and Bob are in the same reference frame is not difficult to understand via conservation of spin angular momentum, because Alice and Bob's measured values of spin angular momentum cancel directly when $\hat{a} = \hat{b}$ (Figure 10.3). But, when Bob's SG magnets are rotated by θ relative to Alice's SG magnets, we need to clarify the situation.

We have two subsets of data, Alice's set (with SG magnets at angle α) and Bob's set (with SG magnets at angle β). They were collected in N pairs (data events) with Bob's(Alice's) SG magnets at $\alpha - \beta = \theta$ relative to Alice's(Bob's). We want to compute the correlation function for these N data events which is

$$\langle \alpha, \beta \rangle = \frac{(+1)_A(-1)_B + (+1)_A(+1)_B + (-1)_A(-1)_B + \ldots}{N} \tag{4}$$

Now partition the numerator into two equal subsets per Alice's equivalence relation, i.e., Alice's $+1$ results and Alice's -1 results:

$$\langle \alpha, \beta \rangle = \frac{(+1)_A(\sum \text{BA+}) + (-1)_A(\sum \text{BA}-)}{N} \tag{5}$$

where $\sum$ BA+ is the sum of all of Bob's results (event labels) corresponding to Alice's $+1$ result (event label) and $\sum$ BA− is the sum of all of Bob's results (event labels) corresponding to Alice's -1 result (event label). Notice this is all independent of the formalism of QM. Next, rewrite Eq. (5) as

$$\langle \alpha, \beta \rangle = \frac{1}{2}(+1)_A\overline{BA+} + \frac{1}{2}(-1)_A\overline{BA-} \tag{6}$$

with the overline denoting average. Notice that to understand the quantum correlation responsible for Bell state entanglement, we need to understand the origins of $\overline{BA+}$ and $\overline{BA-}$ for the Bell states. We now show what that is for the spin singlet state [38], the spin triplet states are analogous in their respective symmetry planes [10].

In classical physics, one would say the projection of the spin angular momentum vector of Alice's particle $\vec{S}_A = +1\hat{a}$ along $\hat{b}$ is $\vec{S}_A \cdot \hat{b} = +\cos(\theta)$ where again θ is the angle between the unit vectors $\hat{a}$ and $\hat{b}$. That's because the prediction from classical physics is that all values between $+1\left(\frac{\hbar}{2}\right)$ and $-1\left(\frac{\hbar}{2}\right)$ are possible outcomes for a spin measurement (Figure 10.2). From Alice's perspective, had Bob measured at the same angle, i.e., $\beta = \alpha$, he would have found the spin angular momentum vector of his particle was $\vec{S}_B = -\vec{S}_A = -1\hat{a}$, so that $\vec{S}_A + \vec{S}_B = \vec{S}_{Total} = 0$. Since he did not measure the spin angular momentum of his particle at the same angle, he should have obtained a fraction of the length of $\vec{S}_B$, i.e., $\vec{S}_B \cdot \hat{b} = -1\hat{a} \cdot \hat{b} = -\cos(\theta)$ (Figure 10.4; this also follows from counterfactual spin measurements on the single-particle state [39]). Of course, Bob only ever obtains $+1$ or -1 per NPRF, but suppose that Bob's outcomes *average* $-\cos(\theta)$ (Figure 10.5). This means

$$\overline{BA+} = -\cos(\theta) \tag{7}$$

Likewise, for Alice's $(-1)_A$ results we have

$$\overline{BA-} = \cos(\theta) \tag{8}$$

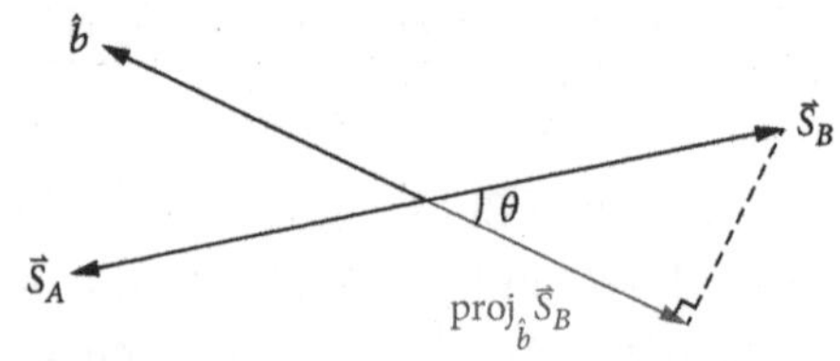

Figure 10.4 The spin angular momentum of Bob's particle $\vec{S}_B = -\vec{S}_A$ projected along his measurement direction $\hat{b}$. This does *not* happen with spin angular momentum.

Putting these into Eq. (6), we obtain

$$\langle \alpha, \beta \rangle = \frac{1}{2}(+1)_A(-\cos(\theta)) + \frac{1}{2}(-1)_A(\cos(\theta)) = -\cos(\theta) \tag{9}$$

which is precisely the correlation function given by QM for the spin singlet state. Notice that Eqs. (7) and (8) are mathematical facts for obtaining the quantum correlation function, we are simply motivating these facts via conservation of spin angular momentum in accord with the SU(2) Bell state invariances. Of course, Bob could partition the data according to his equivalence relation (per his reference frame) and claim that it is Alice who must average her results (obtained in her reference frame) to conserve angular momentum (Figure 10.5).

We posit that the reason we have average-only conservation in different reference frames is ultimately due to NPRF per Information Invariance & Continuity. To further motivate NPRF for the Bell states, consider the empirical facts. First, Bob and Alice both measure $\pm 1\left(\frac{\hbar}{2}\right)$ for all SG magnet orientations, i.e., in all reference frames. In order to satisfy conservation of spin angular momentum for any given trial when Alice and Bob are making different measurements, i.e., when they are in different reference frames, it would be necessary for Bob or Alice to measure some fraction, $\pm\cos(\theta)$. For example, if Alice measured $+1$ at $\alpha = 0$ for an $S = 1$ state (in the plane of symmetry) and Bob made his measurement (in the plane of symmetry) at $\beta = 60°$, then Bob's outcome would need to be $\frac{1}{2}$ (Figure 10.6). In that case, we would know that Alice measured the "true" spin angular momentum of her particle (and therefore the "true" value of Planck's constant) while Bob only measured a component of the "true" spin angular momentum for his particle. Thus, Alice's SG magnet orientation would definitely constitute a "preferred reference frame."

Figure 10.5 Average View for the Spin Singlet State. Reading from left to right, as Bob rotates his SG magnets relative to Alice's SG magnets for her +1 outcome, the average value of his outcome varies from −1 (totally down, arrow bottom) to 0 to +1 (totally up, arrow tip). This obtains per conservation of spin angular momentum on average in accord with no preferred reference frame. Bob can say exactly the same about Alice's outcomes as she rotates her SG magnets relative to his SG magnets for his +1 outcome. That is, their outcomes can only satisfy conservation of spin angular momentum *on average* in different reference frames, because they only measure ±1, never a fractional result. Thus, just as with the light postulate of SR, we see that no preferred reference frame leads to a counterintuitive result. Here it requires quantum outcomes $\pm 1 \left(\frac{\hbar}{2}\right)$ for all measurements and that leads to the mystery of "average-only" conservation.

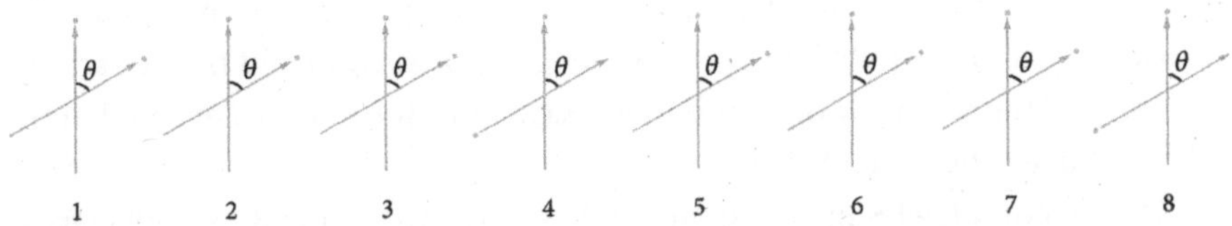

Figure 10.6 A spatiotemporal ensemble of 8 experimental trials for the spin triplet states showing Bob's outcomes corresponding to Alice's +1 outcomes when $\theta = 60°$. Spin angular momentum is not conserved in any given trial, because there are two different measurements being made, i.e., outcomes are in two different reference frames, but it is conserved on average for all 8 trials (six up outcomes and two down outcomes average to $\cos 60° = \frac{1}{2}$). It is impossible for spin angular momentum to be conserved explicitly in any given trial since the measurement outcomes are binary (quantum) with values of +1 (up) or −1 (down) per no preferred reference frame and explicit conservation of spin angular momentum in different reference frames would require a fractional outcome for Alice and/or Bob.

But, this is precisely what does *not* happen. Alice and Bob both always measure $\pm 1 \left(\frac{\hbar}{2}\right)$, no fractions, in accord with NPRF for the measurement of Planck's constant. And, this fact alone distinguishes the quantum joint distribution from the classical joint distribution [37] (Figure 10.2). Therefore, the average-only conservation responsible for the correlation function for the Bell states is actually conservation resulting from NPRF (or conservation per Information Invariance & Continuity in quantum information-theoretic terms).

If Alice is moving at velocity $\vec{V}_a$ relative to a light source, then she measures the speed of light from that source to be c (= $\frac{1}{\sqrt{\mu_o \epsilon_o}}$, as predicted by Maxwell's equations). If Bob is moving at velocity $\vec{V}_b$ relative to that same light source, then he measures the speed of light from that source to be c. Here "reference frame" refers to the relative motion of the observer and source, so all observers who share the same relative velocity with respect to the source occupy the same reference frame. NPRF in this context means all measurements produce the same outcome c.

As a consequence of this constraint we have time dilation and length contraction, which are then reconciled per NPRF via the relativity of simultaneity. That is, Alice and Bob each partition spacetime per their own equivalence relations (per their own reference frames), so that equivalence classes are their own surfaces of simultaneity. If Alice's equivalence relation over the spacetime events yields the "true" partition of spacetime, then Bob must correct his lengths and times per length contraction and time dilation. Of course, the relativity of simultaneity says that Bob's equivalence relation is as valid as Alice's per NPRF.

This is completely analogous to QM, where Alice and Bob each partition the data per their own equivalence relations (per their own reference frames), so that equivalence classes are their own +1 and −1 data events. If Alice's equivalence relation over the data events yields the "true" partition of the data, then Bob must correct (average) his results per average-only conservation. Of course, NPRF says that Bob's equivalence relation is as valid as Alice's, which we might call the "relativity of data partition" (Table 10.1).

Thus, the mysteries of SR (time dilation and length contraction) ultimately follow from the same principle as Bell state entanglement, i.e., no preferred reference frame. So, if one accepts SR's principle explanation of time dilation and length contraction, then they should have no problem accepting conservation per NPRF as a principle explanation of Bell state entanglement. Thus, the relativity principle (NPRF) is a unifying principle for non-relativistic QM and SR per QIT's principle of Information Invariance & Continuity in general, thereby answering Bub's question specifically.

Despite the fact that this principle explanation supplies a unifying framework for both non-relativistic QM and SR, some might demand

Table 10.1 Comparing SR with QM according to no preferred reference frame (NPRF).

Special Relativity	**Quantum Mechanics**
Empirical Fact: Alice and Bob both measure c, regardless of their motion relative to the source	Empirical Fact: Alice and Bob both measure $\pm 1 \left(\frac{\hbar}{2}\right)$, regardless of their SG orientation relative to the source
Alice(Bob) says of Bob(Alice): Must correct time and length measurements	Alice(Bob) says of Bob(Alice): Must average results
NPRF: Relativity of simultaneity	NPRF: Relativity of data partition

a constructive explanation with its corresponding "knowledge of how things in the world work, that is, of the mechanisms (often hidden) that produce the phenomena we want to understand" [40, p. 15]. This is "the causal/mechanical view of scientific explanation" per Salmon [40, p. 15]. Thus, as with SR, not everyone will consider our principle account to be explanatory since, "By its very nature such a theory-of-principle explanation will have nothing to say about the reality behind the phenomenon" [41, p. 331]. As stated by Brown [33, p. 76]:

> What has been shown is that rods and clocks must behave in quite particular ways in order for the two postulates to be true together. But this hardly amounts to an explanation of such behaviour. Rather things go the other way around. It is because rods and clocks behave as they do, in a way that is consistent with the relativity principle, that light is measured to have the same speed in each inertial frame.

In other words, the assumption is that the true or fundamental "explanation" of Bell state entanglement must be a constructive one in the sense of adverting to causal mechanisms like fundamental physical entities such as particles or fields and their dynamical equations of motion. Notice that while our account of SR is in terms of fundamental principle explanation, that does not necessarily make it a "geometric" interpretation of SR. For example, nothing we've said commits us to the claim that if one were to remove all the matter-energy out of the universe there would be some geometric structure remaining such as Minkowski spacetime. Furthermore, there is nothing inherently geometric about our principle explanation of Bell state entanglement in particular or of NPRF in general.

Of course we do not have a no-go argument that our principle explanation will never be subsumed by a constructive one. However, especially in light of the unifying nature of our principle explanation, we think it is worth considering the possibility that principle explanation is fundamental in these cases and perhaps others [8, 31, 10]. We think this is especially reasonable in light of the current impasse in attempts at constructive interpretations of QM. Essentially, we are in a situation with QM that Einstein found himself in with SR [42, pp. 51–52]:

> By and by I despaired of the possibility of discovering the true laws by means of constructive efforts based on known facts. The longer and the more despairingly I tried, the more I came to the conviction that only the discovery of a universal formal principle could lead us to assured results. The example I saw before me was thermodynamics.

Thus we are offering a competing account of quantum entanglement for any interpretation that fundamentally explains entanglement in the constructive sense. As Einstein said, this gives us the advantage of "logical perfection and security of the foundations" as our principle account could be true across a number of different constructive interpretations. And, the principle we offer, NPRF, is a unifying principle for non-relativistic QM and SR that holds throughout physics [9]. As Pauli once stated [43, p. 33]:

> 'Understanding' probably means nothing more than having whatever ideas and concepts are needed to recognize that a great many different phenomena are part of a coherent whole.

Per Hicks [30], NPRF is a principle that is accessible ("because it is simple") and whence we can "infer lots of truths." Inferring "lots of truths" implies a unifying principle is superior to its subsumed constituents, since it implies (at minimum) more truths than any proper subset of its subsumed constituents. The point is, we are hypothesizing that the SO(3) symmetry with average-only conservation as an explanation of Bell state entanglement, and Lorentz symmetry with relativity of simultaneity as an explanation of length contraction and time dilation, are expressions of a deeper truth, NPRF, with seemingly disparate multiple physical consequences. It has been suggested that perhaps other unresolved phenomena in physics might be explained in a similar fashion [8].

The bottom line is that a compelling constraint (who would argue with conservation per NPRF?) explains Bell state entanglement without any obvious corresponding 'dynamical/causal influence' or hidden variables to account for the results on a trial-by-trial basis. By accepting this principle explanation as fundamental, the lack of a compelling, consensus constructive explanation is not a problem [85]. This is just one of many mysteries in physics created by dynamical and causal biases that can be resolved by constraint-based thinking [8].

4. Principle Explanation of Wigner's Friend

We introduce the quantum mechanical gedanken experiment called "Wigner's friend" using Healey's version [44] of Frauchiger and Renner's version [45] of Wigner's original version [52]. The whole point of the Wigner's friend scenario is that someone (Wigner in the original story) makes a quantum measurement of someone else (Wigner's friend) who made a measurement of some quantum system. For that to be possible, Wigner's friend must be isolated (screened off) from Wigner and the rest of the universe. Being screened off means Wigner's friend cannot share any classical information with the universe. This introduces crucial (but often ignored) technical and conceptual difficulties.

Technically, one would have to keep Wigner's friend and his entire lab from interacting with the universe, e.g., no exchange of photons. That's certainly beyond anything we can do now, but more importantly, this means the classical information possessed by the friend[1] with lab while screened off is not accessible to Wigner or anyone else in the universe. Since the universe is precisely all shared, self-consistent classical information, the friend's classical information while screened off, being unshared/unaccessible to everyone else, is not even part of the universe and therefore not a part of objective reality. And that means, among other things, that there is no way to establish relative coordinate directions between Wigner's frame and his friend's frame. The alignment of Cartesian frames and synchronization of clocks between observers is already a problem that must be overcome to transfer information via quantum systems [47]. That is because we must first

[1] The classical information could just as well be contained in a computer with no human agent involved.

be able to relate Hilbert space vectors involving, say, polarizer orientations between the two isolated frames in order to use the formalism of QM. Thus, after being measured by Wigner, it is impossible for the friend's classical results to contradict the shared classical information of the universe, as required to refute objective reality.

Conceptually, if Wigner's friend is measuring $\hat{x}$ with the eigenbasis $|heads\rangle$ and $|tails\rangle$ (a "quantum coin flip"), it must be possible for Wigner to measure $\hat{w}$ with the eigenbasis $|heads\rangle - |tails\rangle$ and $|heads\rangle + |tails\rangle$, even though we cannot imagine what that means in terms of a coin flip[2]. In QM, every Hilbert space basis rotated from the eigenbasis of some measurement operator is the eigenbasis of some other measurement operator and therefore constitutes something we can measure, e.g., Stern-Gerlach magnets or polarizers rotated in space giving rise to transformed eigenbases in Hilbert space.

As we will show, the tacit introduction of these inconsistencies is what leads to the inconsistencies (inconsistent, shared classical information in the form of shared measurement outcomes) associated with the Wigner's friend experiment. And, as Baumann and Wolf (BW) show [48] and we will introduce, there are other approaches to QM which do not suffer such inconsistencies.

Thus, the bottom-line question for the Wigner's friend experiment is whether or not a person (such as Wigner's friend) making a quantum measurement can themselves be treated consistently as a quantum system by someone else (such as Wigner). As we will see, the answer is "yes," as long as the friend does not share classical information with the universe while screened off, i.e., while being treated as a quantum system. So, there is no size limit to the applicability of QM as some have asserted based on FR. The problem is, the way Wigner's friend is typically cast, the friend employs a measurement-update rule[3] while Wigner assumes the friend and his lab (to include their measurement records and memories) continue to evolve unitarily per the Schrödinger equation. BW call this "subjective collapse" and show that even in the simplest version of Wigner's friend, the sharing of inconsistent classical information can result. We will provide a similar example below. According to our constraint-based explanation, since this possibility allows for self-inconsistent shared classical information, the answer to this bottom-line question is "no" under these circumstances. The

[2] This is another crucial point that is ignored in analyses of Wigner's friend, but as we will see in section 5, Proietti et al. [4] have shed some light on it.

[3] This is sometimes associated with "wavefunction collapse."

whole point of the constraint in the universe (such as Einstein's equations) is to ensure that the shared classical information composing the universe is self-consistent. We begin with a presentation of Wigner's friend per Healey showing how subjective collapse leads to the sharing of inconsistent classical information. Then we will show how proper treatments per the "standard" and "relative-state" formalisms for QM contain no such inconsistency.

There are four agents in this story—Xena who makes a quantum measurement $\hat{x}$ on quantum state c in her lab X and then sends a quantum state s to Yvonne, Yvonne who makes a quantum measurement $\hat{y}$ on s in her lab Y, Zeus who makes a quantum measurement $\hat{z}$ or $\hat{x}$ on X pertaining to Xena's $\hat{x}$ measurement, and Wigner who makes a quantum measurement $\hat{w}$ or $\hat{y}$ on Y pertaining to Yvonne's $\hat{y}$ measurement. So, we have two "Wigners," i.e., Zeus and Wigner, and two "Wigner's friends," i.e., Xena (Zeus's friend) and Yvonne (Wigner's friend). The first assumption of FR is that it is possible for Xena and Yvonne to behave as quantum systems for Zeus and Wigner to measure, i.e., there are no size restrictions on what can behave quantum mechanically. Again, we agree that it is possible to screen off Xena and Yvonne with the caveat that, in Bub's language, those systems being treated quantum mechanically per non-Boolean algebra are not exchanging classical information with the universe per Boolean algebra. We now explore the implications of violating that caveat via subjective collapse as in FR. The starting state c for Xena is

$$\frac{1}{\sqrt{3}}|heads\rangle + \frac{\sqrt{2}}{\sqrt{3}}|tails\rangle \tag{10}$$

The eigenbasis for Xena's $\hat{x}$ measurement is simply $|heads\rangle$ and $|tails\rangle$ with eigenvalues heads and tails, respectively. If the outcome of her measurement is heads, she sends state s

$$|-\rangle \tag{11}$$

to Yvonne. If the outcome of her measurement is tails, she sends state s:

$$\frac{1}{\sqrt{2}}(|+\rangle + |-\rangle) \tag{12}$$

to Yvonne. Again, if $|+\rangle$ and $|-\rangle$ refer to orientations in space, e.g., polarizers or Stern-Gerlach magnets, their meaning between Xena and Yvonne is problematic, since their labs are isolated from one another. The sharing of such classical information is crucial for the modeling of the spacetime universe as we have defined it per the very meaning of the Hilbert space structure, so ignoring this point is to introduce an inconsistency. But, as with the original papers, we will proceed to find its implications.

The second assumption of FR is that there is only one outcome for a quantum measurement. So, for example, Xena does not measure both heads and tails and send both versions of state s. Now, assuming the subjective-collapse model, Xena and Yvonne's labs are behaving quantum mechanically (evolving unitarily) according to Zeus and Wigner, so they are entangled in the state

$$|\Psi\rangle = \frac{1}{\sqrt{3}}(|heads\rangle|-\rangle + |tails\rangle|+\rangle + |tails\rangle|-\rangle) \tag{13}$$

per Eqs. (10), (11) and (12) for Zeus and Wigner (this is Eq. (13) in Healey's paper). Assuming Xena and Yvonne's labs evolve unitarily means Zeus and Wigner can make measurements of Xena and Yvonne's labs in any rotated Hilbert space basis they choose. In addition to the measurement $\hat{x}$ with eigenbasis $|heads\rangle$ and $|tails\rangle$, Zeus has the option of measuring $\hat{z}$ with eigenbasis:

$$\begin{aligned} |OK\rangle_Z &= \frac{1}{\sqrt{2}}(|heads\rangle - |tails\rangle) \\ |fail\rangle_Z &= \frac{1}{\sqrt{2}}(|heads\rangle + |tails\rangle) \end{aligned} \tag{14}$$

And, in addition to the measurement $\hat{y}$ with eigenbasis $|+\rangle$ and $|-\rangle$, Wigner has the option of measuring $\hat{w}$ with eigenbasis:

$$\begin{aligned} |OK\rangle_W &= \frac{1}{\sqrt{2}}(|+\rangle - |-\rangle) \\ |fail\rangle_W &= \frac{1}{\sqrt{2}}(|+\rangle + |-\rangle) \end{aligned} \tag{15}$$

Again, this introduces another inconsistency with the universe of shared, self-consistent classical information, i.e., the classical information must be intelligible. First, we have already pointed out how the relative orientations of Stern-Gerlach magnets or polarizers are needed to make sense of Wigner's measurements of Yvonne's lab and that this constitutes shared classical information forbidden by the assumption that Yvonne's lab is screened off from Wigner. Second, we also have a problem of intelligibility between Zeus and Xena as mentioned above. That is, the eigenbasis $|heads\rangle - |tails\rangle$ and $|heads\rangle + |tails\rangle$ makes no sense and without that understanding, it is impossible to create a spacetime universe experimental configuration of classical information corresponding to this Hilbert space basis.[4] Now let us continue and show how these inconsistencies play out.

In any given trial of the experiment, since Xena and Yvonne have definite outcomes duly recorded and memorized, there is classical information as to which of the three possible outcomes in Eq. (13) was actually instantiated, i.e., Xena obtained heads and Yvonne obtained -1, Xena obtained tails and Yvonne obtained $+1$, or Xena obtained tails and Yvonne obtained -1. But, it is easy to show that this classical information is not consistent with the entangled state $|\Psi\rangle$. Essentially, as Bub points out [49], this is just to say that mere ignorance about classical information does not constitute a quantum system. Again, in Bub's language, classical information is Boolean while quantum information is non-Boolean.

Suppose Zeus measures $\hat{z}$ and obtains OK (eigenvalue for $|OK\rangle_Z$), which can certainly happen since Xena measured either heads or tails. In other words, since we are assuming $|\Psi\rangle$ is the quantum state being measured by Zeus and Wigner for any of the definite configurations for Xena and Yvonne, and the projection of $|\Psi\rangle$ onto $|OK\rangle_Z$ is non-zero, it must be possible for Zeus to obtain OK for a $\hat{z}$ measurement for any prior definite configuration for Xena and Yvonne. That is, each of the three individual possibilities of heads and -1 or tails and $+1$ or tails and -1 is compatible with an OK outcome for Zeus's $\hat{z}$ measurement. But when Zeus obtains OK, Wigner must obtain $+1$ for a $\hat{y}$ measurement, since $\langle \Psi \mid OK\rangle_Z = -\frac{1}{\sqrt{6}}\langle +|$, which rules out the possibility that the prior configuration of Xena and Yvonne was heads and -1 or tails and -1, respectively, and that contradicts our assumption that $|\Psi\rangle$ is the quantum state for Zeus and Wigner for any of the definite configurations for Xena and Yvonne prior to Zeus and Wigner's measurements. Obviously, our mistake is to assume classical definiteness for a quantum system. You can see that we have a QM interference effect by rewriting Eq. (13) as

$$|\Psi\rangle = \frac{1}{\sqrt{3}}\left(\sqrt{2}|fail\rangle_Z|-\rangle + |tails\rangle|+\rangle\right) \tag{16}$$

[4] See section 5 concerning progress on this question from Proietti et al. [4].

Likewise, suppose Wigner first measures $\hat{w}$ and obtains OK (eigenvalue for $|OK\rangle_Y$), which can certainly happen since Yvonne measured either +1 or −1, i.e., each of the three individual possibilities of heads and −1 or tails and +1 or tails and −1 is compatible with an OK outcome for Wigner's $\hat{w}$ measurement. But, when Wigner obtains OK, Zeus must obtain heads if he measures $\hat{x}$, since $\langle\Psi \mid OK\rangle_Y = -\frac{1}{\sqrt{6}}\langle heads|$, which rules out tails and +1 or tails and −1 as possible prior configurations for Xena and Yvonne, respectively, contrary to our assumption. Again, QM interference is at work, which you can see by rewriting Eq. (13) as

$$|\Psi\rangle = \frac{1}{\sqrt{3}}\left(|heads\rangle|-\rangle + |tails\rangle|fail\rangle_W\right) \tag{17}$$

Now let us show explicitly what kind of shared, self-inconsistent classical information can result for the inconsistent assumptions involved in this subjective-collapse experiment.

Suppose Xena measures the state

$$\frac{1}{\sqrt{2}}\left(|heads\rangle + |tails\rangle\right) \tag{18}$$

in the basis $|heads\rangle, |tails\rangle$. Further, suppose that if Xena obtains heads, she sends the state $|heads\rangle$ to Wigner who likewise does a measurement in the basis $|heads\rangle, |tails\rangle$. Conversely, if Xena measures tails, she sends the state $|tails\rangle$ to Wigner who again does a measurement in the basis $|heads\rangle, |tails\rangle$. This constitutes sending classical information of course, which is essentially what is done in FR's approach. Now, Zeus passes Xena and her lab through a 'heads-tails polarizer' ($|heads\rangle + |tails\rangle$) and then does a measurement in the the basis $|heads\rangle, |tails\rangle$. Of course, it is entirely possible that Zeus's final measurement will yield tails. Keep in mind that Zeus's outcome is classical information about Xena's entire recorded history to include Xena's memories of the entire process. Therefore, Xena's records and memories will show that she measured tails and sent $|tails\rangle$ to Wigner. Consequently, Wigner must have a classically written record of tails for his outcome. But, of course, Wigner's classical information is heads, so we do not have a self-consistent collection of classical information being shared between Xena, Zeus, and Wigner. BW call this a "scientific contradiction" and state that it "must not arise for a scientific theory." We concur of course. So, how would the standard formalism of QM deal with this?

In the standard formalism per BW, everyone agrees that all measurements produce a collapse, i.e., measurement update is objective. Therefore, Xena and Yvonne's measurement outcomes constitute classical information that cannot be treated quantum mechanically if it is going to be shared as part of the universe. [Of course, one could argue that "unshared classical information" is an oxymoron, but that is a semantic point with ontological implications that we will not engage here.] Recall, a screened-off system is not part of the universe/objective reality by definition. However, that does not mean we cannot screen off an entire lab and then treat it quantum mechanically. It is simply the case that the screened-off lab cannot interact with the universe, since that creates the possibility of inconsistent, shared classical information in the universe which constitutes a scientific contradiction. Therefore, the exchanges needed in FR to create the inconsistencies are the cause of the inconsistencies. This is where our constraint-based explanation would stand, since all classical information is to be represented in the universe in self-consistent fashion. But, there is no collapse on our view nor any question about how or when or why collapse happens. As we said at the beginning, we construct a realist psi-epistemic account of QM [15].

As we suggested in our initial discussion about the relationship between the classical and the quantum, the presumably classical experimental setup (or the many analogs of that in a natural setting) cannot be reduced away and plays an absolutely essential role in explaining so-called quantum outcomes; that is, the experimental setup must be treated as classical in order to use QM. So again, for us the fundamental explanatory role goes to what we call an adynamical global constraint applied to the spatiotemporal distribution of outcomes. Because of our hunch about an adynamical global constraint being fundamental, we based our account on the path integral formalism and seek a realist account with a single history. The point is that for us, the very idea that something could be both truly screened-off from the rest of the universe and also meaningfully treated classically, makes no sense.

Suppose there exists a "quantum entity" with some set of causal properties traversing the space between the source and detector to mediate a quantum exchange. According to environmental decoherence, this quantum entity cannot interact with its environment or it will cease to behave quantum mechanically, for example, it will act as a particle instead of a wave and cease to contribute to interference patterns. Now if this quantum entity does not interact in any way with anything in the universe (it is screened off), then it is not exchanging bosons of any sort with any object in the universe of classical physics. Thus, it does not affect the spacetime geometry along any hypothetical worldline (except at the source and detector) according to Einstein's equations (or it would be interacting gravitationally, i.e., exchanging gravitons). Accordingly, there cannot be any stress-energy tensor associated with the worldline of this quantum entity any place except at the source and detector. So, practically speaking, this posited "screened-off quantum entity" is equivalent to "direct action." Thus, according to our view, QM is simply providing a probability amplitude for the spatiotemporal distribution of outcomes in the QM experiment. That is, in our view there are no "QM systems" such as waves, particles, or fields that exist independently of spatiotemporal, classical contexts. As Feynman puts it [50, p. xv]:

> In the customary view, things are discussed as a function of time in very great detail. For example, you have the field at this moment, a different equation gives you the field at a later moment and so on; a method, which we shall call the Hamiltonian method. We have, instead [the action] a thing that describes the character of the path throughout all of space and time.... From the overall space-time point of view of the least action principle, the field disappears as nothing but bookkeeping variables insisted on by the Hamiltonian method.

Given our realist psi-epistemic account of QM with unmediated exchanges/direct action (i.e., there are no worldlines of counterfactual definiteness that connect the source and the detector), there is no "screened off quantum entity" that must decohere to behave classically. So on our view, environmental/dynamical decoherence is really just a dynamical take on what is in fact spatiotemporal contextuality. What QM is really telling us is that to exist (i.e., to be a diachronic entity in space and time), is to interact with the rest of the universe creating a consistent, shared set of classical information constituting the universe.

Consistent with this quantum-classical ontology is the distinction between quantum and classical statistics, as characterized by the Born rule. That is, one adds individual amplitudes then squares to get the probability, rather than squaring each amplitude then adding as in classical probability, e.g., statistical mechanics. The quantum exchange of energy-momentum between classical objects requires cancellation of possibilities a la the path integral whereby the spacetime path of extremal action (classical trajectory) is obtained by interference of non-extremal possibilities which contribute with equal weight [51, p. 224–225]. Classical statistics does not provide for this so-called quantum interference. In fact, on our view the reason classical statistics works for classical objects is precisely because a classical object is a set of definite (high probability) quantum exchanges in the context of all other classical objects, as we just explained. That is, classical statistics assumes a distribution of classical objects and any given classical object can be viewed as obtaining from quantum statistics in the context of all other classical objects, having removed non-extremal possibilities via quantum interference. Therefore, classical statistics follows from quantum statistics as the quantum exchange of energy-momentum must be in accord with the classical objects of the universe. Again, since our model of objective reality is based on spatiotemporal-global self-consistency, the quantum/classical is not more fundamental than the classical/quantum—they require each other. Notice again, anything, any piece of equipment, an elephant, etc., could in principle be screened-off and treated quantum mechanically. But again, the thing in question could only be so analyzed in some classical context. TTOs are defined relationally via their interactions with other TTOs, so no interactions means no TTO. This means the entire collection of TTOs cannot be decomposed quantum mechanically "at once." The so-called quantum system is in fact the totality of the entire experimental set-up [20, p. 738], such that different set-ups or configurations are not probing some autonomous quantum realm, but actually constitute different "systems." As with all the other mysteries of QM, the Born rule itself is vexing only if one assumes there is an ontology (created from quantum information) fundamental to that of the classical objects of the universe. Moving to the spacetime perspective allows one to consider an entirely new fundamental ontology, a quantum-classical ontology based on an adynamical global constraint. Indeed, per the Feynman path integral for QM, the most probable path is the classical path and per the transition amplitude for quantum field theory, the most probable field configuration is the classical field configuration [8, Chap. 5].

In the Bohmian account (a relative-state formalism) of FR, Lazarovici and Hubert write [52]:

> [T]he macroscopic quantum measurements performed by [Zeus] and [Wigner] are so invasive that they can change the actual state of the respective laboratory, including the records and memories (brain states) of the experimentalists in it.

Per Lazarovici and Hubert, memories and records change, but the history of those memories and records (along their worldlines prior to measurement) remain intact, so nothing in the past is changed. It is exactly analogous to passing vertically polarized light through a polarizer at 45° then measuring it horizontally. The light incident on the first polarizer at 45° has no horizontal component, but it does after passing through the polarizer at 45°. Consequently, it can now pass through the horizontal polarizer. Thus, for the photons that are now passing through the horizontal polarizer, the polarizer at 45° can be said to have changed them from vertically polarized to horizontally polarized. Likewise, Zeus and Wigner's $\hat{z}$ and $\hat{w}$ measurements can literally change Xena and Yvonne's records and memories of their $\hat{x}$ and $\hat{y}$ measurement outcomes.

This does not necessarily constitute scientific contradiction. If Xena and Yvonne's classical information prior to being measured by Zeus and Wigner is not shared (so that their worldlines are not part of the block universe of objective reality), and the classical information that exists at the end of the experiment that is shared by all participants is not self-contradictory, then there is no scientific contradiction. Of course, it may be impossible to tell a self-consistent dynamical story about how the initial self-consistent set of classical information evolved into the final self-consistent collection of classical information, but again that is not a problem for adynamical explanation.

Healey also formulates a relative-states approach to FR, which we might infer from his statement [53]:

> So one could argue that whatever Wigner says about his outcome (more carefully, whatever Zeus measures Wigner's outcome to be) is not a reliable guide to Wigner's actual outcome. In particular, even if Zeus takes Wigner's outcome to have been OK (because that's what he observes it to be in a hypothetical future measurement on W) Wigner's actual outcome might equally well have been FAIL. That is, Zeus and Wigner's outcomes for measurements on Xena and Yvonne's records can contradict those records.

This violates BW and FR's assumption of consistency only if you subscribe to the belief that QM probabilities apply to an objective reality. Healey and other relative-state approaches simply deny that assumption. Per Healey's pragmatic account of QM [54], the job of QM is simply to provide the probabilities/correlations for outcomes in a quantum experiment given the experimental context for each observer, i.e., QM is not providing a physical model or interpretation of what happens between experimental initiation and termination—in Bub's wording, "the non-Boolean link" between the Boolean initial conditions and the Boolean outcomes. Whereas our constraint-based account of QM provides a physical model that includes direct action, Healey's pragmatic account embraces metaphysical quietism about what happens between the initiation and termination of an experimental set-up. Hence his denial that QM probabilities describe an objective reality.

Assuming the existence of a true quantum system $|\Psi\rangle$ represented by Eq. (13) and the standard formalism, Wigner and Zeus share a common classical context for making their measurements of $|\Psi\rangle$. In the relative-state formalism, Wigner/Zeus must treat Zeus/Wigner as a third quantum system resulting in a new version of Eq. (13) if Zeus/Wigner makes his measurement first. In the standard formalism, Eq. (13) is used by both Zeus and Wigner to determine distributions in their common spacetime context for whatever measurements they decide to make, since their measurements act on different parts of $|\Psi\rangle$ (Zeus on Xena's lab and Wigner on Yvonne's lab). Thus, the order of their measurements does not affect the predicted probabilities. For example, per the standard formalism, regardless of what Zeus measures, the probability that Wigner will get an OK outcome for a $\hat{w}$ measurement if Xena got tails in her $\hat{x}$ measurement is zero. That is because the tails part of Eq. (13) is

$$\frac{1}{\sqrt{3}}|tails\rangle|\,fail\rangle_W \tag{19}$$

But, in the relative-state formalism, this same probability depends on whether or not Zeus makes his measurement first and what measurement Zeus makes, since the functional form of $|\Psi\rangle$ will be different for Wigner if Zeus makes an intervening measurement.

For example, suppose Zeus measures $\hat{x}$ first. After Zeus's measurement per the relative-state formalism, Eq. (13) reads

$$|\Psi\rangle = \frac{1}{\sqrt{3}}\left(|heads\rangle_X|heads\rangle_Z|-\rangle + |tails\rangle_X|tails\rangle_Z|\,fail\rangle_W\right) \tag{20}$$

[Notice we must now distinguish Xena from Zeus even though they are measuring the same thing. The same must be done with Yvonne and Wigner.]

In this case, as in the standard formalism, the probability that Wigner will get an OK outcome for a $\hat{w}$ measurement if Xena got tails in her $\hat{x}$ measurement is zero because the tails part of Eq. (20) is

$$\frac{1}{\sqrt{3}}|tails\rangle_X|tails\rangle_Z|fail\rangle_W \tag{21}$$

after Zeus's $\hat{x}$ measurement. But, suppose Zeus makes a $\hat{z}$ measurement instead. To use the relative-state formalism, we must first cast Eq. (13) in the OK-fail basis as [49]:

$$\begin{aligned}|\Psi\rangle &= \tfrac{1}{\sqrt{12}}|OK\rangle_X|OK\rangle_Y - \tfrac{1}{\sqrt{12}}|OK\rangle_X|fail\rangle_Y \\ &+ \tfrac{1}{\sqrt{12}}|fail\rangle_X|OK\rangle_Y + \tfrac{\sqrt{3}}{2}|fail\rangle_X|fail\rangle_Y\end{aligned} \tag{22}$$

Now, after Zeus makes his $\hat{z}$ measurement, Eq. (22) reads [49]:

$$\begin{aligned}|\Psi\rangle =& \frac{1}{\sqrt{12}}|OK\rangle_X|OK\rangle_Z|OK\rangle_Y - \frac{1}{\sqrt{12}}|OK\rangle_X|OK\rangle_Z|fail\rangle_Y + \\ & \frac{1}{\sqrt{12}}|fail\rangle_X|fail\rangle_Z|OK\rangle_Y + \frac{\sqrt{3}}{2}|fail\rangle_X|fail\rangle_Z|fail\rangle_Y\end{aligned} \tag{23}$$

Thus, the tails part of Eq. (23) is [49]:

$$\begin{aligned}|tails\rangle_X\Big[&\tfrac{\sqrt{5}}{\sqrt{12}}\left(\tfrac{3}{\sqrt{10}}|fail\rangle_Z + \tfrac{1}{\sqrt{10}}|OK\rangle_Z\right)|fail\rangle_Y \\ &+ \tfrac{1}{\sqrt{12}}\left(\tfrac{1}{\sqrt{2}}|fail\rangle_Z - \tfrac{1}{\sqrt{2}}|OK\rangle_Z\right)|OK\rangle_Y\Big]\end{aligned} \tag{24}$$

And since

$$\left(\frac{1}{\sqrt{2}}|fail\rangle_Z - \frac{1}{\sqrt{2}}|OK\rangle_Z\right)|OK\rangle_Y = |tails\rangle_Z|OK\rangle_Y \tag{25}$$

we now have a non-zero probability for Wigner obtaining an OK outcome for a $\hat{w}$ measurement when Xena obtains a tails outcome for her $\hat{x}$ measurement (it is $\frac{1}{6}$ actually [49]). If Zeus does not make a measurement, Eq. (22) becomes

$$|\Psi\rangle = \frac{1}{\sqrt{12}}|OK\rangle_X|OK\rangle_Y|OK\rangle_W - \frac{1}{\sqrt{12}}|OK\rangle_X|fail\rangle_Y|fail\rangle_W + \frac{1}{\sqrt{12}}|fail\rangle_X|OK\rangle_Y|OK\rangle_W + \frac{\sqrt{3}}{2}|fail\rangle_X|fail\rangle_Y|fail\rangle_W \tag{26}$$

after Wigner's $\hat{w}$ measurement. The $|OK\rangle_W$ part of this is

$$\frac{1}{\sqrt{12}}\left(|OK\rangle_X + |fail\rangle_X\right)|OK\rangle_Y|OK\rangle_W = \frac{1}{\sqrt{6}}|heads\rangle_X|OK\rangle_Y|OK\rangle_W \tag{27}$$

which has no tails piece for Xena, so the probability of Wigner obtaining an OK outcome for a $\hat{w}$ measurement when Zeus has not made a measurement and when Xena obtains a tails outcome for her $\hat{x}$ measurement is again zero.

You can see why Zeus's intervening $\hat{z}$ measurement per the relative-state formalism can cause possible contradictions between records (as in Healey's version of the relative-state formalism) or changes to records and memories (as in Lazarovici and Hubert's version of the relative-state formalism). In the relative-state formalism, Zeus's measurement outcome must match Xena's hypothetical or recorded measurement outcome in the basis used by Zeus, e.g., $|OK\rangle_X|OK\rangle_Z$. Everything is fine as long as Zeus and Xena make the same measurement, but if Zeus measures in a rotated Hilbert space basis relative to Xena, his possible measurement outcomes will contain cross terms in Xena's possible measurement outcomes, e.g., $|heads\rangle_X|tails\rangle_Z$ and $|tails\rangle_X|heads\rangle_Z$, which implies a contradiction between what Zeus measures for Xena's measurement outcomes and what Xena actually measured and recorded. So that contradiction stands (as in Healey) or Xena's outcomes change (as in Lazarovici and Huber). Neither of these radical responses is required in our single, self-consistent, spatiotemporal model of objective reality.

5. Experimental Evidence for Wigner's Friend?

In "Experimental test of local observer-independence," Proietti et al. [4] claim to have an experimental result which "lends considerable strength to interpretations of quantum theory already set in an observer-dependent framework and demands for revision of those which are not." As their depiction of their experiment (Figure 10.7) clearly shows, all the experimental

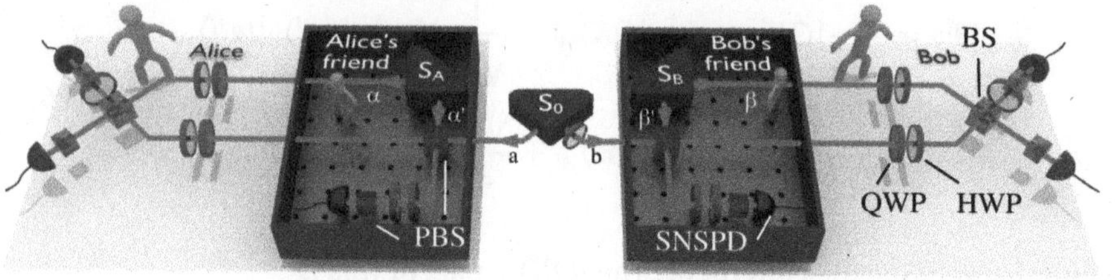

Figure 10.7 Figure from Proietti et al. [4] showing and explaining their experimental set-up.

measurements and outcomes for their experiment occur in a single objective reality, i.e., in the self-consistent, shared classical information of the spacetime model of objective reality. And, as they show in their paper, all these outcomes are in accord with QM. Thus, as Carroll notes [55]:

> What they have not done is to call into question the existence of an objective reality. Such a reality may or may not exist (I think it does), but experiments that return results compatible with the standard predictions of quantum mechanics cannot possibly overturn it.

In Carroll's take on the experiment per Many-Worlds, Proietti et al. did not cause a branching since [55]:

> Rather than having an actual human friend who observes the photon polarization—which would inevitably lead to decoherence and branching, because humans are gigantic macroscopic objects who can't help but interact with the environment around them—the "observer" in this case is just a single photon. For an Everettian, this means that there is still just one branch of the wave function all along. The idea that "the observer sees a definite outcome" is replaced by "one photon becomes entangled with another photon," which is a perfectly reversible process. Reality, which to an Everettian is isomorphic to a wave function, remains perfectly intact.

Maudlin agrees, saying [55]:

> The experiments in question are done on a system composed of only six photons. Obviously the photons do not experience anything at all, much less conflicting realities.

And Crowther says [55]:

> But, on the other hand, the fact remains that these devices are not conscious, and so Wigner could stand resolute in his interpretation. If anything, he could point out that—in the same way that an observation of a non-black, non-raven provides a negligible sliver of confirmation for the claim that 'all ravens are black'—the success of the experiment even provides inductive support in favour of his interpretation: the 'observers' in this experiment are able to record conflicting facts only because they do not experience these facts.

In other words, Proietti et al. did not in any way screen off a macroscopic measurement device and outcome recording—all measurement devices are visible in Figure 10.7 and all the experimental outcomes at all times are accessible to all observers in spacetime. Thus, the experiment constitutes a self-consistent (per QM) collection of observations in the spacetime model of objective reality. Indeed, again, one cannot even employ the formalism of QM without coordination of Cartesian frames and synchronization of clocks throughout the spacetime region of the experiment. As Lazarovici notes [55]:

> A group of physicists claims to have found experimental evidence that there are no objective facts observed in quantum experiments. For some reason, they have still chosen to share the observations from their quantum experiment with the outside world.

Maudlin agrees on this point as well, saying [55]:

> If there is no objective physical world then there is no subject matter for physics, and no resources to account for the outcomes of experiments.

Thus, the Proietti et al. experiment neither establishes the claim in the title of their paper nor provides a true instantiation of the Wigner's friend experiment. However, we believe their experiment does contain an interesting hint of what we pointed out earlier is otherwise ignored in Wigner's friend scenarios.

That is, their experiment does hint at what it might mean for Wigner to measure his friend's lab and measurement results in a rotated Hilbert space basis. Wigner's non-rotated Hilbert space basis (the direct measurement of

Wigner's friend's result) is achieved in Proietti et al. by removing the beam splitter(s) (BS) in Figure 10.7 thereby measuring A_0 and/or B_0. That means Wigner (here represented by Alice and Bob) is measuring directly both the friend's "measurement system" (lower exiting red beam on either side) and the friend's "outcome recording" (upper exiting red beam on either side). Inserting the beam splitter(s) then mixes this information, as represented by a rotated Hilbert space basis.

Of course, in such a case Wigner's friend's result would not be in conflict with Wigner's measurement for two reasons. First, Wigner's result is a quantum conflated measurement of his friend's outcome and measurement, so there could be no contradiction in such a result even if there was some way to make sense of the friend's measurement and outcome when screened off. That is the case in Proietti et al., since the entire experimental set-up exists in the spacetime model of shared classical information. Second, as we stated earlier, when Wigner's friend and his lab are screened off from the rest of the universe (contrary to Proietti et al.) there is no common classical context in which to interpret the friend's measurement and outcome. So, it would be impossible for any contradiction to be observed, i.e., there is no violation of the consistency of shared classical information constituting spacetime.

Simply put, the results that violate Bell's inequality in the Proietti et al. experiment imply, at worst, that there is no counterfactual definiteness for screened-off quantum systems contributing to the spacetime of shared classical information, just as in any other violation of Bell's inequality. The experiment Proietti et al. should have claimed to instantiate is the quantum liar experiment of Elitzur and Dolev [56, 57].

In the quantum liar experiment, an experimental configuration leads to the creation of a quantum state which then violates the Bell inequality. But, the violation of the Bell inequality by this state denies the very counterfactual definiteness responsible for creating the state to begin with. Again, this does not violate the consistency of shared classical information constituting the spacetime model of objective reality [57]. In Proietti et al. the Bell-inequality-violating states represent quantum information about a measurement and its outcome. As with any other combined quantum information, quantum interference can then erase various individual contributions. This interference does not change the friend's measurement and outcome, it just changes the information concerning the friend's measurement and outcome, and it does so without jeopardizing the self-consistency of shared classical information constituting the spacetime model of objective reality. This is precisely what happened in Proietti et al.

In "A strong no-go theorem on the Wigner's friend paradox," Bong et al. also claim to have a "proof-of-principle" experiment for such "extended Wigner's friend scenarios" (EWFS) [5]. Specifically, as we said at the beginning, they derive an inequality that when violated by any physical theory entails the violation of at least one of the following assumptions in the context of EWFS (worded colloquially here by Cavalcanti) [58]:

1. When someone observes an event happening, it really happened.
2. It is possible to make free choices, or at least, statistically random choices.
3. A choice made in one place can't instantly affect a distant event.

They refer to these assumptions collectively as Local Friendliness (LF). Their LF theorem is

> If a superobserver [Wigner or Zeus above] can perform arbitrary quantum operations on an observer and its environment [Xena or Yvonne above], then no physical theory can satisfy Local Friendliness.

Assumption 1 is the "Absoluteness of Observed Events (AOE)): An observed event is a real single event, and not relative to anything or anyone." As they point out, this is a tacit assumption made in deriving Bell inequalities, e.g., the CHSH inequality. And as we showed above, this assumption is necessary for using the formalism of QM, i.e., if one violates this assumption when using the formalism of QM, contradictions and absurdities can arise. So, while it may certainly be true that violating Assumption 1 leads to the violation of their LF inequality, one cannot use QM to check their LF inequality while violating Assumption 1. That means a violation of their LF inequality by QM entails the violation of either or both of "Assumption 2 (No-Superdeterminism (NSD)): Any set of events on a space-like hypersurface is uncorrelated with any set of freely chosen actions subsequent to that space-like hypersurface" or "Assumption 3 (Locality (L)): The probability of an observable event *e* is unchanged by conditioning on a space-like-separated free choice *z*, even if it is already conditioned on other events not in the future light-cone of *z*." Of course, NSD and L are just the assumptions for the Bell inequality, so other than an new inequality nothing new has been introduced in the Bong et al. paper regarding QM.

As with Proietti et al., the spacetime region for the Bong et al. experiment resides entirely in the spacetime of self-consistent, shared classical information, so we can use QM to check their experiment for the violation of their LF inequality. Again, if the LF inequality is violated by QM (as it is for this experiment), we only know that we must abandon NSD or L, since AOE is required to map the QM formalism to experimental arrangements. In this experiment, the "superobservers" are Alice and Bob and their friends are Charlie and Debbie, respectively. No measurement outcomes for Charlie and Debbie are involved in the LF inequality, so we may dismiss them immediately; again, this is certainly not an EWFS (which they readily admit in the paper). Their source produces a mixture that they can tune via μ:

$$\rho_\mu = \mu|\Phi^-\rangle\langle\Phi^-| + \frac{1-\mu}{2}(|HV\rangle\langle HV| + |VH\rangle\langle VH|) \qquad (28)$$

where $|\Phi^-\rangle$ is the spin singlet state. They find the maximum violations of the Bell and LF inequalities occur when the state is tuned entirely to the spin singlet state ($\mu = 1$), so we'll briefly review that case.

Alice and Bob choose between three spin measurements labeled $\{A_1, A_2, A_3\}$ and $\{B_1, B_2, B_3\}$, respectively, in each trial of the experiment as shown schematically in Figures 10.8 and 10.9. In the context of these measurements, the LF inequality is

$$\begin{aligned} -\langle A_1\rangle - \langle A_2\rangle - \langle B_1\rangle - \langle B_2\rangle - \langle A_1B_1\rangle - 2\langle A_1B_2\rangle - 2\langle A_2B_1\rangle \\ +2\langle A_2B_2\rangle - \langle A_2B_3\rangle - \langle A_3B_2\rangle - \langle A_3B_3\rangle - 6 \leq 0 \end{aligned} \qquad (29)$$

By comparison, the CHSH (Bell) inequality for these measurements is

$$\langle A_2B_2\rangle - \langle A_2B_3\rangle - \langle A_3B_2\rangle - \langle A_3B_3\rangle - 2 \leq 0 \qquad (30)$$

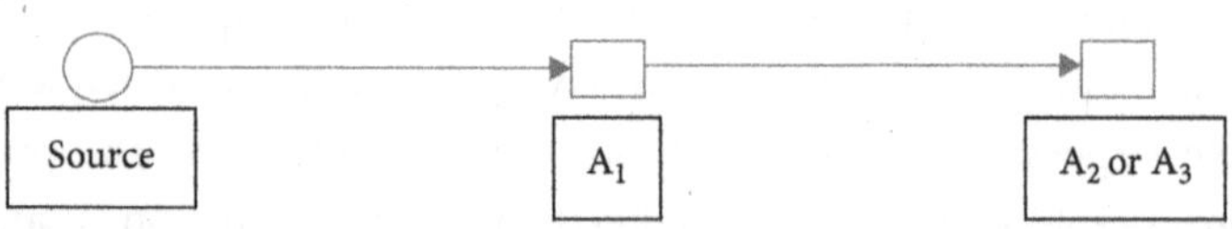

Figure 10.8 Alice's side of Bong et al. experiment. Bob's side is the same with B_j replacing the A_i.

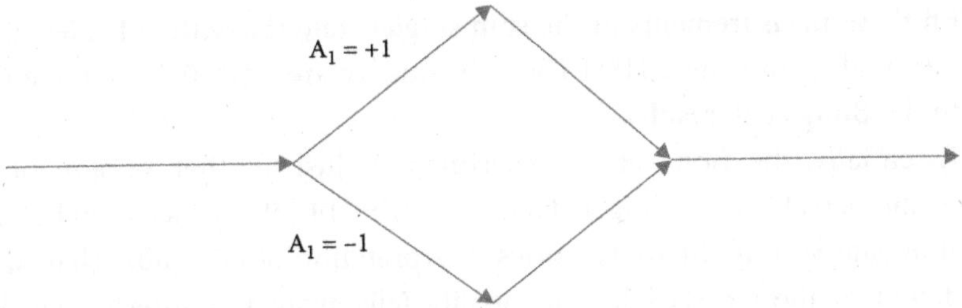

Figure 10.9 If the measurement A_1 is done, the particle is intercepted on the upper path ($A_1 = +1$) or the lower path ($A_1 = -1$). Otherwise, the particle is allowed to continue to either an A_2 or A_3 measurement. In order to promote this to a "proof-of-principle" EWFS experiment, the path inside here is taken to represent Charlie (Debbie on Bob's side) having made his(her) measurement and obtained his(her) result.

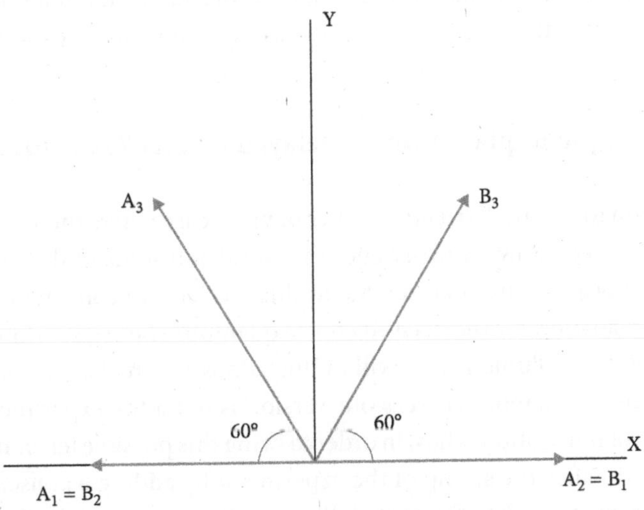

Figure 10.10 Spin measurements A_i and B_j in the xy plane.

These inequalities are saturated for the counterfactually definite set $\{A_1, A_2, A_3, B_1, B_2, B_3\} = \{1, -1, 1, -1, -1, -1\}$, for example. In order to evaluate them for the spin singlet state, we choose the measurements shown in Figure 10.10. These choices are close enough to those used by Bong et al. so as to reproduce their results within experimental error. For the spin singlet state we have $\langle A_i \rangle = \langle B_j \rangle = 0$ and $\langle A_i B_j \rangle = -\cos\theta$ where θ is the angle between the measurements A_i and B_j as shown in Figure 10.10 [10].

With these measurements of the spin singlet state the lefthand sides of the LF inequality and the CHSH (Bell) inequality are both 0.5, in agreement with the Bong et al. results.

Essentially, the Bong et al. experiment is just another version of the Greenberger-Horne-Zeilinger (GHZ) experiment [59], which Dowker uses to motivate Sorkin's Many Histories interpretation of QM [60]. That is, the violation of the CHSH (Bell) inequality tells against counterfactual definiteness (no "instruction sets" per Mermin [61]) which entails no definite path for the photon through the experimental arrangement unless explicitly measured, e.g., A_1 (B_1). But, if the photon's path through region A_1 (B_1) is not definite under measurements A_2 or A_3 (B_2 or B_3), then what does that imply about Charlie or Debbie's measurement outcomes? You can see how this would bear on the mystery of the EWFS, if in fact they had actually screened off a classical measurement and recording device, but they did not, so they're left with another (clever) quantum entanglement experiment.

6. Principle Explanation of Delayed Choice Quantum Eraser

In addition to Wigner's friend, another obvious case where there is a possible tension between how we experience the world and some QM experiment is delayed choice quantum eraser. So, in this section, we consider constraint-based explanation for the delayed choice quantum eraser experiment. In his chapter in this volume Hardy will outline a possible real experiment of the sort we discuss herein, whereas our version is but a toy experiment for the purposes of illustration. Thus, in order to bring this possible tension out most fully we will alter the set-up of the experiment by adding a conscious agent who attempts to violate the probabilities of QM, as one might think a truly free conscious agent ought to be able to do. Let us start with a description of the experiment.

Using pictures from Hillmer and Kwiat [62] we start with a particle interference pattern (Figure 10.11) then we scatter photons off the particles after they have passed through the slits(s) (Figure 10.12) and finally we erase the which-way information obtained by the scattered photons by inserting a lens (Figure 10.13).

In the Hillmer and Kwiat article the lens (eraser) is inserted after the particles have passed through the slits, but experiments have been done where the 'lens is inserted' after the particles have hit the detector. This is

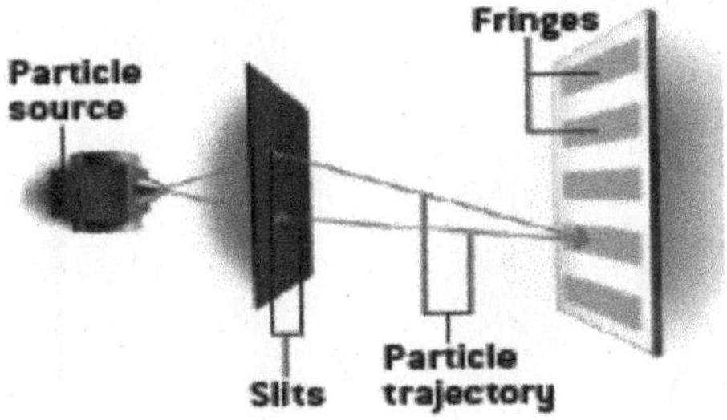

Figure 10.11 Particles create an interference pattern when proceeding through the double slits (figure from Hillmer and Kwiat [62]).

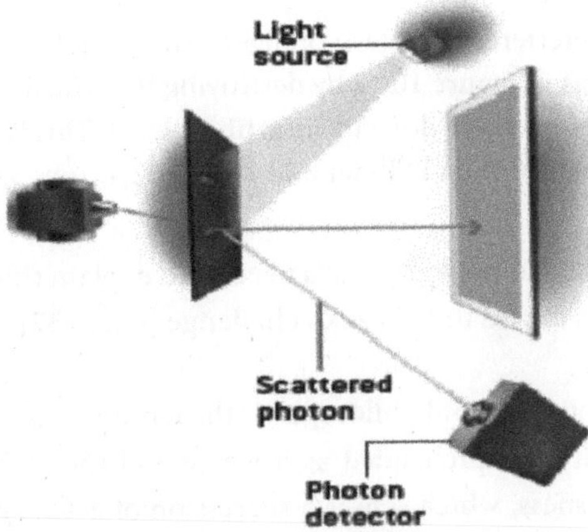

Figure 10.12 The interference pattern of Figure 10.11 can be destroyed by scattering photons and using those scattered photons to determine which slit the particle went through on each trial (figure from Hillmer and Kwiat [62]).

called a "delayed choice quantum eraser experiment" [63]. The question from our dynamical perspective is, How do the particles 'know' whether or not the lens will be inserted? And, if they do not 'know' whether the lens will be inserted or not, how do they 'know' whether or not to create the interference pattern? These questions assume temporally sequential, causal explanation, i.e., we are playing chess.

If we rather seek an adynamical, spatiotemporal constraint-based explanation in crossword puzzle fashion, we are content with the fact per QM that the distribution of particles on the screen is consistent with the presence or absence of the lens in spacetime. The insertion of the lens does not 'cause' the interference pattern any more than the interference pattern 'causes' the

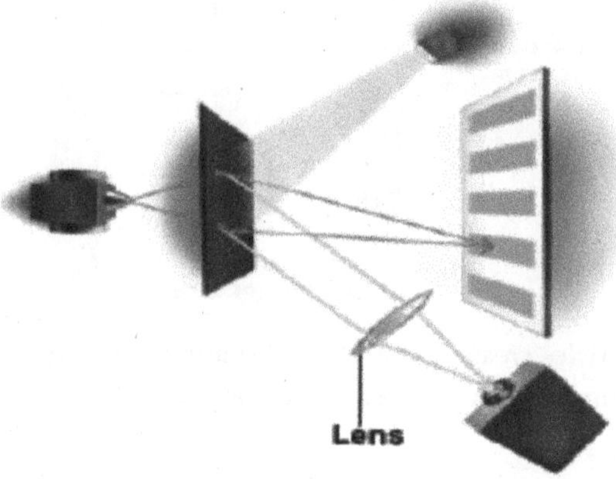

Figure 10.13 The interference pattern of Figure 10.11 can be restored after scattering photons as in Figure 10.12 by destroying the which-way information in the scattered photons (here done by inserting a lens). This is known as "quantum eraser" (figure from Hillmer and Kwiat [62]).

insertion of the lens. No new physics is needed to explain this phenomenon, just the willingness to rise to Wilczek's challenge [64, p. 37]:

> A recurring theme in natural philosophy is the tension between the God's-eye view of reality comprehended as a whole and the ant's-eye view of human consciousness, which senses a succession of events in time. Since the days of Isaac Newton, the ant's-eye view has dominated fundamental physics. We divide our description of the world into dynamical laws that, paradoxically, exist outside of time according to some, and initial conditions on which those laws act. The dynamical laws do not determine which initial conditions describe reality. That division has been enormously useful and successful pragmatically, but it leaves us far short of a full scientific account of the world as we know it. The account it gives—things are what they are because they were what they were—raises the question, Why were things that way and not any other? The God's-eye view seems, in the light of relativity theory, to be far more natural.... *To me, ascending from the ant's-eye view to the God's-eye view of physical reality is the most profound challenge for fundamental physics in the next 100 years* [italics ours].

Let us now bring the conscious agent into the picture by imagining it is a conscious agent inserting the lens (or not) in the experimental set-up. The question from our dynamical perspective is, What will I experience if I am the agent deciding whether or not to insert the lens? If the predictions of QM are to hold, then my decision must always be in accord with the particle's behavior at the detection screen and that event occurred before I made my decision. Assuming QM holds, will I feel mentally 'coerced' into making the appropriate choice? Will I feel some 'physical force' moving my hand against my will? Most people do not like the idea that our "freely made" decisions can be the result of a single particle striking a distant detector. It would seem that QM does not care about choice at all, delayed or otherwise.

While most people predict that a conscious agent will not violate the probabilities of QM anymore than a classical measuring device, Hardy has proposed an experiment to test this fact. Concerning such an experimental test, he states [65]:

> [If] you only saw a violation of quantum theory when you had systems that might be regarded as conscious, humans or other animals, that would certainly be exciting. I can't imagine a more striking experimental result in physics than that. We'd want to debate as to what that meant. It wouldn't settle the question, but it would certainly have a strong bearing on the issue of free will.

What explains the agreement between the agent's decision and the particle's pattern if it is not "spooky action at a distance" or "backwards causation?" Why does the conscious agent always (statistically at least) make the "right" choice in accord with QM? One doubts there is some special new physical or mental force acting on the hand or mind of the conscious observer. For us the answer is simple—we instead ignore our anthropocentric bias and allow for the possibility that objective reality is fundamentally the spacetime of shared, self-consistent classical information whose various patterns/distributions are determined fundamentally by adynamical global constraints, not by dynamical laws/processes acting on matter/mind to make it move/decide. We can then accept that there are some constraint-based explanations that do not allow for dynamical counterparts, at least dynamical counterparts without serious baggage, such as those discussed earlier. The constraint-based explanation here is the distribution of quantum energy-momentum exchanges in the spacetime context for the experimental set-up and procedure according to the adynamical global constraint of QM, as in section 3.

The point is, adynamical global constraints in spacetime also constrain the choices of conscious agents. Thus, physics is already part of psychology in that it places real constraints on what can be experienced to include memories (classical records) and choices. Conscious agents attempting to override QM do not experience any weird forces acting on them because there are no such forces. It is simply the case that their choices will be made in accord with the relevant adynamical global constraints per spacetime. Such agents feel like they have libertarian free will (that the future is open) because they experience reality from the "ant's-eye" view.

7. Re-Thinking the World with Neutral Monism: Removing the Boundaries Between Mind, Matter, and Spacetime

Let us now conclude by putting it all together, that is, our axiomatic principle constraints, our constraint-based explanation of the experiments herein, and neutral monism. How did we get to the point where every decade or two we feel compelled to try and relate the hard problem of consciousness, the measurement problem and mystery of free will? We get there by making certain assumptions. We believe it is high time we jettisoned these assumptions and start again with our best science as our guide. In particular we think the offending assumptions are: 1) physicalism, 2) fundamentalism, and relatedly 3) dualism about conscious experience, 4) the notion that fundamental explanation is always constructive, causal or dynamical, and relatedly, 5) realism about the wavefunction. Together these assumptions force us into the hard problem, they force us into the measurement problem, and they force us to seek the solutions in or add the solutions to fundamental physics, e.g., "panpsychist fusion" and all the rest [9]. Herein we have shown you an account of QM, relativity and their relationship to conscious observers that rejects all these assumptions. To appreciate fully how it all hangs together one must really appreciate neutral monism, so we will begin there.

In its most general form neutral monism is the idea that mental and material features are real but in some specified sense, reducible to or constructable from a neutral basis in a non-eliminative sense of reduction. The neutral basis is not a substance. Mental and material features are not separable or merely correlated, they are non-dual; indeed, they are not essentially different and distinct aspects. Thus, experience isn't inherently or essentially 'inner' or mental and the 'external' world isn't inherently non-mental. The particular brand of neutral monism we want to defend herein is most closely associated with William James and to a lesser degree Bertrand Russell. To hopefully help the reader to appreciate the idea here are some passages from James that we believe captures its character:

> Subjectivity and objectivity are affairs not of what an experience is aboriginally made of, but of its classification. [66, p. 1208]
> "Subjects" knowing, "things" known, are "roles" played. Not "ontological" facts. [67, p. 110]
> The neutral "in itself, is no more inner than outer. . . . It becomes inner by belonging to an inner, it becomes outer by belonging to an outer, world." [68, p. 217]
> A given undivided portion of experience, taken in one context of associates, play[s] the part of the knower, or a state of mind, or 'consciousness'; while in a different context the same undivided bit of experience plays the part of a thing known, of an objective 'content.' In a word, in one group it figures as a thought, in another group as a thing. [69, p. 533]

This idea can also be found in Hinduism and Buddhism long before it appears in the West. Take the following from Evan Thompson for example [70, p. 61]:

> Take a moment of visual awareness such as seeing the blue sky on a crisp fall day. The ego consciousness makes the visual awareness feel as if it's 'my' awareness and makes the blue sky seem the separate and independent object of 'my' awareness. In this way, the ego consciousness projects a subject–object structure onto awareness. According to the Yogacara philosophers, however, the blue sky isn't really a separate and independent object that's cognized by a separate and independent subject. Rather, there's one 'impression' or 'manifestation' that has two sides or aspects—the outer-seeming aspect of the blue sky and the inner-seeming aspect of the visual awareness. What the ego consciousness does is to reify these two interdependent aspects into a separate subject and a separate object, but this is a cognitive distortion that falsifies the authentic character of the impression or manifestation as a phenomenal event.

Let us now relate this all back to physics. In neutral monism, what we call spacetime is nothing but the events therein and those events are neither inherently mental nor inherently physical. Russell calls such "neutral" events "unstructured occurrences," such as "hearing a tyre burst, or smelling a rotten egg, or feeling the coldness of a frog" [71, p. 287]. How do physical phenomena relate to these neutral events? As Russell puts it, "Matter and motion . . . are logical constructions using events as their material, and events are therefore something quite different from matter in motion" [71, p. 292]. How we ultimately taxonomize those events, is as he says "a mere linguistic convenience to regard a group of events as states of a 'thing', or 'substance', or 'piece of matter', or a 'precept' " [72, p. 284]. Going further he says, "electrons and protons . . . are not the stuff of the physical world" [71, p. 386]. Again, "bits of matter are not among the bricks out of which the world is built. The bricks are events, bits of matter are portions of the structure to which we find it convenient to give separate attention" [73, p. 329].

Therefore given neutral monism, the world is not made of or realized by essentially physical entities such as the QM wavefunction. Rather, what we call physical entities are contextually given manifestations of the neutral base. We believe this idea comports well with QM contextuality and relativity. Once one accepts such neutral monism and contextuality, it ought to lead one to question other things as follows. First, the notion of beables as hidden, distinct entities with metaphysical autonomy that are responsible for all observables is questionable. Per neutral monism, sometimes reality (i.e., spacetime) manifests as particle-like, field-like or wave-like, etc., depending on multiscale context, e.g., the twin-slit experiment. There are no context-independent beables, multiscale contextuality itself is fundamental. Second, constructive, constitutive, dynamical, and causal mechanical explanations are not always fundamental. Sometimes principle explanations a la spatiotemporal adynamical global constraints are fundamental, e.g., the light postulate, conservation laws, least action principles, etc. This isn't surprising since the contextuality in question is spatiotemporal.

What is the neutral base you ask? The neutral base in question is what James calls "unqualified actuality" and "the instant field of the present." We call it Neutral Pure Presence (i.e., "pure being") or "Nowness." Philosophers and physicists such as Einstein have long noted that there is something special about the experience of Nowness which is "outside the realm of science," as he put it. Neutral monism holds that Presence is fundamental and universal. To paraphrase Hawking, it's "what puts the fire in the equations."

This is not panpsychism. Panpsychism by definition is the view that whatever fundamental physical entities are, their intrinsic nature and essence is proto-qualia or proto-subjectivity. Given neutral monism, "physical entities" are manifestations of Presence and Presence is neutral, not mental. Unlike panpsychism which merely moves the mysterious dualism from brains to fundamental physical entities, neutral monism is a complete rejection of the primary/secondary property distinction. Panpsychism likes to fancy itself as a kind of dual-aspect monism, but that's just another name for property dualism. Panpsychism claims as an advantage over strong emergence that the origins of consciousness or proto-consciousness is in fundamental physics. The panpsychist thinks it should be comforting to us naturalists to associate consciousness with fundamental physics. We are deeply puzzled by this intuition. For us, panpsychism doesn't make proto-qualia/proto-subjectivity any less weird, on the contrary. At least associating conscious minds with brains makes some intuitive and empirical sense; after all, there are many important dynamical and causal relationships between brain states and conscious states. Rather, panpsychism only makes matter weirder and seemingly less natural. It's like learning that there are fairies in the world, but then being told to relax because we have decided they are just brute features of fundamental physics. How does this help us feel better about either physics or fairies? The truth is, if what one appreciated about physicalism, materialism or ontological reductionism was the beauty and simplicity of explanatory and ontological unity, strong emergence and panpsychism as forms of dualism are both gross disruptions to that picture of reality, just at different scales. Frankly, either view disconfirms the idea that matter traditionally understood and physics alone is fundamental. Substitute immortal souls for fairies and the point is clear, the mere act of putting conscious experience or subjectivity into fundamental physics doesn't magically turn "qualia" or subjectivity into a physical property like momentum. Remember, the whole idea behind physicalism and materialism is to reduce or identify mental properties with biological or physical ones, not give them equal billing. For a detailed critique of panpsychism and dual-aspect theories see [74, 9].

What of consciousness and the hard problem? Given neutral monism, qualia is no longer the right way to conceive of conscious experience. While there are no doubt many important neural and information-theoretic correlates of conscious experience, with neutral monism, we do not tell a story about how fundamental quantum or neural processes dynamically or causally give rise to full-blooded conscious experience (i.e., panpsychism or strong emergence). We tell a story that starts with the nonduality of the so-called mental and physical (neutral monism), and then explain why we incorrectly perceive them as essentially distinct. How does that story go? Think of Kant's unity of apperception: a minimal subject in a world in space and time go hand-in-hand, two sides of the same coin. As James puts it, "[N]ot subject, not object, but object-plus-subject is the minimum that can actually be. The subject-object distinction meanwhile is entirely different from that between mind and matter, from that between body and soul. Souls were detachable, had separate destinies; things could happen to them" [69, p. 535].

However, unlike Kant's a priori transcendental condition (i.e., categories) for this "object-plus-subject" experience, neutral monism is proposing an a posteriori transcendental condition, not some cognitive or neural lens through which the noumenal is filtered. Neutral monism is a kind of direct realism. As James says, "As 'subjective' we say that the experience represents; as 'objective' it is represented" [75, p. 480]. Indeed, fundamentally speaking, there is only "the instant field of the present.... It is only virtually or potentially either object or subject as yet. For the time being, it is plain, unqualified actuality or existence, a simple that" [75, p. 482]. What then is the self on this view? The self is simply subjectivity, a conscious PO to use the language from the beginning of the this chapter. As James puts it, "The individualized self, which I believe to be the only thing properly called self, is a part of the content of the world experienced. The world experienced comes at all times with our body as its centre, centre of vision, centre of action, centre of interest. Where the body is is 'here', where the body acts is 'now'; what the body touches is 'this'; all other things are 'there', and 'then' and 'that' " [76].

What then is the source of the illusion that self or subject is essentially distinct from the "physical," "external" world? James says that the reification of subject/self only arises when, "a given 'bit' is abstracted from the flow of experience and retrospectively considered in the context of different relations, relations that are external to the experience taken singly but internal to the general flow of experience taken as a whole" [69, p. 535]. It is only the discursive intellect, in an inductive act of interpretation, that later creates or projects these dualisms between subject and object, inner/outer, self/world, etc. Keep in mind this discursive intellect is not some a priori cognitive category through which noumena is filtered. The mind's inference to the dualism of knower/known, subject/object, etc., happens after the fact. And again, this is direct realism, there is no noumena.

Given neutral monism, the mind and the world are one, just as Kant suspected. For Kant, given his unity of apperception, time is an a priori condition for experience, no subjectivity means no time or space. Kant here is providing a transcendental analysis in mentalistic terms. This means that the dynamical character of thought/experience and the world are two sides of the same coin. James puts it like this, "According to radical empiricism, experience as a whole wears the form of a process in time" [69, p. 540]. Kant's transcendental arguments from *The Critique of Pure Reason* are supposed to show that we must conceive of the world in a certain way, structure it internally according to certain categories such as time, space, and causation. Those arguments are fraught with many interpretative perils and controversies, but the basic idea is that experience is possible only if some experiences are conceptualized as being of enduring objects, enduring through time and space. Likewise, to experience a world of enduring objects there must be some sense of an enduring self. You cannot have one without the other. Again, however, as James notes, Kant was wrong that the structure of experience is a product or projection of mental filters or categories. Neutral monism takes the world of experience out of the head and also rejects the very idea of noumena or as some people call it, beables. Kant is right however that neither subject nor object alone, but only subject-object is the basic unit of experience. Nothing is mind dependent on this view in the subjective idealist sense. All entities and their properties are extrinsic or interdependent (not mind dependent!), not just colors, tastes and sounds, but mass, charge and spin as well. In short, mind and world are just two interdependent sides of the same coin. You can't have one without the other.

It should be clear that given neutral monism physics is not about the pursuit of some noumenal world or hidden magical beables, it is about the world of experience. It is common to make a distinction between metaphysical things-in-themselves and mere appearances or observations. Neutral monism rejects the very idea of the former, but the alternative is not anti-realism or sense data theory, it's radical empiricism. The idea that realism demands noumena or beables hiding behind the world of observables is simply a question begging misnomer. For us there is no inaccessible, noumenal QM realm hiding behind the world of experience. This is not to deny that under certain conditions, in certain contexts, reality behaves in ways that are best described by QM, e.g., as QM particles, fields or waves in spacetime. This is very much in keeping with the kind of thinking that led Einstein to relativity. As Jim Baggott puts it, "In developing his theory of relativity, Einstein sought to banish from physics the entirely metaphysical concepts of absolute space and time. One consequence is that the observer is put firmly back into the reality that is being observed" [77, p. 109]. Notice that there is nothing inherently anti-realist or instrumentalist in such a move. After all, Einstein is often described as the realist to Bohr's instrumentalist. We are not advocating for a "shifty split" as Bell put it, no magical line between the QM and the classical. However, for us, as evidenced by environmental decoherence, the QM and the classical are co-fundamental and co-dependent. This might violate a certain kind of fundamentalism or reductionism, but it doesn't violate commonsense realism.

Just as the work of David Hume, Immanuel Kant, and Ernst Mach was essential for Einstein's epiphanies behind SR [78, pp. 82–83], so for us neutral monism (i.e., radical empiricism) not only shows us how to properly situate subjective experience in the world formally and metaphysically, but it shows how to move forward in physics as well. Indeed, in many ways these turn out to be the same project. Here is what Einstein said he gleaned from Hume and Mach, one must eliminate concepts that "have no link with experience such as absolute simultaneity and absolute speed" [78, p. 131]. As James noted every scientific theory and all scientific explanations are, however opaquely, rooted in some metaphysical picture of the world such that "the juices of metaphysical assumptions leak in at every joint" [79, p. 112]. The radical empiricism (neutral monism) of James is just an extension of the empiricism of Hume, Einstein and others. In the words of Eugene Taylor [79, p. 130]:

> Radical empiricism was, nevertheless, psychological; that is to say, it placed immediate experience at the center of everything we have to say about the universe. Consciousness, therefore, knower-and-known, subject-and-object, person-and-world, formed the basis of all science and all knowledge-getting. Positivistic science had to conform as much to the dictates of such psychology, as psychology was trying to conform to such a science.

Radical empiricism is thus a metaphysics and epistemology of science. Taking it on board allows us to reconceive scientific explanation just as Einstein did with his principle explanation in relativity. This is precisely what we have done with QM as well.

In Einstein's words, "the totality of our sense experiences . . . can be put in order" [29]. We then use this model to explore regularities and patterns in the events we perceive. We mathematically describe these regularities and patterns and explore the consequences (experiments). In Einstein's words again, "operations with concepts, and the creation and use of definite functional relations between them, and the coordination of sense experiences to these concepts" [29]. We then refine our model of physical reality as necessary to conform to our results. This allows us to explain the past, manipulate physical reality in the present (to create new technology, for example), and to predict the future. While defining physics all the way down to individual "sense experiences" may seem unnecessarily detailed, it is crucial to understanding the relationship between subjective experience and physics being proposed here. In turn, we think that understanding will help the reader see how it could be the case that often the best explanations in physics are principle explanations, or explanations in terms of adynamical global constraints on sense experiences.

Putting this all together again, we can say the following. From our take on neutral monism we understand that each subject is just a conscious point of origin (PO) of Neutral Pure Presence. The perceptions of each PO form a context of interacting trans-temporal (enduring) objects (TTOs) for that PO. Since TTOs are "bodily objects" with worldlines in spacetime, TTOs are coextensive with space and time. When POs exchange information about their perceptions, they realize that some of their disparate perceptions fit self-consistently into a single spacetime model with different reference frames for each PO. Thus, physicists' spacetime model of the "Physical" represents the self-consistent collection of shared perceptual information between POs, e.g., perceptions upon which Galilean or Lorentz transformations can be performed. Here is how Hermann Weyl himself put it, physics is the "Construction of objective reality out of the material of immediate experience" [80, p. 117].

Coordinate transformation is important because relativity states that there is not one reference point (or perspective) in the universe that is more favored than another. As Weyl put it, "The explanation of the law of gravitation thus lies in the fact that we are dealing with a world surveyed from within" [80, p. 117]. Keep in mind that the beauty of neutral monism is that talk about POs and their perceptions should be understood not as some sort of positivism, or some brand of idealism (subjective or otherwise), or sense data theory, or merely as bracketed phenomenology, but in terms of James' "instant field of the present" and what Russell calls "events." That and that alone is what spacetime is. And spacetime is the subject of physics.

This brings us to our axioms that we stated at the beginning of our chapter. Given neutral monism, it should be clear why we chose those particular principles as the basis for all of physics. That is, the universe (as experienced) is the self-consistent collection of shared classical information regarding diachronic entities (classical objects), which interact per QM. The consistency of shared classical information of the universe is guaranteed by the divergence-free (gauge invariant) nature of the adynamical global constraints for classical and quantum physics. In the case of Wigner's friend per Healey, the self-consistent collection of classical information would include all shared classical information between Xena, Yvonne, Zeus, and Wigner. In the case of the delayed choice quantum eraser experiment, conscious choices are equally constrained. Here we see a profound connection between QM, relativity, and conscious experience. But, we do not think this is any weirder than the fact that conscious experiences and choices are constrained by other adynamical global constraints, such as conservation laws, the light postulate, and the relativity principle. Due to limits of time and space, we cannot recapitulate the work along these lines, but for those interested, we showed how to derive QM and relativity from our axioms [9], which in turn gives us the tools to address the experiments herein.

Let us recall our goal for this chapter was to provide a take on QM that explains why there is and must always be determinate and intersubjectively consistent experience about all experimental outcomes (absoluteness of observed events). A take that accepts the completeness of the theory and requires no invocation of relative states (e.g., outcomes being relative to branches, conscious observers, etc.). And finally, a take that requires no allegedly hybrid models such as claims about "subjective collapse." We wanted a take on QM that yields a single world wherein all the observers (conscious or otherwise) agree about determinate and definite outcomes, because those outcomes are in fact determinate and definite. We wanted a realist psi-epistemic take on QM that saves the absoluteness of observed events and the completeness of QM, without giving up free will or locality. We also wanted to show how our realist psi-epistemic account eliminates the measurement problem and, coupled with our take on neutral monism, also eliminates the hard problem of consciousness. We believe we have done all of the above. The key to achieving these goals was to let go of the following offending assumptions: 1) physicalism, 2) fundamentalism, and relatedly 3) dualism about conscious experience, 4) the notion that fundamental explanation is always constructive, causal or dynamical, and relatedly, 5) realism about the wavefunction. Together these assumptions force us into the hard problem, they force us into the measurement problem, and they force us to seek the solutions to these problems in fundamental physics, e.g., by trying to relate these problems to one another directly, with very little success. Once again, sometimes, when a problem is deeply intractable the best move is to jettison the offending assumptions that led to the problem in the first place. This is precisely what we did herein.

References

[1] K. McQueen, "Niels Bohr's interpretation and the Copenhagen interpretation—are the two incompatible?" Philosophy Now **121**, 17–20 (2017).

[2] H. Atmanspacher, "Quantum approaches to consciousness," (2015), https://plato.stanford.edu/archives/sum2015/entries/qt-consciousness/.

[3] S. Gao, *The Meaning of the Wave Function: In Search of the Ontology of Quantum Mechanics* (Cambridge University Press, 2017) https://arxiv.org/abs/1611.02738

[4] M. Proietti, A. Pickson, F. Graffitti, P. Barrow, D. Kundys, C. Branciard, M. Ringbauer, and A. Fedrizzi, "Experimental test of local observer-independence," Science Advances **5**, eaaw9832 (2019), https://arxiv.org/abs/1902.05080.

[5] K. Bong, A. Utreras-Alarc'on, F. Ghafari, Y. Liang, N. Tischler, E. Cavalcanti, G. Pryde, and H. Wiseman, "A strong no-go theorem on the Wigner's friend paradox," Nature Physics (2020), https://doi.org/10.1038/s41567-020-0990-x.

[6] L. Hardy, "Proposal to use humans to switch settings in a Bell experiment," (2017), https://arxiv.org/abs/1705.04620.

[7] C. A. Fuchs and B. C. Stacey, "Some negative remarks on operational approaches to quantum theory," in *Quantum Theory: Informational Foundations and Foils*, edited by G. Chiribella and R. Spekkens (Springer, Dordrecht, 2016) pp. 283–305.

[8] Michael Silberstein, W. M. Stuckey, and Timothy McDevitt, *Beyond the Dynamical Universe: Unifying Block Universe Physics and Time as Experienced* (Oxford University Press, Oxford, UK, 2018).

[9] Michael Silberstein and W. M. Stuckey, "Re-thinking the world with neutral monism: Removing the boundaries between mind, matter, and spacetime," Entropy **22**, 551 (2020), https://doi.org/10.3390/e22050551.

[10] W. M. Stuckey, Michael Silberstein, Timothy McDevitt, and T. D. Le, "Answering Mermin's challenge with conservation per no preferred reference frame," Scientific Reports **10**, 15771 (2020), https://www.nature.com/articles/s41598-020-72817-7.

[11] E. Taylor and R. H. Wozniak, *Pure Experience* (Thoemmes Press, Bristol, UK, 1996).

[12] R. Carnap, "Carnap's intellectual biography," in *The Philosophy of Rudolf Carnap* edited by P Schilpp (Open Court, Chicago, IL, USA, 1963) pp. 3–84.

[13] Z. Merali, "This twist on Schr¨odinger's cat paradox has major implications for quantum theory," Scientific American (2020), https://www.scientificamerican.com/article/this-twist-on-schroedingers-cat-paradox-has-major-implications-for-quantum-theory/.

[14] H. Price and K. Wharton, "Dispelling the quantum spooks: A clue that einstein missed?" in *Time of Nature and the Nature of Time: Boston Studies in the Philosophy and History of Science* edited by C. Bouton and P Huneman (Springer, 2017) pp. 123–137, http://arxiv.org/abs/1307.7744.
[15] W. M. Stuckey, Michael Silberstein, and Timothy McDevitt, "Relational blockworld: Providing a realist psi-epistemic account of quantum mechanics," International Journal of Quantum Foundations **1**, 123–170 (2015), http://www.ijqf.org/wps/wp-content/uploads/2015/06/IJQF2015v1n3p2.pdf.
[16] W. M. Stuckey Michael Silberstein, and Timothy McDevitt, "An adynamical, graphical approach to quantum gravity and unification," in *Beyond Peaceful Coexistence: The Emergence of Space, Time and Quantum* edited by I. Licata (Imperial College Press, London, 2016) pp. 499–544.
[17] C. W. Misner, K. S. Thorne, and J. A. Wheeler, *Gravitation* (W. H. Freeman, San Francisco, 1973).
[18] D. K. Wise, "*p*-form electromagnetism on discrete spacetimes," Classical and Quantum Gravity **23**, 5129–5176 (2006).
[19] A. Kheyfets and J. A. Wheeler, "Boundary of a boundary principle and geometric structure of field theories," International Journal of Theoretical Physics **25**, 573–580 (1986).
[20] R. Gomatam, "Does consciousness cause quantum collapse?" Philosophy of Science **74**, 736–748 (2007).
[21] P. Ball, *Beyond Weird* (University of Chicago Press, Chicago, 2018).
[22] Lucien Hardy "Reconstructing quantum theory," in *Quantum Theory: Informational Foundations and Foils* edited by G. Chiribella and R. Spekkens (Springer, Dordrecht, 2016) pp. 223–248, https://arxiv.org/abs/1303.1538.
[23] L. Felline, "Scientific explanation between principle and constructive theories," Philosophy of Science **78**, 989–1000 (2011).
[24] A. Einstein, "What is the theory of relativity?" London Times, 53–54 (1919).
[25] W. Van Camp, "Principle theories, constructive theories, and explanation in modern physics," Studies in History and Philosophy of Science Part B: Studies in History and Philosophy of Modern Physics **42**, 23–31 (2011).
[26] S. Weinberg, "The trouble with quantum mechanics," (2017), http://quantum.phys.unm.edu/466-17/QuantumMechanicsWeinberg.pdf.
[27] N. D. Mermin, "Bringing home the atomic world: Quantum mysteries for anybody," American Journal of Physics **49**, 940–943 (1981).
[28] B. Dakic and C. Brukner, "Quantum theory and beyond: Is entanglement special?" (2009), https://arxiv.org/abs/0911.0695.
[29] A. Einstein, "Physics and reality," Journal of the Franklin Institute **221**, 349–382 (1936).
[30] M. T. Hicks, "What everyone should say about symmetries (and how humeans get to say it)," Philosophy of Science **86**, 1284–1294 (2019).
[31] W. M. Stuckey, Michael Silberstein, Timothy McDevitt, and Ian Kohler, "Why the Tsirelson bound? Bub's question and Fuchs' desideratum," Entropy **21**, 692 (2019), https://arxiv.org/abs/1807.09115.
[32] Harvey Brown, *Physical Relativity: Spacetime Structure from a Dynamical Perspective* (Oxford University Press, Oxford, UK, 2005).
[33] H. Brown and O. Pooley, "Minkowski space-time: A glorious non-entity," in *The Ontology of Spacetime*, edited by D. Dieks (Elsevier, Amsterdam, 2006) p. 67.

[34] J. Norton, "Why constructive relativity fails," British Journal for the Philosophy of Science **59**, 821–834 (2008).
[35] T. Menon, "Algebraic fields and the dynamical approach to physical geometry," Philosophy of Science **86**, 1273–1283 (2019).
[36] L. Felline, "Quantum theory is not only about information," Studies in History and Philosophy of Science Part B: Studies in History and Philosophy of Modern Physics (2018), doi.org/10.1016/j.shpsb.2018.03.003, https://arxiv.org/abs/1806.05323.
[37] A. Garg and N. D. Mermin, "Bell inequalities with a range of violation that does not diminish as the spin becomes arbitrarily large," Phys. Rev. Lett. **49**, 901–904 (1982).
[38] C. S. Unnikrishnan, "Correlation functions, Bell's inequalities and the fundamental conservation laws," Europhysics Letters **69**, 489–495 (2005).
[39] S. Boughn, "Making sense of Bell's theorem and quantum nonlocality," (2017), https://arxiv.org/abs/1703.11003.
[40] W. C. Salmon, "The value of scientific understanding," Philosophica **51**, 9–19 (1993).
[41] Y. Balashov and M. Janssen, "Presentism and relativity," British Journal for the Philosophy of Science **54**, 327–346 (2003).
[42] A. Einstein, "Autobiographical notes," in *Albert Einstein: Philosopher-Scientist*, edited by P. Allen Schilpp (Open Court, La Salle, IL, USA, 1949) pp. 3–94.
[43] W. Heisenberg, *Physics and Beyond: Encounters and Conversations* (Harper & Row, New York, 1971).
[44] R. Healey "Quantum theory and the limits of objectivity," Foundations of Physics **48** 1568–1589 (2018).
[45] D. Frauchiger and R. Renner, "Quantum theory cannot consistently describe the use of itself," Nature Communications **9**, 3711 (2018).
[46] E. Wigner, "Remarks on the mind-body question," in *The Scientist Speculates*, edited by I. J. Good (Heineman, 1961) pp. 284–302.
[47] S. D. Bartlett, T. Rudolph, and R. W. Spekkens, "Reference frames, superselection rules, and quantum information," Reviews of Modern Physics **79**, 555 (2007), https://arxiv.org/abs/quant-ph/0610030.
[48] V. Baumann and S. Wolf, "On formalisms and interpretations," Quantum **2**, 99 (2018).
[49] J. Bub, "'Two Dogmas' redux," in *Quantum, Probability, Logic: The Work and Influence of Itamar Pitowsky* edited by M. Hemmo and O. Shenker (Springer Nature, Switzerland, 2020) pp. 199–215, https://arxiv.org/abs/1907.06240.
[50] L. M. Brown, *Feynman's Thesis: A New Approach to Quantum Theory* (World Scientific Press, New Jersey 2005).
[51] R. Shankar, *Principles of Quantum Mechanics 2nd ed.* (Plenum Press, New York, 1994).
[52] D. Lazarovici and M. Hubert, "How quantum mechanics can consistently describe the use of itself," (2018), https: //arxiv.org/abs/1809.08070.
[53] R. Healey, "IJQF Wigner's friend workshop," (2018), https://www.ijqf.org/groups-2/workshop-on-wigners-friend-2018/forum/topic/is-there-an-inconsistent-friend/.
[54] R. Healey *The Quantum Revolution in Philosophy* (Oxford University Press, Oxford, 2017).

[55] R. Healey, "Philosophers on a physics experiment that 'suggests there's no such thing as objective reality,'" (2019), http://dailynous.com/2019/03/21/philosophers-physics-experiment-suggests-theres-no-thing-objective-reality/.
[56] A. Elitzur and S. Dolev, "Quantum phenomena within a new theory of time," in *Quo Vadis Quantum Mechanics*, edited by A. Elitzur, S. Dolev, and N. Kolenda (Springer, Berlin, 2005) pp. 325–349.
[57] W. M. Stuckey, M. Silberstein, and M. Cifone, "Reconciling spacetime and the quantum: Relational blockworld and the quantum liar paradox," Foundations of Physics **38**, 348–383 (2008), http://arxiv.org/abs/quant-ph/0510090.
[58] E. Cavalcanti, "A new quantum paradox throws the foundations of observed reality into question," The Conversation (2020), https://theconversation.com/a-new-quantum-paradox-throws-the-foundations-of-observed-reality-into-question-144426.
[59] M. D. Greenberger, M. Horne, and A. Zeilinger, "Going beyond Bell's theorem," in *Bell's Theorem, Quantum Theory, and Conceptions of the Universe* edited by M. Kafatos (Kluwer Academic Press, Dordrecht, 1989) pp. 69–72.
[60] F. Dowker, "Are there premonitions in quantum measure theory?"(2014), https://www.youtube.com/watch?v=jETvuojQ2qs.
[61] N. D. Mermin, "Quantum mysteries revisited," American Journal of Physics **58**, 731–734 (1990).
[62] R. Hillmer and P. Kwiat, "A do-it-yourself quantum eraser," Scientific American **296**, 90–95 (2007).
[63] Y. Kim, R. Yu, S. P. Kulik, Y. H. Shih, and M. O. Scully, "A delayed choice quantum eraser." Physical Review Letters **84**, 1–5 (2000).
[64] F. Wilczek, "Physics in 100 years," Physics Today **69**, 32–39 (2016).
[65] A. Ananthaswamy, "A classic quantum test could reveal the limits of the human mind," New Scientist , 427 (2017), https://www.newscientist.com/article/2131874-a-classic-quantum-test-could-reveal-the-limits-of-the-human-mind/.
[66] W. James, "The notion of consciousness," in *Sciousness* edited by J. Bricklin (Eirini Press, 1905) pp. 87–111.
[67] W. James, *William James: Writings 1902–1910* (Library of American, 1905).
[68] W. James, "Manuscript lectures," (1988).
[69] W. James, "A world of pure experience," Journal of Philosophy Psychology and Scientific Methods **1** 533–543 (1904).
[70] E. Thompson, *Waking, Dreaming, Being: Self and Consciousness in Neuroscience, Meditation, and Philosophy* (Columbia University Press, New York, 2015).
[71] B. Russell, *An Outline of Philosophy* (George Allen and Unwin, London, 1927).
[72] B. Russell, *The Analysis of Matter* (Kegan Paul, London, 1927).
[73] B. Russell, *Logical Atomism* (George Allen and Unwin, London, 1956).
[74] Michael Silberstein, "Review of Lee Smolin's 'Einstein's Unfinished Revolution,'" International Journal of Quantum Foundations **6**, 133–159 (2020).
[75] W. James, "Does 'consciousness' exist?" Journal of Philosophy, Psychology and Scientific Methods **1**, 477–491 (1904).
[76] W. James, *A Pluralistic Universe* (Anodos, England, 2019).
[77] J. Baggott, *Quantum Reality: The Quest For The Real Meaning of Quantum Mechanics A Game of Theories* (Oxford University Press, Oxford, UK, 2020).
[78] W. Isaacson, *Einstein* (Simon and Schuster, New York, New York, 2007).

[79] E. Taylor, *William James on Consciousness beyond the Margin* (Princeton University Press, Princeton, New Jersey 1996).
[80] T. Ryckman, *The Reign of Relativity* (Oxford University Press, New York, New York, 2005).
[81] C. Brukner and A. Zeilinger, "Information Invariance and Quantum Probabilities," Foundations of Physics **39**, 677 (2009).
[82] W. M. Stuckey, Timothy McDevitt, and Michael Silberstein, "No Preferred Reference Frame at the Foundation of Quantum Mechanics," Entropy **24**(1), 12 (2022).
[83] L. Hardy, "Quantum Theory From Five Reasonable Axioms," https://arxiv.org/abs/quant-ph/0101012 (2001).
[84] P. Höhn and M. Müller, "An operational approach to spacetime symmetries: Lorentz transformations from quantum communication," New Journal of Physics **18**, 063026 (2016). https://arxiv.org/abs/1412.8462
[85] Michael Silberstein and W. M. Stuckey, "Beyond Causal Explanation: Einstein's Principle Not Reichenbach's," Entropy **23**(1), 114 (2021).

11

The Roles Ascribed to Consciousness in Quantum Physics

A Revelator of Dualist (or Quasi-Dualist) Prejudice

Michel Bitbol

1. Introduction

Let's ponder two simple questions: "How can consciousness interact with physical systems, thus imposing a reduction of their quantum state?" Or conversely, "How can consciousness be produced by a process involving physical systems described by quantum theory?" The purpose of this chapter is not to offer an answer to these questions that were raised by some of the best physicists of the 20th century, from Eugen Wigner to Roger Penrose, but rather to ask more questions. Why did so many serious thinkers consider that they were legitimate questions at all? What are the implicit ontological and epistemological presuppositions that underpin this sense of legitimacy? Can we move upstream such presuppositions and adopt a standpoint from which these questions would no longer be taken at face value, but rather as symptoms of a conceptual and cultural bias?

2. Sense, Non-Sense, and Philosophy

This strategy that consists of asking questions about questions, instead of addressing them straightaway, can be perceived as a dodge. But it is in line with one of the most specific tasks of philosophy. Indeed, "what truth and falsity is to science, sense and non-sense is to philosophy" [1]. Even before trying to solve a problem by scientific methods, one should try to inquire philosophically into whether this problem makes sense at all, or at least under which intellectual (or existential) conditions one is prone to *believe*

Michel Bitbol, *The Roles Ascribed to Consciousness in Quantum Physics: A Revelator of Dualist (or Quasi-Dualist) Prejudice* In: *Consciousness and Quantum Mechanics.* Edited by: Shan Gao, Oxford University Press.
 DOI: 10.1093/oso/9780197501665.003.0012

that it makes sense. If pushed far enough, and with sufficient boldness, this inquiry may lead to overturning the system of presuppositions that made the problem acceptable in the first place. In the present case, as we will soon realize, such overturn is tantamount to adopting the standpoint of *phenomenology*, a philosophical discipline that deliberately favors a first-person approach of any issue dealing with what there is and what can be known.

3. Tacit Presuppositions

Thus, what presuppositions are hidden under the former questions about the relation between consciousness and quantum mechanics? In fact, these questions encapsulate virtually *all* of the presuppositions that guided the research effort of the West from the 17th century until now. They do so with such compactness that they turn out to be precious revelators of our common cultural background, provided we do not fall immediately under their spell.

To start with, our two initial questions involve a pair of terms: consciousness and physical systems (or processes). These terms having the grammatical status of substantives, they irresistibly call for the vague intuition that there are two "substances" that relate to each other, or interact with each other. Isn't it natural "to try to find a substance for a substantive," as Wittgenstein suggested in the very first page of his *Blue Book* [2]? One may refrain from falling immediately in this elementary trap, but the structure of the question makes the mental attractor of dualism almost irresistible in the long run. Even the archetypal opponent of dualism in the contemporary debate, namely physicalist monism, is not immune from this. Indeed, physicalist monists deny that consciousness *exists* independently of the neural processes. By saying so, physicalist monists implicitly accept that consciousness is something that may exist or not exist in the same sense as physical objects, and they thereby render themselves guilty of virtual dualism.

To sum up, the two major protagonists of the present debate in the philosophy of consciousness, namely (property or substance) dualists, and physicalist monists, share two presuppositions.

The first presupposition is that consciousness is either something or a property of something. Dualists consider that such something has an independent form of existence, and that it can act somehow on physical systems and processes; whereas physicalists try to understand how the property "consciousness" can emerge somehow from physical systems and processes.

The second presupposition (that is, the reciprocal of the first one) is that there exist physical systems and processes apart from consciousness.

4. Phenomenology beyond Presuppositions

Are these two presuppositions, however, unavoidable? The phenomenological tradition has disposed of them long ago.

According to phenomenology, consciousness is no thing or property that may exist or not exist. "Consciousness" is the misleading name we give to the precondition for any ascription of existence or inexistence. What makes this remark obvious for phenomenologists and almost incomprehensible for physicalists is that phenomenologists are settled in the first-person standpoint, whereas physicalist researchers explore everything from a third-person standpoint. From a first-person standpoint, anything that exists (thing or property) is given as a phenomenal content of consciousness. Therefore, consciousness *de facto* comes before any ascription of existence.

Instead, from a third-person standpoint, nothing else than objects of perception and handling is to be taken seriously. Now, the behavioral or neurobiological correlates of consciousness are possible objects of perception and handling. They can be said to exist (if a subject is alive and awake) or not to exist (in other cases). Then, from this standpoint, saying that the neural correlate of consciousness (often taken as its "neural *basis*") may exist or not exist, amounts to saying that consciousness itself may exist or not exist in the same sense.

Let's now turn our philosophical attention on the second presupposition, which is the keystone of physicalist monism, but that is shared by dualism. What is the status of the tacit assumption that there *exist* physical systems and processes apart from consciousness? From the third-person standpoint, this is just a fact that is so glaring that it hardly needs arguments apart from an evocation of common sense. But from the first-person standpoint of phenomenology, this is a bold metaphysical assumption that stems from the "natural attitude" of common sense and extrapolates far beyond it. We have already mentioned that, according to the phenomenological approach, it is *prima facie* obvious that physical objects are given as nothing else than correlates of conscious experience. They arise as poles of stability and identity within a flowing consciousness; they are meanings constituted through consciousness, or intentional correlates of acts of consciousness [3].

Physical objects can still be said to "transcend" consciousness according to phenomenology, but in a sense that is itself rooted in certain features of consciousness. This is what Husserl called "the immanent transcendence" of the objects of our experience in §48 of his *Cartesian Meditations.*

The first feature that evokes transcendence is the fact that physical objects are presented *incompletely* to consciousness at each moment: one just perceives one facet of an object at a time and expects other facets, or one just measures one variable pertaining to a physical system and predicts (deterministically or probabilistically) the values of other variables. This gives rise to the impression that there is still more to come, that a physical system has always something additional in store beyond what appears of it at this very moment. But one must not forget that this feeling of incompleteness is itself generated by an even more primitive act of consciousness: the act of *identifying* past, present, and expected appearances as moments of one and the same object [4]. Once this act of identification has been performed, our ever-developing expectations are understood as a sign of the incompleteness of our knowledge *about this enduring object.*

The second "transcendent-like" feature is that expectations may be disappointed; that a surprise may occur. This is usually expressed by saying that physical systems are *given* to us, since we do not "control" how they manifest to us. But translating this "givenness" into standard ontological terms, asserting that physical systems are "external" entities of a "reality out there" that is completely foreign to their appearances in conscious experience, is a dubious inference. From a phenomenological standpoint, this ontologization of the "givenness" and transcendent-like behavior of certain patterns of experience is just a verbal trick used to favor the intersubjective stabilization of the poles of identity and intentional directedness that structure the field of consciousness.

5. A Phenomenological Critique of the Concept of "Physical System"

Let's notice at this point that the lack of ontological import of the concept of physical system in phenomenology makes the redefinition of the putative objects of physical theories much easier. It invites us to see the existence of physical objects as an open problem rather than an uncontrovertible fact. And it then prepares us for the most radical scientific revolution of all: a revolution in which not even the former ontological furniture of the world can subsist. To understand this, we must come back to the (mostly tacit) criteria we use in order to convince ourselves that a certain phenomenon is underpinned by a permanent entity *of which* it is the appearance. These criteria were described in exquisite details by Husserl [4] from a phenomenological standpoint, and also by Piaget [5] from the standpoint of developmental psychology. Piaget summarized his criteria thus: "[A child] does not believe in the permanence of an individual object as long as she cannot find it again and again by coordinated actions." The key criterion for believing in an entity is active *reidentification*. But in microphysics, reidentifying a localized object with certainty is usually impossible (as can be inferred from considerations about Heisenberg's "uncertainty" relations, or about quantum statistics). It then turns out that the class of objects "localized particles," and more generally "spatio-temporal continuants," is highly problematic in microphysics [6].

Yet, physicists still speak in terms of independent "physical systems" having "states," *on which* various (mostly incompatible) measurements are performed. Can the general concept of "physical system" truly survive its most common variety; can it survive the disappearance of the (more or less) localized individual particles? Probably not. Some good reasons to think that not even the general concept of "physical system" can be left untouched by the quantum revolution were given recently [7]. These reasons pertain to the structure of sets of experimental phenomena: certain sets of measurement outputs are *not* such that they can be *ascribed to* single physical systems. The provocative conclusion is that "physical theory may contain no physical systems" [7]. It thus turns out that a physical theory such as quantum mechanics might well be averse to the second common presupposition of dualists and physicalist monists. Physics no longer supports physicalism (at least not without contrived attempts to save it).

6. Quantum Physics without Physical Systems

But then, what status can we ascribe to quantum mechanics, if we cannot even say that it provides us with a non-standard *description* of *physical systems*? Some suggestions to that effect were made long ago by Bohr, when he wrote that "physics is to be regarded . . . as the development of methods of ordering and surveying human experience" [8], and that quantum mechanics is "a purely symbolic scheme permitting only predictions . . . as to results obtainable under conditions specified by means of classical concepts" [9]. The two key words here are "human experience" and "predictions": a physical theory like quantum mechanics provides us with probabilistic predictions bearing on a fraction of human experience that measurement devices help to shape into elements of experimental information. Similar ideas were further developed in neo-Bohrian circles. Some authors explicitly suggested that quantum mechanics bears on nothing else than pure information [10]. Other authors considered that quantum mechanics is just a probabilistic "user's guide" for agents confronted with the outcomes of their own experimental and technological activities [11].

In other terms, according to the latter authors, experience comes first, and a physical theory is a coherent bundle of expectations about its later developments. This puts physics in line with elementary animal and human cognition, thereby weakening the common belief that physical theories have something exceptional, that they represent a historical leap due to their innovative use of a combination of rationality and technology. Indeed, according to phenomenology, the essence of the human condition is to live in the perspective of one's own *possibilities* of being/becoming, and never to remain trapped in flat factual actualities. The essence of the human condition thus implies a permanent anticipation of what will come next, a projection of oneself onto an expected future [3]. Similarly, according to some significant naturalistic views, the function of cognition is to anticipate by certain bodily capacities and behaviors those features of the environment that are relevant for the survival of cognizing organisms. In particular, the function of the nervous system is to minimize the disruption of expectations, that is, to attenuate the "surprises" of an organism confronted with accidental variations of its environment [12, 13]. This being granted, the previously mentioned neo-Bohrian approaches turn out to be an epistemological golden standard for quantum mechanics. Indeed, according to them, quantum mechanics no longer appears as a maverick theory, but, on the contrary, as one of the purest expressions of a central principle of knowledge, and one of the most straightforward formal extensions of a basic function of elementary cognition.

7. What Must Be Assumed for the Measurement Problem to Make Sense?

Now, how does this relate to the measurement problem, which was the main motivation for the disconcerting introduction of consciousness into physics? As a preliminary to answering this question, we must make an inventory of the *conditions* under which the outcome of von Neumann's quantum theory of measurement is seen as an enigma or a paradox. Let's remember that, according to von Neumann's theory of measurement, the global "state vector" of the large system made of an object and a measurement apparatus becomes an *entangled superposition* after the measuring interaction has taken place. The enigma or paradox then derives from the apparent contradiction between this superposed "state" and the sharp observational state of the measurement chain.

However, for this apparent contradiction to arise, three assumptions must be made tacitly or overtly [14]:

1. Quantum mechanics *describes* the *state* of physical systems.
2. The state of *every* physical system is ruled by quantum mechanics.
3. There *is* nothing but physical systems.

The first assumption is tantamount to adopting a *scientific realist* reading of quantum mechanics; the second assumption asserts the *universality of quantum mechanics* in the domain of physics; and the third assumption asserts the *universality of physics* in the domain of what there is (this is "physicalism").

8. Solving or Dissolving the Measurement Problem?

Three main strategies to solve or dissolve the measurement problem can be elaborated by renouncing each one of the three assumptions in turn. These strategies almost exhaust the propositions that have been made during the history of quantum physics.

1′. If one renounces the realist interpretation according to which certain vectors in Hilbert spaces describe the "(quantum) state" of something, the measurement problem is automatically *dissolved*. Indeed, there is no immediate contradiction between a sharp observational state and a (superposed) symbol that does *not* represent the state of anything. There can be no conflict of this kind if the latter symbol is just taken at face value, namely as a mathematical tool used to calculate the probabilities of sets of measurement outcomes arising in various experimental contexts. This is the reason why Bohr was unable to see a problem in the so-called measurement problem. And this is also how several contemporary physicists dissolve the measurement problem: by considering "state" vectors as relational rather than monadic symbols, or as mathematical expressions of the best strategy for gambling about interrelated sets of measurement outcomes.

Thus, according to Carlo Rovelli [15], it makes no sense to ask what the state of a system *is* in absolute terms, since each quantum state reflects the *relationship* between a physical system and another system playing the role of an observer. There is no state *of* a system, but only relative states. Relative to someone who has made no observation yet, the state of the large system made of an object and a measurement apparatus is a superposition; and relative to someone who has observed the screen of the apparatus, the state is sharp. But there is no fact of the matter as to which of these two "states" (superposed or sharp) *is* the *intrinsic* state of the large system.

The same dissolution occurs when vectors in Hilbert space are supposed to have nothing to do with states, not even relational states, and are rather taken as "user's guides" for making coherent bets [11].

2′. The measurement problem can be *solved* if one considers that standard quantum mechanics somehow lacks universality in the physical domain. Indeed, in this case, some physical systems may escape the general process of entanglement and dissemination of state superpositions that is typical of the quantum theory of (measuring) interactions. And they can accordingly impose the sharp definition of their own properties to quantum physical systems. In history, Bohr's insistence that measuring apparatuses should (to a certain extent) be described by classical concepts was the first variety of this claim of non-universality of quantum mechanics. Later on, the same kind of claim has been systematized as a clause of distinction between theoretical and meta-theoretical entities [16]. But the claim of non-universality has also taken more conservative forms, in approaches that extend the domain of physical entities, properties, or processes beyond standard quantum mechanics. This is the case of Bohm's hidden variable theory and Ghirardi-Rimini-Weber spontaneous collapse theory.

3′. Another strategy to *solve* the measurement problem is to accept that there exists something non-physical that automatically eludes the laws of quantum physics, and might therefore be able to break the unended chain of entanglement and superposition of "states." It is at this point that consciousness may irrupt, and we will then concentrate on the meaning and scope of this third line of thought.

9. An Interlude: Decoherence

Before we proceed on the theme of consciousness as a non-physical *deus ex machina* for solving the measurement problem, however, a brief mention of decoherence must be made. For decoherence seems to fall outside the previous classification of strategies used to address the measurement problem. But is it so?

The decoherence solution to the problem raised by the gap between the quantum domain of superpositions and the classical domain of sharp properties is called "environment-induced superselection" [17]. The principle of this solution consists of showing that the phase coherences of the state vector (or density operator) of an apparatus correlated to a micro-system are rapidly diluted in its environment. Indeed, the virtually complete disappearance of the interference terms is equivalent to a superselection rule that only retains the eigenstates of a given observable.

This looks like a completely new kind of solution. But there are aspects of decoherence that resonate with each one of the three former strategies. We could even say that the reason why decoherence does not fit squarely with any one of these strategies is that it borrows something from all of them. Let's examine this threefold resonance in a different order with respect to the listed strategies.

First, to derive a decoherence process from the evolution of a global state vector, one must split the latter into three sub-states ascribed, respectively, to a micro-system, an apparatus, and an environment. But this neat separation into sub-systems does not arise *a priori* from quantum mechanics; it may become approximately acceptable only *a posteriori,* at the end of the decoherence process. Trying to derive decoherence from quantum mechanics by postulating such separation from the outset thus looks like *a petitio principii.* But it's no longer a *petitio principii* if the postulate of separation is ascribed a meta-theoretical, rather than theoretical, status. Decoherence thus borrows something from the second strategy: it implicitly denies the exhaustivity of *quantum theory* by making use of meta-theoretical assumptions in the physical domain.

Second, unlike the standard "reduction of the state," the decoherence process does not yield a single eigenstate of some observable (corresponding to a single measurement outcome) but an (improper) statistical mixture of eigenstates [18]. Selecting a particular eigenstate is arguably an additional extra-physical act, since its only justification is one's awareness of the actual measurement outcome. Therefore, the decoherence approach is not completely immune from the third strategy that consists of denying the exhaustivity of *physics*.

Third, if one wants to avoid the latter consequence, an option is to suspend any reference to the actual measurement outcome. But this can be done only by considering that the status of quantum symbols is exclusively probabilistic; that they always bear on possibilities and never on actuality. In this case, the decoherence process must be interpreted as nothing more than a transformation of the predictive probabilistic structure, from a quantum interferential structure to a classical Kolmogorovian structure. This is tantamount to renouncing a realist reading of the state vector as the description of something, and accordingly a realist reading of decoherence *qua* "emergence of a classical world from the quantum world." Something of the first strategy is here creeping in: dissolving the measurement problem by a non-realist interpretation of quantum symbols. Indeed, decoherence deals with a purely probabilistic aspect of the measurement problem and discards the rest of the problem, which is the issue of how a unique measurement outcome is actualized. It concerns the predictive structure of quantum mechanics and has no relevance for its alleged descriptive status.

10. Consciousness as an Extraphysical Gimmick, and the Phenomenological Deflation

Let's now come back to the third strategy for addressing the issue of actualization: suspending physicalism. To begin with, what about the (loose) words we have used until now to state what physicalism is, and what it would be like to suspend it? Physicalism has been roughly characterized as the claim that physics is universally valid in the domain of *what there is*. Suspending physicalism then means assuming that there is *something* that is not physical. Hence the dualist idea that, in addition to physical entities, there might be some vague non-physical stuff called "mind" or "consciousness"; and the correlative claim that the irruption of this non-physical stuff might explain the disruption of the law of evolution of quantum physics by way of "state reduction." The trouble is that, in addition to its well-known metaphysical weakness, such a dualistic option is a non-starter from a phenomenological standpoint.

To understand this, remember that phenomenology goes upstream from established ontologies to identify the elementary criterion that allows us in practice to believe that something belongs to the domain of *what there is*. This phenomenological criterion of *being* is that the pattern of expectations that shapes our concept of something is *fulfilled by a perceptual content of experience*. In phenomenology, *to be* is to appear, or at least to have the possibility of appearing. According, for example, to Heidegger [19], "being means appearing." From this criterion, it may easily be inferred regressively that experience, the most specific component of consciousness, *is not something*; experience *does not belong to the domain of what there is*. This sounds paradoxical, but becomes almost obvious after a little reflection. Experience is not presented in experience, for it coincides with presentation itself. Experience fulfills no expectation, for both fulfillments and expectations are experienced. Therefore, experience is not something and does not belong to the domain of what there is. Does it follow that experience is nothing at all? By no means, since experience is the universal precondition for anything to be considered as existent! We could summarize what has just been said by diverting a remark of Wittgenstein [20]: "[Experience] is not a *something*, but not a *nothing* either!" However, this remark is weaker than the conclusion of phenomenology, which could be expressed thus: "[Experience] is not a something, but it's more than anything else, since it is *in it* and *by it* that the existence of anything is ascertained."

11. Transcendental Ego and Introspection

This exterritorial status of conscious experience, which phenomenology insistantly brings out, played an important role in the reflection of some quantum physicists of the past. It is often said that a whole lineage of physicists, von Neumann, London and Bauer, Wigner, Stapp, etc., advocated the idea that the state vector reduction is triggered by consciousness. But in this list of physicists, only the two last were unambiguously dualists, dealing with consciousness as if it were something non-physical. Von Neumann and

London and Bauer were much more nuanced, sometimes coming remarkably close to a phenomenological vocabulary and approach. Even Everett, as we will see later on, can be understood as a crypto-phenomenologist.

Let's start with von Neumann [20]. His key sentence is: "But in any case, no matter how far we calculate—to the mercury vessel, to the scale of the thermometer, to the retina, or into the brain, at some time we must say: and this is *perceived* by the observer." According to von Neumann, the measurement problem would not be solved by just invoking some physical event that occurs in the brain of the observer at the end of a measuring interaction. For such an event would remain "inside the (quantum) calculation" and would therefore do nothing to break the chain of entanglement and superposition. But von Neumann does not make use of some non-physical entity either. What he mentions is only a change in the level of description, between the superposition and the sharp eigenstate. From a neutral mode of description, one switches to a situated mode of description. To a view from nowhere, one substitutes a view from somewhere (or rather for someone). A quantum entangled superposition (involving the system and anything correlated with it) holds for anyone who would like to anticipate probabilistically a measurement outcome, whereas a sharp state holds for someone who has observed this outcome and wants to take it into account for anticipating the outcomes of future measurements. No miracle occurs here, but only a change in one's self-ascribed epistemological status: from anonymous predictor to specific observer, from a neutral stance to a situated view. Both state vectors (superposed and sharp) can be used alternatively by one and the same person, according to her needs: either providing a weighted list of possible experiences available to anyone, or indicating the actual experience of someone who happens to be oneself.

This non-substantialist construal of observers and their consciousness is confirmed by von Neumann's use of the quasi-Husserlian expression "abstract ego" (Husserl would have written "transcendental ego"). According to von Neumann, the divide between the observer and the observed system can be moved back further and further until nothing (not even a brain) is left on the observer's side. It can be moved until the observer is represented only by her "abstract ego," whereas all the rest is treated as a global (quantum) system. This procedure clearly precludes any reification of the observer's residue. What is left on the observer's side is no *thing*, even though it is not *nothing*. In other terms, the expressions "abstract ego" or "transcendental ego" do not refer to some non-physical entity. They play the role of the indexical "I" that does not refer to anything or anyone [21], but indicates the source of every act of reference.

A similar conclusion can be drawn from London and Bauer's famous analysis of the measurement problem [22]. London and Bauer give priority to the act of becoming aware, not to some reified concept of consciousness. According to them, the transition from a superposition to a reduced state vector expresses a change of perspective. The entangled superposition holds from an external standpoint, whereas the reduced state holds from the internal standpoint of an observer who partakes of the measurement chain. The said observer does not need to make a measurement (which would be a sort of self-measurement) in order to know her own state. It suffices for her to resort to the privileged relationship she maintains with herself through her "faculty of introspection." By realizing immediately and non-observationally (through this faculty of introspection) that she is in a definite state, she can "constitute by virtue of her (self-)observation a new objectivity by attributing a new state to the object: (an eigenstate of the observable)." To sum up, it is not necessary for an observer to approach herself indirectly from outside (*qua* brain or reified consciousness) to get *knowledge* of herself. Some kind of direct self-knowledge is available, and this radical change of approach and angle of view is sufficient to break the quantum superpositions that hold in the standard approach, namely from the usual (external) angle of view. This alternative approach is what London and Bauer try to convey with their use of the term "introspection." But this term is ambiguous insofar as it connotates the "inspection" of some inner realm, thus assuming a kind of dualism of the subject and object of inner knowledge. It would be better to use the expression "knowledge by acquaintance" as opposed to "knowledge by description." One knows the superposed entangled state vector of a measurement chain by description, but one finally knows the observed outcome of a measurement by self-acquaintance. Here again, both modalities of knowledge hold for the same concrete situated person, but with two different stances: the stance of an anonymous predictor, and the stance of an individual observer becoming aware of the outcome of a particular measurement.

12. The Phenomenological Flavor of Everett's Interpretation

Surprisingly, the same ideas were suggested (though cryptically) by a physicist who declared that his interpretation of quantum mechanics is "realist," and who insisted that neither consciousness nor the "abstract ego" have any role to play in it. This physicist is Hugh Everett. Yet, his crucial move consists not so much in adding one more element to the measurement chain (an observer or, maybe, a recording robot), but rather in appending a new symbol to it. The new symbol is a "memory bracket" that contains a list of measurement outcomes observed and recorded in the past. There are as many memory brackets as there are terms in the entangled superposition of the global state vector of the measurement chain. So, each memory bracket is supposed to hold *relative to* the corresponding term. In the many-world meta-interpretation of Everett's interpretation, this relativity is made even more concrete by a daring reification. There, each memory bracket holds *within* the world that corresponds to this term.

However, the mere addition of a symbol to each term of the superposition is not sufficient by itself to solve the measurement problem, since no collapse is triggered by it. What really does the trick in this (or these) interpretation(s) is the *situated meaning* ascribed to the symbol "memory bracket." The measurement problem is arguably solved when one endorses one of the two following statements:

a) "In *this* universe (where *I* live), a sharp outcome *appears* to be obtained, even though in the multiverse there *is* a superposition."
b) "From *my* observer's point of view, it *appears* that a sharp outcome has been obtained, even though from the standpoint of distantiated predictors, the initial superposed state still holds."

So, the solution of the measurement problem here arises from full awareness that one occupies an idiosyncratic situation, and that this situation self-manifests in one's conscious realization of some particular measurement outcome. Here again (as in von Neumann and London and Bauer), consciousness *does nothing* to the physical world. Instead, the physical world is *reinterpreted as* what is either predicted or observed from the standpoint of a conscious agent. And this dual reinterpretation of the physical domain, as (i) that about which predictions are made after having been triggered according to the prescriptions *of conscious agents* and (ii) what is observed *by conscious agents,* is entirely encoded in a superposed state vector with memory brackets.

Turning an alleged "realist" interpretation of quantum mechanics such as Everett's into a phenomenological interpretation may sound surprising to some. But even John Bell [23] found a phenomenological reading of Everett compelling in view of the latter's "replacement of the past by memories" (with a strong critical undertone, however, since Bell accordingly accused Everett of "radical solipsism": solipsism of *present experience*).

13. QBism and Phenomenology

However, the most consistent phenomenological approach of quantum mechanics is presumably QBism [24]. QBism is an acronym for "Quantum Bayesianism" or "Quantum Bettabilitarianism." In QBism, "state" vectors are probabilistic valuations, in a Bayesian sense. They are not statements about what is the case, but statements about what each agent can reasonably expect to be the case. Ultimately, they are just expressions of subjective guesses; they express subjective agent's willingness to place *bets* about each outcome. Hence the expression "Quantum Bettabilitarianism."

What makes QBism so close to phenomenology is that it adopts a deliberately first-person standpoint (be it first-person singular or first-person plural). The project of both phenomenology and QBism is to reconstruct the so-called objective knowledge, starting everything anew from the first-person standpoint of knowers and agents. Just as any good phenomenologist, a QBist thinker suspends judgment about a presumably external domain of objects. Her act of suspension resembles what Husserl called the *epoché*. Indeed, in QBism, the symbols of quantum theories do not refer to objects nor do they denote predicates of objects. And, since the attention of QBists is no longer absorbed by claims about an object, it is reflectively redirected towards the epistemic function and the practical use of the symbols of quantum mechanics. QBists then point out that the symbols of quantum mechanics are primarily used by agents to assign probabilistic weights to the outcomes of experiments so that such agents can make consistent bets. Pictures of objects can still play a role in QBism, but only as ancillary mental scaffoldings helping researchers to determine the best possible use of probabilistic valuations. This further reflective move is similar to what Husserl called the *phenomenological reduction*.

14. QBism without Measurement Problem

The measurement problem is addressed in this spirit. About Wigner's friend's so-called "paradox," QBists do not invoke consciousness (neither Wigner's nor his friend's) to reduce the state vector of the measurement chain. However, the reason why they don't need to invoke consciousness as a *deus ex machina* is not that they believe the state vector reduction describes some completely autonomous "physical process out there," thus making consciousness irrelevant for it, but rather the opposite. According to them, the quantum "state" has no direct bearing on physical processes; it is a symbolic tool within "a calculus for gambling *on each agent's own experience*" [25]. This entails that (i) the reduction of the "state" represents no "physical" process, and (ii) conscious experience is the universal presupposition of the quantum "gambling," rather than some additional ingredient. Then, in QBism, there is no "objective reduction of the physical state," triggered by the allegedly non-physical consciousness. Even less is there the possibility of generating consciousness by this non-existent "objective reduction." In QBism, there is only a change in expectations (the dispositions to bet) that takes into account previous experiences of measurement outcomes.

We can go as far as saying that in QBism, quantum symbols bear exclusively on experiences. They bear on their being expected or their being felt, on their being conceived as possible or their being sensed as real, and nothing else. Therefore, the fact that Wigner (who is outside the laboratory) does not use the same state vector as his friend (who is inside the laboratory) is not due to some spooky action of the friend's consciousness on the physical furniture of the laboratory. It is due to a difference in the informational basis on which those two researchers endowed with conscious experience rely for elaborating their optimal bets about future experiences. "One statement refers to the friend's potential experiences, and one refers to Wigner's own" [25]. Since nothing else than conscious experience is involved in the symbols of quantum physics, no action of conscious experience on something else must be called for to account for sudden changes in these symbols.

15. QBism beyond Idealism and Instrumentalism

Does this necessarily imply some sort of solipsism, or subjective idealism? By no means. No particular subject is able to create the experienced outcome of a measurement. And no subjective preferences are involved in the probabilistic anticipation of an outcome (except as an initial guess). Each outcome comes as a partial surprise; it is *given*. And each anticipation, each bet, is framed by rational rules of coherence that ensure (in virtue of the Dutch book theorem) its effectiveness in the long run.

Claiming that "nothing else than conscious experience is involved in physics" would be shocking only for those who believe that conscious experience restrictively concerns some inner realm distinct from the outer realm of physical objects and events. In other terms, it is shocking only under the latent presupposition of dualism. But within the framework of phenomenology, the same claim is almost trivial and does not have the (absurd) consequence that only the inner realm exists, whereas the outer realm is inexistent. For in phenomenology, what there is, namely *appearance* or *experience*, is neither inner nor outer, but present. From the first-person standpoint of phenomenology, it is obvious that the so-called outer objects are always tinged of experience: they are either manifest or imagined, or dreamed, or conceived. In experience, in its anticipatory thrust and in its surprises, one finds all the resources to develop physics, with no need to take the metaphysical concept of a "real world behind the veil of appearances" too seriously. For even this metaphysical concept can be seen as one more aspect of experience: it affords representations that guide and motivate felt expectations, and it offers a background against which one can evaluate the meaning of unexpected experiences. The only particularity of physics, compared to ordinary knowledge, then lies in a tight feedback loop connecting its expectations to the technological activity that allows one to test them, through these heuristic representations.

To avoid the pitfall of pure instrumentalism or idealism, Christopher Fuchs has adopted a metaphysical position that fits quite well with the low level of ontological commitment of QBism. This metaphysical position has been called "participatory realism," [26] thus immediately displaying that it is a (non-conventional) variety of realism. The idea behind it is that the insuperable dependence of the symbols and propositions of quantum theories on our situation and experience *indirectly* reveals the nature of reality. Here, reality is so deeply entangled and holistic that our knowledge of it can only be participatory rather than representational, predictive rather than descriptive. Reality is so entangled that it cannot be described directly by a physical theory, but only suggested indirectly by the failure of descriptions and the use of pure predictions instead.

But how can such an alternative metaphysical picture of reality "from nowhere" fit with the primacy of experience, and first-person standpoint (from somewhere), that were advocated in the epistemological presentation of QBism?

16. "Participatory Realism" and Merleau-Ponty's Embodiment

Quite surprisingly, the two standpoints can easily be put in agreement. This was shown long ago by a lineage of phenomenologists stemming from Maurice Merleau-Ponty. These Merleau-Pontian phenomenologists start from first-person experience, in line with the principle of their discipline. However, when they have performed the *epochè* (or suspension of judgment), they dig into their experience to reach a particular level of it that is especially relevant for the issue of participancy. This level is the sense of *embodiment,* the experience of being-in-a-body. It provides one with full awareness of a most remarkable item called my "own body," whose "flesh" is simultaneously perceived and perceiving. A celebrated example, developed by Husserl and Merleau-Ponty, is that of my hands: my right hand can be touched and thereby perceived (say, by my left hand), but it is also capable of touching, and thereby perceiving, anything else. The latter experience of a bifacial flesh is both commonplace and stunning. It is commonplace because each one of us was born with it. And it is stunning because, through it, we witness what it is like to be coextensive with a fraction of the world. Since we are coextensive to a fraction of the world, we can but oscillate between detachment (when we perceive our own body) and radical commitment (when we realize that our own body is perceiving everything, including itself).

According to Merleau-Ponty, far from being a marginal epiphenomenon in a massively non-sentient universe, the experience of embodiment is paradigmatic. It is the only fact that fully illustrates the nature of this universe of which we partake: a highly integrated and entangled universe endowed with a potential for self-realization and self-objectification that are fully accomplished in us and through us, human beings. Merleau-Ponty then does not shy from endowing this experience of participancy with a bold cosmological significance. Drawing the consequences of his deep *epoché,* Merleau-Ponty suspends even the process of positing boundaries between bodies, and accordingly describes one's own flesh in continuity with the rest of what appears. This particular bodily flesh is then considered as a locus of intense self-revelation of what Merleau-Ponty calls the "flesh of the world": "Where should we locate the boundary between the body and the world, since the world *is* flesh?" [27]. It turns out that embodiment is precisely where the cosmological concept of participancy meets the phenomenological concept of consciousness. Feeling embodied is the most immediate experience of universal participancy we can have.

17. Illusory Separations in a Non-Separate World

But then, how did we come to believe in a separate world? And what happened in quantum physics to remind us so strongly of our participation in the universe?

We came to believe in a separate world through a process that is coextensive to our life. I called it "self-objectification," but it is even more primitive than the act of objectification that is performed by language and science. Heidegger considered that self-separation *defines* our existence, and he called it the "ek-stasis": the fact for us to be always out of ourselves, trying to project into the future or to reflect upon the past. Sartre insisted that, for us, "existing (from the latin *ex-sistere*, to stand out of oneself)" means that we never content ourselves with being what we are, but constitutively tend towards what we have to be and what we hope to reach [28]. As for Michel Henry, he bluntly asserts that "[c]onsciousness is none other than the form of existence that arises in the *tearing of Being*", or that "[c]onsciousness is nothing but the self-alienation of Being" [29]. In other terms, consciousness is consciousness-of-something, not because the universe is made of things and consciousnesses (nor things producing consciousnesses), but because consciousness arises *qua* self-splitting of what there is.

Now, how did quantum physics weaken this far-reaching process of separation and revive the idea of participancy? Imagine, Bohr writes [30], that we attempt to "orient oneself in a dark room by feeling things with a stick. When the stick is held loosely, it appears to the sense of touch to be an object. When, however, it is held firmly, we lose the sensation that it is a foreign body, and the impression of touch becomes immediately localized at the point where the stick is touching the body under investigation." According to the latter option, the blind cane becomes part of our own body when it is held firmly. The locus of our separation from the rest of the world is then pushed far from our skin, at the tip of the stick. Something similar happened with experimental devices in the era of classical science. An experimental apparatus could easily be taken as a sort of prosthesis that extends our bodies (either as an extended hand or as an extended eye) and helps us reach new types of objects. Indeed, the phenomena taking place at the tip of such apparatus obeyed the rules Kant assigned for constituting a (close or remote) domain of objective knowledge.

But these rules are no longer in effect in the domain of microphysical phenomena. Deterministic causation is suspended in the spatio-temporal domain, and the concept of "substance" (or spatio-temporal continuant) is no longer applicable. Then, experimental devices can no longer be taken as prostheses that open up a new realm of objects before us; they have not managed to achieve the neat splitting operation that is necessary for complete objectification. With no complete separation between our devices and the environment they are meant to explore, the concept of participancy imposes itself. It is true that one could still maintain a separation between our bodies and the device (by due analogy with the stick when it is held loosely), or between a proximal and a distal part of the device. However, this alternative separation now sounds arbitrary. Bohr soon noticed this arbitrariness when he insisted that part of measurement apparatuses *must* be described classically, to allow partial objectification and sufficient grip for our language: but which part, at which level? Participancy then becomes the norm, and the subject-object cut appears as a problem, a relic of the past, or an expedient tool.

18. Conclusion

To finish with a very short epilogue, I will ask two dualist questions and propose two non-dualist answers.

Do we need consciousness to reduce the state of physical systems? *No*, since "reduction" is the name we give to a revision of conscious expectations, and "physical system" is the synthetic name we use for a coherent set of consciously expected phenomena.

Can consciousness arise from a physical process such as the alleged "objective reduction"? *No*, since physical processes are nothing else than objects of consciousness, and consciousness is the flux of the self-splitting of what there is into perceiving and what is perceived, expecting and what is expected, subjective existence and its objective targets.

References

[1] M. R. Bennett & P. M. S. Hacker, *Philosophical Foundations of Neuroscience*, Blackwell Publishing, 2003, p. 6.
[2] L. Wittgenstein, *The Blue and Brown Books*, Harper Torchbooks, 1965.
[3] J. Kockelmans, *Ideas for a Hermeneutic Phenomenology of the Physical Sciences*, Kluwer, 1993.
[4] E. Husserl, *Experience and Judgment*, Northwestern University Press, 1975.
[5] J. Piaget, *Introduction à l'épistémologie génétique, 2-La pensée physique*, Presses Universitaires de France, 1974.
[6] S. French & D. Krause, *Identity in Physics: A Historical, Philosophical and Formal Analysis*, Oxford University Press, 2006.
[7] A. Grinbaum, "How device-independent approaches change the meaning of physical theory," *Studies in History and Philosophy of Science, Part B: Studies in History and Philosophy of Modern Physics* 58, 22–30, 2017.
[8] N. Bohr, "The unity of human knowledge," in N. Bohr, *Essays 1958–1962 on Atomic Physics and Human Knowledge*, Ox Bow Press, 1987.
[9] N. Bohr, *Essays 1933–1957 on Atomic Physics and Human Knowledge*, Ox Bow Press, 1987.
[10] Č. Brukner and A. Zeilinger, "Information invariance and quantum probabilities," *Foundations of Physics* 39, 677–689, 2009.
[11] C. Fuchs, N. D. Mermin, & R. Schack, "An introduction to QBism with an application to the locality of quantum mechanics," *American Journal of Physics* 82, 749–754, 2014.
[12] K. Friston, "The free-energy principle: A unified brain theory?," *Nature Reviews Neurosciences*, 11, 127–138, 2010.
[13] M. Bitbol, "Neurophenomenology of surprise," in N. Depraz & A. Celle (eds.), *Surprise Surprise! Surprise at the Crossroads of Phenomenology and Linguistics*, John Benjamins, 2019.
[14] M. Bitbol, *Physique et philosophie de l'esprit*, Flammarion, 2000.
[15] C. Rovelli, "Relational quantum mechanics," *International Journal of Theoretical Physics* 35, 1637–1657, 1996.
[16] P. Mittelstaedt, *The Interpretation of Quantum Mechanics and the Measurement Process*, Cambridge University Press, 1998.
[17] W. H. Zurek, "Decoherence, einselection, and the quantum origins of the classical," *Review of Modern Physics* 75, 715–775, 2003.
[18] B. d'Espagnat, *Conceptual Foundations of Quantum Mechanics*, Westview Press, 1999.
[19] M. Heidegger, *Introduction à la métaphysique*, Gallimard, 1967.
[20] J. von Neumann, *Mathematical Foundations of Quantum Mechanics*, Princeton University Press, 1955.
[21] G. E. M. Anscombe, "The first person," in G. E. M. Anscombe, *The Collected Papers of G. E. M. Anscombe, Volume 2: Metaphysics and the Philosophy of Mind*, Blackwell, 1981.
[22] F. London & E. Bauer, *La théorie de l'observation en mécanique quantique*, Hermann, 1939.
[23] J. Bell, *Speakable and Unspeakable in Quantum Mechanics*, Cambridge University Press, 1987.

[24] C. A. Fuchs, N. D. Mermin, & R. Schack, "An introduction to QBism with an application to the locality of quantum mechanics," *American Journal of Physics* 82, 749–754, 2014.
[25] C. A. Fuchs, "QBism, the perimeter of quantum bayesianism," arXiv: 1003.5209 [quant-ph].
[26] C. A. Fuchs, "On participatory realism,", in T. Durham & D. Rickles (eds.), *Information and Interaction*, Springer, 2018.
[27] M. Merleau-Ponty, *Le visible et l'invisible*, Gallimard, 1964.
[28] J.-P. Sartre, *L'être et le néant*, Gallimard, 1943.
[29] M. Henry, *L'essence de la manifestation*, Presses Universitaires de France, 1963.
[30] N. Bohr, "Atomic theory and the description of nature," in N. Bohr, *The Philosophical Writings of Niels Bohr*, Volume 1, Ox Bow Press, 1987.

12

Proposal to Use Humans to Switch Settings in a Bell Experiment

Lucien Hardy

1. Introduction

In this proposal I discuss performing an experiment to test Bell's inequalities [8] wherein humans are used to change the settings at the two ends. The basic idea is that we perform a Bell experiment over a scale of about 100 km and have, at each end, about 100 humans who intervene on the settings via electrical brain activity obtained by electrodes placed on their scalps to intervene on the settings (as is done in recording an electroencephalogram (EEG)). We want to have a large number of cases where the setting has been changed by human interventions at both ends while a signal as to the new value of the setting cannot have yet reached the other side. We suggest using EEG brain activity (rather than, for example, pressing a button by hand) to minimize delays. We need the experiment to be over a large distance scale and to have many humans at each end to get a sufficiently high rate that we could expect a significant effect. Having more humans at each end would increase the rate. Making the experiment longer is also good as long as we have a high enough coincidence count rate that we can still get a significant effect. The parameters suggested above (100 km with 100 people at each end) may be insufficient or more than sufficient for our purposes.

The radical possibility we wish to investigate is that when humans are used to decide the settings (rather than various types of random number generators) we might then expect to see a violation of Quantum Theory in agreement with the relevant Bell inequality. Such a result, while very unlikely, would be tremendously significant for our understanding of the world. A violation of Quantum Theory under these circumstances would, of course, be very important in and of itself—it would teach us that the world

Lucien Hardy, *Proposal to Use Humans to Switch Settings in a Bell Experiment* In: *Consciousness and Quantum Mechanics*. Edited by: Shan Gao, Oxford University Press. © Oxford University Press 2022.
DOI: 10.1093/oso/9780197501665.003.0013

was, after all, fundamentally local as well as having implications for determinism as we will discuss. But the real importance of such a result would be the demonstration that humans have a special role when compared with computers, machines, and random number generators. As we will discuss later, a natural explanation of such a result would be that it demonstrates some sort of Cartesian mind-matter duality (though one could seek out other types of explanation).

Possible radical implications aside, performing an experiment like this would push the development of new technologies. The biggest problem would be to get sufficiently high rates wherein there has been a human induced switch at each end before a signal as to the new value of the setting could be communicated to the other end and, at the same time, a photon pair is detected. This would require us to distribute entangled pairs of systems at a high rate over a scale of, at least, kilometers and possibly hundreds of kilometers. Additionally, we would have to develop fast electronics applicable to these kinds of experiments. The objective of performing this experiment would act as a stretch goal pushing us beyond what we might otherwise attempt and building on what has already been experimentally achieved [21, 5, 39, 42, 33] without attempting to have human input into the switching. Even in the most likely scenario that Quantum Theory emerges unscathed, there could still be technological payoffs to performing such an experiment. Such possible technological payoffs include: (i) developing even more robust and higher rate entanglement distribution schemes over greater distances; and (ii) opening up the field of coupling human choices to quantum systems using EEG technology (the use of EEG signals to input human choices in computers and robotics is already an active area of research [45]). We can investigate possible applications of such technology. In particular, there may be security advantages to coupling humans directly to the apparatus in quantum cryptography [44, 10, 18]. In particular, in device independent cryptography [6, 3] in which we assume there cannot be signalling faster than the speed of light, there could be security payoff to implementing locality conditions with respect to human interventions.

2. Previous Discussion

In 1989 I wrote two papers [27, 26] on the idea of using humans to choose the settings in Bell experiments. These papers did not, of course, get past the referees (and this was before quant-ph on the arXiv). Such ideas were certainly too speculative at that time. Nevertheless, my PhD supervisor, Euan Squires summarized my idea in his beautiful book *Conscious mind in the physical world* [35] (a physicist's take on the issue of consciousness):

> An... idea being studied by a research student here in Durham, L Hardy, is that there might exist genuine free agents which are outside the physically determined world. Such free agents could be responsible for "mind-acts" affecting the settings in the EPR experiment. Assuming these are constrained by the Bell inequality, they would give rise to violations of quantum theory. (Experiments along these lines would be precise tests of a well defined type of dualism. Unfortunately, the time scales involved suggest they would be very difficult to perform).

The free agents in question being, of course, humans. In 1990, when Squires wrote these words, it was clear that such an experiment was well beyond available technology. However, there now exist much more efficient sources of entangled systems and Bell experiments have been performed over kilometers (and even hundreds of kilometers). Further, fast switching techniques have been developed. It seems that, by now, an experiment like this could be achieved by sufficiently determined experimentalists. For the first iteration of such experiments it would sufficient to attempt to implement human switching without also closing the detection efficiency loop hole. More recently I wrote a paper on using humans to switch the settings [28]. That chapter focussed on deriving Bell inequalities with retarded settings (see Sec. 7) which would be useful in such an experimental test while this chapter focuses on how to actually implement an experiment.

In 1990 John Bell published his Nouvelle Cuisine paper. In this he considers a Bell-type experiment and says

> Then we may imagine the experiment done on a such a scale, with the two sides of the experiment separated by a distance of order light minutes, that we can imagine these settings being freely chosen at the last second by two different experimental physicists, or some other random devices.

Bell, however, did not appear to explore the idea that such an experiment would relate to the mind-matter duality debate. He says

> I would expect a serious theory to permit "deterministic chaos" or "pseudorandomness", for complicated subsystems (e.g. computers) which would provide variables sufficiently free for the purpose at hand.

Experimental physicists, from Bell's point of view, are an example of such subsystems.

In the last decade the idea of using humans to do the switching in Bell experiments has been mentioned in passing a number of times. Aside from [27, 26, 26, 28] there has been no discussion about the importance of such an experiment for the issue of mind-matter duality—one exception being the recent paper "Quantum and Qualia" by Adrian Kent [30] who mentions this kind of experiment in the context of a broader discussion of about consciousness. As described in Sec. 4.1 and Sec. 6, Weihs et al. [42] performed the first experiment in which random number generators were used to choose the settings. At the end of the paper they say

> Further improvements, e.g. having a human observers choose the analyzer directions would again necessitate major improvements of technology as was the case in order to finally, after more than 15 years, go significantly beyond the beautiful 1982 experiment of Aspect et al.

The possibility of using experimental physicists to choose the settings is also mentioned in the Canary Islands paper [33] by Scheidl et al. More recently, Hensen et al. [29], who performed one of three recent experiments that simultaneously closed the detector efficiency and switching loopholes, say at the end of their paper

> Even so, our loophole-free Bell test opens the possibility to progressively bound such less conventional theories: by increasing the distance between A and B (testing e.g. theories with increased speed of physical influence), using different random input bit generators (testing theories with specific free-will agents, e.g. humans).

Very recently there was a public outreach initiative—The Big Bell Test—run by a number of experimental groups around the world [1]. In this members of the public were encouraged to provide input over the internet which was used to switch the settings in various Bell experiments. No attempt was made to impose locality conditions (these human choices were clearly in the backward light cone of both ends of the Bell experiment). However, the idea resonates with the ideas being discussed here.

Since I first attempted to publish ideas along these lines in the late 1980s attitudes have clearly changed so that there may be interest in doing this kind of experiment. Further, the technology has developed tremendously so that such an experiment may be feasible. Of course, if we are going to do an experiment of this sort it is worth being very careful up front as to what the idea is we are really trying to test and that the conditions for a genuine test are realized in any actual experiment.

3. The Need for Interventions to Determine Settings While Particles Are in Flight

When we derive Bell inequalities we assume that the outcome at ends A and B are given by some result functions

$$A(a,\lambda) \qquad B(b,\lambda) \tag{1}$$

respectively. Here a and b are the settings and λ are the hidden variables. It has long been appreciated that the settings in Bell experiments need to be chosen while the particles are in flight to ensure that the choice cannot be communicated to the other end by non-superluminal signals.

Consider an experiment in which the settings at each end are static. In this case it is possible that, in the underlying physics, the setting is broadcast from each end to the other end so the outcomes at each end can depend on both settings. Then we would have result functions $A(a, b, \lambda)$ and $B(b, a, \lambda)$. This would block the derivation of Bell inequalities and, indeed, we can easily construct a local model that reproduces Quantum Theory.

We could imagine having some machine decide the setting at each end. But in this case it is possible that the machine runs according to deterministic rules. Then the earlier state of the machines can be communicated to the other end at non-superluminal speeds and from this earlier state the setting can be determined. In this case we can still have $A(a, b, \lambda)$ and $B(b, a, \lambda)$ dependencies. What we need is an *intervention* at each end that changes the setting from the value it would have taken had there been no intervention in a way that is spacelike separated from the measurement event at the other end (see Fig. 12.1).

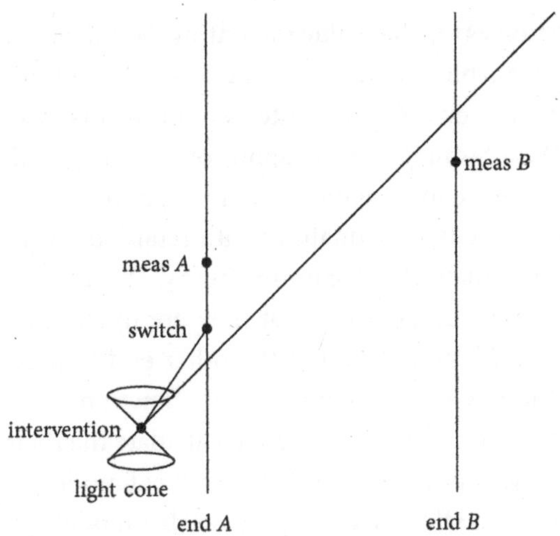

Figure 12.1 This figure shows an intervention changing the setting at end *A* such that no signal carrying information about this intervention can reach end *B*.

For clarity, it is worth mathematically elaborating this simple idea. Let the state of machine *A* at time *t* be α_t. This state may consist of hidden variables that appear in the fundamental theory (that are not directly accessible to experimentalists). Further, this state can describe any physical systems that can locally influence the setting (so the term "machine" is potentially broader than just referring to the box in the laboratory that appears to determine the setting). According to our assumptions, in the absence of interventions, this state is given by some deterministic rules from the state at time 0::

$$\alpha_t = f_t(\alpha_0). \tag{2}$$

We chose time $t = 0$ to be the last time that a light speed signal can communicate the state of machine at end *A* to the measurement event at end *B*. If an intervention happens at a time $t' > 0$ then

$$\alpha(t') \neq f_{t'}(\alpha_0). \tag{3}$$

This is, simply, what we mean by an intervention. We can write the *retarded setting* as

$$a_r = a(\alpha_0). \tag{4}$$

This is the prediction as to what value the setting would take according to the last information available at end B. We can have a result function $B(b, a_r, \lambda)$ for end B. However, when there is an intervention we may have $a_r \neq a$ where a is the actual setting. Similar remarks apply for the state, β_t, of the machine at end B. Hence, if we have interventions then we can recover Bell inequalities (actually, we can derive Bell inequalities with retarded settings as discussed in Sec. 7) which are violated by Quantum Theory. The important point about these interventions is not so much that they are freely chosen but, rather, that they "wrong foot" the attempt at the other end to predict the setting. It could be that there is another deterministic function that can predict the interventions but if it is not the one used to calculate the retarded settings at the other end (in the supposed underlying physical model) then we can still expect a violation of the Bell inequalities in such a model.

The pertinent question now is what are suitable candidates for interventions of this type. Here are some possibilities:

Random number generators: By now there are many experiments using various types of random number generators including quantum random number generators [42, 33, 29, 23, 34]

Signals from distant galaxies: It is possible, depending on what cosmological model one adopts, that distant galaxies have never been in causal contact. Hence, light from such galaxies would be a good candidate for interventions.

Humans: This is the possibility we discuss in this chapter.

The problem with random number generators is that the underlying physics that describes them may actually be deterministic. Even quantum random number generators of the kind used in recent experiments may be governed by an underlying deterministic model. Furthermore, very convincing experiments have now been performed in which Bell's inequalities were violated with random number generators. Whether signals from distant galaxies are a good way to implement interventions really depends on what cosmological model one adopts. However, it is generally believed that there are causally disconnected regions of the universe and so it is certainly it is worth pushing this direction of research. One experiment has been performed using light signals from distant stellar sources in our galaxy [25]. This is taken to be the first step in a series of experiments that could use more and more remote cosmological systems to decide the settings. The last option, using humans, is the most radical. We will discuss the interpretation of this in Sec. 8—in particular we will consider that such interventions might be related to Cartesian mind-matter duality.

4. Components of an Experiment

The basic proposal is that we perform a Bell experiment over some distance d_{sep} with N_A humans doing the switching for end A and N_B humans for end B (see Fig. 12.2). These humans would wear EEG headsets and this electrical activity would be used to change the setting of the measurements at each end of the Bell experiment. The humans do not need to be in the immediate vicinity of the given end. The humans responsible for the switching at end A, for example, could be located some distance away (in the opposite direction from end B) with the switch signal transmitted via radio frequency to the switching device.

For a certain fraction, α, of the coincidences collected, there will be a human induced switch at each end while no light speed signal can have transmitted this information to the other end. We can imagine two strategies for analyzing the data: either (i) we can consider all the data and expect to see a shift in the violation of Bell's inequalities proportional to $-\alpha$ if there

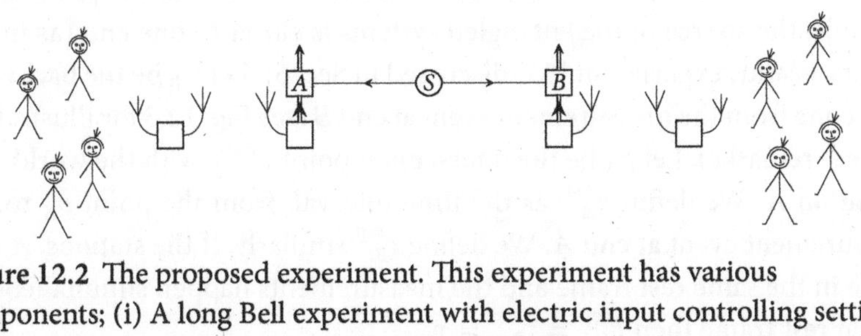

Figure 12.2 The proposed experiment. This experiment has various components; (i) A long Bell experiment with electric input controlling settings at each end; (ii) A large number of humans providing input at each end via EEG headsets; (iii) Radio frequency receivers and transmitters communicate the signal from the headsets to the switching devices at each end. In the figure we show a booster station at each end to receive the signal from the headsets and retransmit a composite signal that is used as input into the setting at each end.

is a real effect; or (ii) we can consider a template that filters for certain types of of EEG activity that we suppose are associated with human interventions and test Bell's inequalities for those cases when the EEG signal satisfies this template at each end.

We will discuss each of the basic components of the proposed experiment.

4.1 A Bell Experiment

In a Bell experiment a source of entangled pairs of systems (system 1 and system 2) are sent to two stations, *A* and *B*, where measurements are made. At each station there is a choice between two (or more) settings for the measurements. The source might, for example, be photons entangled in their polarizations. In this case the measurement would be of polarization and the settings along one of two (or more) different angles.

For the purposes of the current proposal, we need to implement a Bell experiment that, under ordinary operation (without humans performing the switching), violates the Bell inequalities by a significant amount. We require that this experiment is capable of fast switching between the two (or more) settings of the measurement at each end where these fast switches have an electrical input that determines the setting.

The two stations will be a certain distance, $d_{\text{sep}} = c\tau_{\text{sep}}$, apart. It is is possible that, in the earth rest frame the measurement at end *A* happens at a different time than the measurement at end *B*. This could happen if, for example, the source of the entangled systems is closer to one end (as in the Canary Islands experiment [33] discussed in Sec. 6). Let C_B be the backward light cone from the measurement event at end *B* (see Fig. 12.3 for illustration of these remarks). Let p_A be the intersection point of C_B with the world-line of station *A*. We define τ_A^{sep} as the time interval, from the point p_A to the measurement event at end *A*. We define τ_B^{sep} similarly. If the stations, *A* and *B*, are in the same rest frame and the measurements happen simultaneously in that rest frame then $\tau_A^{\text{sep}} = \tau_B^{\text{sep}} = \tau_{\text{sep}}$.

Other key parameters of this experiment are: (i) the time, T_{exp}, required to obtain a violation of the Bell inequalities to some given significance; and (ii) the time delay from the time the switch signal is inputted into the fast switch till the time the measurement setting is changed at each end (these will contribute to the overall delay defined in Sec. 4.2).

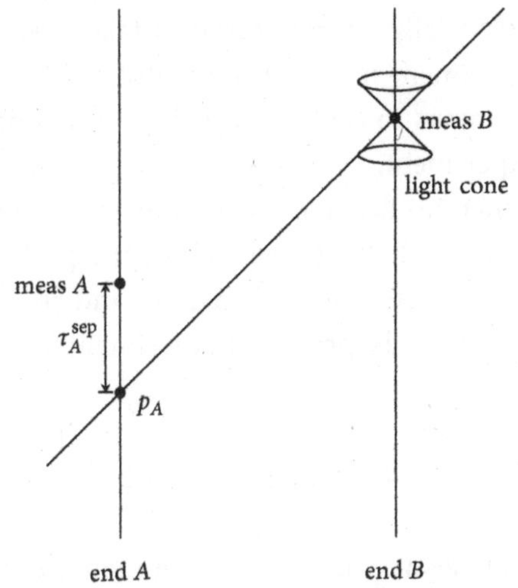

Figure 12.3 Definition of τ_A^{sep}. We define τ_B^{sep} similarly.

There have been very many experiments to test Bell's theorem beginning with Freedman and Clauser's experiment [21] in 1972 which saw a violation of Bell's inequalities by 6 standard deviations. The first experiment to switch the settings while the systems are in flight was performed by Aspect, Dalibard, and Roger in 1982 [5]. A problem with this experiment was that the switching was periodic and, furthermore, the period was unfortunately chosen just such that the setting was back to its original value in the time taken for a signal to go from one end to the other [47, 28]. The first experiment in which a random number generator was used to determine the settings was performed in in 1998 over 400 m across the campus of the University of Innsbruck [42]. There have been many such experiments since then [33, 29, 23, 34]. In particular, in 2010 an experiment was performed over a distance of 144 km in the Canary Islands between La Palma and Tenerife using a free space link [33]. There have also been some long experiments that did not attempt to choose the settings randomly. Of particular note is Gisin's team's experiment using optical fibers with the source in Geneva and the two ends in the neighboring towns of Bellevue and Bernex (a distance of 10.9 km apart) [39].

Very recently experiments have been performed that performed fast switching at the same time as also closing the fair sampling loophole [29, 23, 34]. For our present purposes we do not propose attempting to close the fair sampling loophole (as important as this is to close, it does seem that nature would have to be especially conspiratorial to take advantage of it). Future versions of this experiment might also attempt to close the fair sampling loophole and use humans to do the switching—this might be regarded as the ultimate Bell experiment.

So far, no experiment to date has used humans to determine the settings while the systems are in flight. Nevertheless, the performed experiments have many of the features that we will require and so they can be used to investigate the feasibility of the proposed experiment (as discussed in Sec. 6).

4.2 Human Input

We consider N_A humans at end A and N_B humans at end B. Each human would wear a EEG headset (to collect electrical brain activity). The signals from these headsets would be transmitted (possibly via radio frequency) to an electronic device at each end that inputs a signal into the Bell experiment to switch the setting. We could consider, instead, instructing the humans to press a button but this would operate at mechanical speeds and we would, correspondingly, require a Bell experiment on a much larger scale to hope to get useful results. Indeed, EEG analysis can predict human choices about one tenth of a second before buttons are pressed [12]. In this time a light speed signal will transverse several times the radius of the earth. Since it is not practical to have two groups of humans separated by several times the radius of the earth we propose, instead, using EEG headsets. A concern here is that we want be sure that EEG signals that are associated with human choices are not too delayed by the time they register in the EEG headset (i.e., we need to study how long it takes such signals to get from the part of the brain in which they originate to the outer surface of the skull). This, however, appears not to be a problem. According to [19] "The transmission of a signal through a volume conductor occurs nearly at the speed of light. Therefore, similar appearing waveforms that occur in different or non-adjacent scalp locations without any difference in their time of appearance most likely arise from the same cortical generator."

We do not require that the humans involved deliberately and intentionally switch the settings to this or that value but rather that they engage in appropriate behaviour that we get a high rate of human interventions as defined in Sec. 4.3 (wherein we see certain features of the EEG signals that are conjectured to correspond to interventions—see Sec. 4.3).

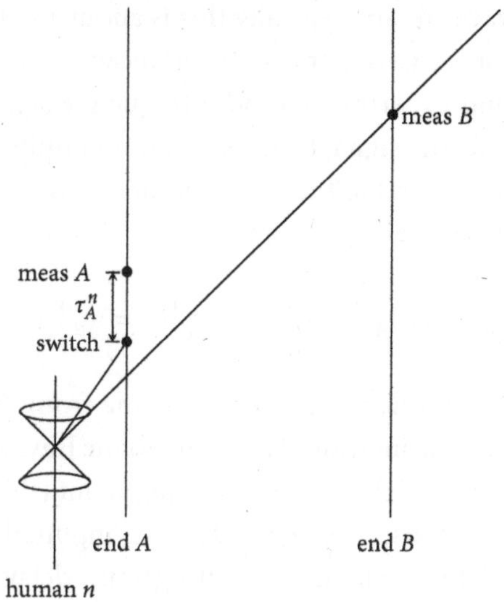

Figure 12.4 Definition of τ_A^n. The signal from the intervention to the switch may be at subluminal speeds.

For a human, *n*, at end *A* we define τ_A^n (as illustrated in Fig. 12.4) as the time interval measured at end *A* during which a human intervention has caused the setting at end *A* to be changed and the latest subsequent time at end *A* that a measurement could occur before a light speed signal could arrive at the corresponding measurement event at end *B* carrying information from the location of human *n* about this intervention. During any such time interval, we will say that the human intervention is *internal* (as it has influenced side *A* but cannot have influenced side *B*). We define τ_A as the average of τ_A^n taken over all the humans at end *A*. We define the average τ_B for humans at end *B* in a similar way. We want τ_A and τ_B to be as big as possible. We will see that the effect scales as $\tau_A \tau_B$.

The humans who switch the setting at one end, *A* say, should be on the opposite side of station *A* to station *B* (i.e., not between *A* and *B*). If we communicate the signal from these humans using radio frequencies then the N_A humans who switch the setting at end *A* could, in fact, be quite far from station *A*. In particular, we could imagine the Bell experiment being performed in a remote location where conditions allowed for free space transmission or the easy laying down of optical fibers while the humans could be located in two towns or cities some kilometers away on either side of the Bell experiment. The limiting factors here are: (i) the speed of light of radio frequency waves in air (typically this is about 1.0003 [7]), and (ii) the speed of the electronics used to transmit and receive radio frequency signals. In the Canary Island experiment a radio frequency signal was successfully used to communicate the signal from the random number generator used to chose the setting at one end of the Bell experiment over a distance of 1.2km.

We define the delay at at ends A and B as

$$\tau_A^{\text{delay}} = \tau_A^{\text{sep}} - \tau_A \qquad \tau_B^{\text{delay}} = \tau_B^{\text{sep}} - \tau_B, \tag{5}$$

respectively. Contributions to these delays come from: (i) delay getting the electrical signal from the interior of the brain to the EEG headset (this should be negligible): (ii) electrical delays in the equipment (this includes delays in any pre-processing of the EEG signal before it is inputted into the switch and delays in the device that switches the setting), (iii) delay in radio frequency signals used to connect headsets to switch due to transmission in air being slower than the speed of light; and (iv) geometric effects (if the humans are not on exactly behind the appropriate station). These geometric effects will kick in if the humans are not on the line subtended from the line AB behind the station they are giving input into.

4.3 Neurological Analysis

Electroencephalograms were invented by Berger in the 1920's (see [11]). Usually a power spectrum analysis is performed with the output in different frequency ranges being given names (for example, alpha is the frequency range 8-15Hz, beta is the frequency range 16-31Hz, . . .). Different frequency ranges are associated with different mental functions and their analysis can help diagnose different pathologies. For our purposes we are only interested in using the signal as a source of candidate interventions. We need to optimize our use and analysis of the signal with this goal in mind.

If we adopt the strategy of conditioning on certain EEG signals taken to be associated with human choice then we would need to develop an understanding of this. There is an active field of research setting up a biological computer interface (BCI). Certain types of EEG signals are strongly correlated with different choices. For example, EEG analysis can predict which key on a computer keyboard is going to be pressed in typing tasks $\frac{1}{10}$th of a second before the key is actually pressed with greater than 96% accuracy [12]. Since we are just interested in human "interventions" that can be used to decide the setting in a Bell experiment, we are open to using signals associated with any kind of mental process. If, for example, we suppose that such interventions are involved in, say, mathematical thinking then we could use EEG signals associated with this activity.

We can subject the EEG signals to a template analysis where these templates are supposed to pick out candidates for cases where there has been a human intervention. For a given template choice, we are interested in the cases that some EEG brain activity passes this template test. There is a question as to whether we (i) do this template analysis in real time (designing our apparatus to switch only when the EEG signal passes the given template) or (ii) simply use the EEG signal as the input into the device switching the setting and only subsequently do the template analysis which can be used to pick out the subensemble for which there are internal candidate interventions at both ends. The first strategy would introduce additional delays into $\tau_{A,B}^{\text{delay}}$ so the second strategy seems better. Another advantage of the second strategy is that we can filter on different templates. An advantage of the first strategy is that we can then more easily deduce the value of the retarded setting and directly evaluate the Bell inequalities with retarded settings discussed in Sec. 7.

Let r_{human} be rate at which an individual human (on average) is able to communicate signals passing the given template test to the switching device of the Bell experiment. Since the relevant EEG activity seems to be at a frequency of order 10Hz, we will suppose $r_{\text{human}} = 10\text{Hz}$. Further, let the rate at which N_A humans at end A can communicate free choices to the switch be r_A (and, similarly, at end B we have r_B). We suppose that $r_A = N_A r_{\text{human}}$ and $r_B = N_B r_{\text{human}}$ (strictly this will only be true up to a point as the electronics will fail to separately implement these switches for sufficiently high rates).

We may also want to design the EEG headsets to maximize the rate of such cases. There are numerous EEG headsets available on the market now (many of them bluetooth enabled) for recreational use. These are relatively cheap and easy for the wearer to put on. However, it is not clear whether the electronics is fast enough for our purposes. Also, we may gain a cost and logistical advantage using fewer humans with more effective headsets. Since we want to optimize the use of these headsets with a different purpose in mind than the usual medical use of EEG headsets, it may make more sense to custom build them.

4.4 Fast Electronics and Radio Transmission

We would need to develop fast electronics and radio transmission (if this is used) to get the signal from the brain activity to the switch at each end on a sufficiently fast time scale so that $\tau_{A,B}^{\text{sep}} > \tau_{A,B}^{\text{delay}}$. Existing experiments [42, 33, 29, 23, 34] have successfully implemented fast electronics from a random number generator. The additional challenge here is to do this for EEG signals from a large number of humans.

4.5 People Management

A new feature of such an experiment is that we would have to manage large numbers of people, fit them with headsets and engage them in some appropriate activity to maximize the value of r_{human}. As we will see, the rate at which we can collect useful data scales as $N_A N_B$ so the payoff of having a large number of people is very significant. It seems practical to have $N_A = N_B \approx 100$ for a time period on the order of an hour. We could imagine being more ambitious and have a thousand people at each end for time on the order of a day but then the people management problems become significantly larger. We could imagine arranging outreach events to attract people to attend. Or we could ask people to wear headsets during their normal work day. Use of wireless EEG headsets (e.g., bluetooth) would be less of an imposition on the humans involved in this experiment.

4.6 Data Analysis

We would need to develop statistical methods to analyze the data from this type of experiment. We could searching on different EEG signal templates for an effect. Further, we could attempt to record the *retarded settings* at each end (this will be discussed below) and use Bell inequalities with retarded settings.

Ideally, we would collect a data stream from each EEG headset as well as recording the setting at each end and also, of course, the outcome of the measurement. All of this information would have to be time stamped so we know what events are cross correlated. It might be unmanageable to to collect this volume of data. In this case we would have to find ways to record, in real time, the salient information. Recording all the data would allow subsequent data analysis. For example, we could investigate different candidate templates on the EEG signal.

4.7 Controls

We can implement various controls on this experiment. In the event that a violation of Quantum Theory were seen we would need, first, to see if there were some instrumental effect causing this anomalous result. For example, the EEG signals passing our templates could induce a spurious behaviour of the apparatus that switches the measurement (after all, it is easy to introduce noise that degrades any violation of Bell's inequalities). We could test for such effects by recording the EEG signals and running the experiment again using the recordings instead of humans. Additionally, we could put a delay between the EEG headsets and the switching. This could be used to set τ_A and τ_B to zero. If the observed effect were really due to having humans switching while the systems were in flight then any effect ought to disappear when such recordings or when long enough delays are used.

5. Fraction of Cases with Switching

To be able to run a successful test we need to have a sufficiently large number of coincidence detection events for which a candidate human intervention has happened at each end and there has not been sufficient time for the new setting to have been communicated to the other end. The fraction of coincidences, α, for which this is the case is

$$\alpha = r_A r_B \tau_A \tau_B, \tag{6}$$

assuming that $r_A\tau_A << 1$ and $r_B\tau_B << 1$ (so that the time intervals during which a candidate human intervention is internal do not overlap too much). If these ratios are close to one or bigger than 1 then α is close to 1. The quantity α tells us how much harder it is to perform a Bell experiment in which locality conditions are imposed using human switching. We can write this as

$$\alpha = N_A N_B r_{\text{human}}^2 \tau_A \tau_B. \tag{7}$$

We see that having many humans at each end gives us a big payoff.

The coincidence rate of useful events (where a signal about neither candidate human intervention can have reached the other end) is

$$r_{\text{coinc}}^{\text{human}} = \alpha r_{\text{coinc}}, \tag{8}$$

where r_{coinc} is the rate of coincidence detections for the Bell experiment when we do not worry about imposing locality conditions for free choices.

The time taken to collect significant statistics is

$$T_{\text{exp}}^{\text{human}} = \frac{T_{\text{exp}}}{\alpha} = \frac{T_{\text{exp}}}{N_A N_B r_{\text{human}}^2 \tau_A \tau_B}, \tag{9}$$

(where T_{exp} was defined earlier as the time taken to collect enough data to get a violation of Bell's inequality to some given significance in the raw Bell experiment).

6. Feasibility

It is instructive to insert some numbers into equation (9). We will put $N_A = N_B = 100$ and $r_{\text{human}} = 10$ Hz. We assume that $\tau_{A,B} \approx \tau_{A,B}^{\text{sep}}$ (so delays are small). In previous experiments involving fast switching (but not humans) it was claimed that this condition was met for the electronics [42, 33, 29, 23, 34]. We will take $\tau_{A,B}^{\text{sep}}$ and T_{exp} from three previous experiments to assess feasibility. The relevant data from these experiments are shown in Fig. 12.5. The bottom line is to know the time, $T_{\text{exp}}^{\text{human}}$, it would take to perform a statistically significant experiment with humans doing the switching.

An experiment reported in 1997 had the source in Geneva [39] and the two ends in the neighboring towns of Bellevue and Bernex (a distance of 10.9 km apart). This experiment did not implement switching but it provides some numbers we can use to look at feasibility. It took 20 s to collect each data point. After analyzing data from hundreds of such data points and performing a best fit, a violation of Bell's inequalities of about 10 standard deviations was seen. With $N_A = N_B = 100$ and $r_{\text{human}} = 10$ Hz we get $\alpha \approx 10^{-3}$ for $d_{\text{sep}} = 10.9$ km. Thus we would require about 5 hours per data point in an experiment with humans. Hundreds of such data points were

Experiment	d_{sep}	T_{exp}(significance)	α	T_{exp}^{human}(significance)
Geneva 1997	10.9 km	1 hour(10)	10^{-3}	40 days(10)
Innsbruck 1998	400 m	10 s(30)	10^{-6}	4 months(30)
Canary 2010	50 km	10 min(16)	10^{-2}	16 hours(16)

Figure 12.5 The first three columns of this table provide information pertaining to three experiments: The 1997 Geneva experiment [39], the 1998 Innsbruck experiment [42], and the 2010 Canary Islands experiment [33]. The distance between the two ends is d_{sep} and time taken to collect the data is T_{exp}(significance) where the significance is given in parenthesis as the number of standard deviations violation of Bell's inequalities. The last two columns provide an estimate for the faction α is provided (with 100 humans at each end assuming $r_{human} = 10$ Hz) and the time taken, T_{exp}^{human}(significance), to perform an experiment with humans providing the switching (with the number of standard deviations given in parenthesis again).

collected in this experiment. In principle, we only require four data points (as we have two settings at each end) so this would require about 20 hours. However, that the 10 standard deviations mentioned above was a best fit with hundreds of data points giving $T_{exp} \approx 1$ hour. Consequently, we would actually require $T_{exp}^{human} \approx 1000$ hours ≈ 40 days.

The 1998 Innsbruck experiment [42] had $d_{sep} = 400$ m. With the above choices we obtain $\alpha \approx 10^{-6}$. In this experiment a violation of Bell's inequalities was observed to 30 standard deviations using data (comprising about 15,000 coincidences) collected over only 10s. A similar experiments with humans would require 10×10^6 seconds per data point (or about 4 months). It is clearly not feasible to run an experiment like this with 200 people in place for 4 months.

In the 2010 experiment performed between in the Canary Islands between La Palma and Tenerife [33], the distance between the two stations was 144km. The source was located at La Palma where one photon from each entangled pair was sent through a 6km coil of optical fibre before measurements and the other photon was transmitted through 144km of free space to a telescope located at Tenerife where measurements were made. The distance between the two ends of the experiment in the frame of reference in which the measurements were simultaneous was calculated to be 50 km. By considering the geometry of the setup we have $c\tau_A = 6$ km and $c\tau_B = 144$ km. With 100 humans at each end and a rate, $r_{human} = 10$ Hz, as above we get $\alpha \approx 0.01$. This is clearly much better as we would only have to collect data for about 100 times as long to get similarly significant figures. In this experiment it took 600 s to collect enough data for a violation of Bell's inequalities of 16 standard deviations so we should be able to get a similar significance in 16 hours.

Sixteen hours is, perhaps, too long to keep 200 people wearing EEG headsets (in addition to keeping the experiment stable). However, it is not too far from being feasible. If we accept a lower significance then we could get a result more quickly. The error scales as $\frac{1}{\sqrt{N}}$, where N is the number of coincidence counts. Hence, the number standard deviations seen scales as $\sqrt{N}$ which scales as $\sqrt{\frac{T}{\alpha}}$, where T is the time we collect data. Then, all other things being equal, we can get a violation of the Bell inequalities (with the human switching locality condition imposed) of about 6 standard deviations in about two hours (or, a similarly significant violation of quantum theory if the effect we are looking for actually exists with the given value of r_{human}). Of course, many other improvements could be made. For example, in recent years photon detectors with much higher efficiencies have been built. The coincidence rate depends on the square of the detector efficiency so this could make a big difference. We could also imagine a more symmetric geometry in which a central source is viewed by two telescopes each at a distance on the scale of 100 km. Additionally, we could consider using more people at each end.

These feasibility calculations all used $r_{\text{human}} = 10$ Hz. The main justification for this is that this is, roughly, the frequency of EEG signals. Even in the case that humans can make interventions that would violate Quantum Theory in a Bell experiment, it is possible that r_{human} is much smaller. Since the effect scales as r^2_{human} we are particularly sensitive to this. It is also possible that r_{human} is bigger than 10 Hz (there is fine grained structure on the EEG scans at higher frequencies than 10 Hz).

One advantage of this experiment is that we can use the massive volume of data that is collected even when the locality conditions are not satisfied (i.e., when there has not been a human induced switch at each end) to stabilize the experiment. These data will violate Bell's inequalities.

7. Bell Inequalities with Retarded Settings

In [27, 26, 28] I derived Bell inequalities with retarded settings for the purpose of this type of experiment. The idea is to take into account the retarded values of the settings. The retarded value, b_r, of the setting at end

B as seen at end A is the value the setting at end B is predicted to take on the basis of information that can be locally communicated to A (see Sec. 3 for more details). If there is a human intervention changing the setting at end B happening after the time a signal could have been sent from end B to end A, then the actual setting, b, and retarded setting, b_r, will be different. We can define a and a_r as the actual and retarded settings at end A similarly.

We assume that the outcome at end A can depend on a, a_r, b_r, and some local hidden variables (but not on b). We write this as

$$A(a, a_r, b_r, \lambda), \tag{10}$$

where this is the result of the measurement. Similarly, at end B we have

$$B(b, b_r, a_r, \lambda). \tag{11}$$

We will assume that the results at each end are equal to ± 1. Here $\lambda \in \Gamma$ is a hidden variable with some distribution $\rho(\lambda) \geq 0$ such that

$$\int_\Gamma \rho(\lambda) d\lambda = 1. \tag{12}$$

We define the correlation function

$$E(a, b|a_r, b_r) = \int_\Gamma A(a, a_r, b_r, \lambda) B(b, b_r, a_r, \lambda) d\lambda. \tag{13}$$

We use the mathematical result that

$$X'Y' + X'Y + XY' - XY = \pm 2. \tag{14}$$

where $X, X', Y, Y' = \pm 1$. We put

$$X = A(a, a_r, b_r, \lambda) \tag{15}$$

$$X' = A(a', a_r, b_r, \lambda) \tag{16}$$

$$Y = B(b, a_r, b_r, \lambda) \tag{17}$$

$$Y' = B(b', a_r, b_r \lambda) \tag{18}$$

If we substitute these into (14) and integrate over λ, we obtain

$$-2 \leq E(a', b'|a_r, b_r) + E(a', b|a_r, b_r) + E(a, b'|a_r, b_r) - E(a, b|a_r, b_r) \leq +2. \tag{19}$$

These are the Bell inequalities with retarded settings in Clauser Horne Shimony Holt form (i.e., based on the CHSH [15] of the Bell inequalities). We can also also obtain such inequalities in Clauser Horne [14] form (see [27, 26, 28]). Note that (unlike in previous papers [27, 26, 28]) we have allowed the result at end A to depend on the retarded setting, a_r at this end (and similarly at end B).

It is shown in [28] that it is easy to write down a local model (in which dependence on retarded values is allowed) that reproduces the predictions of Quantum Theory when the actual and retarded settings are equal. The model works as follows. Let

$$0 \leq \lambda < 2\pi \qquad \rho = \frac{1}{2\pi}, \tag{20}$$

where $\rho(\lambda)$ is the distribution function (so we have a flat distribution). Define

$$A(a, b_r, \lambda) = \begin{cases} +1 & \text{for } \theta_L \leq \lambda < \theta_L + \pi \\ -1 & \text{for } \theta_L + \pi \leq \lambda < \theta_L + 2\pi \end{cases} \tag{21}$$

and

$$B(b, a_r, \lambda) = \begin{cases} +1 & \text{for } \theta_R \leq \lambda < \theta_R + \pi \\ -1 & \text{for } \theta_R + \pi \leq \lambda < \theta_R + 2\pi \end{cases} \tag{22}$$

It is easy to prove that

$$E(a, b|a_r, b_r) = 1 - \frac{2|\theta_R - \theta_L|}{\pi}. \tag{23}$$

If we set

$$\theta_L = -\frac{\pi}{4}(1 + \cos(a - b_r)) \quad \theta_R = \frac{\pi}{4}(1 + \cos(a_r - b)), \tag{24}$$

we obtain

$$E(a, b|a_r, b_r) = -\frac{1}{2}(\cos(a - b_r) + \cos(a_r - b)) \tag{25}$$

When the retarded settings are equal to the actual settings, we get

$$E(a, b|a, b) = -\cos(a - b), \tag{26}$$

in agreement with Quantum Theory. Note, incidentally, that with this model we get a violation of Quantum Theory even when the actual and retarded settings are different for only one end. This is not demanded by the Bell inequalities with retarded settings. It should be possible to build a model that gives the quantum predictions as long as the actual and retarded settings are equal for at least one of the two ends.

If we have some candidate as to what counts as an intervention then we can, in principle, determine the value of the retarded settings. Then we can test the retarded inequalities directly. We can only be sure that a local hidden variable model will violate Quantum Theory when the actual and retarded settings are different at each end. It is, however, worth investigating what happens when this locality condition is only obtained at one end. Indeed, it is much easier to implement these conditions experimentally. Furthermore, as we have just seen, there do exist models having the property that Quantum Theory is violated even when the actual and retarded settings are different for only one end.

One way to be sure that we can determine the retarded settings is to only allow the setting to be changed by those EEG signal that have been identified as candidate interventions (these are our EEG signal templates). Then the retarded setting is equal to the setting at the earlier time (intersected by the past light cone from the measurement event on the other side). However, this requires us to put electronics in place up front that can implement this conditional switching. Such electronics would introduce additional delays reducing the overall effect we see. Further, this would only allow us to investigate one candidate EEG signal template. It is better to allow the all of the EEG signal to switch the settings. Then we can analyze the data for different candidate EEG signal templates. We can only expect to marshal hundreds of people to participate in this experiment for limited time and so it is better to be able to investigate multiple EEG signal templates.

If we are not able to determine the retarded settings then we should think about whether the distribution over them is even or not. In [28] it is shown how to go from Bell inequalities with retarded settings to standard Bell inequalities under certain assumptions about the distribution over the retarded settings. Failure to investigate this might allow local hidden variable models that take into account the retarded settings to produce a violation of the standard Bell inequalities. An extreme example of this type is the original Aspect et al. experiment [5] which used periodic rather than random switching [28, 47].

8. Connection to Mind Matter Debate

8.1 Consciousness

Humans are conscious. There is no generally agreed scientific account of this. One viewpoint is that consciousness is something that will emerge in systems that are sufficiently complex in the right way. For example, a computer program that is functionally indistinguishable from a human (as in Alan Turing's "imitation game" [40]) might be argued to be conscious. Another possibility is that certain sorts of physical process give rise to consciousness. Roger Penrose has suggested that quantum superposition may play a role in mental processes [32]. The idea is that objective reduction events (collapses of the wave function) in the brain demanded from arguments from Quantum Gravity are non-computable and provide the missing link to explain consciousness. For such quantum effects to play a role in cognition, we would need quantum superposition to be possible in the brain (otherwise we would expect a fully classical account to be possible). Penrose and Hameroff proposed that quantum superpositions can be maintained in microtubules in the brain [24]. Others have weighed in on this debate. Max Tegmark argued that decoherence effects are too strong in the brain to allow quantum superposition in microtubules [38]. Matthew Fisher proposed, instead, that nuclear spins could maintain quantum superpositions allowing quantum effects to play a role in cognition [20].

It is difficult to conceive of how something as distinctive and singular as consciousness could emerge once something becomes complex. Similarly, it is hard to see how consciousness could be a result of certain types of physical process. Why, for example, would non-computability bring about the sensation of being conscious? Why are some systems (humans for instance) capable of having experiences? This was called the "hard problem of consciousness" by David Chalmers [13].

8.2 Cartesian Dualism

One viewpoint is that that consciousness is due (at least in part) to some sort of non-physical mind outside regular physics that is endowed with the property of consciousness. This is the Cartesian duality point of view put forward by Descates in 1641 (translated in [16]). As more and more aspects of human mental functioning have been accounted for by studying the brain, the case for this kind of dualism has receded. Furthermore, dualism has echoes of a pre-scientific attitude towards nature that is now largely discredited in scientific circles. However, if we were able to make some empirical progress of the sort suggested in this chapter, then these criticisms would be mute—experiment is, after all, the final arbitrator in science.

Even without taking into account issues arising from Bell's theorem, dualism ought to have some empirical consequences. An argument for this in [35] is to note that the word "consciousness" appears in ink in the dictionary. The fact that the atoms in the ink got into that particular configuration must have been influenced by whatever is responsible for the attribute of consciousness. A community of non-conscious robots would, presumably, not invent and consistently use the term "consciousness." In other words, there seems to be little support for a notion of duality in which the minds merely passively observe without influencing matter.

From the dualistic viewpoint we can tell a story in which the mind acts on the brain and thereby imparts information into the physical world (that, somewhere down the line, can lead to certain configurations of ink in a dictionary). We could imagine looking at the detailed behavior of atoms inside the brain searching for a violation of the standard rules of physics. However, it is not really clear what we would be looking for and, besides, it would be very hard to verify a violation of Quantum Theory (or whatever the prevailing scientific theory is) in such a messy environment. The proposal described in this chapter offers a way forward without having to make any kind of precision measurements inside the brain.

From our present point of view such "mind-acts" provide candidate interventions that could be used to determine the settings in a Bell experiment. The important point is not whether humans have free will as such, but rather whether the effects of mind on matter can be "anticipated" by the laws of physics.

We will describe two types of model that would lead to a violation of Quantum Theory in accord with Bell's inequalities when humans are used to determine the settings. These are *local (super)-deterministic dualistic theories* (LDD) [27, 26] and *local (super)-deterministic interventionist brain theories* (LDFB). It is not entirely clear that the second type of theory can be consistently formulated. We will discuss this.

Although there has been no discussion of mind-matter duality in the context of the Bell experiment (aside from [27, 26, 28]) there has been much discussion of it in other areas of Quantum Foundations—in particular in attempted resolutions of the measurement problem. In 1932 von Neumann [41] argued that it makes no difference whether the projection postulate is applied on the system immediately prior to measurement, at the measurement apparatus, or at the level of the brain (or anywhere in between). In 1939 London and Bauer proposed that it happens at the point when observer becomes conscious of the measurement outcome (their article is reprinted in [43]). Wigner pushed this point of view also coming up with his well-known Wigner's friend example. In more recent years, Stapp has developed a point of view along these lines [36]. A variant of the many worlds interpretation is the many minds interpretation [46, 4, 31, 17]. Gisin asserted that the Copenhagen interpretation is dualistic in nature [22].

The application of dualism proposed in this chapter is quite different. In particular, the proposal in this chapter does not relate to the measurement problem of Quantum Theory as such. Rather, it is an attempt to provide interventions in determining the settings of a Bell experiment.

While the existence of consciousness is motivation to posit dualism, it is not so clear that we can claim that dualism *explains* consciousness. In the end, in science, we are only able to offer a place for the most fundamental properties we see (less fundamental properties are explained in terms of these more fundamental properties). Dualism, at least, offers place for consciousness understood as a fundamental property.

8.3 Super-Deterministic Theories

Before we bring discussion of humans and minds into the picture we want to recap the idea of super-determinism. It is well known that it is possible to maintain locality in the face of the observed violations of Bell's inequalities if we have determinism [27, 26, 9, 33, 37]. The idea being that the settings at each end are taken to be determined by conditions in the past light cone of both ends. Hence the information about the setting can be communicated to the other end without violating locality. This is usually called super-determinism [9]. One way to do this is to (extravagantly) imagine that information about the initial state of the entire universe is encoded into every spacetime point.

We could attempt to guard against this super-determinism by performing experiments wherein signals from causally disconnected parts of the universe are used to determine the settings at each end (see discussion in Sec. 3).

8.4 Local (Super)-Deterministic Dualistic Theories

Now we introduce humans into the picture. We suppose some sort of mind-matter duality in which the physical universe is local and super-deterministic. In the absence of minds, then, it is possible to violate Bell inequalities. Now we suppose that minds act on the physical universe locally introducing "new information" into the physical universe through the brain. This information spreads out locally such that, one second after such a mind-act, this new information can be available 3×10^8 m away. It is clear that we can derive Bell inequalities for this kind of situation in which humans choose the settings at each end (so long as we ensure that the new information cannot have reached the measurement event at the other end). We call theories of this kind local (super)-deterministic dualistic theories (LDD). Such theories were originally put forward in [27, 26].

8.5 Local (Super)-Deterministic Interventionist Brain Theories

To examine whether we would be forced to dualism were Quantum Theory violated (in accord with Bell's inequalities) when humans are used to choose the settings, we can attempt to outline a non-dualist class of theories that would also lead to this result. It could be the case that physics is super-deterministic except for systems that are complex in the kind of way that happens inside brains. Were the world described by such a theory then settings chosen by humans will count as interventions. We call such theories local (super)-deterministic interventionist brain theories (LDIB).

It is difficult to see how to build these type of theories especially if we want to have local microphysical laws (a reasonable demand if we are interested in constructing a local theory). For then the local microphysical laws would have to apply to small collections of atoms in the brain as they apply to small collections outside. For outside atoms these laws would have to be super-deterministic so as to account for already performed Bell experiments in a local way. But theories with local microphysical laws satisfy reductionism—we can account for the behavior of a composite system in terms of the behavior of its parts. But then it would seem that brains, taken as a whole, should be super-deterministic. This is not a rigorous proof against the possibility of LDIB theories but it does point to some serious difficulties with constructing them.

If we imagine it were established fact that we only get a violation of Quantum Theory in accord with Bell's inequalities when humans (and, presumably, other animals) are used to choose the settings then we would have to explain this in terms of local physics. In particular, we would have to explain why (in this imagined scenario) other similarly complex physical systems do not produce a similar violation of Quantum Theory.

We could turn this round and attempt to outline a theory in which only systems with certain types of complexity lead to the ability to make interventions and that these become natural candidates for conscious systems. An analogue for this idea is to imagine the surface of a lake where ripples and waves follow the usual laws of wave physics. However, a fish poking its nose at the surface from below, or a bird skimming the surface from above could not be anticipated by the laws of wave physics and would appear as interventions. In this analogue, however, we will not see a violation of Bell inequalities as such. Consequently, much of the motivation for the foregoing discussion is absent. Further, robotic fish and flying drones would produce the same effects as fish and birds in this system whereas the hypothetical result we are considering is wherein humans (and other animals) produce different results than robots, random number generators, and other machines.

It is doubtful that we can consistently build LDIB theories for the above reasons. The main problem is that there may be breakdown of reductionism. If we consider extracting our random choices from just a few atoms (inside or not inside a brain) then this, surely, would not lead to a violation of Quantum Theory in a Bell experiment.

8.6 Turing-Style Test

Turing was interested in the question of whether programable computers can simulate humans. He proposed a test [40] in which a human interrogator can ask questions to a mystery system. This mystery system is either a human or some artificially intelligent device (a computer with a suitable program). The interrogator's questions are communicated by a keyboard and he reads the answers off a screen. The objective for the interrogator is to guess whether he is talking to a human or an artificial intelligence device (see Fig. 12.6). If such artificial intelligence devices can consistently fool human interrogators then we have evidence that human brains are effectively equivalent to a computer program. This would provide evidence against mind matter dualism—if a computer is functionally equivalent to a human then what need is there for some sort of non-physical mind? The Loebner Prize is an annual competition running since 1991 to write a computer program that can pass the Turing test. This has shown a steady improvement in the ability of computers to fool human interrogators [2]. However, humans still outperform computers in these tests.

One problem with the Turing test is that it depends on the subjective judgement of interrogator. Here we propose an alternative test, utilizing a Bell experiment, to distinguish humans from artificial intelligence devices. Now we need two humans or two AI devices (or whatever type of system we want to use to compete with humans; the "I" in AI may more appropriately stand for "intervention"). The role of the interrogator is now played by the Bell experiment. The two humans or two artificial intelligence devices are placed so as to provide input to switches at each end of the Bell experiment (see Fig. 12.7). The objective is to cause a Bell experiment (which under ordinary operation, violates Bell inequalities by a significant amount) to satisfy the Bell inequalities only by providing the input to this switching. In the case that the world is described by a local (super)-deterministic dualistic theory, humans will be able to do this but artificial intelligence devices will

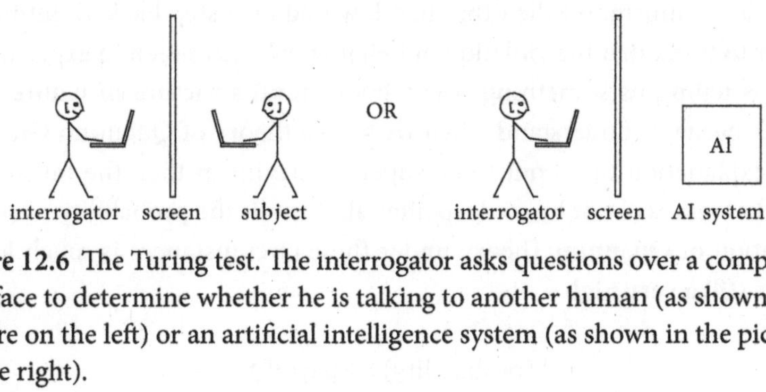

Figure 12.6 The Turing test. The interrogator asks questions over a computer interface to determine whether he is talking to another human (as shown in picture on the left) or an artificial intelligence system (as shown in the picture on the right).

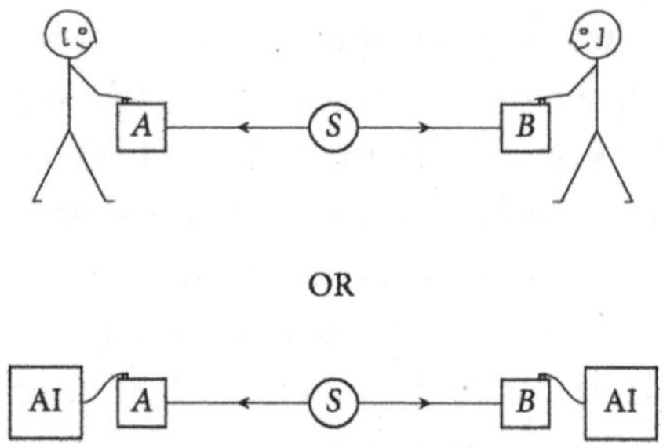

Figure 12.7 A Turing style test but using a Bell experiment. Pairs of humans compete against pairs of AI systems.

not. Unlike the Turing test, this test is objective. Were humans able to pass the test while artificial intelligence machines were not then this would provide evidence for mind-matter dualism.

9. Discussion

Quantum Theory has been tremendously successful empirically. It seems very unlikely that a Bell experiment using humans as described in this chapter, this would lead to a violation of Quantum Theory. On my more optimistic days I would put the probability at about 1–2% though I suspect many of my colleagues would give a much lower figure. To have a violation of Quantum Theory in agreement with Bell inequalities under these circumstances is best motivated if we have (i) mind-matter duality and (ii) super-determinism. Personally I think that we need some radical change to the scientific world view to deal with the hard problem of consciousness and so I would give a relatively high weighting to dualism (perhaps 30%). Super-determinism, on the other hand, would be a step back scientifically. I prefer to think that the violation of Bell inequalities as seen in experiments so far is telling us something deep about causal structure in nature (that we will, perhaps, understand when we have a theory of Quantum Gravity). If the explanation turns out to be super-determinism then the message of Bell's theorem will not be so deep after all. Even if the probability of seeing a violation of Quantum Theory under these circumstances is much lower, there is still a very high

$$(\text{probability}) \times (\text{payoff}) \tag{27}$$

Indeed, the payoff, in scientific terms, would be tremendous both in terms of importance for our understanding of Quantum Theory and, even more significantly, our understanding of mind.

Even if this experiment does not lead to a violation of quantum theory it shows how we can talk scientifically about mind-matter dualism. There exists a class of scientifically testable theories invoking duality that are open to falsification.

Acknowledgments

I am especially grateful to Mike Lazaridis for suggesting I write up a proposal for this experiment. This chapter is a direct consequence of that discussion. I am also grateful to Hilary Carteret, Adrian Kent, Matthew Leifer, Markus Mueller, Rob Spekkens, Steven Weinstein, and Elie Wolfe for discussions. I would like to thank the staff at Perimeter Institute Black Hole Bistro where a substantial fraction of this work was done.

This project/publication was made possible through the support of a grant from the John Templeton Foundation. The opinions expressed in this publication are those of the author(s) and do not necessarily reflect the views of the John Templeton Foundation.

Research at Perimeter Institute is supported by the Government of Canada through the Department of Innovation, Science and Economic Development Canada and by the Province of Ontario through the Ministry of Research, Innovation and Science.

References

[1] The big bell test. http://bigbelltest.org, 2016.
[2] The loebner prize. http://www.loebner.net/Prizef/loebner-prize.html, 2016.
[3] Antonio Acin, Nicolas Gisin, and Lluis Masanes. From Bell's Theorem to Secure Quantum Key Distribution. *Physical review letters*, 97(12):120405, 2006.
[4] David Albert and Barry Loewer. Interpreting the Many Worlds Interpretation. *Synthese*, 77(2):195–213, 1988.
[5] Alain Aspect, Jean Dalibard, and Gérard Roger. Experimental Test of Bell'S Inequalities Using Time-Varying Analyzers. *Physical Review Letters*, 49(25):1804, 1982.
[6] Jonathan Barrett, Lucien Hardy, and Adrian Kent. No Signaling and Quantum Key Distribution. *Physical Review Letters*, 95(1):010503, 2005.
[7] Bradford R Bean and John D Horn. Radio Refractive Index Climate Near the Ground. *J. Res. NBS*, 63(3):259–271, 1959.

[8] John S. Bell. On the Einstein-Podolsky-Rosen Paradox. *Physics*, 1(3):195–200, 1964.
[9] John S. Bell. La Nouvelle Cuisine. In A. Sarlemijn and P. Kroes, editors, *Between Science and Technology*, page 97. Elsevier, 1990.
[10] C. H. Bennett. Quantum Cryptography: Public Key Distribution and Coin Tossing. In *International Conference on Computer System and Signal Processing, IEEE 1984*, pages 175–179, 1984.
[11] Hans Berger. Über Das Elektrenkephalogramm Des Menschen. *European Archives of Psychiatry and Clinical Neuroscience*, 87(1):527–570, 1929.
[12] Benjamin Blankertz, Gabriel Curio, and Klaus-Robert Müller. Classifying Single Trial eeg: Towards Brain Computer Interfacing. *Advances in Neural Information Processing Systems*, 1:157–164, 2002.
[13] David J. Chalmers. Facing up to the Problem of Consciousness. *Journal of Consciousness Studies*, 2(3):200–219, 1995.
[14] John F. Clauser and Michael A. Horne. Experimental Consequences of Objective Local Theories. *Physical Review D*, 10(2):526, 1974.
[15] John F. Clauser, Michael A. Horne, Abner Shimony, and Richard A. Holt. Proposed Experiment to Test Local Hidden-Variable Theories. *Physical Review Letters*, 23(15):880, 1969.
[16] René Descartes. *Meditationes de Prima Philosophia*, translated by John Cottingham, Robert Stoothoff, and Dugald Murdoch, volume 2. Cambridge University Press, 1985.
[17] Matthew J. Donald. Progress in a Many-Minds Interpretation of Quantum Theory. *arXiv preprint quant-ph/9904001*, 1999.
[18] Artur K. Ekert. Quantum Cryptography based on Bell's Theorem. *Physical Review Letters*, 67(6):661, 1991.
[19] Bruce J. Fisch and Rainer Spehlmann. *Fisch and Spehlmann's EEG Primer: Basic Principles of Digital and Analog EEG*. Elsevier Health Sciences, 1999.
[20] Matthew P. A. Fisher. Quantum Cognition: The Possibility of Processing with Nuclear Spins in the Brain. *Annals of Physics*, 362:593–602, 2015.
[21] Stuart J. Freedman and John F. Clauser. Experimental Test of Local Hidden-Variable Theories. *Physical Review Letters*, 28(14):938, 1972.
[22] Nicolas Gisin. Collapse. What Else? *arXiv preprint arXiv:1701.08300*, 2017.
[23] Marissa Giustina, Marijn A. M. Versteegh, Sören Wengerowsky, Johannes Handsteiner, Armin Hochrainer, Kevin Phelan, Fabian Steinlechner, Johannes Kofler, Jan-Åke Larsson, Carlos Abellán, et al. Significant-Loophole-Free Test of Bell'S theorem with Entangled Photons. *Physical Review Letters*, 115(25):250401, 2015.
[24] Stuart Hameroff and Roger Penrose. Orchestrated Reduction of Quantum Coherence in Brain Microtubules: A Model for Consciousness. *Mathematics and Computers in Simulation*, 40(3–4):453–480, 1996.
[25] Johannes Handsteiner, Andrew S Friedman, Dominik Rauch, Jason Gallicchio, Bo Liu, Hannes Hosp, Johannes Kofler, David Bricher, Matthias Fink, Calvin Leung, et al. Cosmic Bell Test: Measurement Settings from Milky Way Stars. *Physical Review Letters*, 118(6):060401, 2017.
[26] Lucien Hardy. Local Determinism and Mind-Matter Duality. *preprint*, 1989. Available on request.
[27] Lucien Hardy. Proposal for Experiment to Investigate Local Deterministic Models Which Incorporate Non-Material Mind. *preprint*, 1989. Available on request.

[28] Lucien Hardy. Bell inequalities with Retarded Settings. In *Quantum [Un] Speakables II*, pages 261–272. Springer, 2017. arXiv:1508.06900.
[29] Bas Hensen, H Bernien, AE Dréau, A Reiserer, N Kalb, MS Blok, J Ruitenberg, RFL Vermeulen, RN Schouten, C Abellán, et al. Loophole-Free Bell Inequality Violation Using Electron Spins Separated by 1.3 Kilometres. *Nature*, 526(7575):682–686, 2015.
[30] Adrian Kent. Quanta and Qualia. *arXiv preprint arXiv:1608.04804*, 2016.
[31] Michael Lockwood. 'Many Minds' Interpretations of Quantum Mechanics. *The British Journal for the Philosophy of Science*, 47(2):159–188, 1996.
[32] Roger Penrose. The Emperor's New Mind. *RSA Journal*, 139(5420):506–514, 1991.
[33] Thomas Scheidl, Rupert Ursin, Johannes Kofler, Sven Ramelow, Xiao-Song Ma, Thomas Herbst, Lothar Ratschbacher, Alessandro Fedrizzi, Nathan K. Langford, Thomas Jennewein, et al. Violation of Local Realism with Freedom of Choice. *Proceedings of the National Academy of Sciences*, 107(46):19708–19713, 2010.
[34] Lynden K. Shalm, Evan Meyer-Scott, Bradley G. Christensen, Peter Bierhorst, Michael A. Wayne, Martin J. Stevens, Thomas Gerrits, Scott Glancy, Deny R. Hamel, Michael S. Allman, et al. Strong Loophole-Free Test of Local Realism. *Physical Review Letters*, 115(25):250402, 2015.
[35] Euan J. Squires. *Conscious Mind in the Physical World*. CRC Press, 1990.
[36] Henry P. Stapp. *Mind, Matter, and Quantum Mechanics*. Springer, 2004.
[37] Gerard T. Hooft. The Cellular Automaton Interpretation of Quantum Mechanics. *arXiv preprint arXiv:1405.1548*, 2014.
[38] Max Tegmark. Importance of Quantum Decoherence in Brain Processes. *Physical Review E*, 61(4):4194, 2000.
[39] Wolfgang Tittel, Jürgen Brendel, Hugo Zbinden, and Nicolas Gisin. Violation of Bell Inequalities by Photons More than 10 km Apart. *Physical Review Letters*, 81(17):3563, 1998.
[40] Alan M. Turing. Computing Machinery and Intelligence. *Mind: A Quarterly Review of Psychology and Philosophy*, LIX(236), pages 433–460, 1950.
[41] John Von Neumann. *Mathematical Foundations of Quantum Mechanics*. Number 2. Princeton University Press, 1955.
[42] Gregor Weihs, Thomas Jennewein, Christoph Simon, Harald Weinfurter, and Anton Zeilinger. Violation of Bell's Inequality Under Strict Einstein Locality Conditions. *Physical Review Letters*, 81(23):5039, 1998.
[43] John Archibald Wheeler and Wojciech Hubert Zurek. *Quantum Theory and Measurement*. Princeton University Press, 2014.
[44] Stephen Wiesner. Conjugate Coding. *ACM Sigact News*, 15(1):78–88, 1983.
[45] Jonathan R. Wolpaw, Niels Birbaumer, Dennis J. McFarland, Gert Pfurtscheller, and Theresa M. Vaughan. Brain–Computer Interfaces for Communication and Control. *Clinical Neurophysiology*, 113(6):767–791, 2002.
[46] H. Dieter Zeh. On the Interpretation of Measurement in Quantum Theory. *Foundations of Physics*, 1(1):69–76, 1970.
[47] Anton Zeilinger. Testing Bell's Inequalities with Periodic Switching. *Physics Letters A*, 118(1):1–2, 1986.

PART III

QUANTUM APPROACHES TO CONSCIOUSNESS

13

New Physics for the Orch-OR Consciousness Proposal

Roger Penrose

1. Conscious Understanding, as Opposed to Computation

In this section, I discuss various early motivational issues underlying my viewpoint concerning the scientific nature of the phenomenon of consciousness, which was subsequently developed in conjunction with Stuart Hameroff as the Orch-OR proposal [1, 2, 3]. My own early viewpoint on this issue dated back to some of my initial interests in the early 1950s. I relate, here, how they have played out and become significantly modified in view of some very recent developments in physical understanding.

My own motivations began with some of my early concerns, when I was a young Cambridge graduate student working in pure mathematics (algebraic geometry), where I spread my interests by attending lecture courses in General Relativity (by Herman Bondi), Quantum Mechanics (Paul Dirac), and Mathematical Logic (by S. W. P. Steen). In Steen's course, I learned about the notion of computability and the famous (but frequently misunderstood) theorem of Kurt Gödel [4]. This led me to a firm conclusion that the human quality of conscious *understanding* could not be a computational process. Many years later, I described my early ideas about this in my 1989 book *The Emperor's New Mind* [5].

It is worth spelling out, here, the basis of this conclusion. Gödel's Theorem was concerned with the issue of our beliefs with regard to how we establish the truth, or otherwise, of mathematical propositions. How, indeed, does one ascertain whether or not a mathematical proposition is true, or whether it is actually false? The normal idea about how one addresses such a question rests on the notion of mathematical *proof*. In the considerations that follow, I shall restrict attention to propositions in *number theory* (although mathematics itself is far broader than this). Number theory is concerned with propositions

Roger Penrose, *New Physics for the Orch-OR Consciousness Proposal* In: *Consciousness and Quantum Mechanics.* Edited by: Shan Gao, Oxford University Press. © Oxford University Press 2022.
DOI: 10.1093/oso/9780197501665.003.0014

about natural numbers (non-negative integers 0, 1, 2, 3, 4, 5, . . .). Examples of such propositions would be Lagrange's theorem that every natural number is the sum of four square numbers, or, perhaps, the yet unestablished Goldbach conjecture, which asserts that every even number larger than 2 is the sum of two prime numbers or, as another example, the currently unresolved assertion that there are infinitely many pairs of prime numbers differing by 2.

Now, what is a mathematical proof? One may consider that the normal notion of rigorous proof consists of the correct following of appropriate steps within a well-defined formal system of axioms and rules of procedure, where it is to be considered that this system would formalize reasoning available to a human mathematician in providing actual proofs of suggested theorems. Suppose we have such a formal system Σ of such axioms and rules of procedure. Then, by their very nature, the correct following of the operations of Σ would be *algorithmic*, i.e., it would be possible to program a computer to follow all the specific procedures provided by Σ so that any suggested Σ-proof of a given proposition *P* could be checked by a general-purpose computer (theorized to have an unlimited storage capacity).

Accordingly, for any number-theoretic proposition *P*, if it had a proof (or disproof) by the Σ-procedure, then this proof (or disproof) could actually be discovered by an entirely algorithmic action (the action of a general-purpose computer with unlimited storage capacity), where the computer simply enumerates all possible strings of operations of Σ getting successively longer and longer as the computer's search continues until it finds a string that proves (or disproves) the proposition. In practice, this might well be far too lengthy to be feasible, even with current technologies, because of the explosive lengths of actual search procedures for a proof or disproof of *P*, by the procedures of such a Σ.

However, Gödel's Theorem is much more interesting than practical issues of this nature. It tells us that there is a matter of deep principle underlying how we ascertain the truths of number-theoretic propositions, irrespective of the impractical nature of such search procedures. It tells us that our understandings that allow us to ascertain certain truths of number theory cannot actually be formulated in tens of any algorithm whatsoever! Whatever Σ-system may be put forward as encompassing the capabilities of human reasoning for ascertaining the truth (or falsity) of propositions in number theory, this will necessarily fall short. Let us try to see, in outline, how this works.

In effect, what Gödel does is to encode the actual procedures of any such Σ into curious-looking number-theoretic statements. This is where all the hard work lies, but in effect what this encoding procedure does is to allow Gödel to construct a well-defined number-theoretic proposition $P_{G\Sigma}$, clearly of the type that can be addressed by the procedures of Σ, that can be interpreted as *asserting* that there is, in fact, no Σ-proof of the proposition $P_{G\Sigma}$. This proposition $P_{G\Sigma}$ would indeed be a rather odd-looking number-theoretic assertion, from the standpoint of ordinary number theory, but it is all within the rules. Maybe $P_{G\Sigma}$ is true, or alternatively, maybe it is false. Let us suppose that it is actually false. But then $P_{G\Sigma}$'s interpretation simply tells us that it must actually have a Σ-proof, and if it has a proof, then it ought to be true nevertheless. Indeed, if the Σ-proof procedures are to be trusted at all, then Σ-provability certainly ought to establish $P_{G\Sigma}$'s actual truth. So $P_{G\Sigma}$ must indeed be true, but by what it says about itself, it cannot be established by the Σ-procedures themselves! By use of our understanding, we have actually established the truth of $P_{G\Sigma}$, but also that establishing this truth is beyond the capabilities of Σ!

Some people might assert that $P_{G\Sigma}$'s truth is "unprovable." For sure, it is unprovable by the procedures of Σ, but by its very construction, it is seen to be actually true! Some might argue that it has not been properly proved, so we can't count it as a "theorem," but to me this is an absurdity. The whole point of using Σ as a way of establishing mathematical truths depends on our belief that it is actually *sound*—i.e., that it establishes actual number-theoretic truths. What Gödel shows is that if our understandings do lead us to believe that Σ is actually sound, and therefore safe to use as a theorem-proving procedure, then our understanding actually extends *beyond* the use of Σ itself!

Although I have not gone into the tricky details of Gödel's arguments, nor of the simplifications that can be achieved by framing the arguments in terms of Turing machines (i.e., idealized computers, where the storge capacity is taken to be unlimited), the conclusion is clear: the procedures whereby mathematicians can rigorously establish the truths (or falsehoods) of various mathematical assertions cannot be encapsulated in algorithmic procedures that we actually understand to be sound. The crucial issue here is that we can go beyond the capabilities of any algorithmic procedure that we know, trust, and understand (here, Σ). Human understanding reaches beyond *any* suggested algorithmic procedures, whatever these procedures might be!

The moral of Gödel's argument, as I see it, is that when we apply the procedures of some algorithmic system, such as Σ, which asserts that certain number-theoretic statements can be established as true by the use of these Σ-procedures, then we need to have some reason to trust Σ in this regard, and this requires our understanding of what the rules of Σ actually *mean*. The remarkable thing is that this understanding also gives us the power to go *beyond* Σ and ascertain truths that Σ itself is unable to achieve. It is our *understanding* of Σ that allows us actually to transcend the power of Σ, in a way not achievable simply by slavishly following Σ's rules.

My conclusion from all this is that conscious understanding itself must be a non-computable process. We can certainly use our understandings to construct devices that are far more reliable, accurate, and enormously faster in actually applying our understandings than we are ourselves. In this respect, they are able to apply our understandings in ways that far exceed our own abilities, but it would be wrong to assign any actual understanding to the devices themselves. After all, this is what we *use* electronic computers for. They can vastly exceed what we can do simply by applying these procedures "by hand." But a device like an electronic computer, able to carry through these procedures vastly more quickly, efficiently, and accurately than can be done by hand, does not, by virtue of these capabilities, need to have any "understanding" whatsoever of what it is actually achieving. It would indeed be a serious mistake to conflate these two abilities, and to take the view that what we achieve by the use of conscious understanding has much to do with the mindless activity of a computer. This was the underlying theme of my 1989 book [5], the *The Emperor's New Mind*—henceforth abbreviated ENM.

2. Why Unknown Algorithms Don't Underlie Our Understandings

Needless to say, many objections have frequently been raised against my contention that conscious understanding must be a non-computable procedure. The most significant complaint was the very reasonable objection that we cannot apply this argument to ourselves if we do not know what algorithm—in effect "our Σ"—actually is. In my subsequent 1994 book *Shadows of the Mind* [6], I addressed all the complaints that had come my way, many of which were misunderstandings of one kind or another. The main (and very reasonable) objection relates to the fact that human conscious reasoning could be controlled by some algorithmic action, say, Σ_H, where *we do not know* what Σ_H actually is, so we would not be able to construct, and believe in, anything like a Gödel statement for Σ_H, so any contradiction with our supposed algorithmic actions would be removed!

In *Shadows of the Mind* (henceforth abbreviated SM), I went through various tortuous arguments to demonstrate that our absence of knowledge of what "our Σ" actually is cannot really get us out of the conundrum that Gödel's argument presents us with. There is still some controversy expressed about this aspect of the argument, as given in SM in purely logical terms, but in my opinion, there is in any case a stronger and more immediate type of argument, against our using an "unknown algorithm" that is supposed to control our ability to know things in mathematics. This basically sidesteps all the aforementioned objections and tortuous logical consideration, by the use of another type of argument (also indicated in SM) against there being an "unknown algorithm" that underlies all our mathematical understanding.

To make this case, it is useful to concentrate on a particular number-theoretic result, presented in 1944, known as *Goodstein's Theorem* [7]. So let me digress a little to describe what this theorem is. It applies to any positive integer N whatsoever, but to illustrate Goodstein's procedure, let us make a particular choice of N, e.g.,

$$N = 1077.$$

and we begin by writing N in binary form

$$1077 = \mathbf{10000110101}$$

which really means (only the **1**'s contributing):

$$1077 = 2^{10} + 2^5 + 2^4 + 2^2 + 1.$$

Now, we write the exponents in this binary form, too, and find:

$$1077 = 2^{2^3+2} + 2^{2^2+1} + 2^{2^2} + 2^2 + 1.$$

We haven't finished yet, because we want to write the "3" in the second-order exponent in the first term also in this binary-type notat1on, to obtain:

$$1077 = 2^{2^{2+1}+2} + 2^{2^2+1} + 2^{2^2} + 2^2 + 1.$$

For other numbers, particularly those much larger than 1077, we may need to go to higher-order exponents than this, but the general idea should be clear from this. We express N in this way so that no numbers greater than 2 appear in any part of the expression.

Now, Goodstein asks us to perform two operations on the number N. The first is

Operation A. *Increase the base number by 1.*

That is, at this stage, replace all the 2's by 3's, so we obtain:

$$3^{3^{3+1}+3} + 3^{3^3+1} + 3^{3^3} + 3^3 + 1,$$

so our number has increased enormously, namely to something around 10^{60}. Now, we perform the following:

Operation B. *Subtract 1, so we obtain:*

$$3^{3^{3+1}+3} + 3^{3^3+1} + 3^{3^3} + 3^3,$$

This makes very little difference. Now we perform **Operation A** again, to obtain:

$$4^{4^{4+1}+4} + 4^{4^4+1} + 4^{4^4} + 4^4,$$

Such increases the number to around 10^{600}, and we again perform **Operation B**, which gives us:

$$4^{4^{4+1}+4} + 4^{4^4+1} + 4^{4^4} + 3 \times 4^3, +3 \times 4^2 + 3 \times 4 + 3$$

which is analogous to what would be obtained subtracting 1 from 10000 to obtain 9999, in ordinary denary notation. The coefficients "3" in this expression are all less than base "4," so they are unaffected by the next **Operation A**, which now gives us:

$$5^{5^{5+1}+5} + 5^{5^5+1} + 5^{5^5} + 3 \times 5^3, +3 \times 5^2 + 3 \times 5 + 3,$$

(around 10^{10000}) to which we again apply **Operation B**, and then **A** again, and then **B**, and so on, alternately, and we see the numbers getting enormously bigger and bigger and bigger, until ultimately—very remarkably—coming down to *zero*! This extraordinary conclusion, for any natural number *N*, is Goodstein's Theorem!

It is an extreme example of a hare and tortoise: The hare **A** is always eventually beaten by the tortoise **B**. It is a good exercise to try this out first with N = 3, for which the sequence rapidly comes down to zero, and then try with $N = 4$. It is not recommended that any form of electronic computer be used as an aid, as the numbers get far too enormous—just pencil and paper—and some understanding—will do!

Goodstein's Theorem is a wonderful example to demonstrate the utter implausibility of the idea that our understanding of mathematics could be the result of some unknown algorithm, working in our brains. For the question arises: How is it that mathematicians can know, and explain convincingly to others, that Goodstein's Theorem is actually true—even for the case of $N = 4$, for which the full underlying mathematical notions are not needed. In fact, Goodstein's Theorem is a rapid consequence of a mathematical principle known as *transfinite induction*, first found by the great 19th-century mathematician Georg Cantor.

If you already know about transfinite induction, you can immediately see that Goodstein's Theorem must be true, simply by replacing the base number (2 3, 4, 5, etc.) by the smallest transfinite ordinal number ω, and then **A** does nothing, but **B** provides a descending sequence of ordinals that must come down to zero. I assume that most readers will not know about transfinite induction, and so will not appreciate the simple magic of Goodstein's proof. Nevertheless, it is sufficient to show the truth of Goodstein's argument. All you need is to be able to follow, and to understand—and to believe—the argument. That can be learned, appreciated, and understood. You don't need to possess the genius of Cantor to appreciate that Goodstein's Theorem is actually true. Even the case $N = 4$ gives us a fair taste of the reason why Goodstein's Theorem is indeed true.

There is something else that is remarkable about Goodstein's Theorem. It cannot be proved by the ordinary procedure that we learn at school, referred to as the principle of *mathematical induction*. This fact was shown in 1983 by Kirby and Paris [8]. Ordinary mathematical induction is a way of proving that some assertion *P*(*n*) is true for *all* natural numbers *n*. What we do is to show that *if* P(*n*) is true for a natural number *n*, then P(*n*+1) must also be true. We must, in addition, show that P(0) is true as well. Then we shall have established that P(n) is true for *all* natural numbers, even though we have established only two things! Ordinary mathematical induction is a powerful and familiar procedure for proving statements about natural numbers. It is the basis of the logical system known as *Peano arithmetic*—basically, what can be achieved by the use of mathematical induction, together with the ordinary rules of logic. All the basic rules of algebra and arithmetic can be formulated and established within the scope of Peano arithmetic. But, as Kirby and Paris have shown, it is not powerful enough to establish Goodstein's Theorem!

Yet, we can still establish that Goodstein's Theorem is actually true. It seems to me that this presents us with a very powerful demonstration that conscious understanding cannot be the result of an algorithm inside our heads. For if it were, then how is it that we can know that Goodstein's Theorem is actually true? As remarked above, we do not need the genius of Cantor to know that Goodstein's Theorem is true. We just need to have the argument properly explained to us (and, incidentally, this requires only the very simplest and most uncontroversial part of Cantor's powerful theory). We can already get something of a feeling for what is involved by trying out the case $N=4$.

The fundamental question that I am raising here is whether there could be some algorithm acting inside our heads that underlies our faculty of conscious understanding. If this were some kind of algorithm somehow "wired" into our brains, then how could it have arisen by natural selection? In Figure 13.1, I depict, in cartoon form (taken from SM), the problems that I have in believing that any algorithmic process capable of dealing with things like Goodstein's Theorem or transfinite induction could have arisen by natural selection. I even have some trouble seeing how ordinary induction (which might indeed be taken as algorithmic) could have arisen in this way, though one could well take the view that the forming of ideas of arithmetic could be of value in farming, domesticating animals, and building shelters and would definitely benefit from the ideas of basic arithmetic and geometry. Maybe a facility to get some general grasp of what the infinite array of natural numbers is could indeed have a selective advantage, and the basic concept of ordinary induction is central to this, allowing us to formulate general properties of these numbers. Yes, it is even conceivable that some kind of inner instinctive trust in the validity and usefulness of Peano arithmetic could perhaps be selectively advantageous. But having the inbuilt

Figure 13.1 Why an ability to do sophisticated mathematics was not what natural selection favored, for our remote ancestors.

algorithmic facility to go to the rules beyond this, so we can perceive the truths of things like Goodstein's Theorem—that's another thing altogether! It simply does not come under the scope of what can be built into our "inner algorithm." If algorithmic at all, it would have to be the effect of another algorithm altogether, having no selective advantage whatsoever.

In Figure 13.1, I have indicated the results of some human skills that could definitely be selectively advantageous. In the distance, we see the domestication of animals, the growing of crops, and the construction of some shelters. At the left the brilliant conception of a mammoth trap is illustrated, duly capturing a mammoth. But in the foreground is a mathematician, developing his subtle theorem, oblivious to the presence of a sabretooth tiger poised to devour him. Clearly, his advanced mathematics serves him no advantage, and there is no favoring of his genes in this regard.

To me, the message is clear. There is no selective advantage whatsoever in promoting algorithmic skills that would be advanced enough to get the correct results for mathematical problems that go beyond the scope of Peano arithmetic. Yet, it is the general quality of *understanding*, not specific to any of the tasks listed above, that could surely have a powerful selective advantage. This is one of the things that consciousness is *for*. Understanding is enormously broadly applicable, and it is what enables us—albeit often with considerable effort—to achieve things that inbuilt algorithms cannot.

The human ability to appreciate Goodstein's Theorem, even if this does not include humanity as a whole, is to my mind a powerful demonstration that the quality of understanding cannot be the result of some inherited algorithm, somehow encoded inside our heads.

3. What Kind of Physical Processes Could Be Non-Algorithmic?

Despite arguments like that presented above, implying that our conscious actions are not of an algorithmic nature, it is natural to ask what kind of physical processes could possibly be regarded as being "non-computational," and which might thereby underlie the brain actions that result in our ability to "understand" things by way of this curious phenomenon of "consciousness." Do we see anything in the actions of the physical world that we need to regard as non-computable? Indeed, what kind of physical action might, in any sense, possibly be regarded as "non-computational" or "non-algorithmic"?

Before we directly address the question of physics, it is important that we appreciate that non-computability is a well-defined concept in the area of pure mathematics. I should make clear here that I do not just mean "random" or anything like that. Accordingly, I do not simply mean "non-deterministic," which is a separate (though related) issue. Let us, for the moment, indeed set aside actions in the physical world, and merely ask what the notion of non-computability means within mathematics itself. It refers to situations where one has an infinite class C of mathematical problems, where for each member of C, an answer might be YES or NO. There might be an algorithm that is able to address this issue, or there might well not be.

Now a very simple example would arise when C refers to a property that might or might not hold for a natural number, i.e., C assigns either YES or NO to each element of the set $\mathbb{N}$ of natural numbers. For example, C might be the question of whether an element n of $\mathbb{N}$ is even or not. This example is, of course, trivially algorithmic (i.e., if n, when written in standard denary notation, ends in 0, 2, 4, 6, or 8 or not; or if written in binary form, whether or not this final digit is 0). More complicated would be for C to ask whether or not an element of $\mathbb{N}$ is prime, but again there is a straightforward algorithm to decide this issue. Nevertheless, there are perfectly well-defined questions that we might ask of an element of $\mathbb{N}$ for which it can be shown that there is no algorithm whatsoever that can decide YES or NO correctly for every element of $\mathbb{N}$.

I do not know of an example of a non-computable C that is simple to state and directly refers to natural numbers, so let us turn to an example where C is a question that does not directly refer to $\mathbb{N}$, but that is much easier to state. Now, C is to be concerned with the family S_P of finite sets of *polyominos*. What is a polyomino? It is a finite connected plane shape that is constructed from equal-sized squares glued together along their edges. The question C that I shall be concerned with is whether or not the entire Euclidean plane can be tiled (i.e., covered without gaps or overlaps) by an element of S_P (i.e., by a given finite set of polyominos). Somewhat remarkably, it follows from a theorem of Robert Berger (proved in 1966 [9], basing his ideas partly on an earlier 1961 result by Hao Wang [10]; see Grunbaum and Shephard [11]) that C is non-computable! That is to say, there is no algorithm that can correctly tell us, for any finite set of polyominos, whether or not they can tile the Euclidean plane!

This result depends on the existence of finite sets of polyominos that will tile the plane only in a non-periodic way (i.e., for which the entire pattern does not ever exactly repeat itself through any translational motion). An example is given in Figure 13.2, where there are just three polyominos involved, as provided at the top. The example is one I constructed myself, based on a tile set due to Robert Ammann (see Penrose [12], Grunbaum and Shephard [11]).

Of course, this example of a non-algorithmic problem is very far from anything that we would anticipate might reflect non-computability in the laws of physics. But, it does illustrate that mathematically well-defined non-computable actions (such as one, like this, that depends on when a given set of shapes will fill space without gaps) are not uncommon in pure mathematics. When it comes to actions that are directly relevant to physics, I can certainly think of one example, in particular, of a physical model that seems to me very likely to be non-computable. This is a physical proposal put forward by Paul Dirac—the great physicist who discovered the equation for the electron (the Dirac equation [13]), and who formulated the general framework of quantum mechanics now adopted by physicists across the world.

In 1938 [14], Dirac explored the issue of the classical evolution of a system of point-like charged particles interacting with each other through electromagnetic forces. There is a well-known problem involved

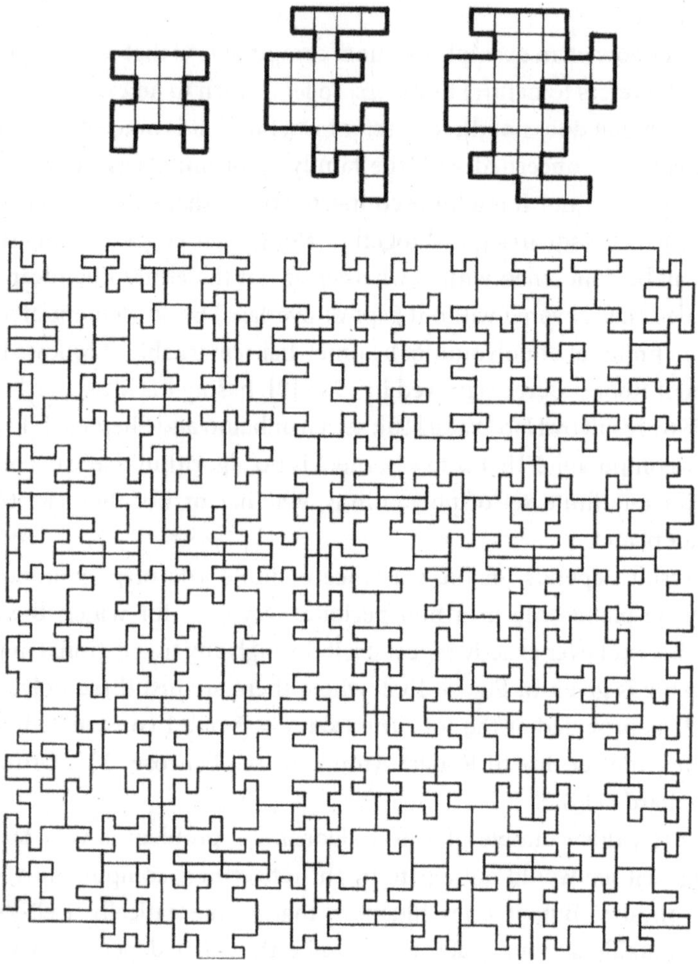

Figure 13.2 A necessarily non-periodic tiling using the three polyomino shapes at the top.

here, concerning the reaction of individual point-like charged particles in the presence of other such particles, each acting in response to the electromagnetic fields of all the other particles. Initially, one thinks just of the Lorentz force, which describes how a charged particle responds to an ambient background electromagnetic field. However, there is a serious additional problem, because the actual field near a point-like charged particle will be completely dominated by *its own* field there, particularly when that particle is itself accelerating due to its motion in the ambient field, and the radiation it emits, when it accelerates, will itself be infinite at the particle's location. It is problematic to consider that the particle can feel out the much weaker background field near its own location, due to the other particles, separating that external part from the far more dominant (and, in fact, divergent) radiating field that it generates itself, where it is located.

Dirac adopts a specific procedure for dealing with this problem, but then he finds that the evolution is governed by a third-order equation for describing the system's evolution, instead of the normal second-order equation. This implies that there are too many solutions, as compared with one's normal expectations, and those that must be discarded involve what are referred to as "runaway" behavior, in which one or more particles are accelerated exponentially out of the system—clearly not physically appropriate! The point of view is adopted that all these runaway solutions are regarded as unphysical, and the true evolution is that in which there are is no such runaway behavior.

Now a curious situation arises here, owing to the fact that, in many cases, this runaway behavior may occur very late in the evolution. Accordingly, in order to know whether or not any chosen solution is to be taken seriously as a truly allowed physical behavior, one may have to wait an indefinite time before knowing whether the solution is indeed acceptable, by virtue of *no* runaway behavior ever becoming involved.

This situation is very much like what happens with the Turing machine action. Problems that are *non-computable* often arise because one cannot tell (according to any algorithmic criterion) whether or when a computation will ever terminate. For example, with the polyomino tiling problem we are presented with a given finite set of polyomino tile shapes, and we can assign a computer the task of simply trying all possible connected arrangements, using the given shapes, and then working its way outwards. This computation halts whenever it is found that every arrangement of a certain diameter cannot be continued any further to a surrounding region. It is when—for a given finite set of polyomino shapes—it is found that no continuation is possible at some stage, then this computation halts, so we know that the answer is NO for that set of shapes. On the other hand, if the computation continues forever, then we would know the answer is YES.

This is logically the same as with the Dirac physical model described above. In each case, we regard the solution to be acceptable when we know that the computation in question continues forever. In Dirac's case, with the charged point-like particles, this means that the computed evolution continues forever without any runaway behavior taking place, where one (or more) of these particles has become exponentially ejected from the system. If the computation finds any runaway behavior, then the computation halts, and we know that this case is unacceptable as a solution.

There are, nevertheless, some technical difficulties with this analogy. In the first place, the polyomino problem is clearly a discrete problem that we can imagine being precisely treated by conventional computational procedures, whereas Dirac's model involves continuous parameters that would normally be treated by approximations. Also one would need a clear-cut procedure for deciding when a "runaway behavior" has actually taken place. More seriously, although it is a known theorem that the polyomino tiling problem is non-computable, it is only my guess that something analogous occurs with the Dirac model.

It should be pointed out that Dirac's model was not intended to provide a realistic picture of what would actually occur for a system of charged particles, since it is a classical rather than a properly quantum model. Moreover, as stated above, I do not know whether his classical model is actually non-computable, though this is certainly a strong possibility, and it shows that non-computability is not at all out of the question when it comes to physical behavior under the known laws of physics.

4. The Collapse of the Wave Function

Since the issues raised by Dirac's model, as just described, refer to *classical* behavior, its relevance to the actual physical laws that govern our world may well be indirect. We need to appreciate that it is *quantum* mechanics that governs the behavior of the particles that constitute the material of our universe at a small scale. Again, I must turn to Dirac for my own basic understandings, for—as indicated in Section 1—it was from attending Dirac's lectures on basic quantum theory at Cambridge, when I was a graduate student in pure mathematics, that I gained my first genuine appreciation of the subject. I believe that it was in Dirac's first lecture that he introduced the *superposition principle*, basic to quantum theory, where he referred to a quantum particle or atom, where it might inhabit one location on his desk, say, A, or it could inhabit another such location, say, B, and then there would be many other possible states for its location, where it could be at A and at B at the same time! Then he produced a stick of chalk, perhaps breaking it into two pieces to illustrate his point, where the chalk might simultaneously inhabit both locations A and B at the same time.

Unfortunately, my mind had wandered at that moment, as he evidently gave his explanation as to why we do not find macroscopic objects like sticks of chalk simultaneously located in two separate places. When my mind returned to the topic at hand, he had moved on to something else, and all I could recall about his explanation of why we do not experience such superpositions of macroscopic objects, was that it had something to do with how much greater energy would be involved when it came to pieces of chalk.

Perhaps it was fortuitous that I missed Dirac's explanation, since I am guessing that he may have provided some reason why we should not really worry about the problem, just to calm us down so that we could get on to the important issue of coming to grips with the serious issues of understanding how we are to make use of quantum mechanics as it applied to sub-microscopic bodies. But, to me, the issue of how the superposition principle appeared not to apply to sticks of chalk remained a profound mystery to me.

When I began thinking further about Steen's lectures, and the non-computability of the processes underlying our understandings of things like Gödel's Theorem, I could not see how there could be serious non-computability in Newtonian mechanics, in Maxwell's amazing theory of electromagnetism, nor did I see how non-computability in Einstein's even more amazing Theory of General Relativity (henceforth referred to as GR), which I had gained much of my understanding about from the wonderful lectures by Hermann Bondi. It may be remarked that, at the time of Bondi's lectures, little had been achieved concerning actual detailed computational simulation of solutions of Einstein's equations. Nevertheless, the possibility was clearly there, at that time, and the much later computational simulations of gravitational wave forms arising from black-hole encounters have been crucial for the LIGO detections of recent times, confirming the basic computability of GR.

Accordingly (and knowing nothing of Dirac's work on classically interacting point-like charged particles referred to in Section 3), it had seemed to me that we must look to quantum behavior, if we are to find any chance of finding a place for non-computability in physical actions. However, as Dirac's lectures later made clear, the evolution of the quantum state occurs via a well-defined differential equation—the Schrödinger equation—and the evolution of this equation is still a computational matter, in the same sense as with GR. But the trouble that I had had with Dirac's stick of chalk remained in my mind, and it seemed to me that there was something seriously missing with regard to the stick of chalk, as raised in Dirac's lecture.

What is normally decreed, in standard quantum mechanics, is that one needs to "make an observation" on the quantum system (e.g., the superposed stick of chalk), and this "observation" would convert the superposition into one or the other. This always seemed very odd to me because surely if the observer (whatever construction of material might be deemed to constitute an "observer") ought also to follow the rules of quantum mechanical evolution, i.e., the Schrödinger equation, and following the Schrödinger equation does not normally "un-superpose" anything. The way that this issue is treated in standard quantum theory is that "upon observation" there is a *collapse of the wave function* at which, instead of the system following the Schrödinger equation, the state seems to "jump" to one of the different possibilities allowed by the type of observation being performed.

It should be explained that in the *Schrödinger picture*, the quantum state of a system is described by the *wave function,* and it is the ordinary time-evolution of this state that the Schrödinger equation describes. The view is taken that somehow it is the act of "observation" that causes the jump in this evolution that is referred to as the "collapse of the wave function." In more general terms, we may refer to the "quantum state" of a system, and the "collapse" is then referred to as the *reduction of the quantum state.*

I became convinced that there must be something seriously missing from quantum theory as it stands—and it was not until a good deal later that I was reassuringly informed that not only Einstein, but also Schrödinger and even Dirac himself, also shared this view. I felt that maybe it was this needed missing completion of quantum mechanics that would supply the non-computability that I felt was needed in order to provide the scope for consciously understanding brains to come about, acting in non-computable ways, but still acting within the constraints that the actual laws of physics would allow.

I can't quite remember when it was that I became convinced that it had to be with *gravity* that the essential changes in the framework of quantum mechanics must come about. I was certainly of that view when I wrote ENM, although at that stage I had not quite found what I currently believe to be the correct criterion governing the conditions for this "wave function collapse" to occur, this being viewed as an *objective* phenomenon that results in some kind of "jump" in the quantum state. According to this view, it is the physics, together with the "jump," that takes place objectively and has nothing to do with some "external observer" somehow becoming aware of the system—that being a type of viewpoint not uncommonly expressed by some of the great theorists of the past, such as Eugene Wigner and John von Neumann. My own viewpoint is almost the opposite of this, namely that the reduction of the quantum state is indeed an *objective* phenomenon, taking place when the gravitational features of the state come to dominate, and the reduction of the state then occurs objectively. On the other hand, conscious experience comes about within suitably organized entities (such as ourselves) who are able to harness the potentially non-computable activity in the objective state-reduction process.

I refer to this state-reduction process as **OR** (where evolution by the Schrödinger equation—or *unitary evolution*—would be **U**), with this acronym **OR** conveniently spelling the word "or" so that Dirac's stick of chalk becomes either in location A *or* in location B, not in some strange superposition of both at the same time. We must also bear in mind that in any ordinary situation there would be many more degrees of freedom involved, and the situation could not just be described in terms of a simple pair of locations A and B, but that numerous additional degrees of freedom would rapidly also get involved in the system, quickly spreading out through the environment, such as in the air or in whatever might be holding the constituents up, so it is not simply a question of a body being in a superposition in A and B at the same time but something vastly more complicated.

In standard quantum-mechanical procedure, one does not attempt to keep track of all these environmental degrees of freedom, but instead one would describe things in terms of a mathematical description introduced by von Neumann (and independently by Lev Landau), referred to as a "density matrix" [15], which very remarkably allows one to ignore all these extra degrees of freedom in the environment and simply to come up with an answer that the original body is not to be considered as being in a superposition of two locations A and B at the same time, but that there is a probability so-and-so of it being in location A and a probability such-and-such of it being at B.

When the environment cannot be tracked in detail, then this is the best that we are able to do, and we may say that the system has come to behave classically because of this *environmental decoherence*, and we "may as well" say that the original body has become "either" located at A "or" located at B, with these calculated probabilities. However, this does not answer the question of what, *in detail*, has actually occurred in nature, with the evolution of the system that, according to the Schrödinger equation—this being a linear equation—would maintain these quantum superpositions right through until a "measurement" is performed, and we would end up with a quantum *superposition* of the following:

"The body is at A and the measuring device says it's at A"

and

"the body is at B and the measuring device says it's at B."

So, we have gained nothing with regard to how standard theory tells us what is, in detail, what is actually going on in the world.

It is not my purpose to go into all the different ways that have been put forward to resolve this issue, such as the "many-words" viewpoint first formally enunciated by Hugh Everett III [16], in which there is an overarching "reality" in which all such alternatives eternally persist, or the many alternative ways in which the rules of quantum theory might be modified, so the time-evolution of the state of a body such as Dirac's superposed stick of chalk becomes "actually at A" or else "actually at B." Instead, I want to concentrate on the more specific proposal (or class of proposals) in which the *gravitational field* of the body is what, in some sense, resists superposition and the body's state becomes actually at A *or* actually at B.

I shall come to this in the next and later sections, but before doing so, it will be useful to return to the issue of environmental decoherence, just considered above. The picture will be that unless we are very careful to prevent the environment from getting involved, we must indeed expect that the system would become very complicated, with many environmental particles indeed becoming involved in the system, so the **OR** scheme would be likely to apply all the time, reducing the state in a way that would be indistinguishable from the normal procedures that are adopted by physicists when they adopt a density matrix description. Nevertheless, it is proposed that there would be a genuinely evolving *reality* in which numerous uncoordinated **OR** events would be taking place all the time. Although Stuart Hameroff and I are proposing that the phenomenon of consciousness actually be explained in terms of **OR** events, according to the physical scheme I shall describe in the following sections, actual consciousness would require that there is some kind of coordination between them that we refer to as "orchestrated." The multitude of **OR** events would have no such coordination, and it would not be reasonable to refer to them as actually evoking consciousness. Instead, we are calling them *proto-conscious* events, providing merely the "building blocks" out of which actual consciousness is constructed.

The line between proto-consciousness and actual consciousness is, as yet, very vague, but our scheme does provide an ontological framework that enables the evolution of ordinary material to take place in such a way that vast numbers of **OR** processes are taking place all the time, thereby allowing something extremely close to classical dynamics to occur, despite the underlying quantum dynamics, without there being any need for conscious beings to be involved. Yet, when circumstances become appropriate, and the pressures of natural selection begin to reveal the power of orchestrating the underlying **OR** processes, actual conscious beings begin to reveal its potential power. The Orch-OR proposal asserts that consciousness is a phenomenon whose "building blocks" are indeed these ubiquitous **OR** events, already present in the actions of the world, thus allowing classicality to emerge from underlying (Schrödinger) dynamics. In this sense, we believe that we are actually addressing the so-called hard problem enunciated by David Chalmers [17].

5. The Diósi-Penrose Criterion for the Rate of OR

Imagine a quantum experiment performed on a tabletop, where Earth's gravitational field is being taken into consideration. We regard Earth's acceleration field to be a 3-vector $\boldsymbol{g}$, which is taken to be constant both in space and in time. How do we bring Earth's field into our calculation? The normal procedure that most physicists would adopt would be "to put a term in the Hamiltonian" to account for Earth's field, this being the standard way of incorporating any force, in quantum mechanics, thereby treating the gravitational force in the normal Newtonian way, no differently from any other force. I use lowercase letters $(\boldsymbol{x}, t)$, $\boldsymbol{x}$ being a 3-vector for the space coordinates and t being the time coordinate, in this system, which I refer to as the *Newtonian perspective*, the wave function being $\psi(\boldsymbol{x}, t)$.

On the other hand, in accordance with the *Galilei-Einstein principle of equivalence*, we might consider it to be more appropriate to describe Earth's gravitational field simply as an acceleration effect, to be cancelled out in a freely falling reference frame. This is what I shall call the *Einsteinian perspective*, according to Einstein's underlying ideas of general relativity, whose foundation stone is indeed this equivalence principle, according to which the gravitational field is really to be treated as simply an acceleration effect, so we can eliminate Earth's gravitational field simply by adopting free-fall coordinates—here with capital letters for the respective 3-space and time coordinates ($\boldsymbol{X}$, T). The relation between the two sets of coordinates is given by

$$\boldsymbol{X} = \boldsymbol{x} - \frac{1}{2}t^2\boldsymbol{g}, \quad T = t,$$

where the two systems of coordinates are taken to agree at time $t = 0 = T$, Since in the freely falling frame, there is no gravitational field, we do not now have any gravitational term in the Hamiltonian. With this Einsteinian perspective, the wave function is $\Psi(\boldsymbol{X}, T)$.

We should bear in mind here that in order to move to more complicated gravitational situations, it is the *Einsteinian* perspective that we try to maintain, so we can fit in with general relativity in an appropriate way. Einstein's general relativity is now confirmed by many experiments over a broad range of phenomena, Newtonian theory giving us only an approximation—albeit an excellent one in most circumstances.

Somewhat remarkably (see [18] and compare [19]), a direct calculation provides the following relation between our tabletop system's wave functions, according to these two different perspectives, where we take into account the lack of a gravitational term in the Hamiltonian in the Einsteinian perspective:

$$\Psi(\mathbf{X}, \mathrm{T}) = \mathrm{e}^{\frac{\mathrm{i}M}{\hbar}\left(\frac{t^3 g^2}{6} - t\boldsymbol{x}\cdot\boldsymbol{g}\right)} \quad \psi(\mathbf{x}t),$$

where M is the total mass of the system. Now, one might take the view that, since there is only a *phase factor* (i.e., a complex number of unit modulus) between them, it makes no difference which perspective is adopted, since the measured physical effects—probabilities, etc.—are obtained, in quantum mechanics, from squaring the moduli of amplitudes, and the phase factor disappears. So, we might well argue that the overall phase factor is of no consequence. However, looking at this phase factor a little more closely, we see that there is a t^3 in the exponent that, according to the rules of quantum field theory (as we may see from performing a temporal Fourier decomposition, if desired), the quantum field theory vacua are different for the two perspectives. Again, one could legitimately say that this is of no consequence. For so long as one sticks consistently to one perspective or the other (i.e., to one notion of the quantum vacuum or to the other), there ought to be a consistent vacuum—and therefore a consistent quantum field theory whichever perspective we adopt.

Now, let us change the physical situation in a crucial way. Let us suppose that, as part of the physical system under consideration, there is a massive body that, like Dirac's stick of chalk of Section 2, is put into a superposition of being in two different locations A and B at the same time. It should be mentioned that there is no particular difficulty in doing this in an actual experiment. We could, for example, arrange that a laser emit a high-energy photon, aimed at a beam-splitter (e.g., a half-silvered mirror), which puts the photon into a superposition of being reflected and being transmitted. Let us suppose that the transmitted part triggers a spring that pushes away the body, whereas the reflected part leaves the body alone. The body is now in a superposition of being moved and not moved, as was intended, so *its own* gravitational field, when brought into the calculation, is far from being a constant, like Earth's gravitational field, and is now in a quantum superposition of *two different* gravitational fields!

A trouble here, in an ordinary experiment of this kind, is that the body itself would be not the only part of the system that is put into a superposition. There would be much in the body's environment that would also be involved, such as the molecules of air in its neighborhood. The proper description of the state would be very complicated, and since one does not want to have to keep track of everything in the detailed state that is involved, one reverts to a description in terms of a *density matrix* **D** (see Section 4), which takes into account all the random features that one has no control of, and providing a *probability mixture* of all the different things that the environment might do. The trouble here is that all we are then presented with, for the description the body, is that there is a certain probability of finding it (with regard to the location of the body) at $\mathcal{A}$ or at $\mathcal{B}$, i.e., there is a probability mixture of it being moved (location $\mathcal{A}$) and of not being moved (location $\mathcal{B}$), and we have lost the possibility of determining whether it is in a *quantum superposition* of being in state $\mathcal{A}$ and state $\mathcal{B}$, as opposed to there being merely a probability mixture of being in one of these states or the other. This is the phenomenon referred to as *environmental decoherence* and, as things stand, there is no way of distinguishing a quantum superposition of these two states from being merely a probability mixture of these two alternatives, which would be the case with a purely classical description.

However, if a very perfect experiment could be carried out, in which environmental decoherence can be eliminated almost completely, then the distinction between a quantum superposition and probability mixture could be ascertained. Indeed, there are various experiments either proposed (e.g., [20, 21]) or actually underway for some time (by Dirk Bouwmeester and colleagues [22, 23]). Such experiments are aimed at testing the implications of the discussion above, where we try to consider the issue of the effect of a gravitational field of a body in a superposition of two locations at the same time, when viewed from the Einsteinian perspective.

This leads us to a serious issue. If we consider the physics in a small region $\mathcal{R}$, fairly close to our superposed body, we find that we are in a bit of a quandary if we choose to adopt the Einsteinian perspective. For, within the region $\mathcal{R}$, we should have to adopt a quantum field theory appropriate to one component $\mathcal{A}$ of the body's location in the superposition, which differs from that appropriate to the component $\mathcal{B}$ (i.e., with different vacuum states). In quantum field theory, we are not permitted to superpose states from quantum field theories that differ in this way so, technically, we are stuck!

However, we should not give up at this point, because the physical world must do something! The point of view that I adopt here is to say that there is some measure of *error*, which I am interpreting as a measure of *uncertainty* about what the quantum state actually is. We can identify this uncertainty in the coefficient of the troublesome t^3 term in the exponent—now for the squared *difference* between the accelerations $\mathbf{g}_{\mathcal{A}}$ and $\mathbf{g}_{\mathcal{B}}$ that gives us a local measure of uncertainty:

$$\frac{m(\mathbf{g}_{\mathcal{A}} - \mathbf{g}_{\mathcal{B}})^2}{6\hbar},$$

where m is the body's mass, and we integrate this quantity over the whole of 3-space. After a little mathematical manipulation (essentially an integration by parts, see [18] (and [20, 21] for further details), we obtain the quantity E_G, which we find can be interpreted as

E_G = the gravitational self-energy of the difference between the mass distributions in each of the two superposed states.

Since we are interpreting E_G to be a fundamental uncertainty in the energy of the system, we may invoke Heisenberg's time-energy uncertainty principle and regard:

$$\tau \approx \frac{\hbar}{E_G}$$

as a *lifetime* of the superposed state, after which it "decays" to one component $\mathcal{A}$ of the superposition, or else to the other component $\mathcal{B}$.

The analogy here is with an unstable nucleus. Such a nucleus might decay one way or another, at a lifetime τ, referred to as a "half-life," at which time there is a 50 percent probability of it having decayed. In accordance with this, Heisenberg's time-energy uncertainty principle tells us that there will be an uncertainty in the nucleus's energy of around $\hbar/\tau$.

Applying this principle in the reverse direction, we may take it that the fundamental energy uncertainty E_G in our superposed system, as described above, is correspondingly inversely related to its lifetime. The lifetime (or half-life) in question is the time τ (or some simple multiple of it), according to which the superposition "decays" to one component of the superposition, or to the other (with perhaps some reasonably small numerical constant multiplying τ).

As a historical comment, the above argument, appearing in various articles, e.g., [18, 21] around 2014–2018, was not my first reasoning for a reduction time τ, as given above. Previously [24, 25], I had argued for it from Einstein's principle of *general covariance* (independence of coordinate description), but I think that the argument given above is more physically compelling. In any case, some years even earlier, Lajos Diósi had already put forward the same proposal for a gravitationally induced **OR** process [26–29], but without the specific motivations from general relativity that I later provided.

Accordingly, the terminology "Diósi-Penrose" is frequently attached to proposals with this specific **OR** lifetime, as described above. Nevertheless, I wish to make clear that the particular proposal that Diósi has developed is very different from what I have in mind myself. In particular, Diósi's proposal predicts a spontaneous heating in a body, and experiments have been performed to detect this heating, and so far this effect has not been found (see [30]). I wish to describe in the following sections the type of scheme that I believe is more likely to accord with the ways of nature, and in which such spontaneous heating is not expected to occur. Nevertheless, the avoidance of this effect demands a very curious viewpoint, and it is this that will be described in Sections 6, 8, and 9 that follow.

6. Special Relativity and Retro-Activity in OR

Let me begin this section by considering a very special system for which E_G can be worked out explicitly. This is the case where a spherical body of radius a and total mass m, with uniform density, is put into a quantum superposition of two locations, where their centers are distance b apart. We find that E_G takes the value

$$E_G = \frac{m^2 G}{a}\left(2\lambda - \frac{3}{2}\lambda^3 + \frac{1}{5}\lambda^5\right) \quad \text{if } 0 \leq \lambda \leq 1$$

and

$$E_G = \frac{m^2 G}{a}\left(\frac{6}{5} - \frac{1}{2\lambda}\right) \quad \text{if } 1 \leq \lambda.$$

where $\lambda = \frac{b}{2a}$ (see [18, 21]; for more general ellipsoidal shapes, see [20]).

We see that most of the value of E_G occurs from coincidence to contact, but still a significant amount $\left(\frac{5}{12}\right)$ is achieved from contact out to infinity. If the mass is not too great, the two locations could well be separated to a large distance before **OR** may be expected to come into play (according to this scheme). Let us consider such a situation and examine it in the perspective of special relativity. In Figure 13.3, I have depicted a space-time diagram, with time going up the page, where the vertical line right at the bottom, up until it reaches O, represents a stationary body that, from O upwards, has been put into a quantum superposition of two locations, gradually separating from one another as time evolves (broken lined indicating superposed states). We suppose that the body's mass, according to the above considerations, is such that it remains in this quantum superposition (as marked with the broken lines) until most of the way up the picture, until the locations Q* and Q are simultaneously reached, at which time **OR** takes place and, in the case depicted, the quantum state of the body becomes entirely located at Q*, so that it disappears from location Q, and the continuing time-evolution follows the left-hand branch from then on, with the right-hand branch terminating at Q.

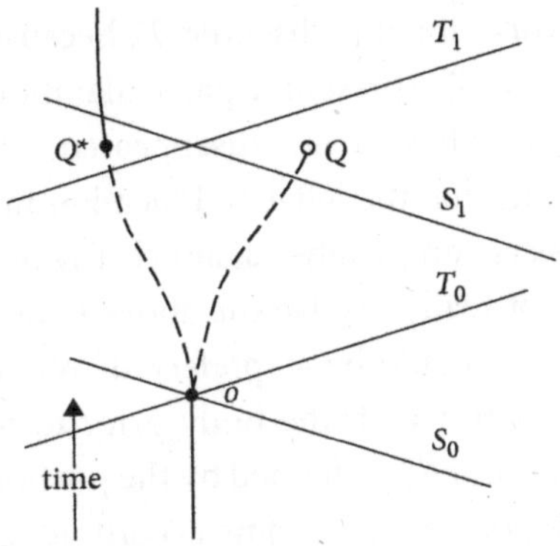

Figure 13.3 Space-time picture of **OR** for a separating quantum superposition.

However, this picture presents a problem with Einstein's special relativity, because it has been considered only from the point of view of the description of an external observer at rest with the original un-superposed body shown at the bottom of the picture. Let us instead consider the description with respect to an observer $\mathcal{S}$ moving uniformly but very rapidly towards the left of the picture. According to special relativity, the lines that are regarded as representing simultaneous events, according to $\mathcal{S}$, are tilted upwards in the direction towards which $\mathcal{S}$ would be traveling. In particular, the origin event O of the superposition of the body would be regarded by $\mathcal{S}$ as being along the line $\mathcal{S}_0$ in the picture. According to a second observer $\mathcal{T}$, moving with the same speed as $\mathcal{S}$, but in the opposite direction, i.e., to the right, the lines perceived as being simultaneous according to $\mathcal{T}$'s reference system would be tilted upwards to the right, and $\mathcal{T}$'s viewpoint as to which events are simultaneous with the separation event O would be tilted along the line $\mathcal{T}_0$.

So far, so good, but now let us consider the picture that $\mathcal{S}$ would be presented with, as regards the occurrence of **OR**. Consider, most particularly, the line $\mathcal{S}_1$ in Figure 13.3. At this particular moment, according to $\mathcal{S}$'s reference frame, the **OR** action would have already resulted in the body being entirely in the left-hand location, since the event Q* would already have passed, whereas since the event Q would have yet to occur, the right-hand possibility for the body's location would still be in play, so there would be some definite possibility that the body might find itself in the righthand location. This is complete nonsense, of course, because it would tell us that there is a definite probability (say, 50 percent) that the body would have magically become two copies of itself!

Things are even worse for the observer $\mathcal{T}$, because, with respect to $\mathcal{T}$'s reference system, the line $\mathcal{T}_1$ represents a particular moment of $\mathcal{T}$'s time when the right-hand possibility has already disappeared, while there remains the chance (say, 50 percent) that the left-hand location may also disappear, i.e., that the body disappears completely—again clearly nonsense!

Although it does not really resolve the above issues, one might consider the possibility that there could be a "preferred" reference system, perhaps that for an observer at rest with the body prior to its bifurcating into its two superposed locations, i.e., as defined by the portion of the body's world-line leading up to the point O (the frame according to which Figure 13.3 is actually drawn). Even this makes no real physical sense, especially because the superposition under consideration could well only be one of several taking place nearby, all with different initial velocities, and no particular one of their initial word-lines would make sense to single out, rather than any one of the others. Nor would it make sense to select some kind of average velocity, especially if such superpositions spread outwards to a great distance, with many different velocities.

Indeed, if we are to consider that **OR** is some kind of "instantaneous" event, at which just one or the other component of a superposition becomes, in some clear sense, the "reality" of the situation, then the only consistent objective picture is what I shall call the *retro-active* viewpoint. This is where, in the example of Figure 13.4, we adopt the strange-seeming point of view that *only* the left-hand branch, in the situation depicted, is what actually represents reality, beyond the point O, and that the right-hand branch "never actually occurred"! Our picture of reality is that the *left*-hand branch was there "all the time," and that we have to take the view that, in a sense, the right-hand branch indeed somehow never actually became a part of this "reality"! Of course, if the **OR** process were to have gone the other way and the reduction had been to the *right* rather than to the left, then we would have to consider that it was the right-hand branch that was "there all the time" rather than the left-hand one. What we are presented with is a strange *retro-active* picture of quantum state-reduction, where the "objective reality" of the situation is that the surviving member of the **OR** process was retro-actively pre-determined by the "choice" that would *later* be made by the **OR** occurrence! (See [31] for an earlier description of this curious retro-active viewpoint.) This notion of "reality?" is what, in Section 8, will be clarified as "classical reality," where there is also a somewhat less robust notion of "quantum reality" that, in some sense, more resembles all of that depicted in Figure 13.3.

7. Gradualist OR: CSL Models

Before we proceed to explore this strange retro-active picture of classical reality more thoroughly, as we shall come to in the following section, there is another possibility that must also be considered, according to which the **OR** process might *not* actually be "instantaneous," as depicted in Figure 13.5, but in some sense a "gradual" process, as is considered in many other state-reduction proposals (CSL models [32]). In order to explain such models, I shall need to enter into the formalism of quantum mechanics a little more thoroughly than before. Returning to Dirac's stick of chalk, introduced in Section 2, I referred to the fact that whereas in classical physics a stick of chalk might be at a point A, or at a point B, in quantum mechanics there are "many other possibilities" for a body's state to be in a superposition of locations: "at A and at B at the same time." What are these "*many* possibilities" for a quantum superposition? Let us suppose that Ψ and Φ are distinct quantum states (necessarily belonging to the same quantum field theory); then we have many possibilities for their superpositions, written as:

$$\alpha\Psi + \beta\Phi$$

with α and β being *complex numbers*, not both zero, referred to as the *amplitudes* of the superposition (where we recall that a "complex number" is a quantity $x + iy$, where i is the "imaginary unit," satisfying $i^2 = -1$, with x and y being ordinary real numbers). Two such state superpositions are considered to represent the same physical situation if one is a complex multiple of the other; thus, the physical situation depends on the *ratio* of α to β (written $\alpha{:}\beta$) rather than to α or β individually. If Ψ and Φ are what are called "orthogonal states"—basically *fully independent* of one another—then one can (in principle, at least) perform a *measurement* on the state $\alpha\Psi + \beta\Phi$, to ask it whether the physical system "is" in state is Ψ or in state Φ. That is to say, after the measurement, the state is "found to be" in state Ψ or "found *not* to be" in state Ψ, in which latter case it has to be in state Φ, each with a certain probability. The rules of quantum mechanics tell us that the ratio of the probabilities that it comes up with the answer "Ψ," as opposed to the alternative "Φ," is the ratio $|\alpha|^2{:}|\beta|^2$ (where this *squared modulus* $|x + iy|^2$ of the complex number $x + iy$ is $x^2 + y^2$).

Now, the class of "**OR**" models that are referred to as "CSL models" (continuous spontaneous localization models; see [32]) are the modifications of standard quantum mechanics where a macroscopic quantum superposition $\alpha\Psi + \beta\Phi$ evolves with time (over and above the standard Schrödinger evolutions of Ψ and Φ), where in CSL the amplitudes α and β change continuously, but in a stochastic manner, eventually becoming so extremely close to either 1:0 or 0:1 that we "may as well" say that the state has *actually* become Ψ or Φ, respectively. This process has been described as a "gambler's ruin" in an ingenious proposal offered by Philip Pearle [32], an originator of the CSL idea.

From such a viewpoint, one would not consider the abrupt termination of the right-hand branch in Figure 13.3 to be appropriate, and instead we might consider that there is a (stochastic) evolution of an amplitude α, attached to the left-hand state, of Figure 13.3, which state we label as Ψ, and also (independently) of an amplitude β attached to the right-hand state, which we label as Φ. In the situation illustrated, the amplitude β attached to Φ, eventually dwindles extremely close to zero (or perhaps even actually *reaching* zero) at Q whereas α, attached to Ψ, remains distinctly non-zero at or near Q*, so we consider that the state has finally collapsed to the left-hand state Ψ.

Now, as opposed to the situation described in Section 6, there does not appear to be a conflict with special relativity, although we need (at least) to abandon the usual convention, in quantum mechanics, that the quantum states are normalized to unity, which would imply:

$$|\alpha|^2 + |\beta|^2 = 1.$$

In my opinion, there is, in any case, no need to insist on this normalization requirement in quantum mechanics, and the entire conventional theory can perfectly well be phrased without it.

It must be admitted that this picture has some definitely attractive aspects to it for the operation of an **OR** scheme. It does indeed appear to resolve the issues raised in the previous section. However, there are certain other severe difficulties inherent in CSL models that lead me, instead, down the retroactive route described in the previous section. One of these difficulties has to do with the state reduction *time* depending on the *initial* ratio $\alpha:\beta$ of the amplitudes of the superposition in question. If, for example, the superposition starts with a value of β, say, which is much smaller than the other amplitude (here, α), then one finds that the state-reduction time would tend to be much shorter than if the two amplitudes were comparable in size. If the reduction time does actually depend on the initial ratio of the amplitudes in such a way, then, following up on an idea put forward by Sandro Donadi [33] (also see [34]), we find that it would be possible to send messages faster than light. This would be in situations involving long-distance quantum entanglements known as Einstein-Podolski-Rosen (EPR) effects [35, 36]—such non-local effects being convincingly confirmed in numerous experiments that have been performed. Although, at first sight, EPR effects would seem to violate the relativity principle of the impossibility of sending signals faster than light, this is *not* actually so in conventual quantum mechanics. But even mild deviations from standard relativistic quantum theory may be in danger of permitting such super-luminal signaling. And, as Donadi has shown [33], **OR** theories for which the collapse time depends on the initial amplitude ratio would actually permit such signaling, in violation of the basic principles of special relativity. Since in CSL modes the expected time for **OR** to take place would indeed appear to depend on the initial ratio of the amplitudes $\alpha:\beta$, we are presented with a severe conflict between such CSL models and this basic non-signaling property of relativity theory [33].

Another difficulty with models of the CSL type, and others like them, is that they appear to give rise to "spontaneous heating," an effect that has been searched for but not as yet observed (e.g., [34]). To understand, in general terms, what the "heating" effect would come from, one must imagine that some form of **OR** would be occurring all the time within a solid body or fluid. We imagine that the wave function of all the particles would begin to spread out, due to the action of the Schrödinger evolution, their locations becoming more and more de-localized as time progresses. The argument would be that this spreading must, from time to time, be cancelled by some type of **OR** process that achieves a "localization" of the particles' positions. This localization involves an effective "jump" in the state of the system, and the continuing jumping would, in effect, provide a heating of the material that, at some level—in detail depending on the theory—ought to be measurable.

Apart from the absence of this effect to be seen in any experiment so far dedicated to detect it (e.g., [30]), there are reasons from Einstein's general relativity (and also other physical considerations) to believe that such an effect, resulting in the spontaneous production of energy, ought not actually to be present. Accordingly, it is important to see how the retro-active picture of Section 6 evades this difficulty. This comes about because although the particle locations begin to spread out as time progresses, in accordance with the Schrödinger equation, when it comes to the point when **OR** is to take place, the "**OR** decisions" retro-actively refer back to an earlier stage, before the quantum spreading has started so that the separations in the alternatives, subsequently to evolve into large quantum separations, are still very tiny (in effect, to the time of the event O in Figure 13.3). The classical evolution remains continuous and without the big "jumps" that would occur if it were not for the retro-activity.

In order to get a better picture of the physical evolution that I believe must be involved, we shall need to find a distinction between two different kinds of "reality," where the "retro-activity" refers to the "classical reality," but where we also need the somewhat different, somewhat less robust, notion of "quantum reality." This is an effect of the tension between the two great revolutions of 20th-century physics: general relativity and quantum mechanics, which we shall have to touch on in the following section.

8. Quantum and Classical Reality

In quantum theory, there has always been a lot of confusion about the ontological status of different parts of the theory. Is the wave function to be considered as "physically real," for example, or is it some artifact of the observer's "knowledge of the system"? I have no desire to get into all the facets of what is involved here, but I would like to take full consideration of a point of view expressed by Einstein in the following quotation [35], which I shall refer to as "Einstein's dictum":

> *If, without in any way disturbing a system, we can predict with certainty (i.e., with probability unity) the value of a physical quantity, then there exists an element of physical reality corresponding to this physical quantity.*

The notion of reality referred to here is what I shall call *quantum reality.* There is also a firmer notion of reality that I refer to as *classical reality,* which is more the kind of reality that we think of as consisting of tangible physical objects like rocks or tables, or even things like magnetic fields or radio waves, so long as we are not worrying about quantum effects, but most importantly, the classical space-time (even black holes), subject to being influenced by macroscopic bodies, in accordance with Einstein's general relativity, is an important aspect of classical reality. I am prepared to accept that the notion of space-time becomes a bit fuzzy when we try to imagine it at extremely tiny scales—what are called the Planck distances of 10^{-33} cm and time intervals of 10^{-43} seconds—but on the much larger scales that we are accustomed to, these classical spatiotemporal notions are extremely well defined.

There is an important ontological difference between these two types of reality, which can be expressed as follows:

CR. With regard to *classical reality*, we can ask a system: "What is your state?" The system can then legitimately respond: "My state is *X*," whichever *X* its state might actually happen to be.

QR. With regard to *quantum reality*, one cannot expect a response of this kind. However, one may have a good theoretical idea of what the quantum state is, and accordingly ask the system: "Is your state *X*?" Then, if you have it right, the system must respond, *with certainty*: "Yes, my state is indeed *X*." Moreover, in standard quantum mechanics, this certainty implies that the state was also *X* immediately before the system had been "asked" (i.e., "measured") so it has not been disturbed in any way, as Einstein's dictum requires.[1]

In the case of quantum reality, if one does not have it quite right, and the quantum-real state is not actually quite the *X* that one ascribes to it, then the system's response may still almost always be "yes" but, accordingly, *not* with certainty. That is to say, one may enact the same physical situation a very large number of times, and if one has calculated one's *X* correctly, then the response will be "yes" every time, but if one's calculation is just slightly off, then there would be an occasional "no" in the system's response. It seems to me that Einstein's dictum does indeed provide a good notion of a kind of reality—namely *quantum* reality. But it is not necessarily the same as the firmer notion of classical reality, as we shall shortly see.

[1] It may be remarked, in relation to the comments made at the beginning of this section, that the quantum state (or wave function) of any system whatsoever may be viewed as being part of "quantum reality" if we take the broadest view as to what kinds of "measurements" are allowed. However, most such "measurements" are very far from what could be carried out in practice.

In this connection, there is one issue that needs to be addressed in relation to the **OR** process as put forward here. Suppose we have a superposition $\alpha\Psi + \beta\Phi$ that, according to our gravitational **OR** theory, we expect to reduce to either Ψ or Φ, after some very long period of time, perhaps $\sim \tau$ Then, if after some much shorter time than τ, we wish to confirm that the quantum reality of the state is indeed $\alpha\Psi + \beta\Phi$, by the use of Einstein's dictum (by performing an experiment to distinguish between $\alpha\Psi + \beta\Phi$ and something orthogonal to $\alpha\Psi + \beta\Phi$), we do not quite succeed, because within this shorter time there will still be some very small probability that the **OR** process has actually occurred by then, so with this tiny probability the state will have become either Ψ or Φ, rather than remaining as $\alpha\Psi + \beta\Phi$, so the "certainty" required by Einstein's dictum will not quite be achieved. Nevertheless, as above, we may well say that the quantum reality is "very close" to being $\alpha\Psi + \beta\Phi$.

In Figure 13.4, I have depicted a space-time cartoon of the evolution of a system in which **OR** manifestly takes place. At the bottom of the picture, we see a laser emitting a single high-energy photon that encounters a beam splitter angled at 45° to its direction, so the photon's state is half reflected downwards and half transmitted horizontally, whereupon its horizontal component impacts on a tiny massive body. Its impact on the body is sufficient to move it slowly towards the left, so that as time progresses—time being now depicted upwards in the diagram—the body is put into a quantum superposition of being moved and not moved, different instances of this separation being depicted as we follow it up the picture. We also see the space-time, slightly warped by the presence of the mass of the body. Since the body's mass location is in quantum superposition, we find that the space-time itself must also be in a superposition of two slightly different geometries, so it is depicted as bifurcating into two slightly differing space-time geometries.

At the top of the picture, we see the **OR** event killing off the right-hand branch of the bifurcation, but in accordance with the discussion of Section 6, we must now (retro-actively) discard the entire right-hand branch of the bifurcation to obtain only the left-hand one, in a smoothly evolving *classical* space-time, right from the bottom of the picture to the top, and that we imagine continuing beyond the top of the picture.

With regard to the *quantum* reality of the situation, on the other hand, we must fairly well retain both branches, at least for a while, because we can conceive of an experiment aimed at confirming the actual quantum

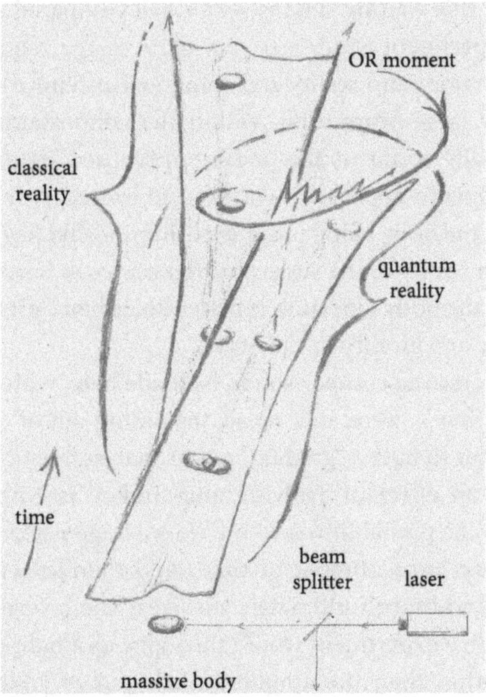

Figure 13.4 Classical and quantum reality in an **OR** event.

reality of the body's superposition. Let us suppose that this confirmatory experiment takes place well after the photon encounters the beam splitter, but still a good deal earlier than the anticipated reduction lifetime τ. We might conceive of such an experiment in which the downward (i.e., reflected) component of the photon's path, after it emerges from the beam-splitter, is put into a route enabling it to be reflected (up in the picture), to allow its entanglement with the body's location to be ascertained in some interference effect. In such a way, one could indeed confirm that the body's *quantum* state is actually in superposition, provided that this occurs well before **OR** has taken place. If we are able to ignore the potential ultimate **OR** event, then we could indeed confirm the quantum reality exhibited in the picture, using Einstein's dictum. However, this quantum reality may be expected to become less secure the later this confirmatory experiment is performed, as it gets closer to the reduction time τ.

We must actually imagine that the whole situation is repeated many times, for each proposed waiting time at which the confirmatory experiment is to be performed. If this waiting time is very small compared with τ, then the confirmatory experiment would get very close to the required "certainty" needed for a firm quantum reality according to Einstein's dictum (i.e., there would be a very large proportion, within the confirmatory experiments, that indeed actually confirms the body's superposed state). But when the waiting time begins to approach τ, or even to exceed τ, the proportion of confirmations of the body being in superposition would begin to drop away, and the quantum reality of the superposition ceases to have significance, so we consider that the body's location is better described by its classical reality of being in either one locality or the other.

There is an interesting comparison to be made here, with the CSL picture described in Section 7, were, in a sense, the fading out of one branch with respect to the other to have a "gradual" rather than sudden character. Rather than this being an effect of evolving amplitudes, as with CSL, here we would see this in the probabilities, where we envisage performing the whole experiment a very large number of times and then select only the cases where the left-hand branch ultimately survives. The probabilities obtained by the confirmatory experiment would then give us a fade-out of the right-hand branch, rather than the amplitude fade-out of CSL. The problems of consistency with special relativity encountered in Section 7 seem to be evaded in a different way, however, as here the confirmatory experiments are all done in a particular Lorentz frame, and the contradiction with special relativity described in that section is evaded.

An important ontological issue needs to be pointed out. Einstein's dictum is here being used in a counterfactual way, which is very much according to the nature of quantum reality. If one were *actually* to perform such a confirmatory experiment, as above, and the answer were to come back "Yes, my state is indeed in the superposition that you surmise," the effect on **OR** is not entirely passive, because if the answer indeed comes back "yes," we need to "reset the **OR** clock" all over again, and the timescale τ must be measured from the time of the measurement that just confirmed the quantum superposition. The act of "measurement" that is made in this confirmatory experiment will have its *own* **OR** event that constitutes this measurement, and this would affect the classical as well as the quantum reality of the situation. This would complicate the situation, if this confirmatory experiment is *actually* performed, rather than being imagined as being performed in this counterfactual way!

The behavior of **OR** according to this scheme closely mirrors what happens with an unstable nucleus. If an observation of the nucleus has been made that confirms the nuclear decay has not yet occurred, then its half-life must again be taken from the time of measurement. Of course, this "stay of execution" is balanced by the probability that the nucleus may already have decayed at the time that it is measured, and this appears also to be consistent with the gravitational **OR** proposal being put forward here. This parallel between nuclear decay lifetimes and this **OR** proposal is, of course, no accident. For the very derivation of the lifetime τ, as given on Section 5, is based on an analogy with nuclear decay, so such similarities are indeed to be expected.

It may seem that, with regard to quantum reality, when gravitational influences must be taken into consideration, things are not so clear-cut, when gravitational **OR** is to be part of the picture. Einstein's dictum does not acquire its necessary "certainty," so the quantum reality becomes a little less robust than in the absence of such effects.

However, the same must be said of *any* quantum-reality determination, quite apart from gravitational **OR** effects, since "certainty" could only be expected as an idealization. Moreover, so long as gravitational **OR** is taken to be a true phenomenon of nature, any quantum superposition in which mass displacement is involved would have a lifetime, even if an enormously long one, according to the Diósi-Penrose criterion of Section 5. As remarked earlier, there could be an extremely tiny probability that the normal rules of quantum theory would upset the "certainty" required for quantum reality.

Of course, such considerations would affect the notion of classical reality also, and these considerations should not get in the way of the idealized notions of "reality" described above. Just as the presence of gravity provides limitations on our notion of "quantum reality," we must conversely recognize that quantum effects provide limits on the notion of "classical reality." As mentioned earlier, there are limits conferred by quantum mechanics on the precision to which classical space-time can be defined—Planck-scale distances and times, as referred to earlier.

As a point of unification concerning these fundamental scales, it may be pointed out that the 4-volume difference between the two superposed branches, such as in Figure 13.4, up until the **OR** event is of the order of unity in Planck-scale units, which is another way of phrasing the Diósi-Penrose state-reduction time τ (see [24])!

9. OR and Consciousness

Let us finally return to our goal of applying these ideas to the phenomenon of consciousness. As already explained in Section 4, the Orch-OR proposal basically reverses the idea, variously put forward by some distinguished quantum theorists of the past, such as Wigner and von Neuman, that the "collapse of the wave function" could be an effect of a conscious observer actually "observing" a quantum system. Instead, the idea of Orch-OR is that there are structures in the brain that take advantage of a physically objective wave function collapse phenomenon—albeit far from fully understood—of a nature that I have referred to in this article as **OR**. In the writing of ENM, I had already formed the view that, in order to perform non-computational acts of conscious understanding, some form of **OR** needs to be involved, in brain action, in an effective way.

Accordingly, in order to understand what structures in the brain might be able to make use of **OR**, I tried to learn enough about neurophysiology that might help me to perceive how there could be, in what was currently understood about brain function, some clue as to what structures in the brain might conceivably be capable of maintaining a quantum superposition, adequately free from environmental decoherence, so that the **OR** process might have some positive effect. But from what I was then able to understand about the Hodgkin-Huxley theory of nerve propagation [37], I could see no hope of maintaining quantum superpositions in nerve signaling, as it would immediately decohere via the electric fields that would necessarily accompany a nerve transmission signal, thus causing uncontrollable environmental decoherence. Nevertheless, I needed to complete my book, and I ended my neurophysiological discussion very weakly, suggesting only some vague analogy with the formation of quasi-crystals, which appear also to require some non-local seeming "foresight" for their formation.

Fortunately for me, however, Stuart Hameroff, an anesthesiologist attached to the University of Arizona in Tucson, read my book, and realizing the limited nature of my physiological understanding, acquainted me with the existence and importance of *microtubules*! Stuart had himself independently come to the conclusion that something beyond mere (Hodgkin-Huxley [37]) nerve propagation needed to be involved, and he had, in particular, been struck by the chemically widely different gases that were able to act as general anesthetics. He had formed the view that such anesthetic gases might act directly (but non-chemically) on the microtubules within neurons. There is, in fact, now some distinctive evidence in support of this proposal (see the accompanying article by Stuart Hameroff [38]).

Upon learning from Stuart about microtubules, I was particularly struck by the remarkable symmetry that these structures possess—although at that time I had not become aware of the two different kinds of structure, referred to as A-lattice and B-lattice microtubules. The A-lattice microtubules have the greater symmetry; the B-lattice ones have a "seam" along the lengths of the microtubules, which interrupts the axial screw symmetries. I don't think that, at that time, there was much appreciation about this distinction. Microtubules have many roles to play in biology, and for many of these, the B-lattice microtubules seem to be preferred, perhaps owing to the presence of this "seam." Nevertheless, it was the greater symmetry of the A-lattice microtubules that had intrigued me most.

Why did this symmetry impress me so much? There is an effect in quantum mechanics whereby structures with high symmetry may enclose a region within. Of the many modes of quantum oscillation allowed, such symmetry would provide a class of lowest-energy quantum modes that are "degenerate," so a certain amount of quantum information can be preserved, by virtue of a comparatively large *energy gap* that shields these modes from immediate decoherence.

As a side remark, it may be mentioned here that there are curious macromolecules called *clathrins*, in the neighborhood of synapses, that also exhibit remarkable symmetry—in this case, icosahedral symmetry—and I could not help wondering whether these play some contributing role. In any case, the overall hope was that such symmetrical structures like clathrins, or more particularly A-lattice microtubules, might provide the necessary protection required for the preservation of quantum information.

In order to have something sufficiently macroscopic, such as an overall conscious experience, much larger well-shielded quantum states would be needed, involving these high-symmetry local ingredients, combining together from many different nerve cells. Clearly, this is a huge challenge, but the Orch-OR proposal would require such to be possible (perhaps interconnecting through the agency of *gap junctions* between cells (see [38])) so that quantum modes on a large scale, straddling large numbers of nerve cells, would be possible, in order to allow the preservation of some large-scale quantum state, with many degrees of freedom of quantum information, involving very large numbers of microtubules acting together in some kind of quantum-mechanical concert. This might allow quantum coherence to be preserved until the limits of **OR** could be reached and the state reduction therefore achieved in some beneficial "orchestrated" way—Orch-OR—so that the resulting proposed non-computability might arise as a basis for the notion of conscious understanding, as well as the many other aspects of conscious experience!

Of course, there is much speculation in this view and, as yet, it is grossly lacking in much of its necessary detail, though there is certainly some significant movement in this direction in recent work, some of this newly understood detail being spelled out in the accompanying article by Stuart Hameroff [38]. But before coming to this, I should also mention several features of brain structure and function that had already made me feel uncomfortable with the conventional "computer-like" action of brain operation, and had already disturbed me in my writing of ENM.

The most obvious of these, to me, was the contrast between the *cerebrum* (the most manifest, convoluted brain structure that presents itself when we think of a "brain") and that other structure, behind and beneath the cerebrum, called the *cerebellum*. As we now know, there are many more nerve cells, and certainly vastly more connections between nerve cells, in the cerebellum than in the cerebrum, yet the action of the cerebellum seems to be entirely unconscious! This enormous contrast would seem challenging to the "computer-like" picture of consciousness.

Moreover, unlike the cerebrum, the cerebellum seems to be much more "sensibly" wired up than the cerebrum, since there is that very curious "crossing over" of cerebral function where, on the whole, it is its left side that is directly concerned with the right side of the body, whereas the right-hand side of the cerebrum is concerned with the left side of the body! The unconscious cerebellum does not suffer from this seeming "mistake" in the organization of the cerebrum. Needless to say, I do not believe that it was a mistake on the part of nature, but this crossing over must be serving some important purpose. Could there be something other than direct nerve signaling that gives advantage to the cerebrum's crossing over, which enables consciousness to benefit from this seeming anomaly in construction?

Related to this is the seeming "accident" that becomes evident when one examines the fissure between the frontal and parietal lobes of the cerebrum. The frontal part is concerned with motor control (active influence) and the parietal (rear) part with the sensory (passive) experiences. What I find particularly striking is that the local regions of the frontal side of this fissure match very closely with the corresponding parts of the parietal side of this fissure, where we seem to see a curious correspondence, in location, between the passive and active aspects of conscious function, where the sensations on the hind part of this fissure (on the parietal lobes) have an uncanny correspondence with the locations for active control on the frontal part on this fissure (see ENM [5], Figures 9.3 and 9.4). What's the point of this local correspondence? If there were direct nerve connections across the fissure, then it might make sense, but there aren't! To trace such a nerve connection, one needs to go right into the central parts of the brain, which would seem to be incredibly inefficient, but there it is! Something other than direct nerve connection must be crucially important in conscious perception and its relation to conscious action, it had seemed to me very early on. It may be mentioned that there are other things that seem to require something more subtle than direct neve transmission, but these are not so clear-cut, and I shall leave them aside in this account.

How might the considerations of the previous sections, most particularly Section 8, relate to these issues? Somehow, with conscious control, there must be a trade-off between the direct and rapid nerve action of the unconscious but superbly trained cerebellum, and the very indirect and curiously organized conscious (to some extent) cerebrum. In my view, some clues are to be found in the tension between classical and quantum reality, as described in Section 8 and illustrated in Figure 13.4. The viewpoint here is that a conscious event has to do with **OR** occurrences, as illustrated at the top part of Figure 13.4. Since, according to this scheme, **OR** occurrences would be happening all the time and, as already pointed out in Section 4, it would be unreasonable to refer to absolutely *all* these occurrences as being "conscious," so the "un-orchestrated" occurrences of **OR** are merely "proto-conscious." There certainly is nothing like a conscious "purpose" (i.e., anything that we could meaningfully refer to as "free-will decisions") in these purely randomly occurring "choices" that the rules of standard quantum theory indeed demand must be purely "random," though with precise probabilities determined by the amplitude ratios, as described in Section 7. For us to be able to assign any "meaning" to these **OR** occurrences, they would need to be set in an appropriate context and "orchestrated" in some appropriate way. For actual conscious experience, these **OR** events would indeed have to be coordinated (or orchestrated) in some way; hence the terminology "Orch-OR" that Stuart Hameroff and I have agreed on.

Does this issue of some kind of "meaningful purpose," attached to these orchestrated **OR** events, provide us with a role for "free will"? There is a good case for this as we shall be seeing shortly, in relation to the changed perspective that the retro-activity of Section 6 provides us with, but I prefer not to hold a dogmatic position on the issue of "free will" just for now. However, the arguments frequently made for the necessary *absence* of free will in rapid reactions would be invalidated by the retro-active proposal being presented here.

The general picture that I am presenting here is that conscious perceptions and conscious actions are indeed to be understood in terms of such orchestrated **OR** events. The long routes that the nerve signaling seems to have to take in the cerebrum would need to be compensated by the "retro-active" nature of the **OR** events, as described in Sections 7 and 8.

There is some striking support for this point of view from experiments carried out by Benjamin Libet in the 1970s and later [39, 40] (these being described by Erich Hath in his book [41], which is where I first came across these ideas and then discussed them in ENM [5], see Figure 10.6). Libet had studied patients who needed a brain operation for some other reason. He would electrically simulate a particular point on a finger of the patient, and also independently electrically stimulate a corresponding point on the patient's parietal cerebral cortex. Although the patient would regard the brain stimulation as something very like a stimulation of that finger at the same associated place, the two would be clearly felt as different sensations by the patient. What was particularly remarkable about these experiments was that although the direct stimulation of the finger would be almost immediately felt by the patient, the direct stimulation of the brain would be felt only about one-half a second later than the actual time of initiation of the brain stimulation. Even more remarkable would be that if the finger stimulation was followed by the brain stimulation, the latter being initiated at around one-quarter of a second *after* the finger had been stimulated, then the finger stimulation would not be felt at all! How could it be that a later event (the brain stimulation) could *prevent* something that would otherwise have been experienced *earlier*? Somehow, the later brain stimulation, occurring around a quarter of a second *after* the time that the finger stimulation would otherwise be felt, rendered the finger stimulation to be "unfelt"! This seems to be a blatant contradiction with ordinary progression of conscious experience.

This seeming paradox can be understood if, as Orch-OR proposes, conscious sensations occur at **OR** events that, according to the images in Figures 12.3 and 12.4, imply a retro-active nature with regard to classical reality.

This may well tie in with Stuart Hameroff's proposal (see [38]) that it is in the *pyramidal* calls that the main location for the production of conscious experience occurs. As he explains in the accompanying article [38], such cells appear *not* to occur in the cerebellum at all (!), but they do occur in a particular layer of cerebral cells. These pyramidal cells contain a large number of microtubules, organized in a different way from in ordinary neurons, and Stuart proposes that they play a key role for conscious experience. It seems to me that this picture could provide an explanation for Libet's very curious results, as described above.

Consider, first, the skin stimulation of the person's finger. In order for it to be felt as an actual conscious occurrence, the nerve transmission would have to reach the pyramidal cells, but, according to Hameroff's picture, this requires a significant fraction of a second before the relevant pyramidal cells are eventually reached—let us say something like one-half of a second or so. However, because the **OR** "experience" of Figure 13.4 is caused retro-actively—perhaps to something again like a half-second earlier—the classical-reality response to an **OR** reaction to this conscious experience could be pushed back by around half a second or more. According to this curious viewpoint, the person involved would have the systematic conscious interpretation that the sensations actually experienced had *always preceded* this actual conscious (**OR**) event, say, by something like a half-second, i.e., the perception would have to be "referred back," in one's interpretations to say at least about half a second *earlier* than the timing of the **OR** event itself.

This would be reasonable, within this scheme, because any subsequent classically real action that would result from this experience would be referred back by the **OR** retro-activity of Figure 13.4, since the person involved would normally "act" in a classically real way. That classically real action would seem not to be retro-active, because the person involved would always refer back to his or her timing to the earlier moment when the classically real action would take place. The delay in nerve signals reaching the pyramidal cells would be compensated by this retro-activity, and the person involved would regard any consciously willed action to occur almost immediately following the willing of that action (in any normally willed action), despite the neuronal remoteness of the pyramidal cells.

Accordingly, in Libet's experiment referred to above, the perceived "moment of sensation" of the finger stimulation would not actually be felt until *after* the following brain stimulation, so it would not be at all impossible for it to be wiped out by the seemingly subsequent brain stimulation, because it would not actually have been "felt" until more than a quarter of a second after that brain stimulation. Such a sensation would normally be "referred back" in accordance with ordinary conscious experience. With the brain stimulation, however, this might well interrupt the normal progression of nerve transmission to the pyramidal cells, so the finger stimulation could consistently *not* be felt following the brain stimulation, because the brain stimulation would be *before* the actual **OR** event that would otherwise have been consciously felt and then "referred back," had it not been for the interrupting brain stimulation!

In many of Libet's experiments, the timing of the patient's experiences would be measured by a special fast-moving *clock* that would register, fairly closely, the actual times that the patient would experience the various conscious events. Surely, one might well think, this would accurately nail the *actual times* that the patient consciously experienced these events, rather than some later time, half a second or more later when the relevant **OR** event would have taken place? *Not so*, because in this proposal, it would take about half a second before one's sensation of a clock's face reaches the subject's consciousness, and that *is* the time according to the "consciousness clock" of the patient, and this has to be referred back by half a second or more to get the actual time, as registered by the clock!

Of course, this is a very radical picture of things, and much more study and experimentation would be needed in order to make full sense of it in all its aspects—or else to refute it! In any case, we are likely to be in dangerous waters with the idea of retro-activity, and we must not fall foul of various paradoxes that are potentially present with any form of "retro-causality," Nevertheless, I do not think that the ideas described in Section 8 actually do run into such paradoxical behavior, strangely enough, because the retro-causality described here is of a very different character, and seems not to be subject to the paradoxes of "time travel." The retro-activity described here is indeed very unlike the "traveling back in time" as described in many science fiction stories. Such postulated "time travel" is indeed subject to serious paradoxes and must be rejected as a feature of our actual universe, but the retro-activity described here is a more subtle thing, and more study is needed to make its implications completely clear.

Nevertheless, the ideas described here provide a new perspective on numerous experiments performed over the years, mainly following Libet's work, but some preceding it (see [38], Section 6). As a result of such experiments, many of which are described in Section 4 of Stuart Hameroff's accompanying article [38], the common view has come about that actions that have to be performed very rapidly, say, within about half a second, cannot be under conscious control. Reactions of this rapidity are common in many sports, such as in tennis, ping-pong, badminton, to name but a few. Is it possible that these actions cannot be under conscious control, despite the clear impressions of the players themselves? Can it be that their clear impressions that they are "consciously" controlling these rapid actions are an "illusion" and there is, accordingly, no actual "free will" directly involved in these actions (apart from pre-plane courses of action that would be hard to implement in response to so many possible alternative unexpected situations)?

The retro-activity involved in the Orch-OR proposal would nullify this uncomfortable conclusion since, as described above, the "actual" timing of the **OR** event that is the conscious experience controlling the action in question is "referred back" in the way that I have described above. It would seem clear that rapid conscious control would have a strong selective advantage, and this supplies an additional reason for nature to evolve beings able to make use of this curious feature of physical action, quite apart from the non-computational abilities of consciousness that formed the motivations behind the earlier sections of this article.

But what about this non-computational nature of conscious understanding that formed this central line of reasoning of the earlier sections of this article? How would such retro-activity be able to result in such non-computability? Here, the reasoning is somewhat less clear, but there are definite possibilities. Consider, for example, the Dirac model of interacting charged particles discussed in Section 3. Here, the issue had to do with eliminating the "runaway solutions," and the actual evolution had to take place when such runaway behavior does not later come about. Somehow, the system needs to "probe the future" in order to ascertain the presence or absence of such unphysical activity. Perhaps, in the retro-activity of the "control" of the classical evolution by the "later" **OR** event, the quantum reality is somehow also "probing potential futures," in order to "decide" what classical action to take. But I suspect the actual answer is a good deal more subtle than this!

Acknowledgment

Discussions with Stuart Hameroff have been invaluable for the development of ideas expressed in this article, and I thank him greatly for his essential input.

References

[1] Hameroff S. and Penrose R. (1996). Conscious events in orchestrated spacetime selections. *J. Consc. Stud.* **3**, 36–53.
[2] Penrose, R. and Hameroff, S. (2011). Consciousness in the universe: Neuroscience, quantum space-time geometry and Orch OR theory. *Journal of Cosmology* **14**, 1–32.
[3] Hameroff, S. and Penrose, R. (2014). Consciousness in the universe: A review of the "Orch OR" theory. *Physics of Life Reviews* **11**(1), 39–78, 104–112.
[4] Gödel, K. (1931). Über formal unentscheidbare Sätze der Principia Mathematica und verwandter Systeme I. *Monatshefte für Mathematik und Physik* **38**, 173–198.
[5] Penrose, R. (1989). *The Emperor's New Mind: Concerning Computers, Minds, and the Laws of Physics*. Oxford: Oxford University Press.
[6] Penrose, R. (1994). *Shadows of the Mind: An Approach to the Missing Science of Consciousness*. Oxford: Oxford University Press.
[7] Goodstein, R. L. (1944). On the restricted ordinal theorem. *J. Symb. Logic* **9**, 33–41.
[8] Kirby, L. A. S. and Paris, J. B. (1982). Accessible independence results for Peano arithmetic. *Bull. Lond. Math. Soc.* **14**, 285–293.
[9] Berger, R. (1966). The undecidability of the domino problem, *Memoirs Amer. Math. Soc.* **66**.
[10] Wang, H. (1993). On physicalism and algorithism: Can machines think? *Philosophia mathematica* (Ser. III), 97–138.
[11] Grünbaum, B. and Shephard, G. C. (1987). *Tilings and Patterns*. New York: W. H. Freeman.
[12] Penrose, R. (1997). Remarks on tiling: Details of a $(1 + \varepsilon + \varepsilon^2)$-aperiodic set. In *The Mathematics of Long-Range Aperiodic Order* (ed. Robert V. Moody, pp. 467–497). Amsterdam: Kluwer Academic.
[13] Dirac, P. A. M. (1928). The quantum theory of the electron. *Proc. Roy. Soc. (Lond.)* **A117**, 610–624. Part II of same, **A118**, 361–361.
[14] Dirac, P. A. M. (1938). Classical theory of radiating electrons. *Proc. Roy. Soc. (Lond.)* **A167**, 148–169.
[15] von Neumann, J. (1955). *Mathematical Foundations of Quantum Mechanics*. Princeton, NJ: Princeton University Press.
[16] Everett, H. (1983). "Relative state" formulation of quantum mechanics. In *Quantum Theory and Measurement* (eds. J. A. Wheeler and W. H. Zurek), Princeton, NJ: Princeton University Press. Originally appeared in *Revs. Mod. Phys.* (1957) **29**, 454–462.
[17] Chalmers, D. J. (1996). *The Conscious Mind: In Search of a Fundamental Theory*. Oxford: Oxford University Press.

[18] Penrose, R. (2014). On the gravitization of quantum mechanics 1: Quantum state reduction. *Found. Phys.* **44**, 557–575.

[19] Greenberger, D. M. and Overhauser, A.W. (1979). Coherence effects in neutron diffraction and gravity experiments. *Rev. Mod. Phys.* **51**, 43.

[20] Howl, R., Penrose, R., and Fuentes, I. (2019). Exploring the unification of quantum theory and general relativity with a Bose-Einstein condensate. *New J. Phys.* **21**, 043047.

[21] Fuentes, I. and Penrose, R. (2018). Quantum state reduction via gravity, and possible tests using Bose-Einstein condensates. In *Collapse of the Wave Function: Models, Ontology, Origin, and Impli-cations* (ed. S. Gao). Cambridge, UK: Cambridge University Press.

[22] Marshall, W., Simon, C., Penrose, R., and Bouwmeester, D. (2003). Towards quantum superpositions of a mirror. *Phys. Rev. Letters* **91**, 13–16, 130401.

[23] Eerkens, H. J., Buters, F. M., Weaver, M. J., Pepper, B., Welker, G., Heeck, K., Sonin, P., de Man, S., and Bouwmeester, D. (2015). Optical side-band cooling of a low frequency optomechanical system. *Optics Express* **23**(6), 8014–8020.

[24] Penrose, R. (1993). Gravity and quantum mechanics. In *General Relativity and Gravitation 13. Part 1: Plenary Lectures 1992. Proceedings of the Thirteenth International Conference on General Relativity and Gravitation held at Cordoba, Spain.*

[25] Penrose, R. (1996). On gravity's role in quantum state reduction. *Gen. Rel. Grav.* **28**, 581–600.

[26] Diósi, L. (1984). Gravitation and quantum-mechanical localization of macro-objects. *Phys. Lett.* **105A**, 199–202.

[27] Diósi, L. (1987). A universal master equation for the gravitational violation of quantum mechanics. *Phys. Lett.* **120A**, 377–381.

[28] Diósi, L. (1989). Models for universal reduction of macroscopic quantum fluctuations. *Phys. Rev.* **A40**, 1165–1174.

[29] Diósi, L. (1990). Relativistic theory for continuous measurement of quantum fields. *Phys. Rev.* **A**42, 5086.

[30] Donadi, S., Piscicchis, K., Curceanu, C., Diósi, L., Laubenstein, M., and Bassi, A. (2021). Underground test of gravity-related wave function collapse *Nature Physics* **17**, 72–78.

[31] Penrose, R. (2018). See https://www.youtube.com/watch?v=n941JWmT_xQ towards the very end, just before the general discussion.

[32] Pearle, P. (1989). Combining stochastic dynamical state-vector reduction with spontaneous localization. *Phys. Rev.* **A39**, 2277–2289.

[33] Donadi, S. (2020). Personal communication.

[34] Penrose, R. (2020). New thoughts on gravitational quantum state reduction, Version 4. Unpublished manuscript. Widely circulated notes.

[35] Einstein, A., Podolsky, P., and Rosen, N. (1935). Can quantum-mechanical description of physical reality be considered complete? *Phys. Rev.* **47**, 777–780. Reprinted in *Quantum Theory and Measurement* (1983, eds. J. A. Wheeler and W. H. Zurek), Princeton, NJ: Princeton University Press.

[36] Bohm, D. (1951). *Quantum Theory* (Chapter 22, Sections 15–19). Englewood Cliffs, NJ: Prentice-Hall. Reprinted as The paradox of Einstein, Rosen, and Podolsky, in *Quantum Theory and Measurement* (1983, eds. J. A. Wheeler and W. H. Zurek), Princeton, NJ: Princeton University Press.

[37] Hodgkin A. L. and Huxley A. F. (1952). A quantitative description of membrane current and its application to conduction and excitation in nerve. *J. Physiol.* **117**, 500–544.
[38] Hameroff, S. (2022). Orch OR and the quantum biology of consciousness. Essay appearing in this volume.
[39] Libet, B., Wright, E. W. Jr., Feinstein, B., and Pearl, D. K. (1979). Subjective referral of the timing for a conscious sensory experience. *Brain* **102**, 193–224.
[40] Libet, B. (1992). The neural time-factor in perception, volition and free will. *Review de Métaphysique et de Morale* **2**, 255–272.
[41] Harth, E. (1982). *Windows on the Mind.* Hassocks, Sussex, UK: Harvester Press.

14

Orch OR and the Quantum Biology of Consciousness

Stuart Hameroff

1. Introduction: Consciousness in the Universe

Consciousness implies awareness, a subjective phenomenal experience of internal and external worlds, composed of what philosophers term "qualia." Consciousness may also imply a sense of self, feelings, choice, control of voluntary behavior, memory, thought, language, and (e.g., when we close our eyes or meditate) internally generated images and geometric patterns, also composed of experiential qualia. Our views of reality, of the universe, of ourselves depend on consciousness that defines our existence. But what consciousness actually *is* remains unknown. Three general possibilities regarding the place of consciousness in the universe have been commonly expressed (adapted from Hameroff & Penrose, 2014).

A) Computational Materialism Consciousness is not an independent quality but arose in terms of conventional physical processes as a natural evolutionary consequence of computation in brains and nervous systems. As an emergent adaptation, consciousness is commonly assumed to be epiphenomenal (i.e., a secondary effect without independent influence) and also illusory (largely constructing reality, rather than perceiving it). In this view, consciousness is purported to have a scientific basis but is *not* an intrinsic feature of the universe.

B) Dualism/Idealism/Panpsychism Consciousness is a separate quality, distinct from physical actions and not controlled by physical laws. Descartes' dualism, idealism (consciousness is "all there is"), religious viewpoints, and approaches in which consciousness causes quantum state reduction ("subjective reduction" or SR)—all place consciousness outside science. Panpsychism attributes consciousness to properties of all matter, or underlying

Stuart Hameroff, *Orch OR and the Quantum Biology of Consciousness* In: *Consciousness and Quantum Mechanics*.
Edited by: Shan Gao, Oxford University Press. © Oxford University Press 2022.
DOI: 10.1093/oso/9780197501665.003.0015

reality, but without scientific identity or causal influence, and generally fails to deal with quantum physics that governs matter and reality at microscopic levels. In these views, consciousness is an intrinsic feature of the universe, but without scientific basis.

C) Orch OR: Quantum Pan-Protopsychism Consciousness results from discrete physical events that have always existed in the universe as non-cognitive, proto-conscious moments, these acting as part of precise physical laws not yet fully understood. Biology evolved a mechanism to orchestrate such events and couple them to neuronal activity resulting in meaningful, cognitive, conscious moments, and hence also to causal control of behavior. These events are proposed specifically to be moments of quantum state "objective reduction" (intrinsic quantum "self-measurement") connected to the structure of space-time geometry. Such events need not necessarily be taken as part of current theories of the laws of the universe, but should ultimately be scientifically describable. This is basically the type of view put forward, in very general terms, by the philosopher A. N. Whitehead and also fleshed out in a scientific framework in the Penrose–Hameroff theory of "orchestrated objective reduction" (Orch OR; Penrose & Hameroff, 1995; Hameroff & Penrose 1996a, 1996b, 2014a, 2014b). In the Orch OR theory, these conscious events are terminations of "orchestrated" quantum computations in brain microtubules reducing by Penrose "objective reduction" (OR), and thus having experiential qualities. In this view, consciousness has a scientific basis and is an intrinsic feature of the universe.

2. Computability and Non-Computability in the Brain

Cortical pyramidal neurons—most likely sites for consciousness The prevailing assumption in science and philosophy is "computational materialism." Indeed, brain function and consciousness have always been likened to contemporary information technology, from the ancient Greek idea of memory encoding as a "seal ring in wax," to the mind as a telegraph switching circuit, and now computers and artificial intelligence (Jaynes, 1977). Several theories of brain function and consciousness are based on computer-like cognitive architectures, e.g., bottom-up sensory processing, and top-down executive action.

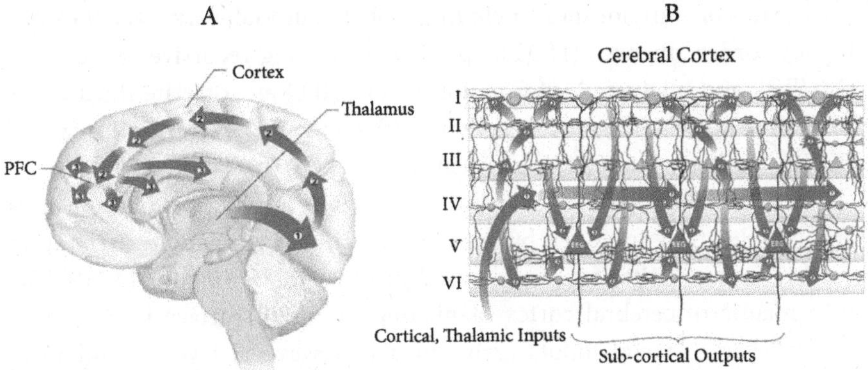

Figure 14.1 A. In the brain, sensory inputs including vision are routed through thalamus and from there in 3 waves through cortex, resulting in consciousness. They are (1) thalamus to primary sensory (e.g., visual) cortex, (2) feed-forward to frontal and pre-frontal regions, and finally, (3) feed-back to other cortical regions, correlating with conscious vision and/or other conscious sensations. **B.** Sensory signals in 6 layers of cerebral cortex also occur in 3 waves, converging on layer 5 giant pyramidal neurons. Cortico- and thalamic inputs (1) project to layer 4 and from there to (2) layers 1, 2, 3, and 6, and then finally, onto (3) layer 5 pyramidal neurons whose apical dendrites arise to cortical surface, giving rise to measurable electro-encephalographic (EEG) activity. Third wave pyramidal neuron activity correlating with consciousness occurs approximately 300 milliseconds after input to sensory organs.

Inputs from sensory organs for vision, hearing, taste, and touch (but not smell) are routed through the thalamus, which then relays signals through the cerebral cortex in three waves (Figure 14.1A). These are (1) the thalamus to primary sensory cortex, e.g., for vision, V1 in the back of the brain; (2) forward from primary cortex to frontal and pre-frontal cortex; and (3) frontal/pre-frontal to other cortical areas and elsewhere. Third wave activity occurs about 300 to 500 milliseconds after sensory impingement and seems to correlate with consciousness. For example, only third wave activity is inhibited by anesthesia in all its forms: propofol, ketamine, inhalational gas (Lee et al., 2013). And yet we often respond to such stimuli within 100 milliseconds, seemingly in conscious control (see Section 6 on free will and retroactivity). Accordingly, the conventional view in science and philosophy is that we respond non-consciously, and have a false illusion of conscious control (Dennett, 1991; Wegner, 2002).

Theories of consciousness including global neuronal workspace (GNW), higher order theory (HOT), predictive coding/recursive processing (PC/RP), and integrated information theory (IIT) all correlate third wave, frontal-to-posterior broadcast activity with consciousness (or in the case of PC/RP, recursive interaction between waves 2 and 3).

What might be special about third wave activity to generate consciousness and be selectively inhibited by anesthetics? Third wave activity arriving in cerebral cortex is itself also processed in three waves (Figure 14.1B). The thin mantle of cerebral cortex blanketing the brain surface is composed of six layers. Axonal inputs arrive in a first wave at layer 4, and from there a second wave from layer 4 to layers 1, 2, 3, and 6. Finally, a third wave from layers 1, 2, 3, and 6 converges on layer 5 giant pyramidal cell neurons (Koch, 2004). These distinctive neurons have extremely large cone-shaped cell bodies/soma that extend basilar dendrites laterally within layer 5 to interact with basilar dendrites of other giant pyramidal neurons, and inter-neurons. Karl Pribram (1991) suggested such dendritic-dendritic webs extending laterally through the cortex mediate consciousness. Each giant pyramidal cell body/soma also projects an apical dendrite vertically to the brain surface (Figure 14.1B). Collective oscillations of parallel, vertical apical dendrites of layer 5 pyramidal neurons give rise to electro-encephalography (EEG) measured at the scalp or brain surface. Finally, layer 5 pyramidal neuron cell bodies/soma also extend outgoing axons "downward" to cortical and sub-cortical areas, e.g., to exert causal action on behavior.

Cortical layer 5 pyramidal neurons are the apex of brain inputs and executor of outputs, a convergent intersection, receiving inputs from the external world via ascending pathways, as well as horizontal cortical-cortical interactions, e.g., default mode networks. Pyramidal cell dendrites and soma contain unique arrays of mixed polarity networks of microtubules, optimal for quantum interference as proposed in Orch OR (Figure 14.2). The cerebellum, which apparently lacks consciousness, has no pyramidal neurons. Networks of mixed polarity microtubules in dendrites and soma of cortical layer 5 pyramidal neurons are likely sites for consciousness in the brain.

The Hodgkin-Huxley integrate-and-fire computable neuron The contemporary view is that the brain is a computer of neuronal threshold logic devices, including layer 5 giant pyramidal cells, each acting as a Hodgkin-Huxley (H-H) model neuron. Hodgkin and Huxley (1952) mathematically described neurons as "integrate-and-fire" devices, their functions mediated

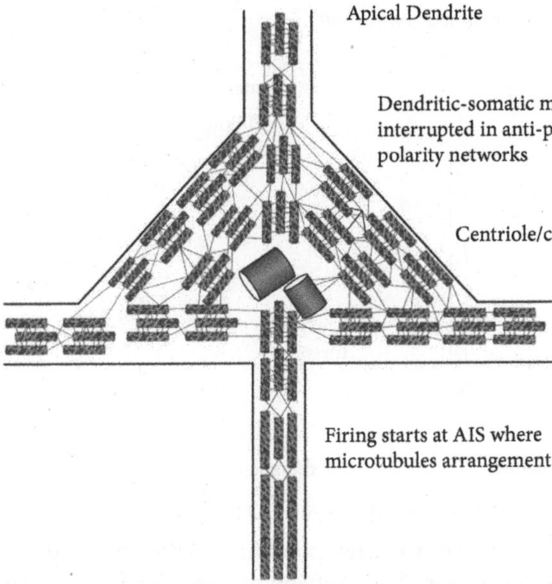

Figure 14.2 Schematic of layer 5 giant pyramidal neuron, with cone-shaped cell body. An apical dendrite arises to cortical surface and is a source of electroencephalography (EEG). Basilar dendrites extend laterally to other pyramidal dendrites (not seen). A single axon extends downward, conveying action potential firings, or spikes that can exert volitional, causal action.

entirely through neuronal surface membrane potentials and synaptic connections. According to H-H, neuronal dendrites and cell bodies/soma receive synaptic inputs from other neurons as membrane potentials. These graded, "analog" potentials are then integrated along the dendritic-somatic membrane surface to a threshold potential at the axon-initiation segment (AIS) on the proximal axon. When a particular threshold potential is reached, at that moment, at the AIS, an axonal *firing* (action potential or "spike") is triggered, which propagates the length of the axon to a subsequent synapse, exerting causal action and behavior. In computer analogies and artificial intelligence, spikes are often taken as digital or bit-like states. The integration and firing of H-H neurons are "computable" algorithmic processes with minimal variability and randomness (Figure 14.3).

Circuits and networks made up of H-H neurons as simple threshold logic devices can "compute" and account for cognitive brain functions. But the "brain-as-computer of H-H neurons" approach has limitations, casting consciousness as:

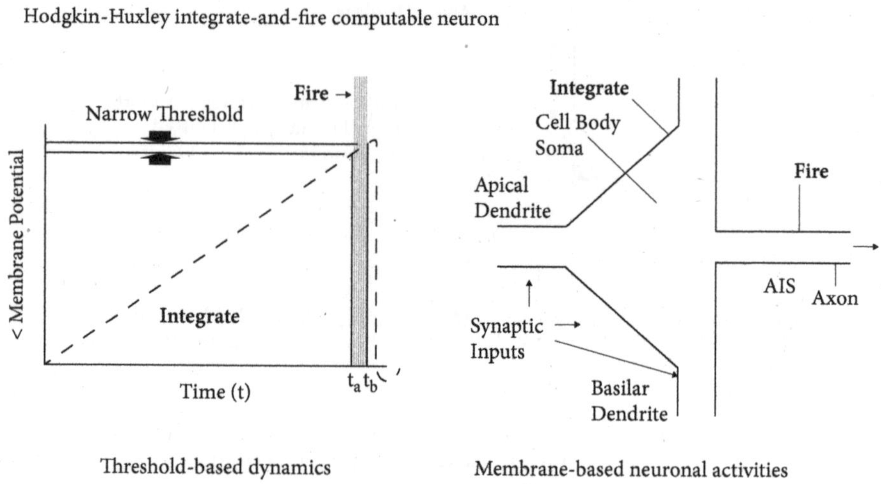

Figure 14.3 *Left:* Diagram of Hodgkin-Huxley (H-H) integrate-and-fire neuronal activity. Narrow threshold and temporal firing windows reflect algorithmic "computable" behavior. *Right:* Schematic pyramidal neuron (apex pointing left, axon to right) showing/utilizing only its surface membrane for integration and firing as prescribed by H-H.

A) Fragmented and unbound, failing to account for binding and unity of conscious experience (Mashour, 2013). For example, various aspects of visual inputs (e.g., shape, color, motion, and meaning) are processed in different regions of visual cortex and at different times, yet we consciously perceive integrated, unified visual objects and scenes.

B) Epiphenomenal and illusory, occurring "too late" for real-time conscious action. Brain activity correlating with sensory inputs occurs 300 to 500 milliseconds after the inputs, and yet we respond to those inputs, seemingly "consciously," at around 100 milliseconds. Accordingly, the conventional view in science and philosophy is that we respond non-consciously and have a false illusion of conscious control (Dennett, 1991; Wegner, 2002).

C) Computable, algorithmic, and predictable ("robot-like"), seeming to preclude insight, creativity, and free will.

D) Non-conscious, "zombie-like," failing to account for phenomenal conscious experience or qualia. Chalmers (1996) described the experiential nature of conscious experience as the "hard problem," in contrast to so-called cognitive "easy problems" like sensory discrimination, information integration, purposeful movement, focusing attention, and others that may be, at times at least, entirely non-conscious, "autopilot" cognitive functions. The H-H brain-as-computer approach can in principle account for "easy problems," non-conscious autopilot cognition, but not the hard problem of conscious experience.

The Orch OR "orchestrate and fire" non-computable neuron The brain-as-computer approach based on H-H neurons paints a bleak picture. However, actual neurons in awake, presumably conscious animals appear to deviate from Hodgkin-Huxley computable behavior. Naundorf et al. (2006) placed electrodes in cortical pyramidal neurons of awake cats and found that axonal firing threshold in individual neurons varies "firing-to-firing," "spike-to-spike" (Figure 14.4). In control experiments, pyramidal neurons in cell culture showed no deviation from H-H computable behavior, and the cause of deviation in neurons of awake animals was debated in the literature for several years (McCormick et al., 2007; Naundorf et al., 2007). Colwell and Brenner (2009) proposed that the deviation from computable

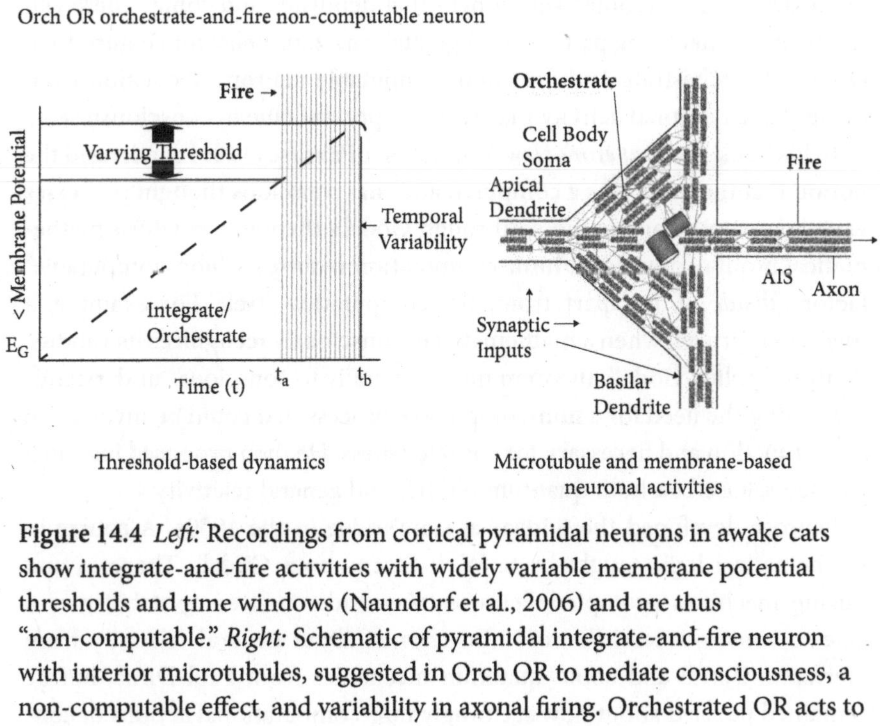

Figure 14.4 *Left:* Recordings from cortical pyramidal neurons in awake cats show integrate-and-fire activities with widely variable membrane potential thresholds and time windows (Naundorf et al., 2006) and are thus "non-computable." *Right:* Schematic of pyramidal integrate-and-fire neuron with interior microtubules, suggested in Orch OR to mediate consciousness, a non-computable effect, and variability in axonal firing. Orchestrated OR acts to modulate integrated membrane potential in actions on firing.

H-H behavior in the neurons of awake animals was caused by "noise" within dendrites and soma from incoming synaptic inputs. But there was no such noise and deviation in cell culture neurons. Some factor other than synaptic inputs and dendritic-somatic membrane potentials, i.e., a factor outside H-H computation, appears to regulate axonal firings in pyramidal neurons of awake, presumably conscious animals.

Deviation from H-H computable behavior could indicate a "non-computable" conscious process modulating non-conscious, cognitive activity. As firings or spikes mediate volitional behavior and causal action, the end of integration in dendrites and cell body/soma of integrate-and-fire neurons (e.g., layer 5 pyramidal cells) is the most strategic time and place for consciousness to occur and regulate otherwise non-conscious, "auto-pilot" cognitive behavior. For example, we commonly perform cognitive tasks non-consciously on "auto-pilot," e.g., while walking or even driving, but when necessary, consciousness suddenly assumes control of our behavior. As we shall see, the Orch OR theory suggests this non-computable deviation from H-H behavior occurs due to consciousness from deeper-level quantum processes in microtubules *within* neuronal dendrites and soma. These can modulate axonal firing patterns to regulate conscious behavior (Figure 14.4; Orch OR "orchestrate-and-fire" non-computable neuron). Deviation from computable neuronal activity may be an important clue to consciousness.

In his book *The Emperor's New Mind*, Roger Penrose (1989) challenged the notion that the brain was a computer, and that conscious thought processes were computational. He did so through Gödel's theorem, in which mathematical proof of any algorithmic computation requires a "non-computable" factor *outside of*, or apart from, the computation itself. For example, a theorem is proven when a mathematician consciously recognizes its validity. Penrose applied Gödel's theorem more generally to conscious "understanding," citing the need for a non-computable process that could be involved in brain function and necessary for consciousness. He then proposed just such a process, based on both quantum physics and general relativity.

Penrose developed these ideas at Cambridge in the 1950s. A course in mathematical logic taught by S. W. P. Steen about Gödel's Theorem and Turing machines prompted Penrose to consider non-computability as a clue to consciousness. Penrose also learned about general relativity from Hermann Bondi and about quantum mechanics from Paul Dirac, but he remained puzzled about how seemingly non-computational actions of conscious understanding could come about, feeling that something was missing. He eventually concluded that the one gap in the computational picture of the world, and the most likely source for non-computability in consciousness, had to lie with the *reduction* of the quantum state by an *objective* threshold occurring in quantum measurement, or "collapse of the wave function" ("objective reduction" or OR). The physics underlying OR would underlie consciousness in the brain. What was that physics? Penrose determined the unknown, non-computable OR physics linked to consciousness was related to the appropriate unification of Einstein's general relativity and quantum mechanics in order to accommodate quantum state reduction. This would require, he realized, a serious modification of quantum mechanics and a solution to the measurement problem.

3. Objective Reduction (OR) and the Nature of Reality

The measurement problem Consciousness and the nature of reality are related through the "measurement problem" in quantum mechanics. That is, at microscopic scales, particles can exist as superpositions of multiple co-existing possibilities, described by a quantum wave function. Yet when amplified, measured, or observed, superpositions seem to reduce or collapse to definite states, the macroscopic world we perceive. Why this occurs is unknown, but there are several proposed explanations. They include:

A) Conscious Observation/Subjective Reduction The very act of measurement or of conscious observation reduces waves to particles. Early quantum theorists John von Neumann and Eugene Wigner, and more recently Henry Stapp (2007), David Chalmers, and Kelvin McQueen (this volume), have suggested that *subjective* conscious observation causes quantum state *reduction* or the "collapse of the wave function," meaning multiple possibilities reduce/collapse to specific states ("Subjective Reduction" or SR).

B) Many Worlds Reduction/collapse does not occur, and all possibilities co-exist eternally (Everett, 1957).

C) Environmental Decoherence The quantum superposition in question mixes with its microscopic environment (which is also quantum in nature) and becomes too complicated for detailed description (Zurek et al., 1993).

D) Objective Reduction (OR) An objective physical process causes reduction of the quantum state to one outcome *OR* another, not both at once, requiring *new physics*, going beyond standard quantum theory.

Some OR approaches (e.g., as occur in GRW; Ghirardi, Rimini, & Weber, 1986; Ghirardi, Grassi, & Rimini, 1990) and also Continuous Spontaneous Localization (CSL) approaches (e.g., Pearle, 1989; Pearle & Squires, 1994) introduce a "new field" or "new process" to reduce the quantum state, but as Penrose suggests, the "new physics" ought to involve some justification from broader aspects of physics. Indeed, as Penrose (1989) has long proposed, the relevant field may be the oldest known field of all, namely the *gravitational* field.

Superposition as separated space-time geometry Gravity reveals a deep tension between the basic quantum mechanical (QM) principle of *linear superposition* and the basic general relativity (GR) *principle of equivalence* (Penrose, 2014)—or the related principle of *general covariance* (Penrose, 1996). This tension led Penrose to conclude that OR physics had to come about through the appropriate unification of Einstein's GR and quantum mechanics, requiring serious modification of the latter to include OR. Einstein's GR describes a space-time metric in which matter relates to curvature in the metric geometry. Penrose observed that when mass is in quantum linear superposition of significantly different locations or states, its corresponding space-time is also put into quantum linear superposition of differing curvatures, a separation in the fine scale space-time geometry of the universe. To illustrate (Penrose, 1989), he used two-dimensional space-time sheets (one dimension of space, one of time) to represent four-dimensional space-time geometry (Figure 14.5A). According to "Subjective Reduction" (SR), conscious observation ("BING!!"; Figure 14.5B) would cause quantum state reduction to one curvature or the other.

With only limited mass-energy superposition separation, then the space-time geometry is minimally affected and remains a classical entity in accordance with general relativity. But when the superposition becomes too de-localized, when the separation becomes too great, then perhaps either the alternate space-time curvatures would completely separate and *many worlds* would result (Figure 14.6A). Or, some resilient property of space-time geometry causes the separation to "snap" and OR takes place! At that moment, one particular mass location or state, and its one particular space-time curvature, are selected, with the previously superpositioned alternatives then ceasing to exist ("ping"; Figure 14.6B).

The specific classical states selected in Penrose OR are thought to not be chosen randomly, as in other approaches, or algorithmically, or a combination of algorithmic and random. Rather, Penrose OR events are thought

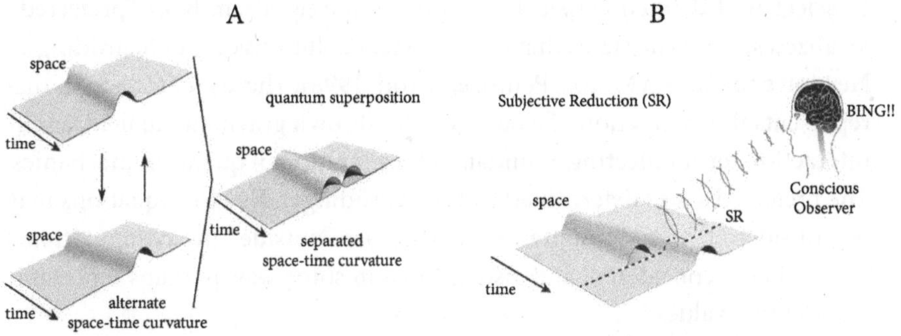

Figure 14.5 A. Two-dimensional space-time sheets representing four-dimensional space-time geometry. Particle locations correspond with specific curvatures in space-time geometry and can oscillate (*left:* top and bottom). Quantum superposition of the particle in both locations corresponds with alternative, separated curvatures in space-time geometry.
B. Subjective Reduction (SR); superposition evolves until conscious (subjective) observation (denoted by "BING") causes quantum state reduction or collapse of the wave function at time *t*. One curvature ceases to exist, and the alternative curvature/mass location exists in the material world. Consciousness causes collapse.

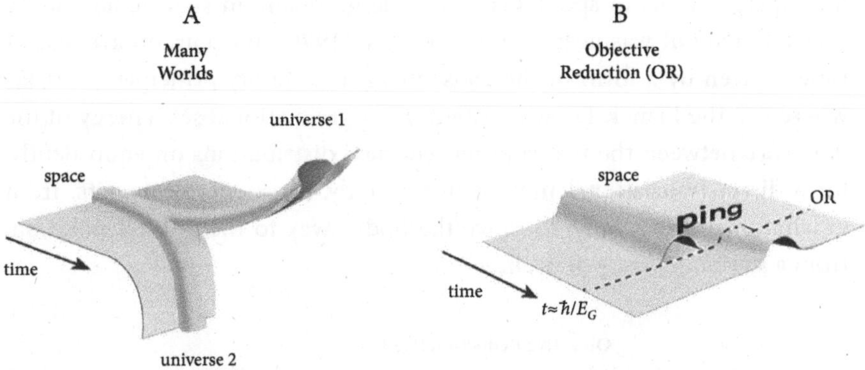

Figure 14.6 A. Quantum superposition as space-time separations that separate into their own universes, i.e., the "many worlds" hypothesis. **B.** Objective Reduction (OR); quantum superposition of a particle in 2 locations in two-dimensional space-time sheet reaches threshold at an average time $t = \hbar/E_G$ at which OR occurs ("ping"). Quantum state reduction occurs due to an objective threshold ("objective reduction," OR). An external conscious observer is not necessary, and "many worlds" is avoided.

to select well-defined classical states from among a family of "preferred," localized space-time states that are stationary solutions of the "Schrödinger-Newton equations" (Moroz, Penrose, & Tod, 1998). These include a term that represents the interaction of a particle with its own gravitational field, a self-interaction term reflecting a fundamental alteration of quantum mechanics. The preferred states determined by the Schrödinger-Newton equations may be considered "non-computable," as they are "outside" the computational system being considered, and also optimal in some way, perhaps representing Platonic values.

Penrose had recognized that non-computability was a subtle but important feature of consciousness, and a clue to its origin. OR could provide non-computability, and also fundamental aspects of conscious experience, a potential solution to the "hard problem" (Hameroff & Penrose, 1996b). As opposed to the "conscious observer" ("Subjective Reduction" or SR) approach in which consciousness causes quantum state reduction (putting consciousness outside science), Penrose proposed instead that spontaneously occurring OR caused, or was equivalent to, phenomenal experience, (proto-) consciousness ("ping"; Figures 14.6B and 14.7).

Objective reduction (OR) and proto-conscious moments The timescale for superpositions to spontaneously undergo quantum state reduction by gravitational OR was described by Penrose (1996) to occur, on average, at time *t*, given by a form of the Heisenberg uncertainty principle: $t = \hbar/E_G$ where $\hbar$ = the Planck-Dirac constant, E_G = gravitational self-energy of the *difference* between the two superposed mass distributions or, equivalently, for a direct translational motion of the body, E_G = energy it costs, from gravitational forces only, to move the body away to the given separation, from a stationary copy of itself.

Objective Reduction (OR)

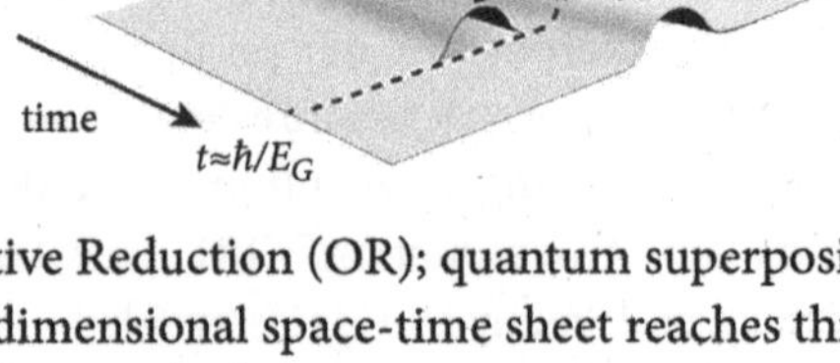

Figure 14.7 Objective Reduction (OR); quantum superposition of a particle in 2 locations in two-dimensional space-time sheet reaches threshold at an average time $t = \hbar/E_G$ at which OR occurs with a moment of (proto-) conscious experience ("ping").

E_G may be calculated for a given superposition from its particular mass, radius, separation distance, and level and degree of separation (Hameroff & Penrose, 1996a). The inverse relationship means that large superpositions E_G will reach threshold quickly, and smaller superpositions E_G will require longer times, potentially providing a spectrum of OR events at time $t = \hbar/E_G$ that can be tested experimentally (Marshall et al., 2003; Eerkens et al., 2015). although detecting or verifying phenomenal experience would be problematic. Penrose OR stands as a testable solution to the measurement problem in quantum mechanics.

In terms of consciousness, Penrose OR is precisely *opposite* to the view put forth by von Neumann, Wigner, Chalmers and McQueen, and others, discussed above, in which consciousness causes quantum state reduction (Figure 14.5B; Subjective Reduction or SR). A fundamental process in the fine scale structure of the universe, Penrose OR *causes* consciousness or may be considered equivalent to consciousness (Figure 14.7).

Superpositions in the general micro-environment at ambient temperature would arise, evolve, entangle, and join neighboring quantum superpositions. E_G would quickly increase and at time t reach the OR threshold. At that moment, OR to classical states would occur, producing a moment of phenomenal experience. Because of the random nature of the microenvironment, the states selected and nature of the experiential qualia would also be random, with minimal non-computable influence. The phenomenal experience would be presumably primitive and isolated, and lack memory, context, and sense of "self." Such proto-conscious moments would be occurring ubiquitously and continuously in the environment, appearing as "decoherence" (and avoiding the need for *many worlds,* or *subjective reduction*).

The ubiquitous occurrence of OR proto-conscious moments is conceptually similar to the philosophical position of panpsychism, in which experience is somehow a property of matter, or of underlying reality. But at microscopic levels, matter is quantum in nature and can exist in superposition states of multiple possibilities. These evolve and entangle to reach OR, which results in repetitive moments of (proto-) conscious experience (perhaps best categorized as "quantum pan-protopsychism"). As opposed to the panpsychist view of consciousness as a state or property, OR casts consciousness as a dynamic sequence of discrete events, akin to philosopher Alfred North Whitehead's (1929, 1933) process of "occasions of experience."

In *The Emperor's New Mind,* Penrose (1989) put forward the case for non-computability in discrete moments of OR (proto-)conscious experience, although without the specific gravitational OR proposal reflected in his current beliefs. OR was, and remains, the only specific scientific mechanism for consciousness yet advanced. However, for the brain to convert proto-conscious OR into full, rich phenomenal experience and non-computable volitional action, some type of organized ("orchestrated") quantum computational process would be required that terminates, or halts, non-computably by OR. Could quantum computing happen in the brain?

4. Anesthesia and the "Quantum Underground"

"Warm, wet and noisy", but oil and water don't mix Technological quantum computers must operate at extreme cold to avoid disruption of quantum information by thermal decoherence (premature OR?). Therefore, the microenvironment in biological brains has generally been considered too "warm, wet and noisy" for organized, and seemingly delicate, quantum processes. However, biological microenvironments are highly heterogeneous, e.g., composed of various different pharmacological solubility compartments in which various drugs dissolve and bind, one particular compartment appearing conducive to organized quantum effects.

However they are ingested, drugs and molecules distribute throughout the body according to their solubility, how easily they dissolve, or bind in different chemical media. A major determinant of solubility is the polarity of both the solute and solvent, polarity being the degree of charge separation within a molecule, leading to positively and negatively charged regions. For example, the water molecule H_2O is polar because its two positively charged hydrogens orient in one direction, and its negatively charged oxygen in the opposite direction. Positive ends of one molecule attract to negative ends of others, and so on. Other polar molecules dissolve in water, and water dissolves in other polar solvents. (In solubility, "like dissolves like.") Most drugs are polar, with positively and negatively charged regions, and thus dissolve in polar compartments like water, blood, and fluids. They may have precise charge interactions, e.g., "lock and key" binding, with receptors. At biological temperatures, microscopic activities in aqueous polar solvent compartments are random and chaotic, and quantum states are generally thought to be extremely short-lived and disorganized.

But there are also non-polar, lipid-like, or oil-like "hydrophobic" compartments in the brain and body, based in organic chemistry. These non-polar regions exclude water ("oil and water don't mix") and support quantum processes. Organic chemistry derives largely from quantum properties of delocalized pi electron resonance clouds of basic benzene ("phenyl") hexagonal rings, found in nearly all biomolecules, most drugs and neurotransmitters. The benzene ring has 6 carbons, 6 hydrogens, and 3 extra electrons that combine to form delocalized pi orbital electron resonance clouds above and around the ring (Figure 14.8). Non-polar, but polarizable, pi electron resonance clouds are fundamental to life and can support collective quantum effects.

Fröhlich coherence In bulk form, benzene is highly flammable. But when properly spaced in organized geometric arrays, benzene rings can have interesting quantum properties due to collective, cooperative pi resonance. Two-dimensional sheets of benzene, known as graphene, as well as fullerene nanotubes, have extended pi resonance clouds that can detect terahertz

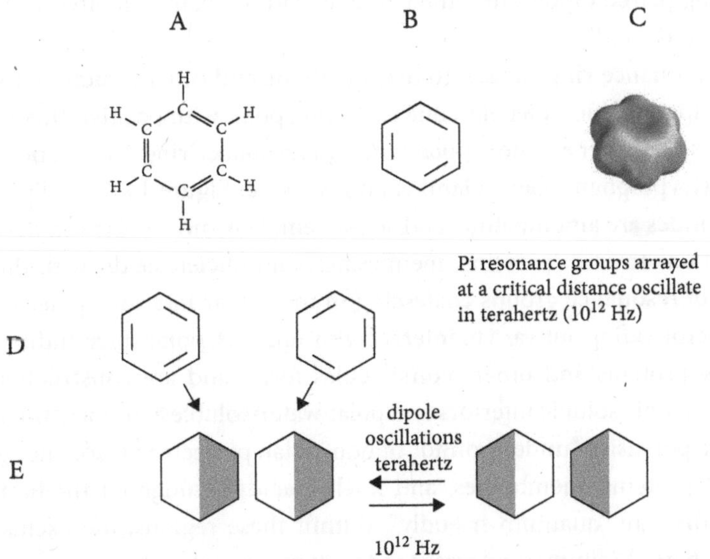

Figure 14.8 *Top:* Three representations of a benzene ring, the fundamental structure in organic chemistry. **A.** Chemical structure: a 6 carbon ring, each carbon binding a hydrogen atom, and 3 carbon-carbon double bonds with 3 extra electrons. **B.** The hexagonal ring with extra double-bond electrons. **C.** Extra electrons form delocalized electron cloud engulfing ring. **D.** Two benzene rings attract by van der Waals forces. **E.** Induce electron cloud dipoles that couple and oscillate in terahertz (10^{12} Hz).

oscillations (Gayduchenko et al., 2021) and do quantum computing at warm temperatures.

Pi resonance plays a key role in quantum mechanisms supporting plant photosynthesis at warm temperatures. To make chemical energy and food, plant protein complexes collect photons and convert them to electronic excitations ("excitons") that are then transported through a protein to reach the chemical energy center. Intra-protein transport occurs by quantum superposition of excitons propagating through seven different pi resonance chromophore pathways (Engel et al., 2007; Scholes, 2010). Without this efficient quantum energy transport, photosynthesis may not have been able to adequately support life on Earth.

Quantum interactions among pi resonance groups drive biological self-organization. At warm temperatures, single neutral benzene molecules attract each other through van der Waals forces by self-induced dipoles. Electronegativity in one cloud repels electronegativity in a nearby cloud, forming paired dipoles that then couple, and oscillate in terahertz (Figures 14.8D and 14.8E).

Pi resonance rings attach to hydrocarbons and other structure, forming "amphipathic" biomolecules, one end being polar with positive and negative charges, the other end non-polar with a pi resonance ring. "Aromatic" amino acids (tryptophan, phenylalanine, and tyrosine; Figure 14.11A), lipids, and nucleotides are amphipathic, and self-assemble by non-polar van der Waals attractions to form proteins, membranes, and nucleic acids, with the non-polar pi resonance groups coalescing in molecular interiors, polar components protruding outward to interact with aqueous, polar surroundings. This is how proteins and other biomolecules form, and are constructed, with non-polar oil-soluble interiors and polar water-soluble exteriors. Thus, there exists a pervasive "underground" of non-polar, pi electron resonance groups within proteins, membranes, and nucleic acids throughout the brain and body that are "quantum-friendly." Within these regions, the exclusion of polar charges minimizes degrees of freedom other than pi resonance dipole oscillations or excitons. Particular symmetries, or geometries (e.g., quasi-crystals, Fröhlich oscillators, Jahn-Teller materials, Schrödinger's "aperiodic lattices"), of quantum-friendly pi resonance regions may approach a "decoherence-free subspace" conducive to quantum processes, i.e., a "quantum underground" (Craddock et al., 2016).

Biophysicist Herbert Fröhlich (1968, 1970, 1975) suggested that pi resonance groups that were spatially arrayed in appropriate geometric symmetry would oscillate and condense to a common quantum coherent vibrational mode. Ambient heat would pump coherent mechanical lattice vibration phonons, coupled to collective pi resonance dipole oscillations in secluded non-polar regions, forming what Fröhlich termed "optical phonons." As the dipole coupling could extend mesoscopic and macroscopic lengths along and among symmetrically arrayed pi resonance loci, he described their coupled vibrations as "giant dipole oscillations" (Figure 14.12).

A connection between Fröhlich dipole oscillations, pi resonance "quantum underground," Orch OR, and consciousness is suggested by the action of anesthetic gases.

Anesthesia In the mid-19th century, certain gases were found that, when inhaled at critical concentrations, caused humans and animals to lose consciousness, and/or cease purposeful behavior. If breathing was adequately maintained, when the "anesthetic" gas was exhaled away, the subjects and animals woke up, returning to consciousness essentially unchanged. Modern studies have shown that anesthesia is selective, preventing consciousness while sparing non-conscious brain activities that continue under anesthesia. The precise mechanism of anesthetic action should pinpoint consciousness in the brain.

Anesthetic gases all have similar actions selectively blocking consciousness in all animals, but have strikingly different types of molecular structures, including halogenated hydrocarbons, ethers, nitrous oxide, and the inert elemental gas xenon. But the anesthetic gases all have one common chemical feature, a surface with complete non-polar electron outer shell exposure. This non-polar but polarizable surface enables van der Waals dipole couplings, the same sort of induced-induced dipoles that occur between pi resonance ring structures (Figure 14.9A). However because of asymmetry, van der Waals couplings between a pi resonance ring and an anesthetic cause "dipole dispersion," disrupting coherent oscillations between pi resonance rings (Figure 14.9B).

Each anesthetic gas was found to have a particular potency based on the gas concentration (equilibrated with lungs, blood, and brain) at which half of a group of animals would move, or not move, in response to a tail clamp or other noxious stimulus. This concentration became known as the "minimal alveolar concentration," or MAC, with potency correlating with 1/MAC; the more potent the gas, the less required to cause loss of responsiveness. Each anesthetic gas had its own MAC, but amazingly, each gas acted at

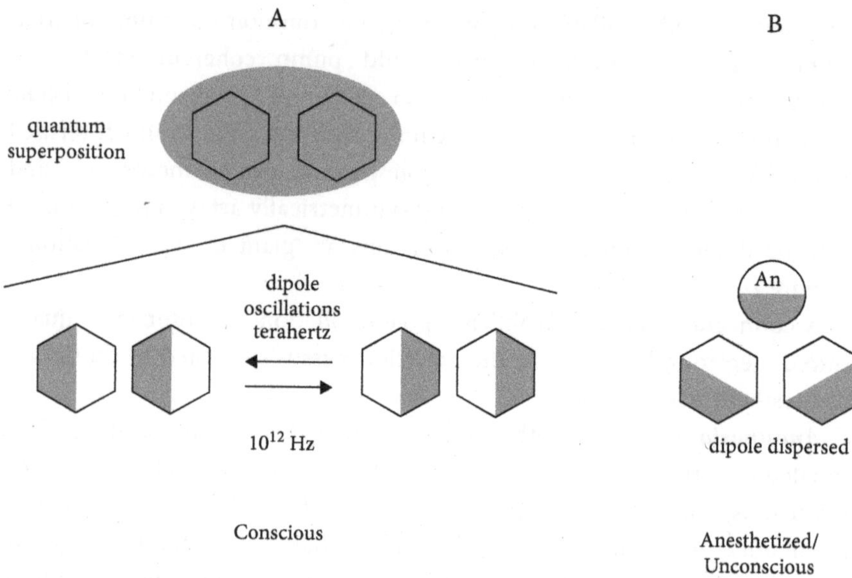

Figure 14.9 Two benzene ("phenyl") rings with pi resonance electron dipoles oscillating in terahertz (10^{12} Hz) by van der Waals forces. **A.** Quantum superposition of both dipole states and possible qubit. **B.** Anesthetic (An) molecule disperses van der Waals dipole oscillation and superposition, resulting in loss of consciousness.

the same concentration—had the same MAC—in all animals, near-identical concentrations anesthetizing mice, salamanders, insects, horses, or humans.

Scientists sought a common property among the gases that might correlate with anesthetic potency, and found the answer in gas solubility in a particular non-polar compartment. Working separately, Germany's Hans Meyer (1899) and Britain's Charles Overton (1901) ranked anesthetic potency for many anesthetics in various animal models and also tested anesthetic solubility in different types of solvents. Both found the same results. Over many orders of magnitude, among many chemically disparate anesthetic gas structures, the potency of all anesthetics in immobilizing all animals correlated nearly perfectly with the anesthetic's solubility in a particular non-polar, lipid-like compartment, characterized by a low Hildebrand coefficient lambda (15.2 to 19.3 SI Units), closely resembling benzene and olive oil.

During anesthesia, large amounts of anesthetic gas are dissolved in the Meyer-Overton compartment in fat stores, membranes, proteins, and nucleic acids all over the body, binding non-specifically by quantum van der Waals interactions. But paradoxically, anesthetics have very few effects other than preventing consciousness. One explanation may be that a subset of the "Meyer-Overton quantum underground" within the brain mediates consciousness by utilizing highly organized quantum processes, easily perturbed by random quantum interactions, there being no such organized quantum processes in membranes, fat stores, and other biomolecules.

Where in the brain's "Meyer-Overton quantum underground" might there be such highly organized quantum processes? Where do anesthetics act to selectively block consciousness? Specific anesthetic sites of action were initially assumed to be lipid membranes, but evidence eventually showed anesthetics acted in non-polar regions of pi resonance groups inside proteins (Franks & Lieb, 1984). The specific proteins involved in anesthetic action (and therefore presumably involved in consciousness) were generally considered to be membrane protein receptors. Indeed, post-synaptic receptors for excitatory neurotransmitters acetylcholine, serotonin, and glycine, and for the inhibitory neurotransmitter $GABA_A$, were found to bind anesthetic gases, having appropriate intra-protein non-polar Meyer-Overton regions. Anesthetics were asserted to inhibit the excitatory receptors and to promote, as an agonist, the $GABA_A$ inhibitory receptor.

However, experiments failed to show correlations between anesthetic potency and effects on such membrane proteins (Eger et al., 2008). Genomic, proteomic, and optogenetic evidence has instead pointed to tubulin and microtubules as the target of anesthetic action (Xi et al., 2004; Pan et al., 2007; Emerson et al., 2013). Computer modeling suggests anesthetics appear to act by dampening collective terahertz quantum oscillations among the 86 pi resonance groups of aromatic amino acid rings in each tubulin in microtubules, the effective dampening proportional to the clinical potency of each anesthetic in rendering humans and animals unconscious and/or unresponsive (Craddock et al., 2017). It remains an open question, but there is more evidence for anesthetic action on microtubules than for anesthetic action on membrane proteins.

Skepticism regarding functional quantum biology related to consciousness in the "warm, wet and noisy" brain may be overblown. Fröhlich coherence suggests "warm" ambient heat *pumps* coherence in biomolecular lattices, rather than disrupting it. Non-polar regions in the Meyer-Overton quantum underground exclude water and thus are not "wet." And coherent pumping and reduced degrees of freedom avoid "noise" in a decoherence-free subspace. In the Meyer-Overton quantum underground, quantum computation terminating by OR to produce consciousness is feasible. Is there a quantum computer in the brain?

5. Microtubules and Orch OR—Orchestrated Objective Reduction

Quantum computers process "quantum information," superpositions of multiple co-existing possible states, e.g., quantum bits ("qubits") of both 1 *and* 0. These then entangle, unify, and compute according to the Schrödinger equation until quantum state reduction ("collapse") occurs, and qubits reduce to classical states. These classical states are the output or solution of the quantum computation.

Microtubules Neurons appear too large, asymmetrical, and susceptible to decoherence (or premature OR) to act as qubits, e.g., of both firing and not firing. Smaller-scale brain structures with a suitable "quantum underground" and proper symmetry could potentially function as entangled qubits at biological temperatures, isolated from the polar surroundings. But they would also have to (1) interface with the polar surroundings for input and output, (2) be intimately involved in sub-cellular cognition, and (3) avoid decoherence (premature, random, superpositions) until reaching Orch OR threshold at time $t = \hbar/E_G$. At that moment, a unified conscious experience would occur, and classical states selected (qubit-to-bit) that "consciously" modulate non-conscious cognitive neurophysiology (e.g., as non-computable deviation from integrate-and-fire H-H neuronal activity). But how and where could quantum computing occur and exert effects on neurophysiology? Microtubules were an appropriate candidate.

Major components of the cell cytoskeleton, and self-assembling lattice polymers of the protein tubulin (along with cytoskeletal actin, the brain's most prevalent proteins), microtubules intelligently organize neuronal shape, growth, movement, and synaptic plasticity, appearing to act as both the nervous system and transport system within each neuron.

Tubulin is a peanut-shaped dimer composed of two tubulin monomers: negatively charged alpha tubulin and positively charged beta tubulin. Tubulins self-assemble such that alpha monomers of one tubulin are attracted and bind to beta monomer ends of another tubulin so that each microtubule has a net polarity, with a beta "plus" end and an alpha "minus" end. Microtubule assembly is facilitated by these electrostatic effects, but is primarily entropy-driven, as groups of unpolymerized tubulins bind and order more water than those same tubulins assembled in a microtubule lattice. Thus, paradoxically, a fully polymerized microtubule has overall less order/more entropy than all of its unpolymerized subunits.

In most cells, microtubules serve multiple purposes, including cell division (mitosis), in which structural microtubules disassemble, their tubulins subsequently reassembling as mitotic spindles that, controlled by centrioles, tease apart duplicated sets of chromosomes and guide them to their new daughter cells. The mitotic microtubules then disassemble, reassemble for their previous function or another, and . . . life goes on.

However, brain neurons don't divide. These neurons, once formed, may grow, form new synapses, and restructure (based largely on microtubule activities), but they don't undergo mitosis. So many neuronal microtubules are stable, although those in axons can "treadmill," assembling at one end and disassembling at the other. But dendritic-somatic microtubules are "capped," and don't add or lose tubulins at their ends. As each tubulin in a brain microtubule may be one of 22 different isoforms, and each may also be modified post-translationally in one of 5 different ways, or phosphorylated, etc., and because dendritic-somatic microtubules are stable lattices, and because there are about 10^8 tubulins per neuron, they offer ideal "memory beds," precisely where consciousness is occurring according to Orch OR.

In long-term potentiation, a model for memory, the hexagonal enzyme calcium-calmodulin-kinase 2 (CaMKII) is activated and deploys 6 kinase domains such that each activated CaMKII can phosphorylate microtubule hexagonal lattices up to 6 "bits" at a time (Craddock et al., 2017). Dendritic-somatic microtubules are ideal sites for memory storage and therefore ideal sites in which consciousness can occur.

Because of their organizational functions and lattice structure, microtubules have long been proposed to process information relevant to cell functions and cognition (e.g., Hameroff & Watt, 1982). Based on Fröhlich coherent oscillations as a clocking mechanism, automata models of "bit-like" tubulins in 2 possible states, dipole coupled between each tubulin and its 6 A-lattice neighborhood tubulins, had shown self-organizing behavior and classical information processing capabilities (Figure 14.10; Smith et al., 1984; Rasmussen et al., 1990). Simulated Fröhlich vibrations in a microtubule A-lattice demonstrate nodes—energy maxima and minima—that precisely match attachment patterns of microtubule-associated proteins (MAPs), patterns determining cell function and structure (Samsonovich

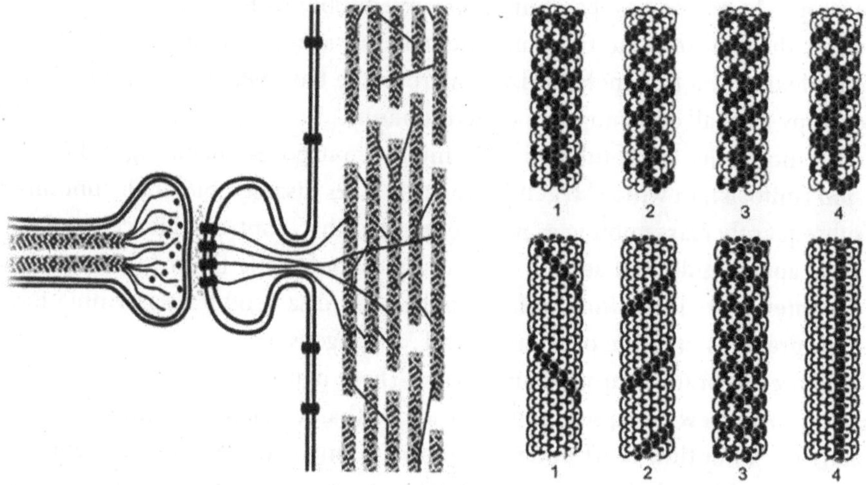

Figure 14.10 A. A synapse showing internal microtubules and other cytoskeletal structures. At left, an axon meets a dendritic spine in which actin filaments contact interrupted (mixed polarity) microtubules in main dendrite. **B.** *Top:* Four-step sequence of A-lattice microtubule automata in which the dipole state of each peanut-shaped tubulin (black or white) interacts with those of 6 neighboring tubulins at discrete time steps. Patterns propagate, conveying and processing information (Rasmussen et al., 1990). *Bottom:* Contiguous pathways following Fibonacci geometry in A-lattice microtubule.

et al., 1992). Microtubules appear to act as the cell's on-board computer system for cognitive functions.

Defining the orchestrated qubit In technological quantum computers, reduction or "collapse" occurs due to external measurement (premature, proto-conscious OR?), introducing randomness, and minimizing non-computability. Quantum computers that could avoid decoherence/premature OR and were able to properly organize, or "orchestrate" quantum information, could instead evolve to terminate or "halt" by "orchestrated" OR at time $t = \hbar/E_G$, resulting in a fully orchestrated conscious moment, i.e., one with maximal non-computable influence.

Orchestration implies integration and evolution of quantum information following the Schrödinger equation, but also dependent on natural symmetry, resonance, and logic of the qubit structure itself, e.g., the A-lattice microtubule with Fibonacci geometry (Figure 14.12). During orchestration, superpositioned qubits entangle with those in other microtubules, increasing E_G, unifying and harmonizing conscious content in accord with the natural properties of the qubit substrate until threshold is met at time $t = \hbar/E_G$. At that moment of Orch OR, a fully integrated multi-modal conscious experience occurs, orchestration having converted experiential proto-conscious OR noise ("ping") to full conscious experience ("BING!!"). And orchestration requires the qubit structure to resist decoherence (premature OR), e.g., through the quantum underground in microtubules as described above.

There are two types of microtubule lattices: the A-lattice microtubule and the B-lattice microtubule. In both types, tubulins align vertically in protofilament columns, "beta plus" end to "alpha minus" end. But lateral side-to-side interactions differ. In the A-lattice, tubulin protofilament columns are aligned in a skewed hexagon, such that winding patterns along adjacent tubulins follow Fibonacci symmetrical geometry. These patterns wind around the cylindrical microtubule to repeat on any vertical protofilament every 3, 5, or 8 tubulins (e.g., 5 helix; Figure 14.12). B-lattice microtubules lack Fibonacci geometry and have a symmetry-breaking seam. A-lattice microtubules are far more favorable than B-lattice microtubule for quantum effects.

Initially, the Orch OR qubit was taken as the dipole state of each individual tubulin, with alternative states and superposition of both states. But tubulin proteins are 110,000 atomic mass units ("daltons"), extremely large compared to quantum particles, highly charged on their surfaces, and seemingly susceptible to decoherence, or premature, random OR. Tubulin qubits seemed unlikely.

In the late 1990s, the atomic structure of tubulin revealed complex clusters of 86 pi resonance rings of aromatic amino acids tryptophan, phenylalanine, and tyrosine within each tubulin (Nogales et al., 1998; Figures 14.11A and 14.11B).

The pi resonance rings within each tubulin are themselves arrayed helically and contiguously around beta sheets, enabling quantum dipole oscillations along one degree of freedom. Pi resonance rings within each tubulin can also couple with those in neighboring tubulins in microtubule lattices, e.g., along helical winding patterns in A-lattice Fibonacci geometry. Orch OR (Hameroff & Penrose, 2014) began to portray microtubule-based qubits as pathways of oscillation and superpositioned dipole orientation within each tubulin, simplified for illustration purposes from 86 helically arrayed rings to 3 rings in one plane (Figure 14.11C), and also bridging by dipole

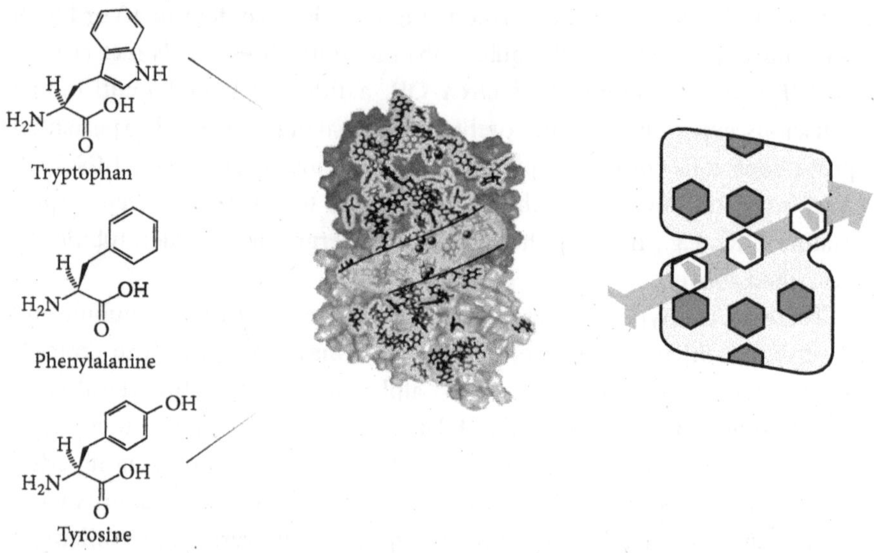

Figure 14.11 *Left:* Structure of pi resonance amino acids tryptophan, phenylalanine, and tyrosine found within (middle) atomic structure of tubulin showing 86 pi resonance rings of the pi resonance rings in non-polar "quantum underground." Spheres show sites of anesthetic gas binding and action. Shaded band going upwards to the right is a sequence of rings aligned with those in neighboring tubulins along the "8-start" helical winding pattern in the microtubule A lattice (see Figure 14.12). *Right:* Schematic of tubulin with only 9 rings to illustrate possible global dipole pathways through the microtubule.

coupling between pi resonance rings in neighboring tubulins, leading to mesoscopic Fröhlich oscillating "giant dipoles" (Fröhlich, 1968, 1970, 1975; Figure 14.12). In A-lattice microtubules, helical winding pathways occur along Fibonacci geometry, e.g., patterns that repeat on any vertical protofilament every 3, 5, and/or 8 tubulins. (Interestingly, the helical qubits are composed of individual tubulins whose internal pi resonance groups are themselves helically arranged.) The giant dipoles are suggested to oscillate and become coherent superpositions of both orientations to function as a pathway qubit (Figure 14.12). These can then interact with other pathway qubits within each microtubule and by non-local entanglement with microtubule qubits elsewhere. Orch OR suggests qubits occur as Fröhlich giant dipoles in microtubules that entangle and can extend throughout the brain. How can qubits lead to consciousness?

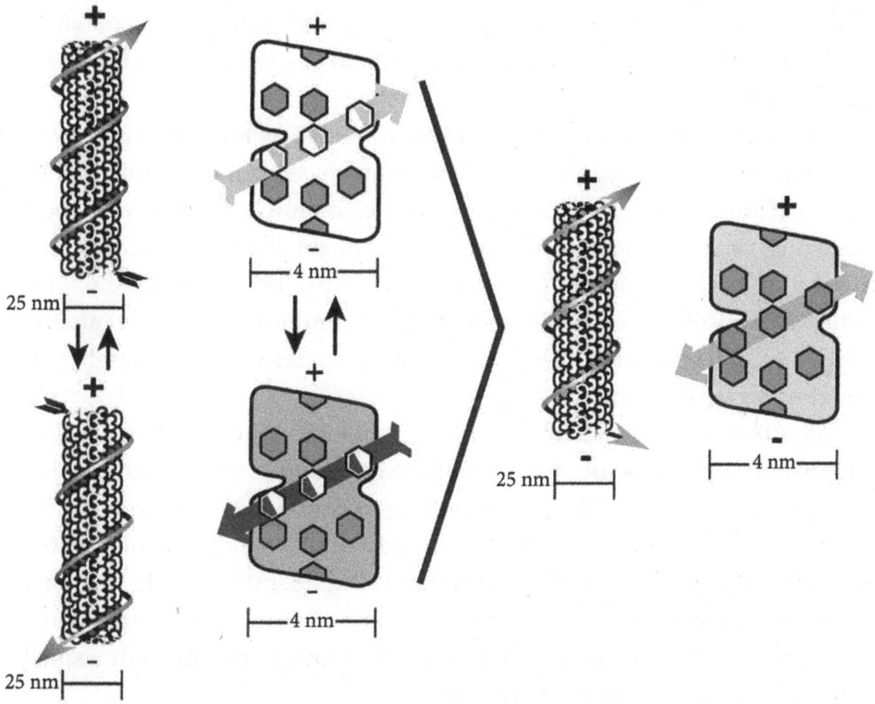

Figure 14.12 The Orch OR qubit in a microtubule. *Left:* Collective dipoles in pi resonance groups in tubulins (schematic versions with only 3, of 86, pi resonance groups shown) oscillate in single tubulin and along a helical pathway in a microtubule A lattice. *Right:* Quantum superposition of both orientations in a tubulin pathway.

Quantifying Orch OR consciousness According to Orch OR, consciousness and non-computability would require orchestrated qubits to reach time $t = \hbar/E_G$ without decoherence, or premature OR, by non-orchestrated processes.

For time t to be reasonably short, thus making it easier to avoid non-orchestrated OR, orchestrated superpositions with significant gravitational self-energy E_G would be required. However, electrons in quantum superposition (as in Fröhlich coherent dipole oscillations) have low mass and E_G (amount of mass/space-time geometry separated from itself). Consequently, time t based on E_G of electrons would be relatively long. Electron movements couple weakly to atomic nuclear movements by the Mossbauer effect, e.g., 1 nanometer electron movement causing displacement and superposition of its carbon atomic nucleus by the nuclear diameter in femtometers. With atomic nuclei included in superposition, E_G and space-time separation increase markedly, time t becomes briefer, and the OR threshold is more easily reached.

We calculated E_G for tubulin in three ways: (1) as a partial 10 percent separation of the whole protein, (2) as complete separation at the level of atomic (carbon) nuclei, or (3) as complete separation at the level of carbon nucleons, i.e., protons and neutrons. The dominant effect was (2), complete separation of atomic nuclei (Hameroff & Penrose, 1996a), consistent with Mossbauer displacement. Using this approach, and with the tubulin atomic weight of 110,000 daltons, we calculated E_G in terms of the number of tubulins for various times t.

Cognitive events and membrane physiological action times occur in tens to hundreds of milliseconds, e.g., gamma synchrony electroencephalography at 40 Hz, corresponding to time $t = 25$ milliseconds. By $t = \hbar/E_G$, setting t equal to 25 milliseconds would require E_G of 2×10^{10} tubulins to reach the Orch OR threshold. But 25 milliseconds is a long time for a quantum state to avoid environmental decoherence (premature, random, unorchestrated OR). And with $\sim 10^8$ tubulins/neuron, only a small number of neurons would be involved.

Figure 14.13 illustrates the evolution of E_G superposition in microtubules in a pyramidal neuron (14.13A), as individual microtubules with increasing superposition E_G (gray tubulins, 14.13B), and its space-time separation (14.13C) at arbitrary t and E_G. When the threshold is met, an Orch OR conscious moment occurs.

Orch OR conscious moments need not coincide directly in frequency with cognitive and electrophysiological times, and could thus occur at larger E_G values and faster quantum vibrational frequencies than cognitive functions. Faster Orch OR frequencies reduce the time needed to avoid decoherence/premature OR, but require more superpositioned tubulins.

But cognitive phenomena occur in tens to hundreds of milliseconds, far slower than the Orch OR time scale. To account for this discrepancy, Orch OR proposes that interference "beats" occur between slightly different fast frequencies (e.g., among mixed polarity, anti-parallel arrayed dendritic-somatic microtubules) and may result in slower oscillations in hertz, including EEG rhythms (Hameroff & Penrose, 2014). Used for nearly a century, the origin and significance of EEG remain unknown (e.g., Cohen, 2017).

The neurons of the human brain contain about 10^{19} tubulins. If these were all entangled in one superposition E_G, time t would be reached in about 10^{-11}

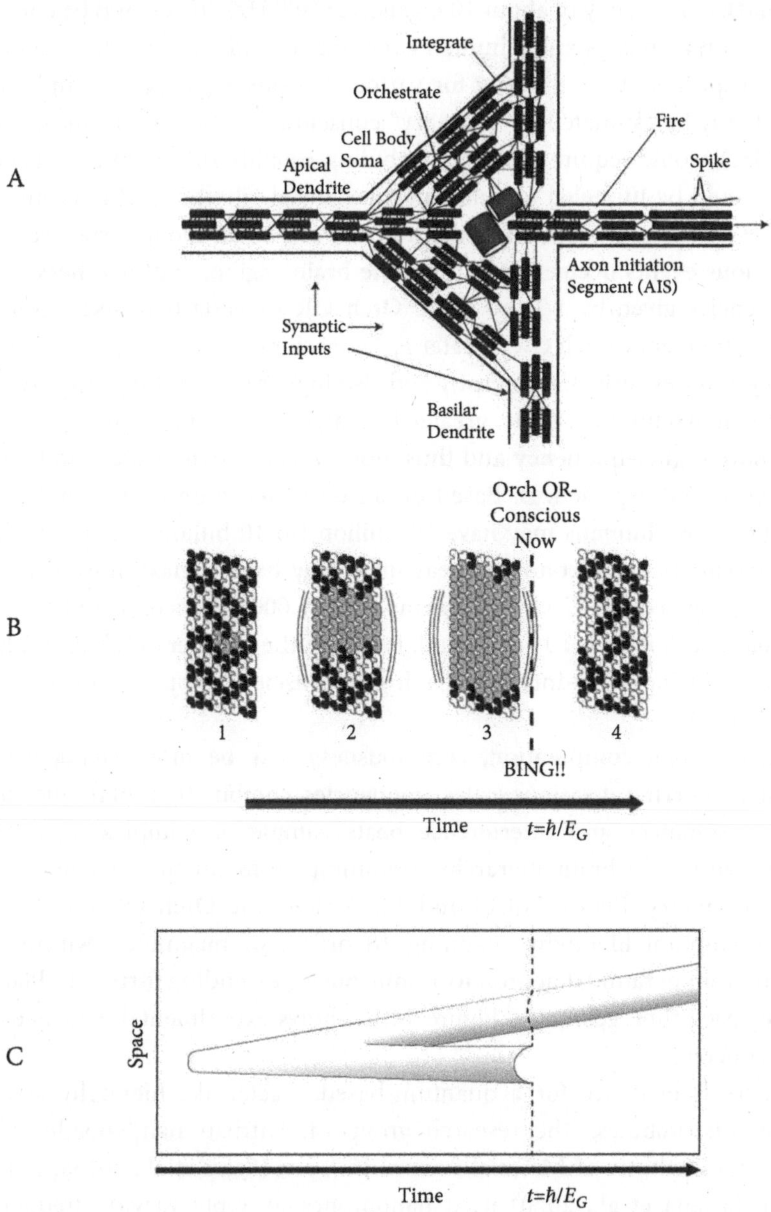

Figure 14.13 A. Schematic of pyramidal neuron in which orchestration and integration occur during **B.** Evolution of quantum superposition states E_G of tubulin (gray) reaches threshold for Orch OR at time $t = \hbar/E_G$. One microtubule fragment is shown, but many entangled microtubules and tubulins throughout brain regions would be involved. **C.** Corresponding space-time curvature at Planck scale.

seconds, a frequency of about 10 gigahertz (10^{10} Hz). This could be considered the maximal possible human consciousness involving 100 percent of brain capacity, but is unlikely for various reasons, e.g., the heat involved. What may be estimated as "ordinary" consciousness, at, say, 10 megahertz (10^7 Hz), would require about 10^{16} tubulins, roughly 10^{-3} of brain capacity. Heat could be funneled to coherent mechanical vibrations of microtubule lattices at 10 megahertz. Indeed, Orch OR can occur along a spectrum of conscious events occurring in the same brain region, and/or others, with frequencies given by $1/t = E_G/\hbar$. Orch OR suggests that, like photons, higher-frequency Orch OR (greater $\underline{E_G}$) conscious events would correspond with greater experiential intensity and also higher information capacity.

On an evolutionary scale, greater E_G (larger brains, more microtubules) can have higher frequency and thus more intense, orchestrated, and more widely varied experiences. Based on an estimated number of tubulins in the brain, we humans may have 10 million (to 10 billion maximal) Orch OR moments per second, whereas mice may have a maximum of about 100,000 per second, *C. elegans* worms about 1,000 per second, and a single cell *paramecium* only 10 per second. Based on the number of tubulins, plant cells might have low-intensity Orch OR conscious moments once every few minutes.

Rather than computation, consciousness may be more analogous to music, occurring at various scales, frequencies, combinations of frequencies, with resonances and interference beats, simple or complex, operating in a multi-scale brain hierarchy, stemming from fluctuations in space-time geometry. Figures 14.14 and 14.15 show the Orch OR multi-scale spatiotemporal hierarchy spanning 18 orders of magnitude within the brain, from pyramidal neuron to atomic nuclei, extending further to Planck scale space-time geometry. Figure 14.16 shows experimental evidence for this concept.

There is evidence for a quantum-based, fractal-like hierarchy arising from microtubules. The research group of Anirban Bandyopadhyay at National Institutes of Material Science in Japan (Sahu et al., 2013a, 2013b, 2014; Saxena et al., 2020) used nanoprobes to apply varying frequency alternating current (AC) at three different biological scales—to whole neurons, microtubules, and tubulin proteins (Figure 14.16). The neurons and proteins were generally insulators, except at certain applied AC frequencies that resulted in efficient electronic ("ballistic") conductance at all three levels. Plots of conductance versus frequency at these levels showed remarkably

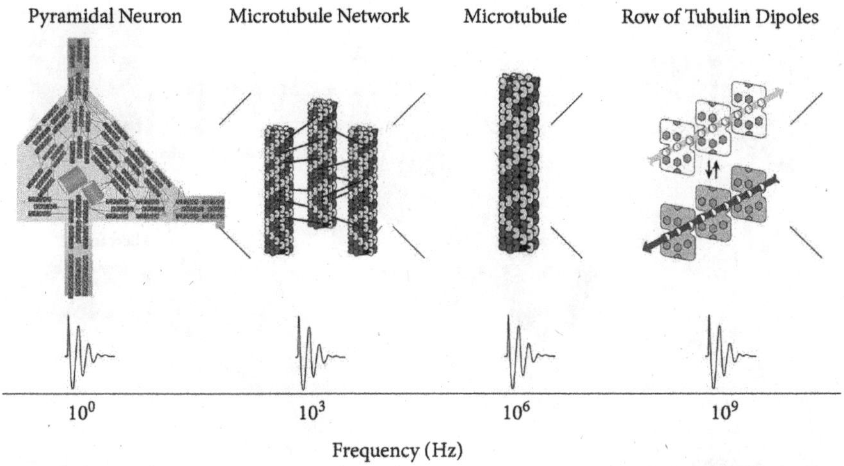

Figure 14.14 First part of a proposed spatiotemporal hierarchy in which Orch OR-mediated consciousness can occur: From left to right, a cortical pyramidal neuron showing internal networks of mixed polarity microtubules, a network of mixed polarity microtubules, a single microtubule, a row of tubulins within a microtubule with schematic display of collective dipoles among pi electron resonance rings.

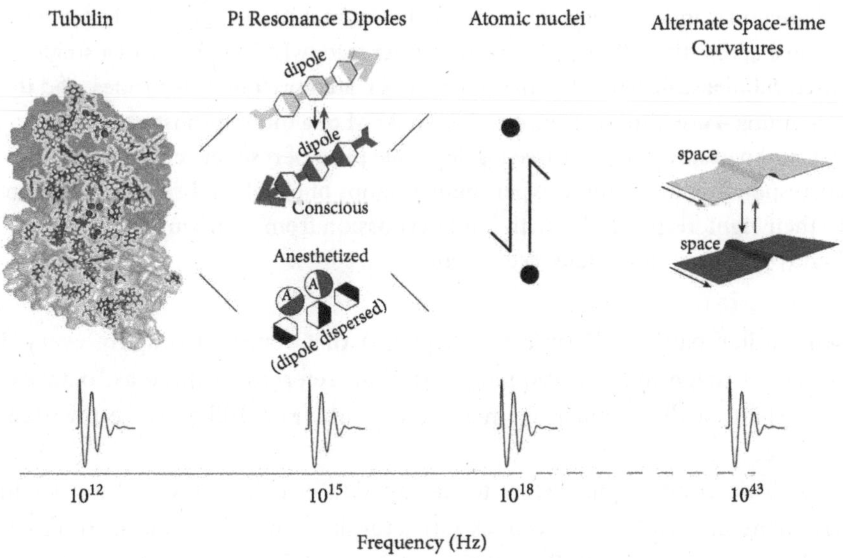

Figure 14.15 Second part of a spatiotemporal hierarchy in which Orch OR-mediated consciousness can occur: From left to right, a single tubulin protein showing internal pi resonance amino acids, pi resonance dipoles with effect of anesthetic gases—dispersing dipoles, atomic nuclei, and (much smaller) curvatures in fundamental space-time geometry. At bottom, dynamical patterns of activity are shown at different spatiotemporal levels (Saxena et al., 2020; Sahu et al., 2013a, 2013b, 2014; Craddock et al., 2012).

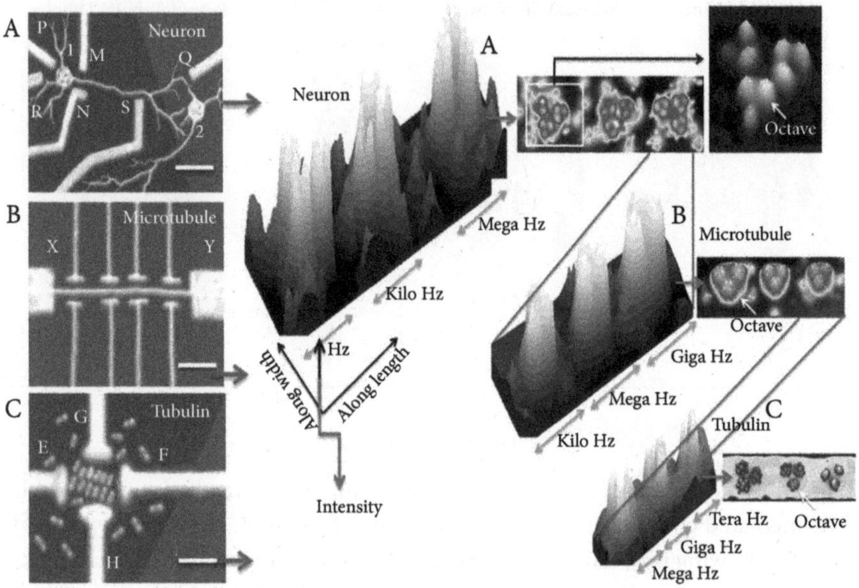

Figure 14.16 Experimental evidence for self-similar resonance bands at 3 scales for neuron, microtubule, and tubulin. **A.** Microscopic image of rat hippocampal neuron with gold electrodes; scale bar = 10 micrometers. PQ and RS are non-contact electrodes to supply horizontal AC signal. MN electrodes are just across the AIS. **B.** Atomic force microscopy (AFM) image of a single microtubule; scale bar = 120 nanometers. XY are horizontal electrodes, and the remaining 4 are transverse electrodes. **C.** AFM of a tubulin substrate; scale bar = 50 nanometers. The three log-log scale plots of resonance frequency corresponding to the three experimental setups of panels A, B, and C are given to their right, respectively. With kind permission from Anirban Bandyopadhyay (for details, see Saxena et al., 2020).

self-similar patterns ("triplets-of-triplets") that repeated roughly every 3 orders of magnitude in frequency. Anirban referred to these as "octaves," consistent with a musical metaphor, and fractal-like, scale-invariant process.

Ballistic conductance refers to unimpeded electron flow and occurs in graphene and carbon nanotubes (structurally similar to tubulin interiors with pi resonance rings). Ballistic conductance differs from superconductivity, lacking the Meissner effect in which current flow continues after the driving forces stops. The ballistic conductance in tubulin and microtubules was suggested to be quantum in nature because it was temperature-independent, and because conductance through microtubules was greater than through individual tubulins. The conductance pathway was not confirmed, but is consistent with excitons and phonons in pi resonance regions in microtubules (the "quantum underground"). Further experiments are underway.

Where in the brain does Orch OR consciousness occur? Although microtubules exist in all cells, they are uniquely arranged in dendrites and cell bodies/soma of neurons. Microtubules each have a net polarity, with positive "plus" ends on beta monomers of the tubulin dimer and negative "minus" ends on alpha monomers (Figure 14.12). In non-neuronal cells and neuronal axons, microtubules are structurally continuous and organized radially, like spokes of a wheel, from the cell center outward to cell periphery, or parallel bundles in axons. And they are unipolar, with beta plus ends at the cell center (centrosome/centriole) and alpha minus ends at the cell periphery. However, microtubules in dendrites and soma of neurons are short and interrupted, and arrayed as mixed-polarity, anti-parallel networks.

The reason for this arrangement has never been explained, although mixed polarity microtubule networks have been proposed to enable recursive processing (Rasmussen et al., 1990). Adjacent dendritic-somatic microtubules of opposite polarity in a common voltage potential (e.g., from their neuronal membranes) would have slightly different energies and therefore oscillate at slightly different frequencies, leading to interference "beats" and slower frequency vibrations, e.g., including cognitive events and electroencephalography (EEG) in tens to hundreds of milliseconds. Slight divergence from parallel alignment in microtubules in pyramidal cell bodies could contribute further to interference. Conscious Orch OR events in dendrites/soma would also be well positioned to regulate axonal firings and control behavior, and thus cause observable deviation from Hodgkin-Huxley behavior.

Computational views of brain function often ascribe the currency of consciousness to bit-like axonal firings, but whether consciousness emerges from axonal firings or dendritic-somatic integration and orchestration remains unknown. But anesthetics that selectively block consciousness while sparing non-conscious activities appear to act on dendrites and soma with little or no effect on axonal firing capabilities. Consciousness may occur primarily in dendrites and soma of brain neurons, specifically in mixed polarity networks of dendritic-somatic microtubules.

Several lines of reasoning suggest cortical layer 5 pyramidal neurons are the most likely sites for consciousness. (1) These neurons are collectively an intersection, sites of convergence of ascending "vertical" pathways from the thalamus (via other cortical layers) with sensory inputs from the external world, and also "horizontal" influences from cortical-cortical interactions and internal inputs, e.g., default mode networks. (2) Pyramidal basilar dendrites, inter-neurons, and gap junction connections form a grid-like dendritic web throughout the entire cerebral cortex layer 5. (3) Apical dendrites of cortical layer 5 pyramidal neurons ascend vertically to the cortical surface, responsible for EEG signals, an important neural correlate of consciousness. (4) Layer 5 pyramidal neuron cell bodies and dendrites have the largest geometric arrays of mixed polarity, anti-parallel microtubule networks, optimally suited for interference. Following the cell body/soma's conical/pyramidal shape, somatic microtubules would be divergent, leading to further interference effects. (5) Cortical layer 5 pyramidal neurons are strategically placed to receive and integrate/orchestrate inputs for global executive actions via descending axonal outputs. (6) Layer 5 pyramidal neurons in awake animals exhibit deviation from Hodgkin-Huxley behavior and appear to implement non-computable firing responses.

The boundary between dendritic-somatic integration (or orchestration) and firing in integrate-and-fire neurons was traditionally viewed as the "axon hillock," the membrane connection between the cell body/soma and axon. However, the axonal firing or "spike" has been found to begin at some distance down the axon, an area now known as the axon initiation segment (AIS; Figures 14.2 through 14.4). Integration/orchestration continues within the axon up to the AIS, as does the configuration of microtubules as interrupted, mixed polarity microtubules. Beyond the AIS, the axon "fires" all-or-none spikes, and the microtubules are continuous and uninterrupted. The AIS within the axon may play a critical role in non-computability and firing regulation, and could help facilitate selection of states that resonate with non-computable preferred states as indicated by the Schrödinger-Newton equation (Moroz et al., 1998). A modification of the Schrödinger equation that considers the self-gravitational effects of the superpositioned mass, the Schrödinger-Newton equation introduces possible Platonic values into non-computability.

There may be other Orch OR modes, systems, and processes in the brain that support consciousness under various circumstances. Consciousness itself may quite literally move around the brain, e.g., within the brain-wide dendritic web of cortical layer 5 pyramidal neurons, and/or other systems (Hameroff, 2010).

6. Free Will and Retroactivity

Conscious "free will" is difficult to address because (1) brain mechanisms underlying consciousness and causal agency are considered unknown (though argued here to depend on Orch OR); (2) determinism, i.e., our actions according to prevalent theories, and the world around us seem algorithmic and inevitable; and (3) measurable brain activity correlating with consciousness apparently occurs too late for real-time conscious response. Consequently, the common assumption in neuroscience and philosophy is that consciousness is epiphenomenal and illusory, and that free will, at least for real-time events, is impossible.

Does sensory conscious perception occur too late for conscious response? Neural correlates of conscious perception occur hundreds of milliseconds after impingement on our sense organs, but yet we respond consciously, it seems, within 100 milliseconds. In tennis, Jeffrey Gray (2004) observed, "The speed of the ball after a serve is so great, and the distance over which it has to travel so short, that the player who receives the serve must strike it back before he has had time consciously to see the ball leave the server's racket. Conscious awareness comes too late to affect his stroke." Nonetheless, tennis players claim to see the ball consciously before they attempt to return it.

Max Velmans (1991, 2000) listed other activities whose correlating brain activity appears to occur too late: sensory and emotional content processing, phonological and semantic analysis of heard speech, preparation of one's own spoken words and sentences, learning and formation of memories and choices, planning and execution of voluntary acts. Consequently, the subjective feeling of conscious control of these responsive behaviors is deemed illusory and free will as nonexistent (Dennett, 1991; Wegner, 2002).

In planned volitional actions, a famous study by Kornhuber and Deecke (1965) asked subjects to move their finger randomly, at no prescribed time, and recorded brain electrical activity over pre-motor cortex. They found activity that preceded finger movement by ~800 milliseconds, calling this the *readiness potential* (RP). Libet et al. (1983) repeated the RP experiment, except they asked subjects to also note precisely when they consciously decided to move their finger. (To do so, and to avoid delays caused by verbal report, Libet et al. used a rapidly moving clock and asked subjects to note when on the clock they consciously decided to move their finger.) This conscious decision came ~200 milliseconds before actual finger movement, hundreds of milliseconds after onset of RP (Figure 14.17A). Libet and many

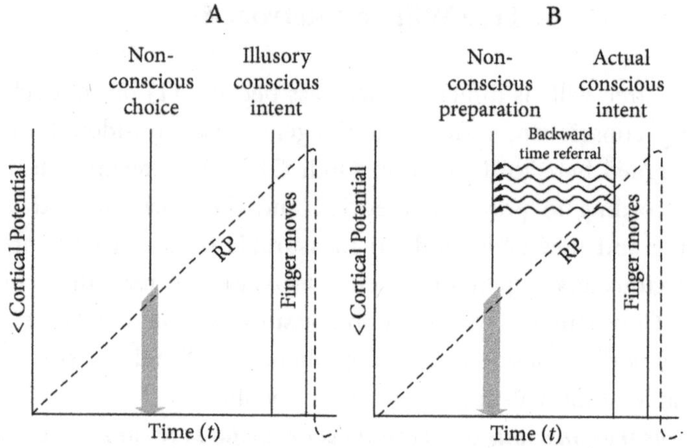

Figure 14.17 The "readiness potential" (RP) (Libet et al., 1983) **A.** Cortical potentials recorded from a subject instructed to move her hand whenever she feels ready, and to note when the decision was made (conscious intent), followed quickly by the finger actually moving. Readiness potential, RP, preceding conscious intent is generally interpreted as representing the non-conscious choice to move the finger, with conscious intent being an illusion. **B.** Assuming RP is necessary preparation for conscious finger movement.
Actual conscious intent could initiate the earlier RP by (quantum) temporal non-locality and backward time referral, enabling preparation while preserving real-time conscious intent and control (Penrose, 1989; Hameroff, 2012).

authorities concluded that the RP represented non-conscious determination of movement, and that many seemingly conscious actions are actually initiated by nonconscious processes. Consciousness apparently comes too late, and conscious control, even of moving a finger, is an illusion.

Unless... aspects of conscious experience are referred backward in what we perceive as a forward flow of time. Figure 14.17B shows referral from the moment of actual conscious intent backward in time to the RP for non-conscious preparation. Could this occur?

Quantum physics has no real preference for a direction of time, e.g., entanglement (or "quanglement" in Penrose terms) can propagate in either direction, into the past or into the future. In his chapter in this volume, Penrose elaborates on "retroactivity" in objective reduction (OR). Aharonov and Vaidman (1990) also proposed that quantum state reductions send quantum information forward and backward in what we perceive as time, "temporal nonlocality." However, it is generally agreed that quantum information going backward (or forward) in time cannot, by itself, communicate or signal ordinary classical information; it is "acausal." However in quantum cryptography and teleportation, acausal quantum information can influence or correlate with classical information, greatly enhancing capabilities of causal, classical processes. This could apply to non-computable conscious quantum brain processes influencing non-conscious classical cognition, as proposed in Orch OR.

Apparent backward-in-time effects in the brain have been known about for many years. Before his famous volitional "move-your-finger" RP experiments, Ben Libet et al. (1979) studied the timing of conscious sensory experience in awake, cooperative patients undergoing brain surgery with local anesthesia. In these patients, Libet was able to record from and stimulate specific areas of somatosensory cortex, e.g., corresponding to the skin of the patient's hand and the hand itself, as well as communicate with the conscious patients. As depicted in Figure 14.18A, a peripheral stimulus to the skin of the hand caused subjects to report conscious experience of the stimulus (using Libet's rapidly moving clock) near immediately, 30 milliseconds after stimulation of the hand, consistent with the time required for a neuronal signal to travel from hand to spinal cord, thalamus, and brain, resulting in an evoked potential (EP) spike in the hand somatosensory cortical area

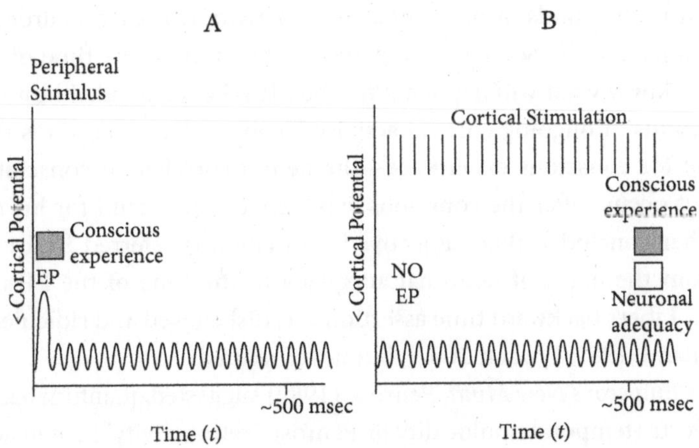

Figure 14.18 Cortical potentials in Libet's sensory experiments. **A.** Peripheral stimulation, e.g., at the hand, results in a near-immediate conscious experience of the hand stimulation at ~30 milliseconds, as well as an evoked potential EP at ~30 milliseconds in the "hand area" of somatosensory cortex, followed by 500 milliseconds or more of ongoing cortical electrical activity. **B.** Without peripheral stimulation, the "hand area" of somatosensory cortex was stimulated continuously. No EP, but there was ongoing electrical activity in that area. Conscious sensation of the hand occurred after 500 milliseconds of stimulation and ongoing activity (to reach what Libet termed "neuronal adequacy").

30 milliseconds after skin contact (e.g., first wave, Figure 14.1A, except to sensory cortex rather than visual cortex), followed by several hundred milliseconds of ongoing cortical activity (waves 2 and 3).

Libet also stimulated the "hand area" of the subjects' brain somatosensory cortex directly (Figure 14.18B). This type of stimulation did *not* cause an EP spike, but did result in ongoing brain electrical activity. Conscious sensation felt in the hand was reported, but only after 500 milliseconds of ongoing brain activity after cortical stimulation. This requirement of ongoing, prolonged electrical activity (what Libet termed "neuronal adequacy") to produce conscious experience was subsequently confirmed by Amassian et al. (1992), Ray et al. (1999), Pollen (2004), and others.

But if hundreds of milliseconds of brain activity are required for neuronal adequacy, how can conscious sensory experience occur at 30 milliseconds? To address this issue, Libet also performed experiments in which stimulation of the hand area of thalamus resulted in an EP at 30 milliseconds, but only brief ongoing following activity, i.e., without reaching neuronal adequacy (Figure 14.19A). Despite the EP, no conscious experience occurred. Libet concluded that for "real-time" conscious perception (at the time of the EP), two factors were necessary: an EP at 30 milliseconds and several hundred milliseconds of ongoing cortical activity (to reach neuronal adequacy) *after* the EP. Somehow, apparently, the brain at the time of the EP seems to know what will happen *after* the EP. Libet concluded hundreds of milliseconds of ongoing cortical activity ("neuronal adequacy") is the *sine qua non* for conscious experience—the neural correlate of consciousness, even if it occurs *after* the conscious experience. To account for his results, he further concluded that subjective information is referred *backwards in time* from the onset of neuronal adequacy to the time of the EP (Figure 14.19B). Libet's backward time assertion was disbelieved and ridiculed (e.g., Churchland, 1981; Pockett 2002), but never refuted.

In his *Emperor's New Mind*, Penrose (1989) suggested quantum backward time effects (temporal nonlocality or Penrose "retroactivity") could account for Libet's results and play a role in consciousness, providing more non-computable influence. Regarding conscious "free will," Orch OR provides mechanisms for consciousness, causality, and backward time effects so that consciousness doesn't occur too late (Hameroff, 2012). As shown in Figures 14.20 and 14.21, backward time referral can contribute to Orch OR, including non-computable influences, and rescue conscious free will.

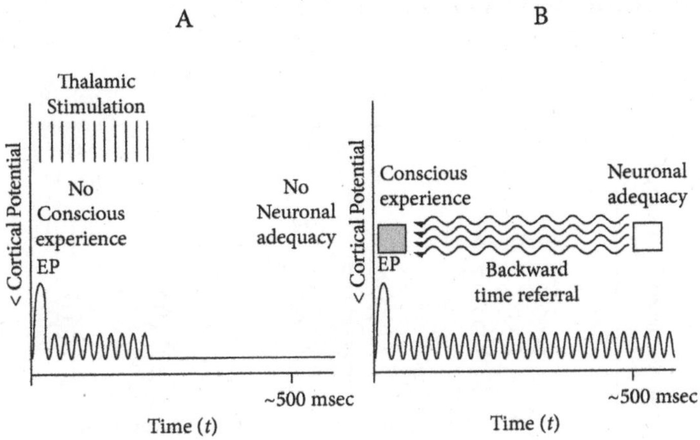

Figure 14.19 Libet's sensory experiments, continued. **A.** Libet et al. stimulated medial lemniscus of thalamus in the sensory pathway to produce an EP (~30 milliseconds) in somatosensory cortex, but only brief post-EP stimulation, resulting in only brief cortical activity. There was no apparent "neuronal adequacy" and no conscious experience. An EP and several hundred milliseconds of post-EP cortical activity (neuronal adequacy) were required for conscious experience at the time of EP. **B.** To account for his findings, Libet concluded that subjective information was referred backward in time from neuronal adequacy (~500 milliseconds) to the EP.

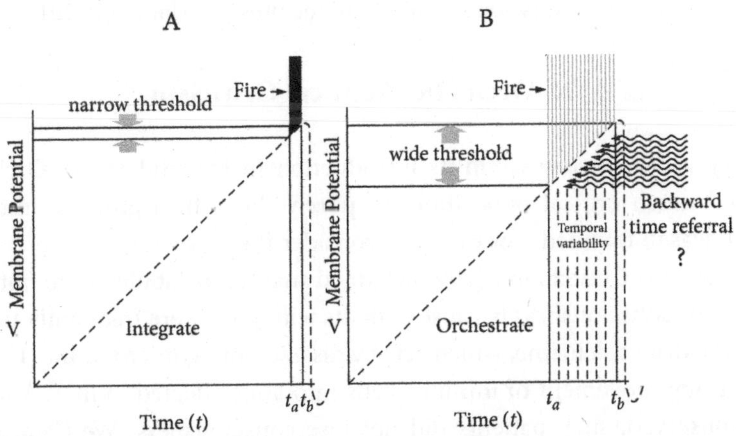

Figure 14.20 Backward time effects and non-computable neuronal behavior. **A.** The Hodgkin–Huxley model predicts that integration by membrane potential in dendrites and soma reach a specific, narrow threshold potential, and fire axonal action potentials for given inputs (Figure 14.1). **B.** Recordings from cortical neurons in awake animals (Naundorf et al., 2006) show a large variability in effective firing threshold and timing. Some non-computable factor(s) cause deviation from H-H behavior. Here, quantum temporal non-locality results in backward time referral to influence Orch OR and firing patterns controlling behavior.

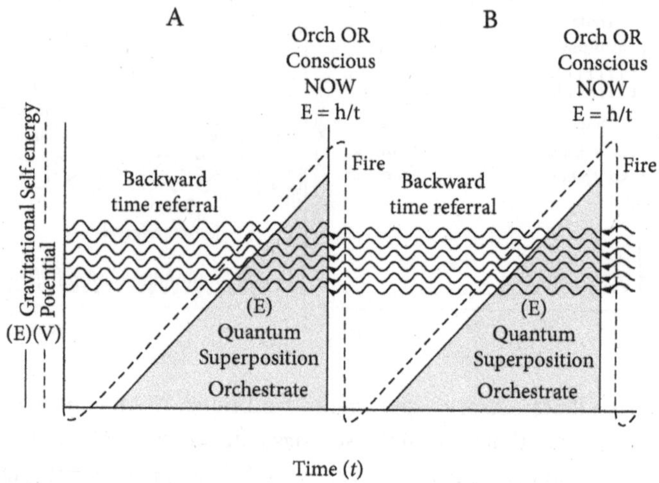

Figure 14.21 Two Orch OR events A and B underlie integrate-and-fire electrophysiology (dotted lines) in neuronal dendrites and soma. Each Orch OR moment occurs with conscious experience, selects tubulin states that can govern axonal firings, and sends quantum information forward and backward in time, in accord with Penrose "retroactivity." Information from Orch OR conscious moments arbitrarily distant in the future may be referred backward in time to Orch OR conscious moments in the present (Hameroff, 2012).

7. Orch OR Skeptical Criticism

Theory versus theory Upon its introduction in the mid-1990s, Orch OR evoked much critical opposition on purely theoretical grounds, most of which was ill founded. For example, consider the following.

Grush and Churchland (1995) claimed that microtubules were not necessary for consciousness because patients with gout were frequently treated with the drug colchicine, which acts by disassembling microtubule (thereby preventing movement of immune cells into gout-affected joints). And yet, they observed, such patients did not lose consciousness. We (Penrose & Hameroff, 1995) replied in the same journal that (1) colchicine does not cross the blood-brain barrier, (2) colchicine only affects microtubules actively assembling and disassembling (unlike brain microtubules that are quite stable over time, not needing to repurpose for mitosis or mobility, and suitable for memory), and (3) colchicine injected into the brains of animals caused severe cognitive dysfunction and loss of wakefulness (Bensimon & Chernat, 1991).

Tegmark (2000) criticized Orch OR on the basis of decoherence, calculating that microtubule quantum superpositions would decohere after only 10^{-13} seconds, far too brief for cognitive events. However, Tegmark based his calculation on a superposition separation distance of 24 nanometers (a soliton separated from itself by the lengths of 3 tubulins in a microtubule). The superposition separation distance stipulated in Orch OR is the diameter of a carbon atom nucleus, a discrepancy of 7 orders of magnitude. Hagan et al. (2002) replied to Tegmark in the same journal, correcting the discrepancy in the superposition separation that occurs in the denominator of the decoherence formula. This lengthened the calculated decoherence time to 10^{-6} seconds, and additional errors extended the calculated microtubule coherence to as long as 10^{-4} seconds, as later found experimentally in microtubules (e.g., Saxena et al., 2020).

Reimers et al. (2014) attacked the Orch OR qubit, which they incorrectly asserted was based on switching and superposition of tubulin states by a single pi resonance cloud within each tubulin. We (Penrose & Hameroff, 2014b) responded that, as described above, the Orch OR qubit is proposed to involve: (1) coupled dipole oscillations among 86 pi resonance clouds within each tubulin, and (2) alignment of multiple tubulins in helical pathways determined by Fibonacci geometry in microtubule lattices, enabling "giant dipoles" as proposed by Fröhlich.

Challenges to Orch OR largely stemmed from the issue of maintaining quantum superpositions and coherence in the "warm, wet, and noisy" brain. Most technological quantum systems required extreme cold to avoid "decoherence" (premature OR), and yet the brain operates at 37.6 degrees C. But Fröhlich proposed that ambient heat would pump coherent phonons in a lattice with proper geometry and common voltage potential. Plants utilize quantum superposition in sunlight, so "warm" is not necessarily an issue. Regarding "wet," polar, liquid regions do exist within the biological microenvironment and do disrupt organized quantum activities. But non-polar, lipid-like regions also exist, deep within proteins and other biomolecules, composed largely of pi resonance groups. These regions in which non-polar anesthetic gas molecules bind and act to selectively prevent consciousness (the "Meyer-Overton correlation") exclude water, being non-polar and hydrophobic, and thus not "wet." Such regions support quantum mechanisms, particularly if they are periodically arrayed and coherently pumped by ambient heat, as proposed in Fröhlich coherence, and are thus not "noisy." Such a set of regions within biomolecules has been characterized by a particular solubility parameter, akin to olive oil, in which anesthetics bind and act, and thus apparently where consciousness originates. Termed the Meyer-Overton "quantum underground" (Craddock et al., 2016), these regions can support quantum coherence pervading the deepest level of living systems. Life itself may be a quantum phenomenon, as first proposed by Schrödinger (1937).

Experimental evidence, testing, and falsifiability As described above, the group of Anirban Bandyopadhyay found self-similar patterns of "ballistic" conductances at specific frequencies in terahertz, gigahertz, megahertz, and kilohertz at ambient temperatures (Figure 14.16; Sahu et al. 2013a, 2013b, 2014; Saxena et al., 2020). The self-similar patterns were described as "triplets of triplets," suggesting a type of fractal-like, scale-invariant process. The conductance was suggested to be quantum in nature because it was temperature-independent, and because conductance through microtubules was greater than through individual tubulins.

Organized biological quantum coherent superposition at ambient temperatures has been demonstrated in proteins involved in photosynthesis. Plants convert photons to "excitons," quasi-particles that are transported in quantum superposition in multiple pathways simultaneously, through pi resonance chromophore groups within a protein (e.g., Engel et al., 2007; Scholes, 2010). In a quantum optical technique called two-dimensional electron spectroscopy, low-intensity laser beams are focused at the intra-protein pi resonance chromophores, and outgoing photons are measured. Sawtooth wave "interference beat" patterns of outgoing photons are considered characteristic of quantum superposition, and such patterns were found with this technique in photosynthesis proteins at ambient temperatures.

Excitons are couplings of separated electrons and electron holes, and thus similar to van der Waals dipoles. But rather than serving energy transport, as excitons do in photosynthesis, in Orch OR van der Waals dipole oscillations (or excitons) among pi resonance groups inside tubulin in microtubules are proposed to process quantum information, perform quantum computation, couple to atomic nuclei by the Mossbauer effect, evolve to Orch OR conscious moments at time $t = \hbar/E$, and regulate neuronal function including axonal firing.

Accordingly, Orch OR predicts tubulin and microtubules should exhibit quantum interference, and experiments are underway using quantum optical techniques to look for quantum interference and superradiance in microtubules (Celardo et al, 2019). If found, the samples will be exposed to anesthetic gases, which Orch OR predicts will inhibit quantum interference, superradiance and other quantum optical processes. The degree of inhibition for each anesthetic gas should correlate with their potency in rendering humans and animals unresponsive and presumably unconscious (Hameroff, 2021). Failure to find quantum interference, superradiance or other quantum optical process in tubulin/microtubules, or, if found, failure to show its inhibition by anesthesia, would tend to falsify Orch OR (though Penrose OR would remain viable as a solution to the measurement problem and for consciousness). Presently, Orch OR has far more experimental support than have other theories of consciousness and remains easily falsifiable.

8. Discussion

Orch OR and the problems of consciousness Most theories of consciousness view brain neurons as Hodgkin-Huxley integrate-and-fire threshold logic devices, and axonal firings as the currency of synaptic computation, leading to consciousness. But as described in Section 2, conventional theories fail to account for certain critical features, rendering consciousness as A) fragmented and unbound; B) epiphenomenal and illusory; C) computable, algorithmic, and predictable; D) non-conscious, "Zombie-like."

Unlike other theories, Orch OR attributes consciousness to quantum computations in microtubules inside brain neurons, most specifically within dendrites and soma of layer 5 cortical pyramidal neurons. These quantum computations are orchestrated by microtubules, and entangle, evolve, and terminate by Penrose "objective reduction" (OR), introducing non-computability and conscious experience. Can Orch OR explain these critical features?

A) Fragmented and unbound, *failing to account for binding and unity of conscious experience* (Mashour, 2013).

Sensory information comprising conscious experience, e.g., in waves 1, 2, and 3 through the brain (Figure 14.1A), is ascribed in conventional theories to neuronal-level axonal firings, and/or synaptic transmissions mediating classical computation. However in Orch OR, sensory information is based on deeper levels, finer-grained states, both classical and quantum. These finer-grained states are pi electron resonance dipoles within tubulin subunits, within microtubules, within neurons, particularly within dendrites and soma of pyramidal neurons, with their mixed polarity microtubule networks. At this smaller, faster scale, quantum coherently oscillating sensory content in the brain may entangle with related states in other tubulins in microtubules in the same way and in other neurons. Entanglement spatiotemporally unifies, and microtubules orchestrate experiential content into fully bound conscious moments when Orch OR threshold is met.

B) Epiphenomenal and illusory, *occurring "too late" for real-time conscious action.*

As described in Section 6, there is evidence that the brain somehow refers subjective information *backward* in what we perceive as the arrow of time (e.g., Libet et al., 1979; Bem, 2011). From the standpoint of physics, Penrose (1989) suggested backward time referral of subjective information can occur in the brain due to OR events sending quantum information both forward and backward in time. If so, sensory inputs reaching "neuronal adequacy" for consciousness 300 to 500 milliseconds after sensory impingement could result in backward referral of an orchestrated conscious perception and response to 30 to 100 milliseconds after impingement, a backward time referral of several hundred milliseconds. This earlier time is when we consciously perceive and respond to that input. Via a process Penrose terms retroactivity, consciousness can thus act in real time, and be neither epiphenomenal nor illusory. Quantum state reduction is a causal action, selecting particular states. With Orch OR, free will is possible.

C) Computable, algorithmic, and predictable *("robot-like"), seeming to preclude insight, creativity, and free will.*

Other theories assume neurons act as Hodgkin-Huxley integrate-and-fire threshold logic devices, resulting in completely algorithmic computation. This may suffice for non-conscious, auto-pilot "easy problems" cognition, but not consciousness. As Penrose has observed, consciousness is to some extent non-computable, requiring extrinsic features augmenting and influencing recognized brain computation serving cognition. Particular feelings, creativity, and insight could reflect "preferred states," or Platonic values, as determined by the Schrödinger-Newton equations, in selection of states in Orch OR events. This influence can occur at time t of Orch OR, after sensory inputs have been processed, but result in orchestrated conscious experience and action at an earlier time—in what is perceived as "real time"—due to backward referral by Penrose retroactivity with each Orch OR event. With

Orch OR and retroactivity, consciousness is neither epiphenomenal nor illusory. Free will is possible.

D) Non-conscious, "zombie-like," *failing to account for the "hard problem" of phenomenal conscious experience or qualia.*

There is a growing consensus among neuroscientists, philosophers, and many others that the experiential nature of phenomenal consciousness, composed of "qualia," is in some way intrinsic to the universe, akin to mass, spin, or charge, or some "fundamental" feature of reality. A solution to the "hard problem" may lie in panpsychism, or pan-protopsychism, ascribing qualia or entities that can manifest qualia, to properties, states, or arrangements of matter or of reality itself. But therein lies a problem. At micro scales, reality is obscure, although space-time geometry is known to extend downward in size to the Planck scale at 10^{-33} centimeters (and 10^{-43} seconds dynamics). At micro scales, matter is quantum in nature and appears to continually go into and out of superposition, e.g., undergoing quantum state reduction, or "collapse of the wave function" to definite, classical states.

The questions of how, why, and if this happens is known as the "measurement problem" in quantum mechanics, discussed in Section 3 above. One proposed solution for the measurement problem also potentially solves the hard problem.

Orch OR and the "hard problem" Penrose used Einstein's general relativity to link superposition possibilities to alternate curvature separations in underlying space-time geometry, each curvature corresponding to a possible micro-reality and experience. The superposition separations were said to be unstable and would collapse, or spontaneously undergo quantum state *reduction* to a particular state at an *objective* threshold (objective reduction, OR). The threshold would occur at a time t related to the Heisenberg uncertainty principle and given by $t = \hbar/E_G$ (where $\hbar$ is the Planck-Dirac constant, and E_G the gravitational self-energy of the superposition). Because superpositions were ascribed to separations in space-time geometry, in Penrose's view, the wave function and its OR collapse were not just mathematical, but actual events, ripples, and rearrangements in the fine scale structure of the universe. In a general context of panpsychism and pan-protopsychism, and as intrinsic causal events in fundamental space-time, OR moments are logical candidates for a spectrum of qualia, units of (proto-) conscious experience.

In a random, polar microenvironment, superpositions would rapidly entangle with adjacent superpositions to quickly reach OR threshold at time $t = \hbar/E_G$, resulting in random selection of classical states. A moment of "proto-conscious" experience would accompany each OR event, presumably random in terms of the type of qualia and lacking in memory or context. But such OR moments would have fundamental units of phenomenal experience ("ping"; Figures 14.6B and 14.7) and be ubiquitous, occurring continually in the micro-environment (and taken as decoherence).

These proto-conscious OR moments would be isolated. How could multiple disparate experiential OR events be unified into a common cognitive consciousness? This is like the "combination problem" in panpsychism, e.g., the question of how qualia associated with individual atoms may be integrated and unified. In this regard, quantum states and processes have an advantage in that they may entangle and literally become one common system.

Metaphorically, proto-conscious OR moments would be like the sounds and noise of individual members of an orchestra tuning their instruments ("ping"), a cacophany. Metaphorically, fully conscious OR moments would be like music, with cooperative resonance, interference, beats, and emotion, playing in the brain's multi-scale hierarchy (BING!!), reaching down to the Planck scale (Figures 14.16 and 14.17). In a special issue of the *Journal of Consciousness Studies* on the "hard problem," we (Hameroff & Penrose, 1996b) specifically proposed that orchestrated OR "space-time events" cause, or are equivalent to, fully conscious moments of phenomenal experience, qualia.

Orchestration implies (1) appropriate symmetry and lattice geometry, e.g., fractal-like helices of pi resonance rings in microtubules, for Fröhlich coherence and Jahn-Teller effects; (2) natural resonance and implicit logic of the qubit structure itself, e.g., Fibonacci geometry in microtubules; (3) a quantum-friendly medium, e.g., the "quantum underground" of pi electron resonance, where anesthetics act inside microtubules and other biomolecules. Microtubules with an A-lattice configuration have Fibonacci geometry whose symmetry and lattice structure enables helical winding pathways suitable for Fröhlich giant dipole qubits and quantum computations (Figure 14.14). Individual superpositioned qubits could entangle with others, increasing E_G and unifying conscious content, in accord with the natural properties of the qubit substrate, until threshold is met at time $t = \hbar/E_G$. Such Orch OR moments may happen, e.g., at 10 megahertz, 10 million events per second, but with interference effects also creating epochs in slower ranges for EEG and cognition.

Orch OR events occur along a spectrum $t = \hbar/E_G$, with higher frequency (smaller t) presumed to provide more intense phenomenological experience, and because of greater E_G also greater content capacity. During orchestration phases, qubits entangle with those in other non-polar regions of the quantum underground in tubulins, microtubules, neurons, and distant brain regions, e.g., across layer 5 pyramidal neuron cell bodies and their basilar dendrites, and gap junctions. When Orch OR threshold is reached, a unified, global moment of orchestrated conscious experience occurs, and microtubule states selected that can "consciously" trigger axonal firings, and otherwise "non-computably" regulate behavior and neuronal activities. Sequences of Orch OR moments give our stream of consciousness.

Orch OR, life, and evolution Which came first: life or consciousness? Most would say life preceded consciousness, which then emerged as an adaptation of biological cognition. But Penrose OR would have been present on Earth long before the advent of life about 4.5 billion years ago. For example, proto-conscious OR moments could have occurred within the "Primordial soup" from which life apparently arose. A simulation of the oily soup done in the 1950s, including sparks for lightning, showed spontaneous formation of amphipathic biomolecules, each with a non-polar, pi resonance group on one end, and a polar, water-soluble group on the opposite end (Miller, 1953).

Amphipathics include aromatic amino acids tryptophan, phenylalanine, and tyrosine, whose non-polar pi resonance rings coalesce to form and fold proteins, resulting in a non-polar pi resonance core and a polar, water-soluble exterior. Billions of years ago, similar amphipathics in the Primordial soup are presumed to have coalesced to form primitive "micelles" or "coacervates" that eventually self-organized into biological cells and life (Oparin, 1938). The drive or motivation for micelles—structurally similar to simple soaps—to self-organize into functional cells, and eventually all of life, is unknown. What was the feedback fitness function for life to have developed and flourished? Was it pleasure?

As micelles with pi resonance interiors grew larger from van der Waals attractions and the hydrophobic effect, Fröhlich coherence could have developed and enabled collective oscillations and superposition E_G to reach threshold for partially orchestrated OR proto-conscious moments. Some of these would have been pleasurable qualia and could have served teleologically as the feedback fitness function for micelles and coacervates to self-organize, develop, and evolve (Hameroff, 2017). Behavior is generally driven, at some level, by seeking pleasure and avoiding displeasure. With OR-mediated pleasure as a teleological feedback fitness function, micelles, coacervates, and simple biological systems would have self-organized to optimize pleasure. Even today, our behaviors are driven to optimize pleasure, be it hedonistic, altruistic, spiritual harmony, intellectual understanding, or delayed gratification. Optimizing conscious pleasure in all its forms may be the purpose of life.

9. Conclusion

Most theories of consciousness view the brain as a computer of neurons, but fail to account for essential features of consciousness, as described above. Moreover, most theories fail to suggest what consciousness might actually *be*. Is consciousness a computation? A critical level of complexity? An entity outside the realm of science? Or, as panpsychists maintain, is consciousness a property of all matter? Of reality itself? But our views of matter and reality depend on consciousness, as evidenced by the "measurement problem" in quantum mechanics.

At small scales, particles can exist in wave-like quantum superposition of multiple co-existing possible states. But when measured or consciously observed, superpositions undergo quantum state reduction, or "collapse of the wave function," resulting in particular, definite states. Why and how this occurs remain controversial, but proposed possibilities (see Section 3) include: (1) conscious observation causes quantum state reduction (e.g., Chalmers and McQueen, this volume); (2) "many worlds," in which superpositions don't reduce and each possibility continues independently; (3) decoherence, in which superpositions mix with their microscopic environment, also quantum in nature, and become too complicated for detailed description: and (4) objective reduction (OR) in which an objective physical process causes reduction of the quantum state to one outcome, "OR" another (Penrose, 1989).

Roger Penrose proposed a particular OR approach through Einstein's general relativity in which superposition, e.g., a particle in two places at once, was considered equivalent to two separated curvatures in fundamental space-time geometry. Such separations were unstable, Penrose suggested, and would undergo *reduction* spontaneously due to an *objective* threshold (OR) in space-time, a basic equation at the most basic level of the universe. Moreover, Penrose further proposed, such OR events would entail phenomenal experience—qualia—that, in the random, polar microenvironment would be random and isolated, without memory, context, or meaning. Such OR experiential events would be occurring spontaneously and ubiquitously, not unlike panpsychism. Because of the random, isolated nature of their qualia, these experiential OR events would be merely "proto-conscious," perhaps akin philosophically to Whitehead's process of "occasions of experience." Penrose OR constitutes the most specific scientific mechanism for an essential feature of consciousness yet proposed.

For the full, rich consciousness we humans experience, the brain would need some kind of widely distributed quantum computer, strategically positioned to support, organize, and logically process ("orchestrate") quantum information until reaching threshold at time $t = \hbar/E_G$. To do so, these devices would need to avoid random processes, be involved in cognition, have suitable symmetry, and be properly interfaced to receive inputs, and consciously process and respond with causal outputs.

Microtubules are self-assembling polymers of "tubulin" proteins that organize activities within brain neurons and other cells. The Orch OR theory attributes consciousness to "orchestrated" quantum computations in microtubules inside brain neurons, quantum computations that terminate by Penrose "objective reduction" (OR), introducing non-computability and phenomenal experience. Orch OR conscious events can select microtubule states that trigger axonal firing and thus causally control behavior.

Key points from the brain biology of Orch OR include the following:

(1) Microtubules encode and process cognitive information including memory as states of tubulins within microtubule lattices inside brain neurons. Orch OR consciousness may then actively modulate cognition. (2) Cortical layer 5 pyramidal neurons, a convergence of sensory and default mode inputs, and executors of essential outputs to control behavior, are likely anatomical brain sites for consciousness. (3) Mixed polarity networks of microtubules in dendrites and soma/cell bodies of cortical layer 5 pyramidal neurons are likely molecular sites for consciousness. (4) Non-polar regions of pi electron resonance rings inside tubulins can connect to those in neighboring tubulins in microtubule A-lattices to form "quantum-friendly" helical pathways following Fibonacci geometry (the "quantum underground"). (5) Collective quantum dipole oscillations among pi resonance groups in these

helical pathways are enabled by lattice symmetry (e.g., Fibonacci geometry) and pumped by ambient heat (Fröhlich coherence) to "orchestrate" quantum vibrations as quantum bits or "qubits" in microtubule quantum computation. (6) Microtubule qubits entangle with those in other neurons, unify, and evolve to meet Orch OR threshold for full, rich conscious experience. (7) Selection of microtubule states with each Orch OR event may regulate axonal firings and behavior in layer 5 pyramidal neurons, collectively well situated to implement conscious control of behavior. Penrose has further described (8) the "retroactivity" inherent in OR and Orch OR, which can resolve Libet's "backward time referral" and rescue conscious free will.

Orch OR is the most complete, ambitious, and detailed theory of consciousness, with significant explanatory power. Orch OR consciousness is proposed to operate, resonate, and interfere over 18 orders of magnitude in a multi-scale brain hierarchy extending through quantum vibrations in microtubules, further downward to the Planck scale of fundamental spacetime geometry (Figures 14.14 and 14.15). Unlike other theories that relegate consciousness to epiphenomenal illusion, Orch OR renders consciousness as a physical process in the structure of the universe, connected to the brain through quantum vibrations in microtubules. Orch OR has both experimental and theoretical support, is currently being tested, and remains unfalsified, but falsifiable.

Acknowledgments

Thanks to Roger Penrose for collaboration and ideas, Anirban Bandyopadhyay for data and Figure 16, Edgar Mendoza for artwork, Abi Behar-Montefiore for manuscript assistance, Harrison Hameroff for support, and Betsy Bigbee for consultation and advice.

References

Aharonov, Y., & Vaidman, L. (1990). Properties of a quantum system during the time interval between two measurements. *Phys. Rev.* A41(1): 11–20.

Amassian, V. E., Eberle, L., Maccabee, P. J., & Cracco, R. Q. (1992). Modelling magnetic coil excitation of human cerebral cortex with a peripheral nerve immersed in a brain-shaped volume conductor. *Electroencephalography and Clinical Neurophysiology/Evoked Potentials* 85(5): 291–301.

Bem, D. J. (2011). Feeling the future: experimental evidence for anomalous retroactive influences on cognition and affect. *Journal of Personality and Social Psychology* 100(3): 407–425.

Bensimon, G., & Chernat R. (1991). Microtubule disruption and cognitive defects: Effects of colchicine on learning behavior in rats. *Pharmacol. Biochem. Behav.* 38(1): 141–145.

Celardo GL, Angeli M, Kurian P, Craddock TJA (2019) On the existence of superradiant excitonic states in microtubules *New J. Phys.* 21 023005

Chalmers, D. (1996). *The Conscious Mind: In Search of a Fundamental Theory.* Oxford University Press, New York.

Chalmers, D., & McQueen, K. (this volume).

Churchland, P. S. (1981). On the alleged backwards referral of experiences and its relevance to the mind-body problem. *Philos. Sci.* 48, 165–181.

Cohen, M. X. (2017). Where does EEG come from and what does it mean? *Trends Neurosci.* 40(4): 208–218.

Colwell L. J., & Brenner, M. P. (2009). Action potential initiation in the Hodgkin-Huxley model. *PLoS Comput. Biol.* 5(1): e1000265.

Craddock, J. A., Hameroff, S. R., & Tuszynski, J. A. (2016). The "quantum underground": Where life and consciousness originate. In *Biophysics of Consciousness: A Foundational Approach*, eds. R. R. Poznanski, J. A. Tuszynski, and T. E. Feinberg. World Scientific, Singapore.

Craddock, T., St. George, M., Freedman, H., Barakat, K., Damaraju, S., Hameroff, S., & Tuszynski, J. A. (2012). Computational predictions of volatile anesthetic interactions with the microtubule cytoskeleton: Implications for side effects of general anesthesia. *PLoS ONE* 7(6): e37251.

Craddock, T. J. A., Kurian, P., Preto, J., Sahu, K., Hameroff, S. R., Klobukowski, M., & Tuszynski, J. A. (2017). Anesthetic alterations of collective terahertz oscillations in tubulin correlate with clinical potency: Implications for anesthetic action and post-operative cognitive dysfunction. *Scientific Reports* 7: 9877.

Dennett, D. C. (1991). *Consciousness Explained.* Little Brown, Boston.

Eerkens, H. J., Buters, F. M., Weaver, M. J., Pepper, B., Welker, G., Heeck, K., Sonin, P., de Man, S., & Bouwmeester, D. (2015). Optical side-band cooling of a low frequency optomechanical system. *Optics Express* 23(6): 8014–8020.

Eger, E. I. II, Shafer, D. E., Hemmings, Jr., H. C., & Sonner, J. M. (2008). Is a new paradigm needed to explain how inhaled anesthetics produce immobility? *Anesth. Anal.* 107(3): 832–848.

Emerson, D. J., Weiser, B. P., Psonis, J., Liao, Z., Taratula, O., Fiamengo, A., Wang, X., Sugasawa, K., Smith, A. B., Eckenhoff, R. G., & Dmochowski, I. J. (2013). Direct modulation of microtubule stability contributes to anthracene general anesthesia. *J. Am. Chem. Soc.* 135: 5389–5398.

Engel, G. S., Calhoun, T. R., Read, E. L., Ahn, T. K., Mancal, T., Cheng, Y. C., Blankenship, R. E., & Fleming, G. R. (2007). Evidence for wavelike energy transfer through quantum coherence in photosynthetic systems. *Nature* 446: 782–786.

Everett, H. (1957). Relative state formulation of quantum mechanics. *Rev. Mod. Phys.* 29: 454–462.

Franks, N., & Lieb, W. (1984). Do general anaesthetics act by competitive binding to specific receptors? *Nature* 310: 599–601.

Fröhlich, H. (1968). Long range coherence and energy storage in biological systems. *Int. J. Quant. Chem.* 2: 641–649.

Fröhlich, H. (1970). Long range coherence and the actions of enzymes. *Nature* 228: 1093.

Fröhlich, H. (1975). The extraordinary dielectric properties of biological materials and the action of enzymes. *Proc. Natl. Acad. Sci.* 72: 4211–4215.

Gayduchenko, I., Xu, S. G., Alymov, G., Tretyakov, I., Taniguchi, T., Watanabe, K., Goltsman, G., Geim, A. K., Fedorov, G., Svintsov, D., & Bandurin, D. A. (2021). Tunnel field-effect transistors for sensitive terahertz detection. *Nat. Commun.* 12: 543.

Ghirardi, G. C., Grassi, R., & Rimini, A. (1990). Continuous-spontaneous-reduction model involving gravity. *Phys. Rev.* A42: 1057–1064.

Ghirardi, G. C., Rimini, A., & Weber, T. (1986). Unified dynamics for microscopic and macroscopic systems. *Phys. Rev.* D34; 470.

Gray, J. A. (2004). *Consciousness: Creeping Up on the Hard Problem*. Oxford University Press, Oxford.

Grush R., & Churchland P. S. (1995). Gaps in Penrose's toilings. *J. Consciousness Studies* 2(1): 10–29.

Hagan, S., Hameroff, S. R., & Tuszyński J. (2002). Quantum computation in brain microtubules: Decoherence and biological feasibility. *Phys. Rev.* E65: 061901.

Hameroff, S. (2010). The "conscious pilot"-dendritic synchrony moves through the brain to mediate consciousness. *J. Biol. Phys.* 36(1): 71–93.

Hameroff, S. (2012). How quantum brain biology can rescue conscious free will. *Front Integr. Neurosci.* (6): 1–17.

Hameroff, S. (2017). The quantum origin of life: How the brain evolved to *feel good,* In *On Human Nature: Biology, Psychology, Ethics, Politics, and Religion,* pp. 333–353.

Hameroff, S. (2021). "Orch OR" is the most complete, and most easily falsifiable theory of consciousness. *Cogn. Neurosci.* 24: 1–3.

Hameroff, S., & Penrose, R. (1996a). Orchestrated reduction of quantum coherence in brain microtubules: A model for consciousness. *Math. Comput. Simul.* 40: 453–480.

Hameroff, S., & Penrose, R. (1996b). Conscious events as orchestrated space–time selections. *J. Conscious. Stud.* 3(1): 36–53.

Hameroff, S., & Penrose, R. (2014a). Consciousness in the universe: A review of the "Orch OR" theory. *Phys. Life Rev.* 11(1): 39–78.

Hameroff, S., & Penrose, R. (2014b). Reply to criticism of the "Orch OR qubit"—"Orchestrated objective reduction" is scientifically justified. *Phys. Life Rev.* 11: 104–112.

Hameroff, S. R., & Watt, R. C. (1982). Information processing in microtubules. *J. Theor. Biol.* 98: 549–561.

Hodgkin A. L., & Huxley A. F. (1952). A quantitative description of membrane current and its application to conduction and excitation in nerve. *J. Physiol.* 117: 500–544.

Jaynes, J. (1977). *The Origin of Consciousness in the Breakdown of the Bicameral Mind*. Houghton Mifflin, Boston.

Koch, C. (2004). *The Quest for Consciousness: A Neurobiological Approach*. Roberts and Company, Englewood, NJ.

Kornhuber, H. H., & Deecke, L. (1965). Hirnpotential andrugen bei Willkurbewegungen und passiven Bewungungen des Menschen: bereitschaftspotential und reafferente potentiale. *Pflug. Arch.* 284: 1–17.

Lee, U., Ku, S., Noh, G., Baek, S., Choi, B., & Mashour, G. A. (2013). Disruption of frontal-parietal communication by ketamine, propofol and sevoflurane. *Anesthesiology* 118(6): 1264–1267.

Libet, B., Gleason, C. A., Wright, E. W., & Pearl, D. K. (1983). Time of conscious intention to act in relation to onset of cerebral activity (readiness potential): The unconscious initiation of a freely voluntary act. *Brain* 106: 623–642.

Libet, B., Wright, E. W. Jr., Feinstein, B., & Pearl, D. K. (1979). Subjective referral of the timing for a conscious sensory experience. *Brain* 102: 193–224.

Marshall, W., Simon, C., Penrose, R., & Bouwmeester, D. (2003). Towards quantum superpositions of a mirror. *Phys. Rev. Lett.* 91: 13–16.

Mashour, G. A. (2013). Cognitive unbinding: A neuroscientific paradigm of general anesthesia and related states of unconsciousness. *Neurosci. & Biobehav. Rev.* 37(10), Part 2: 2751–2759.

McCormick, D. A., Shu, Y., & Yu, Y. (2007). Neurophysiology: Hodgkin and Huxley model—still standing? *Nature* 445(7123): E1–E2.

Meyer, H. (1899). Welche eigenschaft der Anasthetica bedingt ihre Narkotische wirkung? *Naunyn-Schmied Arch. Exp. Path. Pharmakol.* 42: 109–118.

Miller, S. L. (1953). Production of amino acids under possible primitive earth conditions. *Science* 117(3046): 528–529.

Moroz, I. M., Penrose, R., & Tod, K. P. (1998). Spherically-symmetric solutions of the Schrödinger–Newton equations. *Class. Quantum Grav.* 15: 2733–2742.

Naundorf, B., Wolf, F., & Yolgushev, M. (2006). Unique features of action potential Initiation in cortical neurons. *Nature* 440(7087): 1060–1063.

Naundorf, B., Wolf, F., & Volgushev, N. (2007). Hodgkin and Huxley model—still standing? (Reply). *Nature* 445(7123): e2–e3.

Nogales, E., Wolf, S. G., & Downing, K. H. (1998). Structure of the alpha beta tubulin dimer by electron crystallography. *Nature* 391(6663): 199–203. Erratum, *Nature* (1998), 393(6681): 191.

Oparin, A. I. (1938). *The Origin of Life*. Macmillan, New York.

Overton, E. (1901). *Studien uber die Narkose zugleich ein Beitrag zur allgemeinen Pharmakologie*. Verlag von Gustav Fischer.

Pan, J. Z., Xi, J., Tobias, J. W., Eckenhoff, M. F., & Eckenhoff, R. G. (2007). Halothane binding proteome in human brain cortex. *J. Proteome Res.* 6: 582–592.

Pearle, P. (1989). Combining stochastic dynamical state-vector reduction with spontaneous localization. *Phys. Rev.* A39: 2277–2289.

Pearle, P., & Squires, E. J. (1994). Bound-state excitation, nucleon decay Experiments and models of wave-function collapse. *Phys. Rev. Lett.* 73(1): 1–5.

Penrose, R. (1989). *The Emperor's New Mind: Concerning Computers, Minds, and the Laws of Physics*. Oxford University Press, Oxford.

Penrose, R. (1996). On gravity's role in quantum state reduction *Gen. Rel. Grav.* 28: 581–600.

Penrose, R. (2014). On the gravitization of quantum mechanics 1: Quantum state reduction. *Found. Phys.* 44: 557–575.

Penrose, R., & Hameroff, S. (1995). What gaps? Reply to Grush and Churchland. *J. Conscious Stud.* 2: 98–112.

Pockett, S. (2002). On subjective back-referral and how long it takes to become conscious of a stimulus: A reinterpretation of Libet's data. *Conscious. Cogn.* 11: 144–161.

Pollen, D. A. (2004). Brain stimulation and conscious experience. *Conscious. Cogn.* 13: 626–645.

Pribram, K. (1971). *Languages of the Brain: Experimental Paradoxes and Principles in Neuropsychology.* Prentice-Hall, Englewood Cliffs, NJ.

Rasmussen, S., Karampurwala, H., Vaidyanath, R., Jensen, K. S., & Hameroff, S. (1990). Computational connectionism within neurons: A model of cytoskeletal automata subserving neural networks. *Physica D: Nonlinear Phenomena* 42: 428–449.

Ray, P. G., Meador, K. J., Smith, J. R., Wheless, J. W., Sittenfeld, M., & Clifton, G. L. (1999). Physiology of perception: Cortical stimulation and recording in humans. *Neurology* 52: 1044–1049.

Reimers, J. R., McKemmish, L. K., McKenzie, R. H., Mark, A. E., & Hush, N. S. (2014). The revised Penrose-Hameroff orchestrated objective-reduction proposal for human consciousness is not scientifically justified *Phys. Life Rev.* 11(1):101–103.

Sahu, S., Ghosh, S., Fujita, D., & Bandyopadhyay, A. (2014). Live visualizations of single isolated tubulin protein self-assembly via tunneling current: Effect of electromagnetic pumping during spontaneous growth of microtubule. *Scientif. Repts.* 4: 7303.

Sahu, S., Ghosh, S., Ghosh, B., Aswani, K., Hirata, K., Fujita, D., & Bandyopadhyay, A. (2013a). Atomic water channel controlling remarkable properties of a single brain microtubule: Correlating single protein to its supramolecular assembly. *Biosens. Bioelectron.* 47: 141–148.

Sahu, S., Ghosh, S., Hirata, K., Fujita, D., & Bandyopadhyay, A. (2013b). Multi-level memory-switching properties of a single brain microtubule. *Appl. Phys. Lett.* 102 12370.

Samsonovich, A., Scott, A., & Hameroff, S. (1992). Acousto-conformational phase transitions in the cytoskeleton, adaptive resonance networks with nonlinear synapses and trainable intraneuronal pattern recognition. In *Proceedings IJCNN International Joint Conference on Neural Networks.* IEEE.

Saxena, K., Singh, P., Sahoo, P., Sahu, S., Ghosh, S., Ray, K., Fujita, D., & Bandyopadhyay, A. (2020). Fractal, scale free electromagnetic resonance of a single brain extracted microtubule, a single tubulin protein and a single neuron *Fractal and Fractional* 4(11).

Scholes, G. D. (2010). Quantum-coherent electronic transfer: Did nature think of it first? *J. Phys. Chem. Lett.* 1(1): 2–8.

Schrödinger, E. (1937). *What Is Life? The Physical Aspect of the Living Cell.* Cambridge University Press, Cambridge, UK.

Smith, S., Watt, R., & Hameroff, S. (1984). Cellular automata in cytoskeletal lattice proteins. *Physica* D10: l68–174.

Stapp, H. P. (2007). *Mindful Universe: Quantum Mechanics and the Participating Observer.* Springer, Berlin.

Tegmark. M. (2000). Importance of quantum decoherence in brain processes. *Phys. Rev.* E61(4), Part B: 4194–4206.

Velmans, M. (1991). Is human information processing conscious? *Behav. Brain Sci.* 14: 651–669.

Velmans, M. (2000). *Understanding Consciousness.* Routledge, London.

Wegner, D. M. (2002). *The Illusion of Conscious Will.* MIT Press, Cambridge, MA.

Whitehead, A. N. (1929). *Process and Reality.* Macmillan, New York.

Whitehead, A. N. (1933). *Adventure of Ideas.* Macmillan, London.

Xi, J., Liu, R., Asbury, G. R., Eckenhoff, M. F., & Eckenhoff, R. G. (2004). Inhalational anesthetic-binding proteins in rat neuronal membranes. *J. Biol. Chem.* 279: 19628–19633.

Zurek, W. H., Habib, S., & Paz, J. P. (1993). Coherent states via decoherence. *Phys. Rev. Lett.* 70(9): 1187.

15

Can Quantum Mechanics Solve the Hard Problem of Consciousness?

Basil J. Hiley and Paavo Pylkkänen

1. Introduction

The hard problem of consciousness is the problem of explaining how and why physical processes give rise to consciousness (Chalmers 1995). Regardless of many attempts to solve the problem, there is still no commonly agreed solution. It is thus very likely that some radically new ideas are required if we are to make any progress. In this chapter we turn to quantum theory to find out whether it has anything to offer in our attempts to understand the place of mind and conscious experience in nature. In particular, we will be focusing on the ontological interpretation of quantum theory proposed by Bohm and Hiley 1987, 1993 and its further development.

The ontological interpretation makes the radical proposal that quantum reality includes a new type of potential energy which contains active information. This proposal, if correct, constitutes a major change in our notion of matter. We are used to having in physics only mechanical concepts, such as position, momentum and force. The common intuition that it is not possible to understand how and why physical processes can give rise to consciousness is partly the result of the assumption that physical processes (including neurophysiological processes) are always mechanical. If, however, we are willing to change our view of physical reality by allowing non-mechanical, organic and holistic concepts such as active information to play a fundamental role, this, we argue, makes it possible to understand the relationship between physical and mental processes in a new way. It might even be a step toward solving the hard problem.

*About the division of labor between the authors: BJH is responsible for the more mathematical parts (especially sections 7, 10, 15, 16), while PP is responsible for the more philosophical parts (especially sections 8, 9, 11–14). However, much of the chapter is the joint result of our discussions over many years.

Basil J. Hiley and Paavo Pylkkänen, *Can Quantum Mechanics Solve the Hard Problem of Consciousness?* In: *Consciousness and Quantum Mechanics.* Edited by: Shan Gao, Oxford University Press.
 DOI: 10.1093/oso/9780197501665.003.0016

In our earlier work on mind, matter and quantum theory we have explored and further developed different aspects of David Bohm's (1917–1992) pioneering efforts in this field: analogies between quantum processes and thought;[1] the idea that matter and conscious experience have their ground in a new holistic and dynamic "implicate" order (or "holomovement") that is beyond space and time;[2] and the idea that Bohm's 1952 "causal interpretation" of quantum theory ("the Bohm interpretation," "the ontological interpretation," "hidden variable" interpretation) can be understood in terms of a model where a new type of field containing active information literally in-forms (rather than pushes and pulls mechanically) the particle or corpuscle that it accompanies.[3] It is tempting to see such active information at the quantum level as providing the long-sought missing link between the mental and physical sides of reality, and perhaps even as a rudimentary form of the holistic information in living organisms from which conscious experience in a full sense may emerge in suitable conditions.

In this chapter we first revisit the hard problem of consciousness, review the Bohm-Hiley ontological interpretation and the the role of active information, before briefly discussing the notion of implicate order. We then move on to discuss some new developments introduced by Hiley and Callaghan (2012) and Hiley, Dennis and de Gosson (2021) that illuminate the deeper mathematical and physical background of quantum mechanics. What emerges is a more subtle view of quantum objects that is likely to have implications for all attempts to apply quantum mechanics to the mind-matter problem.

2. The Hard Problem of Consciousness and Our Notion of the Physical

As already mentioned, the hard problem of consciousness is the problem of explaining how and why physical processes give rise to consciousness (Chalmers 1995). In a recent article, David Chalmers (2020b: 223) notes

[1] Bohm 1951; Pylkkänen 2014.
[2] Bohm 1989; Hiley 1996; Hiley and Pylkkänen 2001; Pylkkänen 2007.
[3] Bohm and Hiley 1987, 1993; Bohm 1990; Pylkkänen 1992; Hiley 2004; Hiley and Pylkkänen 2005, Pylkkänen 2020.

that one has to take materialism seriously in order to take the hard problem seriously as a problem:

> If one is antecedently a dualist, the hard problem will be unsurprising and not especially worth addressing. The mental and the physical are fundamentally distinct, and that is that. One might like to know how they interact, but that leads us to other aspects of the mind-problem such as the interaction problem. The problem of explaining the mental in physical terms does not really arise.

There have been many attempts to solve the hard problem since Chalmers invigorated it in 1994, but he has not been impressed by these, writing in 2010:

> There have been many neurobiological and cognitive models of consciousness, but few of them have been offered as a solution to the hard problem, and when they have, hardly anyone has been convinced.... That being said, positive nonreductive theories of consciousness have not had a much easier time of it.

What does it mean to take materialism seriously? It does not necessarily require that one believes that materialism is correct, but one certainly has to feel the pull of materialism. The history of science shows that explanations in terms of physical laws and mechanisms have been very successful in physics and biology. Given that there is clearly a close correlation between mental, conscious phenomena and neurophysiological phenomena in the brain, is it not the most obvious and natural thing to assume that conscious, mental processes are neurophysiological processes—or at least somehow emergent from them? Yet as is well known there are a number of plausible anti-materialist arguments to the effect that we have no idea how subjective qualitative conscious experiences could possibly arise from objective, neurophysiological processes (see Levine 2017).

Given that the hard problem is to explain how and why physical processes give rise to consciousness, it might seem obvious that a rigorous scientific and philosophical approach to the question has to consider what "physical" and "consciousness" mean in the light of our best scientific and philosophical theories. Regarding the "physical," would not this require that we give serious attention to what physics has to say about the physical? While this may seem obvious, if one considers the discussions of the hard problem in the philosophical and cognitive neuroscience literature, it is fairly common to completely ignore our best physics, namely quantum theory and relativity, and the theories that build upon them. The reasons for this ignoring are complex, and have partly to do with the difficulties of finding a coherent interpretation of quantum theory. However, as long as we do not tackle the "physical" in the light of our best scientific theories, can we be said to be taking materialism seriously and approaching the hard problem in a truly scientific way?

One of the main challenges underlying our research has been precisely to try to understand the fundamental changes quantum theory and relativity require to our view of the physical world. Our perspective to these issues has been inspired in particular by David Bohm's pioneering efforts. While Bohm worked on different approaches to quantum theory, he was never satisfied with the prevailing instrumentalist tendency to see quantum theory merely as a mathematical tool to predict the results of measurement, but was always trying to develop models of quantum reality which do not presuppose the existence of a classical level or measuring apparatuses. This makes Bohm's various approaches particularly suitable for anyone who is concerned with finding the place of mind and conscious experience in nature. The conventional or "Copenhagen" interpretation of quantum theory gave up early the attempt to provide a model of quantum objects, and thus it has little to say about the nature of reality at the quantum level, and thus about how consciousness might relate to fundamental physical reality.

3. Bohm's Discovery of a Quantum Ontology

In our research we have on the one hand explored Bohm's ontological (hidden variable, causal) interpretation of quantum theory, which he initially proposed in 1952 and developed further since the mid-1970s with one of us (Bohm 1952; Bohm and Hiley 1987, 1993). On the other hand we have explored a more general approach, a new "implicate order" framework in which one can begin to bring together quantum theory and relativity, which Bohm and Hiley began to develop in the early 1960s.

Is there anything in these approaches that might be relevant to tackling the hard problem of consciousness? We have suggested that there is. When

Bohm and Hiley began to re-examine Bohm's 1952 interpretation in the mid-1970s, they soon came up with a radical proposal. As is well known, the Bohm interpretation postulates that a particle (say, an electron) is not a particle OR a field (as in the conventional interpretation of quantum theory), but it is always a particle AND a field. The field guides the particle through a new potential, the quantum potential, and in this way one can give an intelligible explanation of many puzzling quantum phenomena, such as the two-slit experiment, tunnelling, the measurement problem, etc. (see Bohm and Hiley 1987, 1993).

To understand why the Bohm interpretation is relevant to understanding the mind and even conscious experience, let us consider it in some detail (for more extensive presentations see Bohm and Hiley 1987, 1993 and for more recent work see Hiley and Callaghan 2012 and Hiley, Dennis and de Gosson 2021). Bohm had published in 1951 a textbook *Quantum theory*, in which one of his main aims was to provide a more physical interpretation of the conventional "Copenhagen" interpretation, to complement the mathematical formulation. Bohm's book contains many philosophically important ideas, e.g., the idea that a quantum object should be understood as consisting of mutually incompatible potentialities before interactions with systems such as measuring apparatuses. After completing the book Bohm still felt that something was incomplete or missing in the conventional interpretation, namely an account of what happens to quantum objects between measurements, an account of actual movement. Discussions with Einstein in Princeton further prompted him to seek a realistic and deterministic account of quantum reality.

When thinking about the problem Bohm considered the WKB approximation, which is used to show how quantum theory gives rise to classical mechanics. What happens in the WKB approximation is that one writes the wave function in polar form, substitutes this to the Schrödinger equation and obtains an equation that looks very much like a formulation of classical mechanics, the Hamilton-Jacobi equation, except that there appears an extra term which we denote by Q which looks like some form of potential energy.

To show how this works mathematically, we first express the wave function in polar form $\psi(\mathbf{r}, t) = R(\mathbf{r}, t) \exp[iS(\mathbf{r}, t)/\hbar]$.

Here R and S are two physically real fields that describe the time evolution of the energy that constitutes the particle. To see how this works we substitute this expression into the Schrödinger equation and then separating the real and imaginary parts of the resulting equation, we find that the real part gives

$$\frac{\partial S}{\partial t} + \frac{1}{2m}\nabla S^2 + V + Q = 0. \tag{1}$$

This equation, which is an expression for the conservation of energy, would be identical to the single classical particle Hamilton-Jacobi equation except that there is the extra term Q, which is a quantum energy term, taking the form

$$Q = -\frac{\hbar^2}{2m}\frac{\nabla^2 R}{R}. \tag{2}$$

In the WKB approximation if one now assumes that this extra term goes to zero (e.g., by assuming counterfactually that h is zero), one gets classical mechanics out of quantum mechanics. In classical mechanics each particle has well defined position and momentum, so that we have a particle ontology. But, Bohm asked, what happens if we do not assume that Q is zero? After all, h is a constant so it cannot actually go to zero. Could we still have an ontology, albeit a different, quantum ontology? The answer is yes. Q has the dimensions of energy, so it is natural to interpret it as an additional potential acting on the particle, alongside the classical potential V.

Indeed, it is the term Q which produces a behavior in the particle that distinguishes it from a classical particle. Those unfamiliar with the Hamilton-Jacobi theory will more easily recognize the following formula:

$$\frac{dp}{dt} = -\nabla[V + Q], \tag{3}$$

which is just Newton's equation of motion with an additional potential, Q, which is called the quantum potential.

It is well known that Newton's equation produces particle trajectories in the classical case as does the Hamilton-Jacobi equation. This implies that even in the quantum domain we can still regard every particle as having a well-defined position and momentum giving rise to a trajectory even though we are unable to measure position and momentum simultaneously. In order to produce the quantum behavior the "particle" must be accompanied by the R, S coupled field, $\psi(\mathbf{r}, t)$, which satisfies the Schrödinger equation. Thus Bohm's quantum ontology for non-relativistic particle quantum mechanics is not just a particle ontology, it is a particle AND field ontology; it is the field which gives rise to the quantum potential. Indeed quantum trajectories can be (and have been) calculated for many different situations, including the classic two-slit interference experiment (see Bohm and Hiley 1993 and Oriols and Mompart (eds.) 2019 for more details).

Because of the uncertainty principle we cannot observe the trajectory of a single quantum object directly as long as we remain in the domain in which quantum mechanics is valid. So the idea that a single quantum object moves along a trajectory ought to be seen as a hypothesis which has not been empirically verified. However, by making use of measurements of weak values it is possible to measure average trajectories (see Flack and Hiley 2018).

Importantly, Bohm's interpretation also provides a way of thinking about when we should use quantum mechanics, and when classical mechanics is sufficient to provide a good approximate description. The quantum potential is negligibly small in conditions where Newtonian mechanics works for all practical purposes (see Bohm and Hiley 1993, ch. 8).

4. The Quantum Potential and Active Information

Now, in what sense might the Bohm interpretation be relevant to understanding the place of mind and even conscious experience in nature? One philosophically interesting point is that the quantum potential only depends on the *form* (second spatial derivative) of the quantum field, *R*, since

$$Q = -\frac{\hbar^2}{2m}\frac{\nabla^2 R}{R}. \tag{4}$$

After reflecting upon and debating this feature, Bohm and Hiley (1987) proposed that what is going on is that the field enfolds *information* about the environment of the particle (e.g., the precise nature of the slits) and is organizing its behavior by literally IN-FORMING or putting form into its movement, rather than pushing and pulling it mechanically, an example of "formative" causation. This notion is reminiscent of Kant's discussion of phoronomy (see Stan 2021 for details), later taken up by Einstein (1924) in discussing the shift in the perihelion Mercury that has its explanation in general relativity, a shift that has the same origins as the more familiar Coriolis force.

An example of formative causation is the notion of "active information"—something that is operative in many other contexts. The basic idea of active information is that a form having very little energy enters into and directs a much greater energy. In this process the activity of the greater energy is given a form similar to that of the smaller energy.

We can give a useful analogy by recalling that in radio transmission the audio signal modulates the profile of the high frequency carrier wave. Here the audio energy can be quite small, but its form can be amplified to produce a large effect in the radio itself. By analogy the small energy in the quantum wave can be magnified by some as yet unknown internal process so as to produce a large effect on the particle.

As another example, think about a ship guided by radar waves which contain information about the environment of the ship (e.g., rocks at the bottom of the sea). The radar waves are not pushing and pulling the ship. They have a small energy that in-forms the greater energy of the ship.

The computer is another obvious example where we see information being active in this way. Here information is carried in the chip and all of this information is passive until the appropriate software activates some of the information. Thus when the computer is working, some of this "passive" information becomes "active," modifying the input by giving it new form. Hence in the computer there is a continual interplay between passive and active information.

It is interesting to note that Feynman once proposed that every point of space is like a computer processing incoming information and outputting new information (see Finkelstein 1969). For our approach to nonrelativistic quantum mechanics, it is the particle that processes the information although in field theory (which we discuss later) our approach is very similar to the kind of structure Feynman had in mind. In the case of the computer, the significance of the information is decided by outside human activity both in terms of the software we use and the type of information that is stored in the chip.

Artificial neural networks provide another example. In this case the network learns how to function by receiving information from some external source and adjusting its weights, e.g., as a result of backpropagation to produce a relevant output. Once that information has been stored in the net, it then remains passive until the net is activated. Neural nets do not need quantum mechanics in order to function, being essentially based on the Ising model. However this model is itself a classical approximation to a fully quantum version known as the Heisenberg model. It seems clear that the Heisenberg model will have properties that are different from those of the classical model, but what is not clear is whether the properties of the Heisenberg model will have features that will be of direct relevance to artificial neural networks.[4]

It would be more convincing to have an example where there is no direct human intervention, and here we have also used DNA as an analogy. In this case, the genetic code can be regarded as the passive information that has accumulated over the years through the process of evolution. It was not put there by human beings (although this need no longer be the case.) Parts of this code can be accessed by RNA which carries the information to the appropriate part of the cell where the information can become active in the sense that the processes in the cell change to allow it to develop in a new and meaningful way. Here we see how passive information becomes active under appropriate conditions.

Connected to this, Anita Goel (2008) has studied molecular machines that read and write information into molecules of DNA. She writes:

> These nanomotors (matter) transduce chemical free energy into mechanical work as they copy biological information stored in a DNA molecule. These motors can be thought of as information processing machines that use information in their environment to evolve or adapt the way they read out DNA. In ways (as of yet unknown to us), information from their environment can couple into and modulate the dynamics of these nanomachines as they replicate or transcribe genetic information.

Indeed, she has suggested that quantum mechanics may play a role in influencing the motors as they read/write bits of DNA and suggested to us that it is worth exploring whether the notion of active information at the quantum level might be relevant here (private communication, January 2020).

While we have above emphasized that information is an objective commodity that exists and acts independently of the human mind, it is obvious that active information also plays a role in human subjective experience. For example, when reading a map, the information contained in the map typically gives rise to virtual activities (e.g., possible routes) in the mind, and one of these is typically realized as an actual movement in the territory, depending on where one wants to go.

[4] Pylkkänen 1992, 116 discusses briefly in a qualitative way how insight could be instantiated by a quantum neural network. For a discussion of learning in a quantum neurocomputer (including a critical discussion of the idea of active information) see Chrisley 1996.

5. The Quantum Potential as an Essential Feature of Kinematics

As the idea that the quantum potential contains active information is likely to meet with skepticism, let us look at it in more detail. Firstly, what exactly is the role played by the quantum potential Q that appears in equation (1)? A simple but naive way to think about this term is to regard it as generating a force in a similar way as the classical potential V generates forces. However there is a vital difference between the two potentials. V arises from some external field and can be considered as the effect of an external force driving the particle is some way. In contrast, the quantum potential has no external source, but is an internal feature of the kinematic process and so should not be regarded as a driving force. In this sense it can be regarded as a "formative" cause.

To begin to understand the role of the quantum potential energy, it is important to think of it as a new quality of energy, which is not present in the classical domain. To see the reason for this, let us look at equation (1) written in words:

Kinetic energy + quantum PE + classical PE = total energy.

where PE stands for potential energy. Clearly this is simply a conservation of energy equation as we are considering a closed system. Since the total energy is fixed, any change in the motion of the particle comes from a re-distribution of energy between the kinetic energy, the quantum potential energy and the classical external energy. We can also think of this as involving a new process where there is a redistribution of internal energy between the kinetic energy and the quantum potential energy, a process that does not occur in classical mechanics.

In a situation where interference is involved, the kinetic energy of the particle gives up some of its energy to the quantum potential energy, the amount and form of which is conditioned by the experimental environment as Bohr 1961 insisted. This may seem a strange, new idea, but it emerges from the Schrödinger equation itself, with no new mathematical content added. The kinetic energy term in equation (1) uses the Bohm momentum which, as we have argued, is the weak value of the momentum. Thus the kinetic energy term must be the weak value of the kinetic energy (see Hiley 2012).

This redistribution can be regarded as a kind of self-organization involving in the whole process.

The form dependence of the quantum potential energy also helps us to understand why this energy can produce significant effects over large distances. The quantum wave of a particle typically spreads out over a greater and greater distance, so that its amplitude decreases suggesting the energy becomes more spread out. Had the force depended on the amplitude, then it would necessarily decrease with distance, but since the mathematical form of the quantum potential shows that it does not depend on the amplitude, the resultant effect remains regardless of the distance. As we have already emphasized, it depends on the *form* of the quantum wave. This is yet another reason for asserting that it is not meaningful to talk about the quantum potential as generating a force. Thus it is possible to have very long range quantum effects and even non-local effects of the type required to account for the situation described by the Einstein-Podolsky-Rosen (1935) paradox. That gravitational energy is non-local, has long been known in general relativity, especially in gravitational waves (see Bondi, van der Burg, and Metzner 1962 and Penrose and Rindler 1984).

We can take these arguments one stage further. If we consider the quantum potential in particular cases, e.g., in the two-slit experiment, a detailed examination of the mathematical form of the quantum potential energy shows that it contains information about the momentum of the particle, the width of the slits and how far they are apart. That is, the energy carries information of the *whole experimental arrangement*. We can regard this information as being *active* in the sense that it modifies the behavior of the particle. As we remarked above Bohr 1961 came to a similar conclusion, but from a very different point of view. He saw the necessity of talking about the *wholeness of the quantum phenomenon*. He writes:

> As a more appropriate way of expression I advocate the application of the word *phenomenon* exclusively to refer to the observations obtained under specified circumstances, including the *whole experimental arrangement*.

For Bohr such *wholeness* implied that the quantum process could not be analyzed even in principle, but the Bohm interpretation shows that analysis is possible and by carrying out this analysis, we can provide a different way of understanding what is meant by *quantum wholeness* (for Bohr's view, see Bohm and Hiley 1993, ch. 2.2; Pylkkänen 2015).

It was the above types of considerations that led us to the suggestion that the quantum potential should be considered as an *information potential*. Not only does the quantum potential carry information about the experimental set up, but, more importantly, it induces a change of form from within the system itself. It is in this more general sense that we can regard the quantum potential as an information potential. In making this suggestion, we were strongly influenced by the etymological roots of the word "information." In its simplest form to in-form literally means to form from within. As Miller (1987) writes:

> As with many words in the English language, the word "information" has both Greek and Latin roots. The Latin informatio bears direct and obvious structural similarities to our modern "information". The prefix (in) is equivalent to the English "in," "within," or "into"; the suffix (ito) denotes action or process and is used to construct nouns of action. The central stem (forma) carries the primary meaning of visible form, outward appearance, shape or outline. So informo (or informare) signifies the action of forming, fashioning or bringing a certain shape or order into something, and informatio is the noun from which signifies the "formation" thus arrived at.

In other words this information can be either active or passive. Interestingly, one of the most prominent contemporary theories of consciousness, Tononi et al.'s (2016) integrated information theory (IIT) similarly understand information as in-forming, albeit in a different sense. Quantum active information has holistic features which provide a new type of integration which may well be relevant to the kind of unity we find in conscious experience (see Pylkkänen 2016).[5]

[5] Both Tononi and Bohm use the concept of information in a way different than it is used in communication theory. For Tononi, information refers to how a system of mechanisms in a state, through its cause-effect power, gives rise to a form ("informs" a conceptual structure) in the space of possibilities. Such in-forming is needed to account for the fact that consciousness is specific: each experience is the particular way it is—it is composed of a specific set of specific phenomenological distinctions, thereby differing from other possible experiences (Tononi calls this differentiation). For Bohm active information refers to situations when a form (carrying a little energy) enters and literally in-forms a larger energy. He further says (modifying Bateson's definition) that information is a difference of form that makes a difference of content. We have above considered situations where information in-forms the motion of matter, but we can also apply this to perception. For example, when we are accounting for visual experiences, we can say that the form carried by light waves is taken up by the eye and the brain, and that this process involving differences of form results in rich differences of content in our phenomenal experience. Thus the specific structure of the content of our

6. Information and Meaning

A concept closely associated with information is that of meaning. To connect these concepts in a novel way Bohm suggested that an important aspect of meaning is the activity, virtual or actual, that flows out of information. An example of "virtual activity" can be seen by considering a situation where we are reading a map, and all sorts of possible routes arise in our subjective experience of imagination. These are the virtual inward activities that the information contained in the map gives rise to in us. Depending on the context, one of these may be selected and becomes the actual external activity (see Bohm 1989). In this situation, the information in the map is information for us.

However, the information carried by the quantum field is clearly not information for us, it is information for the electron and as such is objective. If we assume that meaning is activity of information, then there is a sense in which meaning is involved, at least in an elementary sense, in the behavior of a quantum particle:

> [I]n the quantum theory, the quantum potential may... be regarded as representing active information.... .In accordance with the suggestion that meaning is the activity, virtual or actual, that flows out of such information, we are led to regard the movements of the self-active particles as the meaning of this information. (Bohm 1989: 58)

Since meaning is involved, we are not using the word "information" in the sense of Shannon 1948. According to Shannon, the information content for the word "coming" as calculated using the expression $H = \sum p_i \ln p_i$ is exactly the same as the word "gnmioc," but one is meaningful and the other is not. The quantum potential always has a kind of meaning for its particle, although it might not have meaning for other particles at the same location.

Floridi (2015) distinguishes between environmental and semantic information; and semantic information can be further distinguished into factual and instructional information. Environmental information is information

phenomenal visual experiences is partly informed by the form carried by the movement of the light waves that was the input. So active information plays a role in making each phenomenal experience into the specific experience that it is. There is thus an interesting similarity between our and Tononi's schemes, although there are subtle differences. We will discuss Bohm's notion of information in more detail later in sections 6 and 8.

as mere correlation, e.g., the way tree rings carry information about age. The quantum active information can be seen as environmental in the sense that the form of the quantum field is correlated with the environment (e.g., whether or not two slits are open in a two-slit experiment). But we can also say that the quantum active information is *about* something (the environment, slits, etc.), it is *for* the particle and it helps to *bring about* something (a certain movement of the particle). This suggests that it is semantic and has both factual and instructional aspects (cf. Floridi 2015). We note here that Dretske (1981) and Barwise and Seligman (1997) have explored the possibility that information in the sense of factual semantic contents (i.e., information as meaningful data that represents facts correctly or incorrectly) can be grounded in environmental information (i.e., information as mere correlation, e.g., the way tree rings carry information about age).

7. The Organic Unity of the Quantum Many-Body System

So far we have just considered the case of a single particle, but the notion of active information takes on a new and potentially more significant meaning in a many-body system. Consider a situation in which we have two sets of particles, A and B. Suppose system A is described by a non-product, or "entangled" wave function $\Psi_A(\boldsymbol{r}_1, \boldsymbol{r}_2, \ldots \boldsymbol{r}_n, t)$ which will produce a quantum potential $Q_A(\boldsymbol{r}_1, \boldsymbol{r}_2, \ldots \boldsymbol{r}_n, t)$ that couples all the particles of A into a coherent group, while the B group of particles are linked by a different quantum potential $Q_B(\boldsymbol{r}'_1, \boldsymbol{r}'_2, \ldots \boldsymbol{r}'_n, t)$ which arises from a different non-product wave function $\Psi_B(\boldsymbol{r}'_1, \boldsymbol{r}'_2, \ldots \boldsymbol{r}'_n, t)$. This implies that we have two independent groups of particles, each group being coordinated into some kind of coherent unit where each particle of the group responds *only* to the coordinated movement of the rest of the particles in its own group.

To help understand this coordinated movement we have likened the group behavior to ballet dancers whose movements are coordinated, not by direct mechanical forces, but rather each individual is responding to a common theme. In the case of the ballet, each dancer responds to the musical score as it develops in time. Thus in the analogy the wave function provides the "score" to which the particles respond. The two independent wave functions correspond in the analogy to two sets of dancers following their own theme. Here the form of the movement in each group can be regarded as unfolding from within and the energy that is needed to bring about these changes is provided by the individuals themselves. Although the analogy has obvious weaknesses, it nevertheless highlights the radical difference between classical forces and the type of effect generated by the quantum potential.

One can see why attempts to continue to regard the quantum potential as producing another kind of mechanical force will fail by considering the two sets of particles discussed above. Members of the two groups can be in the same region of space and provided they have no classical forces between them, they will not experience the quantum potential of the other group. The quantum potential for each group is somehow a "private" experience for only that group. There is no mechanical way of bringing about such behavior.[6]

Since the group behavior is something that is intrinsic to that particular group of particles and to no others, it seems, once again, as if there were some kind of self-organization involved, but a self-organization that is shaped by the environment and mediated by the quantum potential. Thus the system behaves as a *whole* or a *totality* in such a way that the particles appear to be nonlocally linked. This radical approach suggests that nature at its very fundamental level is more *organic* than expected. We will return discuss the connection between quantum mechanics and life further in section 12.

8. Meaning Organizes Matter

Traditional mechanistic materialism—which still prevails as the underlying paradigm of much of natural sciences and mind sciences alike—assumes that everything that exists can be reduced to basic material elements (particles and fields) and their mechanical interactions. By saying that information exists objectively (independently of the human mind) and plays an active organizing role at different levels of nature, we are proposing that we should radically change our world view. Instead of saying that everything is fundamentally matter and energy, we follow Bohm in suggesting that reality is one process which has two sides, a somatic, material/energetic side and a significant, meaningful side (Bohm called this idea the principle of soma-significance). This implies we can treat any process either as somatic or as significant. So we are leaving traditional mechanistic materialism behind, and move toward a version of dual-aspect monism (see Atmanspacher

[6] This example is related to the metaphysical question of whether quantum objects are individuals, see Pylkkänen, Hiley, and Pättiniemi 2016.

2014). Bohm (1986b: 38) describes this view succinctly in an interview with Renee Weber:

> A very elementary case is the printed paper: it's somatic in that it is just printed ink; and it also has significance. . . . [A]ny part of the body or body processes is somatic, it's the nerves moving chemically and physically; and in addition it has a meaning which is active. The essential point about intelligence is the activity of significance, right? In computers, we have begun to imitate that to some extent. . . . [A]ll of nature is organized according to the activity of significance. This, however, can be conceived somatically in a more subtle form of matter which, in turn is organized by a still more subtle form of significance. So in that way every level is both somatic and significant.

So, e.g., when reading a newspaper, light waves (which are somatic) carry the significance contained in the printed ink toward the reader. When the light hits the retina, the information is abstracted and carried by the more subtle somatic processes of the brain until the meaning of the information is apprehended at some even more subtle somatic level of the brain. This example illustrates the significance of the soma, the idea being that each somatic configuration has a meaning (at least potentially) and that it is such meaning that is grasped at more subtle levels of soma. Bohm calls this the soma-significant relation. But there is also an inverse, signa-somatic relation: "This is the other side of the same process in which every meaning at a given level is seen *actively* to affect the soma at a more manifest level" (Bohm 2003: 163). This needs some qualification, for if we consider the significance carried by printed ink, it does not directly affect the soma; it is only when the significance is interpreted by the more subtle levels in the reader that there may be a signa-somatic effect. While much of information is naturally active in the sense that the significance affects the soma (e.g., information in the DNA molecule), there exists also as a special case inactive information (e.g., printed ink), which needs to be interpreted by a suitable system before its meaning becomes active.

The above can be clarified by further considering the notion of information. Gregory Bateson said that information is a difference that makes a difference. Bohm noted that this is too liberal because every difference makes a difference. To restrict the notion he proposed the following characterization: information is a difference of form that makes a difference of content (Bohm

1989). In this sense form and content are two interrelated aspects of one process. Information is thus not mere form.

The point here is similar to the one John Searle (1980) made with his famous Chinese room thought experiment, which can be described (in a simplified form) as follows. There is a monolingual English speaker in a room, and she is given batches of Chinese writing. She has a rule book written in English telling her that when a certain set of symbols comes in, she is to pass another set of of symbols out of the room. The differences in the forms she receives make no difference of content to her. In other words, she is producing answers by manipulating uninterpreted formal symbols. Searle's point was that such manipulation of forms without understanding what those forms mean does not constitute true understanding. As computers process forms in the same way as the person in the Chinese room, computers do not understand anything, although they appear to give meaningful responses.

Our radical hypothesis is that there exist some informational processes in nature taking place without human intervention, which involve both form and content (this is the basic idea of soma-significance). And not only that. The content is not merely an abstract, ethereal quality, it has causal powers in that it can act to organize material processes at lower levels. Or as we already mentioned above, meaning includes the activity—virtual or actual—which information gives rise to. The activity often unfolds or reveals the meaning, as in the case where something means "danger" and there is a signa-somatic response in the body. Our human experience of meaning is thus not unique, but a highly developed instance of a fundamental feature of the universe. The information in the quantum potential has a primitive content which is active, just as information in the DNA molecule has in suitable circumstances.

We underline the importance of the quantum information potential here. It suggests that at a very fundamental level of nature something at least analogous to meaning or content plays a causal, organizing role. The Bohmian electron where a field of information is guiding a particle-like entity can be seen as a *prototype model* which defines a whole class of similar systems, similarly to the way the Watt governor does for the dynamical systems theory (Pylkkänen 1999; cf. van Gelder 1997). We usually assume that meaning cannot play a genuine causal role, because we assume that physics is fundamental and there is no room for such a role of meaning in physics. However, the quantum potential, when seen as an information potential, is a counterexample to this assumption. If you like, this may be one of the best ways of seeing how quantum theory calls into question an entire world view, namely the mechanistic world view. This is why it is so important.

Bohm's favorite example to illustrate a signa-somatic effect was to consider a person who is walking in a park in a dark night, and has heard that a dangerous assailant may be moving about in the area. Suddenly, he sees a suspicious looking shadow. If he interprets that as "the assailant" this signifies "danger," which will typically give rise to a powerful somatic reaction in the body of the person (e.g., adrenalin will flow). If the person then realizes that he is seeing just a shadow of a tree, he will start to calm down. As another example, think of how you would react if someone comes into a room full of people shouting "fire" and you judge this to be correct. In these situations, significance or meaning is not merely an abstract, passive quality, it is fundamentally active in organizing more manifest material levels of the body. And Bohm's proposal is that all of nature is organized according to such activity of significance, from the way information in the DNA molecule organizes biological process, all the way down to the quantum level, where information contained in the quantum potential signa-somatically guides the activity of, say, an electron. This suggests a radical change in the way we are thinking about physical processes and their relation to information and meaning.

Now, to say that all of nature is organized according to the activity of significance, and that such active information applies even at the quantum level may be radical and interesting, but is it relevant to the hard problem of consciousness?

9. The Mind-Like Quantum

Bohm sketched a new theory of mind and matter based on the notions of soma-significance and active information and discussed it extensively with one of us (PP) from the mid-1980s till the early 1990s (see Bohm and Pylkkänen 1991). In the first formulation (1986a) he wrote, "[I]n some sense, a rudimentary consciousness is present even at the level of particle physics." In a later version of the same article (1990) he qualified this by saying: "[T]he quantum theory, which is now basic, implies that the particles of physics have certain primitive mind-like qualities which are not possible in terms of Newtonian concepts (though, of course, they do not have consciousness)." To say that a rudimentary consciousness is present even at the level of particle physics is to endorse panpsychism. To say that particles of physics have certain primitive mind-like qualities but that they do not have consciousness is compatible with a weaker doctrine, panprotopsychism. As David Chalmers (2015: 259) puts it:

> [P]anprotopsychism is the view that fundamental physical entities are proto-conscious. In more detail, let us say that protophenomenal properties are special properties that are not phenomenal (there is nothing it is like to have a single protophenomenal property) but that can collectively constitute phenomenal properties, perhaps when arranged in the right structure. Panprotopsychism is then the view that some fundamental physical entities have protophenomenal properties.

While both panpsychism and panprotopsychism are likely to sound implausible, or even absurd in the context of the prevailing mechanistic materialism, the failure of mechanistic materialism to tackle the hard problem has led more and more researchers to consider pan(proto)psychism as one of the approaches worth pursuing if we are ever to solve the hard problem (see Seager 2020).

To connect our approach to panprotopsychism, we could say that active information at the quantum level has protophenomenal properties. There is nothing it is like to be a single electron; but when electrons and other elementary particles and fields are arranged in the right hierarchical structure (e.g., that of the human brain), phenomenal properties in a full sense emerge from the underlying protophenomenal ground (for discussions of the relation of panpsychism to active information at the quantum level, see Pylkkänen 1995, 2020; see also Skrbina 2005: 202–206). This would be similar to David Chalmers' own very interesting attempt to solve the hard problem, namely his double-aspect theory of information which he summarizes as follows:

> [I]nformation (or at least some information) has two basic aspects, a physical aspect and a phenomenal aspect. This has the status of a basic principle that might underlie and explain the emergence of experience from the physical. Experience arises by virtue of its status as one aspect of information, when the other aspect is found embodied in physical processing.

One trouble with Chalmers' proposal which he himself realized was that his double-aspect principle may be too liberal. There is information in a thermostat—is a thermostat therefore conscious? Many of us would say "obviously not," which undermines the plausibility of Chalmers' proposal. It is not likely that all information has a phenomenal aspect. As far as we know, only certain biological organisms are conscious and can be conscious of the world and their internal states. For those who are seeking a physicalist explanation of consciousness it is thus a reasonable hypothesis to make that some processes that (at least currently) only take place in living organisms are responsible for consciousness—both for the fact that the organism is conscious (rather than unconscious) and the fact that the organism is conscious of some information and not conscious of some other information.

We agree with Chalmers that it is a good intuition that information (and its meaning) is connected to consciousness. However, rather than saying that information has a phenomenal aspect, let us say that the meaning of information has the potentiality of becoming the content of the consciousness of a system. The hard problem then is to say what it is that makes such potentially conscious information into actually conscious information. And assuming that currently only living organisms are conscious, there then is something in living organisms (but not in other systems) which enables them to be conscious of information.

A further advantage of connecting our approach to Chalmers' double-aspect theory of information is that while Chalmers' theory suffers from epiphenomenalism, our scheme, when modified, opens up the possibility of a genuine causal efficacy of phenomenal properties upon the physical domain (see Pylkkänen 2007: 244–247; Pylkkänen 2017; see also Hiley and Pylkkänen 2005). Also, Chalmers himself has suggested that it is an interesting possibility that some sort of activity is required for experience, and that static information (e.g., information in a thermostat in a constant state) thus is not likely to have experience associated with it (1996: 298). If we say that (proto)phenomenal properties are always properties of some kind of active information, we could do justice to the intuition that activity is required for experience.

In the soma-significance scheme we sketched above, it is natural to say that consciousness comes in only at the more subtle soma-significant levels. It seems obvious that a typical soma-significant process (such as reading a newspaper) starts unconsciously, and consciousness only appears at some stage when the significance is consciously experienced, then perhaps giving rise to a signa-somatic response, depending on the meaning of the information (e.g., as we saw, if the meaning is "danger," this typically gives rise to a powerful signa-somatic response). But what is it that makes this non-conscious soma-significant process conscious? This is the hard problem in the scheme of soma-significance. What we have said thus far is that consciousness comes in at the more subtle levels of soma-significance—perhaps when a certain subtle level is related in a suitable way to a more manifest level (this would be a version of a higher-order theory of consciousness, see e.g., Rosenthal 1997 and Gennaro 2012).

But what do we mean by subtle? Quantum mechanics can here give some insight. As we proposed above, the Bohmian model of the electron can be seen as a prototype model of a coupling between a subtle aspect (the quantum field described by the wave function) and a more manifest aspect (the corpuscle or particle aspect of the electron). The subtle aspect enfolds information about the environment of the particle, and then signa-somatically organizes the movement of the particle. We are not saying that the electron is conscious, but we could say that conscious experience arises in a hierachical relationship between a more manifest and a more subtle level, which is analogous to the relation between the quantum field and the particle.

Indeed, it seems obvious that our conscious mental processes involve a hierarchy of levels of active information. We do not merely think about objects in the external world, but we can also become aware of our thinking. Bohm suggested that such meta-level awareness typically gives rise to a new, higher level of information. This higher level gathers information about the lower level. But because its essential nature is active information, it does not merely make a passive representation of the lower level. Rather, the higher level also acts to organize the lower level, a bit analogously to the way the active information in the quantum potential acts to organize the movement of the particle. And of course, we can become aware of this higher level of information from a yet higher level, and so on. So we have a hierarchy of levels of information in the mind. At some point in this hierarchy conscious experience appears. In the quantum model we have considered thus far there are just two levels, the particle and the field. To connect the mental hierarchy to the physical world, is there any way we could find a quantum model which also has a hierarchy of levels of information?

We have thus far discussed the quantum particle theory, but note that the Bohm model can easily be extended to fields (for more details see

Bohm and Hiley (1993)). One important feature is that this field theory can be naturally extended into a hierarchy of levels, each containing active information. Let us thus next consider the Bohm model of the quantum field theory.

10. Extension to Quantum Field Theory

The field, $\phi(\boldsymbol{r}, t)$, and its conjugate momentum, $\pi(\boldsymbol{r}, t)$, replace the position and momentum as beables. These field quantities would then be the appropriate variables that are to be identified with the relevant fields functioning in the brain. These fields would be organized by a generalization of the wave function, namely, the wave functional of the field, $\Psi(\phi(\boldsymbol{r}, t))$, which we will call the superwave function. The time evolution of this superwave function is described by a super-Schrödinger wave equation

$$i\frac{\partial\Psi(\phi(\boldsymbol{r}, t))}{\partial t} = H(\phi(\boldsymbol{r}, t), \pi(\boldsymbol{r}, t))\Psi(\phi(\boldsymbol{r}, t)). \tag{5}$$

The correspondence between field theory and the particle theory is as follows:

$$\boldsymbol{r} \longleftrightarrow \phi(\boldsymbol{r}, t) \qquad p \longleftrightarrow \pi(\boldsymbol{r}, t) \qquad \left(= \frac{\delta L}{\delta\dot{\phi}}\right)$$

$$\psi(\boldsymbol{r}, t) \longleftrightarrow \Psi(\phi(\boldsymbol{r}, t))$$

$$\frac{\partial S}{\partial t} + \frac{(\nabla S)^2}{2m} + V + Q = 0 \longleftrightarrow \frac{\partial S}{\partial t} + \frac{1}{2}\int[(\frac{\delta S}{\delta\phi})^2 + (\nabla\phi)^2]d^3r + SQ = 0$$

$$Q = -\frac{1}{2m}\frac{\nabla^2 S}{R} \longleftrightarrow SQ = -\frac{1}{2}\int\frac{\delta^2 R}{(\delta\phi)^2}d^3r.$$

Here we find there is a super-quantum potential, SQ, that organizes the field through a Hamilton-Jacobi field equation.

Since these equations have the same form as those describing particles, all the qualitative ideas discussed above for the particle model apply equally to quantum fields. However this provides a much richer structure and is far more appropriate for discussing the mind/brain relationship.

One directly significant feature is the emergence of a hierarchy of levels. To motivate this discussion, it should be recalled that in field theory, these fields, $\phi(\boldsymbol{r}, t)$, in turn affect the particles (which are themselves manifestations of these fields) through the quantum potential so that we have two levels operating, one at the particle level and the other at the field level. As far as these ideas have been applied to physical processes to date, only two levels seem necessary. However our theory has within it the possibility of a third, fourth and yet higher levels, producing a rich hierarchy, each level generating subtle effects in the level below.

This structure frees us from all traces of the original mechanical starting point of the quantum particle theory. For example, the particles themselves are merely semi-autonomous manifestations of the fields which can be created and annihilated and are organized by a "pool of information" that is contained in the superwave function and mediated by the super-quantum potential. Thus even the particles themselves are but discrete manifestations of the quantum fields.

We have above suggested that mental processes typically involve a hierarchy, and we just saw that the Bohm model of quantum field theory can be naturally extended into a hierarchy of fields. How are these two hierarchies related? Bohm (1990) proposed that they are the same hierarchy, which provides us with one schematic way of understanding how mind and matter are related. "Matter" corresponds to the more manifest levels, while "mind" refers to the more subtle levels, but each level has both a somatic and a significant aspect.

11. Consciousness and the Hierarchy of Quantum Fields of Information

We can now make the following speculative hypothesis: conscious experience only arises in the context of a hierarchy of levels of information which involves the activity of quantum fields. This does not mean that all information in the brain would be carried by quantum fields. On the contrary, it is likely that a great deal of information in the brain is carried by more stable structures (e.g., neural activity patterns) that for all practical purposes can be described by classical physics. But, we are proposing, the conscious apprehension of the meaning of such "classical" information involves the operation of quantum fields. We are assuming that conscious experience is not possible in a situation where non-trivial quantum effects are negligible.

What is it that makes non-conscious information become conscious information in this scheme? We propose that the best possibility is to make use of some version of higher order theories of consciousness to explain why a given soma-significant level is conscious. For example, we could say that what makes information at a given level conscious is that it is the intentional target of (typically) unconscious information at a more subtle quantum mechanical or quantum-like level. It is important to assume that the information at a more subtle level can be unconscious; otherwise, we get into an infinite regress (cf. Rosenthal 1997).

But there are other possible accounts for what makes non-conscious information conscious. We have already suggested that one way to tackle the hard problem in our scheme is to make use of Chalmers' idea that experience is an aspect of (some kind of) information; e.g., information that is held in suitably subtle somatic levels in the brains of some biological organisms, in conditions where subtle quantum effects can survive as it were (cf. Hameroff and Penrose 2014). Alternatively, we could apply some aspects of Tononi et al.'s (2016) integrated information theory of consciousness to Bohmian active information. Active information at the quantum level is holistic in the way that is likely to have a high value of phi which in Tononi's theory measures the degree of consciousness. Or finally, we could consider whether a Bohmian many-body quantum potential can be seen as a kind of global workspace in the spirit of Baars's (2017) global workspace theory of consciousness.

12. Life, Consciousness and Quantum Mechanics

As we have already hinted above, one important aspect with conscious experience is that it seems to be intrinsically related to the living state of matter, for it is (thus far) only with certain biological organisms that we find conscious experience with some certainty. Creating various artificial information processing systems with our current technology does not seem to enable conscious experience to appear. But what is it about living (as opposed to non-living) systems that enables conscious experience?

We suggest that this has to do with quantum mechanics. We saw above that the many-body system in the ontological interpretation of quantum theory exhibits a kind of organic unity that is very reminiscent of the organic unity of living organisms. Taking a step toward quantum biology, could it be that a necessary condition for a material system to be a living system is that it has a non-negligible quantum potential (or some higher level quantum-like potential) operating within it, providing its organic unity? And could it be that (currently) only certain biological organisms provide the kinds of conditions in which the kind of hierarchy of levels involving quantum fields which according to our proposal enables conscious experience can arise? Quantum mechanics would thus play a crucial role in making both life and consciousness possible. To put it simply: where there are non-trivial quantum effects operative, there can be life, and where there is life, there can be consciousness.

Given that consciousness only appears in living organisms, when searching for the physical preconditions of consciousness it would then be interesting to find processes that only take place in the conditions of the living state of matter, but which disappear or become negligibly small in the non-living state (e.g., in physical systems, such as thermostats, that operate at the macroscopic domain of non-living matter that can be approximately described by Newtonian physics). So for our hypothesis it would be useful to find non-trivial quantum processes that only operate in those conditions where we find consciousness but become negligible in conditions where consciousness is not present.

What is extremely interesting in this context is that Bohm and Hiley's ontological interpretation of quantum theory provides us with a prototype model of quantum information which is active in some conditions but has a negligible effect in other conditions. We pointed out above that the quantum potential is negligibly small in the domain where classical physics provides a good approximation, but has a significant effect in the quantum domain. We can now proceed to make the following speculative hypothesis: consciousness in biological organisms has to do with the non-negligible, non-trivial operation of quantum active information in the brain. It is only when there exists such quantum information in a suitable biological environment that there can be conscious experience. This immediately restricts consciousness to biological organisms, at least as far as the present moment is concerned.

However, the hypothesis also opens up, in principle, the possibility of artificial consciousness. If we knew the relevant structures and processes in biological organisms which make possible for there to operate a non-negligible quantum potential, it is possible in principle that we could replicate them in an artificial system. But if life essentially involves the organic unity characteristic of the quantum potential, then such system might also be a genuinely living system, albeit artificially manufactured. In this hypothesis, the quantum potential accounts for both life and phenomenal properties.

Indeed, one of the advantages of the Bohm approach is precisely that it provides us with an elegant account of the relation between the quantum level and the classical level, which is notoriously difficult to deal with in the usual interpretation of quantum theory. So insofar as a thermostat operates at the classical level, the quantum potential (and the (proto)phenomenal properties) have a negligible effect, and we need not say that a thermostat has (proto)phenomenal properties or that it is conscious. There is information in the thermostat, but it is not the holistic active information that we meet at the quantum level, but rather a more mechanistic information at the classical level. So although such mechanistic information (e.g., printed ink) can carry significance, it is only in situations where the quantum potential plays a non-negligible role in biological organisms that such significance can be consciously experienced. In our approach thus not all information has protophenomenal properties.

In this chapter our discussion has been largely at the general, philosophical level, and we have not considered in detail the possible "quantum sites" in the brain, where a non-neglibigle quantum potential (or a hierarchy of quantum fields) could play a non-negligible role. However, there are currently a number of more concrete proposals to which we can connect our approach, e.g., Vitiello and Freeman's work on quantum field theory of brain states, Beck and Eccles's work on quantum mechanics at the synaptic cleft and Penrose and Hameroff's work on quantum gravity and microtubuli (for a review and references, see Atmanspacher 2020). We propose that when these approaches are viewed from the perspective of our approach, they can become richer and more relevant to the mind-matter issue, because of the new possibilities opened up by the novel notions of levels of active information and quantum wholeness in our approach.

13. From Panpsychism to Cosmopsychism

The proposal that the particles of physics have certain primitive mind-like qualities may sound like a "quantized" version of traditional pan(proto) psychism However, quantum phenomena exhibit some holistic features which suggest a novel way to approach the combination problem of panpsychism, namely the problem of giving an account of how the primitive consciousness of the elements of a system could possibly combine into the full consciousness of the system. The combination problem presupposes what Atmanspacher (2014) calls a compositional approach, i.e., explaining the properties of the whole in terms of its parts. However, there are quite generally instances in quantum theory where the whole is prior to parts (cf. Schaffer 2010).

For example, in Bohm and Hiley's ontological interpretation of the many-body system, the behavior of individual particles cannot be understood in terms of their spatial relationships only, but we need to consider the quantum state of the whole system. So we do not explain the behavior of the whole in a bottom-up way in terms of the behavior of the parts, but we rather explain the behavior of the parts in a top-down way partly in terms of the properties of the whole (cf. Aharonov et al. 2018). More precisely, as we saw above, the particles in a Bohm-Hiley many-body quantum system are guided by a "common pool" of information enfolded in the many-body quantum potential that cannot be reduced to the "private pools" of individual particles. On the contrary, the whole is prior to the parts in the sense that these private pools arise from the common pool in certain circumstances through factorization. Thus, a key issue is to give an account how the whole decomposes into parts, as opposed to the parts combining to constitute the whole.

To summarize, our scheme has a panprotopsychist flavor in that we postulate that elementary particles have mind-like qualities, when the quantum potential for a particle is non-negligible. However, our emphasis on the priority of the whole goes against the spirit of the bottom-up way of explaining consciousness characteristic of traditional panpsychism.

Now consider the entire universe in the light of the Bohm-Hiley interpretation with its idea that active information (a proto-mental quality) is fundamental. It is suggested that the universe is a process which is in some sense simultaneously proto-mental and physical. This view would be called cosmoprotopsychism in the contemporary literature (cf. Chalmers 2020a). What about cosmopsychism? Is there any sense in which the universe as a whole with such dual aspect properties might be said to be conscious—or is conscious experience restricted to certain individual biological organisms, or at most collectives of them, as our folk psychology would have it? While most of us might take it to be obvious that the universe as a whole is not conscious, Bohm had a somewhat different view, as revealed in his discussion with the American philosopher Renee Weber (Weber (ed.) 1986: 95):

BOHM: Hegel took the view that thought was fundamental and that nature was mind showing itself to itself. Marx turned that upside down and said matter is fundamental and consciousness is matter showing itself to itself. Or you could take the view that neither matter nor mind is fundamental, but something unknown, which you could call a deeper or implicate ground. It is this view which I'm inclined toward.

WEBER: Is that ground self-aware?

BOHM: Yes. I would say, since it contains both matter and mind, it would have in some sense to be aware. Let's say it's in the direction of mind, but beyond it. It's not below it, but above it.

Of course, to say that the ground of the universe is "in some sense aware" is not yet to say that the ground is conscious; e.g., it may be conscious only in virtue of containing or enfolding all the minds and conscious experiences of all mental systems, but does not have a separate consciousness of its own. Such idea of enfoldment leads us to briefly discuss the notion of implicate or enfolded order in the next section.

14. The Implicate Order and Consciousness

We have discussed above the new possibilities opened up by the Bohm and Hiley ontological interpretation of quantum theory, which develops Bohm's 1952 theory, generalizing the original proposals so as to include spin and special relativity. This approach introduces some radically new ideas, such as the idea that material particles respond to information, and that the particles themselves are manifestations of a deeper structure. However, there is a sense in which traditional materialism is still retained. For example, we assume an independently existent 3D space *containing* the material particles which constitute material objects, such as tables, chairs and brains. This framework leads us to treat space-time and the material particles and fields as separately existing fundamental entities, while mental states and conscious experience are features that need to be explained in terms of matter. Yet a deeper reflection of quantum and relativity physics suggests that perhaps we should not take space-time as fundamental but abstract its properties from some deeper structure process as suggested by Bohm (1965) and more recently by Penrose (1991).

In the early 1960s Bohm's attention shifted from the causal interpretation to a more general and fundamental approach in which one could bring together quantum theory and general relativity. This approach introduced the notion of structure process, which Bohm developed during a long correspondence with the American artist Charles Biederman and became known as the implicate order (see Pylkkänen (ed.) 1999). The implicate order involves a radical change in our entire concept of reality, which also has profound implications to our attempts to understand the place of mind and conscious experience in nature.

To begin to understand this radical outlook, Bohm noted that in physics the basic order has traditionally been that of the Cartesian rectilinear grid, a description that is suitable for the analysis of the world into separately existent parts (e.g., independent particles or fields in interaction). Indeed, such analysis has been the key aspect of science since the Newtonian revolution. Yet, Bohm realized that both relativity and quantum theory emphasized, not separately existent parts-in-interaction, but rather an undivided wholeness. This suggests that underlying our usual reality of objects in space-time (the Cartesian or "explicate" order) there exists a deeper reality (the holomovement) in which unbroken wholeness prevails (the "implicate" order). The implicate order cannot reveal all aspects of reality simultaneously in one unique explicate order, rather specific partial orders can be revealed, e.g., in different experimental situations. These partial views reveal different complementary explicate orders so that complementarity becomes ontological, not epistemological. In this way the implicate order prevails in a whole range of phenomena, including biological and psychological phenomena.

The basic idea of the implicate order is that each region of space and time contains a total structure or total order *enfolded within it*. This is illustrated by the fact that the movement of light waves in a small region of a room contains information about the entire room; and more radically, when you look at the night sky, the movement of light waves in the region where your eye is placed contains information about structures covering immense stretches of space and time.

Such distinction between explicate and implicate is also crucial in our lives as human beings. Our body and brain live in the explicate order of space and time. But it is not so obvious that our mental and conscious processes can be thought of as existing *in* space and time. To be sure, many of our experiences, especially sensory experiences involve the explicate order. Even when we dream we can find ourselves in a 3D environment, so it is obvious that the brain can, as it were, simulate a 3D reality, a kind of virtual reality which we then consciously experience in our dreams. Many cognitive neuroscientists suggest that even when we are awake, the brain is constructing a simulation of virtual reality. But unlike during dreams, this awake virtual reality is driven by the senses, and we instinctively take it to be the real world.

The notion of implicate order is useful when trying to understand how the brain constructs this 3D virtual reality of consciousness. One useful analogy here is holography. Indeed, Karl Pribram suggested in 1971 that the brain stores information in a somewhat holographic fashion. Just as in holography we are able to produce 3D images from a complex interference pattern, Pribram thought that the brain reconstructs images through a mechanism that resembles holography (Pribram 1977: 162). Holography exemplifies the implicate order in the sense that each region of the holographic plate enfolds information about the whole illuminated object. In other words, in a hologram a total structure is enfolded in each region, which is a key feature of the implicate order. Bohm (1980: 160) described the change between an illuminated object and its hologram as a *metamorphosis* rather than as a transformation. Similarly, we can propose that the change between information as it is stored in the brain and the phenomenal experience that arises from the information is a kind of metamorphosis. It may well be worthwhile to try to describe this metamorphosis mathematically (as an example, see the mathematical description of metamorphosis connected to the hologram given by Bohm 1980: 160). The challenge is to move from the implicate order of the information stored in the brain into the explicate order of our phenomenal experience which contains phenomenal objects with qualia (e.g., colors, sounds) situated in a spatio-temporal structure. In terms of our notion of information, we can say that the differences of form in the information stored in the brain make a difference of phenomenal content which we consciously experience.

There is, however, much in our conscious experience that is not obviously and fundamentally spatial. Think about your thoughts and emotions, the sense of flow in your stream of consciousness. Language clearly is an implicate order in the sense that the meanings of words enfold each other. Philosophers have emphasized that a kind of unity is a key characteristics of conscious experiences, so might there be an aspect of conscious experience that lives at the level of the implicate order rather than the explicate order? Bohm certainly thought so, and suggested that we can understand some key aspects of conscious experience (such as time consciousness, e.g., when listening to music) in terms of the implicate order.

We will not describe this approach in detail here, but point out that our research program does not take the existence of a physical world of particles and field elements in 3D space as fundamental, but rather sees these as manifestations from a deeper level of reality. This may obviously affect the way we understand and approach problems in philosophy of mind and consciousness studies. To understand these problems it may be necessary to take this deeper level of reality into account. Thus instead of asking how can the physical processes in 3D space possibly give rise to conscious experiences (the usual hard problem), we can ask whether conscious experiences arise in the broader context which takes into account the enfolded or implicate ground from which ordinary matter constantly unfolds (for further discussion see Pylkkänen 2007).

That the universe contains much more than known matter in space and time is also suggested by Bohm's discussion of zero point energy:

> [T]he mathematics of the . . . quantum theory . . . treats the particle as . . . the quantized state of the field, that is, as a field spread over space but in some mysterious way with a quantum of energy. Now each wave in the field has a certain quantum of energy proportional to its frequency. And if you take the electromagnetic field, for example, in empty space, every wave has what is called a zero point energy below which it cannot go, even when there is no energy available. If you were to add up all the waves in any region of empty space you would find that they have an infinite amount of energy because an infinite number of waves are possible. However, you may have reason to suppose that the energy may not be infinite, that maybe you cannot keep on adding waves that are shorter and shorter, each contributing to the energy. There may be some shortest possible wave, and then the total number of waves would be finite and the energy would also be finite. (Bohm 1986: 27)

Bohm then suggests that general relativity implies that the shortest possible wavelength is about 10^{-33} cm.

> If you then compute the energy that would be in space, with that shortest possible wave length, then it turns out that the energy in one cubic centimeter would be immensely beyond the total energy of all the known matter in the universe. (Bohm 1986: 28)

This again implies that the world view of mechanistic materialism gives a very limited picture of the universe.

15. What Is Quantum Mechanics?

We have above given a short summary of our previous work on the Bohm theory, active information and connections to the mind and conscious experience but what has this to do with quantum mechanics? In the last few years one of us (BJH) has been exploring a different mathematical approach to quantum phenomena (Hiley 2016) which does not depend on the Cartesian order; rather it emphasizes the algebraic structures defined by the dynamical variables. It involves no wave functions as all the information normally contained in the wave function is already present in the algebra itself in certain of elements of the left ideals. Using these techniques we can arrive at the equations of the Bohm interpretation in a very different way so there is no need to start with the Schrödinger equation.

This suggests that we should *not* automatically give primary relevance to the Schrödinger picture with its emphasis on the wave function as many mathematically equivalent alternatives are possible. Indeed, as we have already pointed out above, the Stone-von Neumann theorem states that the Schrödinger approach is only one of many unitarily equivalent pictures. These alternatives give very different physical insights into the nature of quantum phenomena.

Our focus in the rest of the chapter is to try to bring out this different approach to quantum phenomena in a way that we hope will be relevant to future efforts attempting to tackle the mind-matter problem.

We believe the obsession with the wave function by giving it ontological status has hindered the wider application of the deep and significant changes needed to think about reality. One should realize that von Neumann himself already noted the shortcomings of giving the wave function a primary role. In a letter to Birkhoff he writes (1935), "I would like to make a confession which may seem immoral: I do not believe absolutely in Hilbert space any more." Why? Because: "the vectors ought to represent the physical states, but they do it redundantly, up to a complex factor, only." But there is a further more fundamental reason in that there are situations where probability cannot be defined because the wave function cannot be defined (see Rédei 1996). In spite of this unease and the many resulting paradoxes that this assumption encounters, we have learnt to live with the ambiguity in the wave function and the problem of probability, but in building any ontology on such an assumption, one should proceed with caution.

Using the Schrödinger picture as described above tends to hide the subtle features of the underlying non-commutative algebraic structure. There is another picture that places the operators centre stage by making them time dependent; this is the Heisenberg picture. This picture has its use almost exclusively in high energy physics, a picture that is rarely raised in discussions on the foundations of quantum mechanics. Yet it is this non-commutative structure of the dynamical variables that should lie at the heart of any philosophical discussion of quantum phenomena.

This point was realized in the early days by people like Dirac, Jordan, Feynman and even Schrödinger who regarded the non-commutativity structure as basic. Instead of relying on continuity and the the continuity of derivatives, the essential point of non-commutativity implied that while we may well have continuous functions, derivatives will not be continuous. (See Feynman 1948.) This was a feature that was built into the sum over paths approach by Feynman. What this means is that we must introduce the notion of a "backward" and "forward derivatives" as Feynman makes clear in his 1948 paper.

Let us see how this distinction can arise. We can write

$$\frac{dA}{dx} = \lim_{\delta\to 0} \frac{A(x+\delta) - A(x)}{\delta} \quad \text{or} \quad \frac{dA}{dx} = \lim_{\delta\to 0} \frac{A(x) - A(x-\delta)}{\delta},$$

so that at the point x, we have the possibility of two "derivatives" if the two expressions are not equal. Classical physics assumes that in the limit they become equal. However in Brownian motion they are not equal. There we have a "cause" for this inequality, a collision with another smaller particle. But what if Newton's first law is not valid at the quantum level? Assume at this level these two expressions are not in fact equal, then non-commutativity will be the consequence. To see this, first introduce the notation

$$\vec{D}A(x) = \lim_{\delta\to 0} \frac{A(x+\delta) - A(x)}{\delta} \quad \text{and} \quad A\overleftarrow{D}(x) = \lim_{\delta\to 0} \frac{A(x) - A(x-\delta)}{\delta}.$$

Now let us form

$$A\overleftarrow{D} - \vec{D}A \neq 0 \quad \text{and} \quad A\overleftarrow{D} + \vec{D}A \neq 0.$$

Thus the commutator and anti-commutator need not vanish so that non-commutativity becomes a necessity in the quantum domain. Remember it is the failure of the operators of position and momentum to commute that gives rise to the uncertainty principle.

To quickly see how the Bohm equations emerge from non-commutativity, first note that the most general description of a state will now need a two point function $\rho(x', x, t)$, which is the analogue of the density matrix. The pure state emerges in the coincident limit, $x' \rightarrow x$, when $\rho(x, x, t)^2 = \rho(x, x, t)$. We can write our state function as $\rho(x', x) = |x'\rangle\langle x|$, so if $H(x', x)$ is the Hamiltonian operator, we can define two equations, one being

$$i\frac{\partial |x'\rangle}{\partial t}\langle x| = H(x', x)\rho(x', x),$$

and the other is

$$-i|x'\rangle\frac{\partial \langle x|}{\partial t} = \rho(x', x)H(x', x).$$

If we subtract these two equations we obtain

$$i\frac{\partial \rho}{\partial t} = [H, \rho], \tag{6}$$

which is just the quantum Liouville equation in the limit $x' \rightarrow x$. This accounts for the conservation of probability if $\rho^2(x, t)$ is taken to be the probability in the usual sense.

We form a second equation by adding the two equations and find

$$i\left[\left(\frac{\partial|\psi\rangle}{\partial t}\right)\langle\psi| - |\psi\rangle\left(\frac{\partial\langle\psi|}{\partial t}\right)\right] = \left(\vec{H}|\psi\rangle\right)\langle\psi| + |\psi\rangle\left(\langle\psi|\overleftarrow{H}\right) = [H, \rho]_+ \tag{7}$$

To connect up with the usual approach we now need to see how the wave function appears, so we can write $\psi(x, t) = \langle\psi|x, t\rangle$. If we polar decompose the wave function, we obtain from equation (6):

$$\frac{\partial P(x, t)}{\partial t} + \nabla \cdot \left(P(x, t)\frac{\nabla S(s, t)}{m}\right) = 0, \tag{8}$$

where $P(x, t)$ is the probability of a particle being at (x, t). Equation (7) now becomes

$$\partial_t S(x,t) + \frac{1}{2m}[\partial_x S(x,t)]^2 + Q(x,t) + V(x,t) = 0. \tag{9}$$

Thus we have obtained the two defining equations of the Bohm approach directly from the non-commutative algebra of quantum operators showing that the quantum potential energy, Q, must take the form it does in order to conserve energy in the system.

In other work the operator algebra has been shown to be isomorphic with the Moyal algebra. (For details see von Neumann 1931 and Moyal 1949.) Indeed the motivation of this approach to quantum phenomena emerged from a detailed study of the algebra of quantum operators as detailed in Dirac 1947. This algebra contains the Bohm equations (8) and (7) but appears from what at first sight seems an entirely different non-commutative structure, namely a structure of a non-commutative phase space (x, p) (see Hiley 2015). The non-commutativity arises from the introduction of a new product, the Weyl ★-product. This is defined through the relation

$$a(x,p) \star b(x,p) = a(x,p) \exp\left[i\hbar\left(\overleftarrow{\partial}_x \overrightarrow{\partial}_p - \overrightarrow{\partial}_x \overleftarrow{\partial}_p\right)/2\right] b(x,p).$$

It should be noted that this expression is just the exponentiation of the Poisson bracket. In this way one sees that classical mechanics appears in the first two terms of the expansion of the exponent. Thus we see that classical physics emerges as an approximation to the non-commutative quantum formalism. This is the origin of the approach to quantum mechanics known as deformation quantum mechanics. (See Hirshfeld and Henselder 2002 for a simple account of this approach.)

This structure contains a density distribution function, $F(x,p)$, from which one constructs a marginal momentum through

$$\rho\bar{p}(x) = \int pF(x,p)dp.$$

Here $\bar{p}(x)$ turns out to be identical to the Bohm momentum $p_B(x) = \nabla S(x)$ and the equation of transport of this momentum is just the equation (9) showing that the Bohm equations also arise from a very different structure than the one that depends on the *abstract* non-commutating operator formalism. (See Hiley 2015.) What is common to both approaches is the non-commutative structure of the elements of the algebra.

But we have already shown in section 3 how the pair of equations that Bohm obtained follow directly from the Schrödinger equation so what have we gained? The approach we have outlined in this section is more general and can be applied to the Pauli and Dirac equations, equations necessary to account for spin and relativity. These equations are not discussed in the context of the Bohm approach because it is wrongly believed that that the approach cannot be applied in the relativistic domain. It can and has been applied to the relativistic domain. Here one cannot simply split the wave function into real and imaginary parts. One has to use the non-commutative algebra, indicating the essential need for non-commutative techniques. (For details of this approach see Hiley and Callaghan 2010a, 2010b, 2012.)

To give the reader a feel of how the results emerge more generally, we will sketch some details of the results that arise when spin and relativity are included. The Pauli quantum Hamilton-Jacobi equation is

$$2mE(t) = P^2 + [2(\nabla W \cdot S) + W^2]. \qquad (10)$$

where the quantum potential is

$$Q = (\nabla W \cdot S)/m + W^2/2m. \qquad (11)$$

We can express equation (11) purely in terms of P, ρ and the spin bivector S. After some straightforward but tedious work we find the quantum potential is now

$$Q = \{S^2[2\nabla^2 \ln \rho + (\nabla \ln \rho)^2] + S \cdot \nabla^2 S\}/2m = Q_1 + Q_2. \qquad (12)$$

An expression for Q has been evaluated by Dewdney et al. (1986) in terms of Euler angles. Hiley and Callaghan (2010a) show that

$$Q_1 = -\frac{1}{2m}\frac{\nabla^2 R}{R}, \qquad (13)$$

which is just the contribution of the Schrödinger particle to the quantum potential. However we now have a spin dependent part, Q_2 which takes the form

$$Q_2 = [(\nabla\theta)^2 + \sin^2\theta(\nabla\phi)^2]/8m. \qquad (14)$$

Thus we can write the Pauli quantum Hamilton-Jacobi equation in the form

$$(\partial_t\psi + \cos\theta\partial_t\phi)/2 + P_B^2/2m + Q_1 + Q_2 = 0, \tag{15}$$

which, in this representation, agrees exactly with the expression given in Dewdney et al. (1986).

For the Dirac particle things are far more complicated but the principles are the same. After some work, the energy equation can be written in the form

$$P^2 + W^2 + [J\partial_\mu W^\mu + \partial_\mu W^\mu J] - m^2 = 0, \tag{16}$$

which is to be compared with the relativistic energy equation

$$p_\mu p^\mu - m^2 = 0.$$

Thus we see that the extra two terms must be related to the quantum potential in some way. After some detailed manipulation we find the quantum potential for the Dirac particle can be written in the form

$$Q_D = \Pi^2 + W^2 + [J\partial_\mu W^\mu + \partial_\mu W^\mu J]. \tag{17}$$

In the non-relativistic limit, $\Pi = 0$, and equation (17) reduces to the quantum potential for the Pauli particle. (See Hiley and Callaghan 2010.)

$$Q_P = W^2 + [S(\nabla W) + (\nabla W)S], \tag{18}$$

where $2\rho S = \psi_L e_{12}\psi_R$ is the non-relativistic spin limit of J. W is the non-relativistic limit of W^μ. We have written the expressions for the quantum energy equations and the quantum potential in these two examples simply to show there is a lot more detail to be explored in this approach.

16. The ⋆-product and Classical Physics

We have seen one of the key steps to maintain a (q, p) phase space description in spite of the uncertainty principle was to replace the usual commutative scalar product $f(q,p) \circ g(q,p)$ by a non-commutative star-product, $f(q,p) \star g(q,p)$ for which $q \star p \neq p \star q$. It turns out that the star-product is the exponential of the classical Poisson bracket, once again showing how classical mechanics is related to the quantum formalism. Those familiar with group theory will recognize this procedure as "lifting" the classical group structure on to the covering group. This suggests that it is the covering groups of the classical symmetry groups that provide the link between classical and quantum phenomena. This is the fact that Feynman exploited when he exponentiated the classical action to produce his "sum over paths" approach.

These results have led one of us (BJH) to reexamine the foundations of classical mechanics itself in this new light. Because Hamilton's equations of motion are a pair of first order differential equations, the significance of the Poisson brackets is often overlooked. To appreciate the significance of the Poisson bracket appearing in the exponential form of the $\star$-product, let us write this product in its original form:

$$f(x,p) \star g(x,p) = \frac{1}{\hbar^2 \pi^2} \int dp' dp'' dx' dx'' f(x',p') g(x'',p'') \times \exp\left(\frac{-2i}{\hbar}\left(p(x'-x'') + p'(x''-x) + p''(x-x')\right)\right).$$

(See Groenewold 1946 and Baker 1958.) The expression in the exponential is twice the area of a phase space triangle (mod $2i/\hbar$) formed by the three points (z'', z', z) where $z = (x,p)$

$$A(z'', z', z) = (z' - z) \wedge (z - z'') = z'' \wedge z' + z' \wedge z + z \wedge z''$$

This expression brings out another novel feature of the product, namely its non-local nature since we see the product involves *three points* in phase space. This gives rise to an invariant area of phase space which corresponds to the quantum invariant

$$\oint p dx = nh.$$

Since the $\star$-product can also be written as the exponentiation of the Poisson bracket then classical mechanics must contain some trace of this non-locality, suggesting that classical mechanics itself has some features that we have missed simply by using Newton's or even Hamilton's differential equations of motion.

Indeed the symplectic geometry upon which classical mechanics is based is a geometry that preserves (q, p) areas in phase space. The difference between classical and quantum mechanics lies in the nature of these *phase space areas*. In classical physics it is assumed these areas can be shrunk to a point, whereas in quantum mechanics they can only be shrunk to a cell of area of the order of $\hbar$ under a symplectomorphism. That is what the uncertainty principle is indicating in the operator language. Clearly the key to the difference between the classical and the quantum becomes geometric with the appearance of $\hbar$.

If we explore what happens mathematically as we let $\hbar \rightarrow 0$, we find the quantum structure remains right down to $\hbar = 0 + \epsilon$ no matter how small we make ϵ. If you make it zero, the whole structure literally "blows up" because an infinity appears (see Schempp 1984 for more details). Since, in the physical world, $\hbar$ is not zero it follows that classical physics cannot hold in the very small. But under the appropriate conditions we should expect the classical world to emerge from the quantum world.

This idea is very significant for biological phenomena. Already there is mounting evidence for quantum effects occurring in biological systems. For example, photosynthesis and avian magnetoreception have recently attracted much attention (for details, see Kolli et al. 2012; Klinman and Kohen 2013; Emlen et al. 1976). This is not the place to discuss these proposal in detail but we feel it is essential to draw the reader's attention to this rapidly developing field which has been held back in the belief that "wave-decoherence" is the fatal factor that destroys quantum effects in living systems. This negative outlook traps us in a totally obscure notion of "wave-particle duality" and all the conflicting images that throws up. We should instead direct our attention towards a dynamical structure conditioned by non-commutative geometry. It is through these structures that quantum mechanics is beginning to play a role in biological systems and therefore could play a significant role in addressing the mind-matter question and perhaps even the hard problem of consciousness.

17. Concluding Remarks

It has been fairly common for researchers to say that there is nothing quantum mechanical going on in the brain. However, without quantum mechanics there would be no brain. All molecular structures ultimately depend on quantum mechanics to account for their stability. One tends to think only of the more spectacular quantum effects such as interference or diffraction and, of course, here the phenomena are extremely susceptible to external disturbances. Great care has to be taken to protect these phenomena against such disturbances and this is the source of the worries about the relevance of quantum mechanics in brain processes and the assumption that it is unlikely that effects like these will play any role in the brain.

However, in the brain we already have molecular structures which gain their stability from quantum processes. Recall that the covalent bond is one of the strongest chemical bonds and they are purely quantum mechanical. The brain contains many macromolecules making up the components of the neural network and its synaptic and dendritic interconnections and a lot more. Could there be some long range effect or global obstructions, common in non-commutative mathematical structures, which are not destroyed by temperature? The stability of long-chain molecules has already received some preliminary investigations by Salari et al. (2017) and Collini et al. (2010) and it is too soon to rule out all possibility of long-range quantum coherence processes. With our knowledge today we should not rule out the possibility of quantum effects in the brain based only on decoherence arguments. Perhaps the fragility of the entangled states raises further worries about stability but with the development of new mathematical techniques to investigate quantum effects we feel it is too early to definitely rule out such possibilities (for a recent review, see Adams and Petruccione 2020).

In this chapter we have given a summary of our previous work on how the Bohm interpretation—and the notion of active information in particular—can help to approach the mind-matter problem in a new way. How do the discussions in the previous two sections affect our ideas about active information? If anything they have strengthened our case. By showing how the quantum potential can be generalized to include spin and relativity, we see that active information can appear in all domains.

We have naturally focused on the idea that the quantum potential is the source of the active information, but ultimately its source is a new quality of energy. It is neither classical kinetic energy, nor classical external potential energy. The appearance of an other quality of energy should not surprise us. In thermodynamics we have many notions of energy, internal energy, Helmholtz free energy, Gibbs free energy and even heat energy, all of which arise in general chemical processes, so why would they not appear in brain processes?

Finally, do these new ideas have any relevance to the mind-matter relation? The answer is absolutely everything. Remember Eccles (1986). His ideas were plagued with the horror of non-conservation of energy. The advantage with this new quality of energy is that it is "borrowed" from the kinetic energy to shape the overall process but in such a way that energy is always conserved. We never see anything like this at the classical level. If we assume that this new quality of energy is part of the essence of mental states, we obtain a much better way of understanding how mental states can influence physical processes. Thus active information is an extremely interesting novel concept in physics, relevant in a wide range of phenomena. As we have proposed, it could be an important way to begin to unite mind and matter and even move forward on the hard problem.

Acknowledgments

Sections 4 and 5 of this chapter contain descriptive parts from our earlier joint article Hiley, B. J. and Pylkkänen, P. (2001), Naturalizing the Mind in a Quantum Framework, in *Dimensions of Conscious Experience*, P. Pylkkänen and T. Vaden (eds.), Amsterdam: John Benjamins. These parts have been adapted and updated for the purposes of the present chapter when feasible. The work for this chapter was partially funded by the Fetzer Franklin Fund of the John E. Fetzer Memorial Trust.

References

Adams, B. and Petruccione, F. 2020. Quantum effects in the brain: A review. *AVS Quantum Sci.* 2:022901. https://doi.org/10.1116/1.5135170

Aharonov, Y., Cohen, E. and Tollaksen, J. 2018. Completely top-down hierarchical structure in quantum mechanics. *Proc. Natl. Acad. Sci.* 115:11730–11735.

Atmanspacher, H. 2014. 20th Century Variants of Dual-Aspect Thinking. *Mind and Matter* 3(2):7–26.

Atmanspacher, H. 2020. Quantum Approaches to Consciousness. In (E. N. Zalta, ed.), *The Stanford Encyclopedia of Philosophy* (Summer 2020 ed.), https:// plato.stanford.edu/archives/sum2020/entries/qt-consciousness/

Baars, B. 2017. The Global Workspace Theory of Consciousness: Predictions and Results. In (S. Schneider and M. Velmans, eds.), *The Blackwell Companion to Consciousness* (2nd ed.). Hoboken, NJ: John Wiley and Sons, Inc.

Baker, G. A. J. 1958. Formulation of Quantum Mechanics Based on the Qausi-Probability Distribution Induced on Phase Space. *Phys. Rev* 109:2198–2206.

Bohm, D. 1951. *Quantum Theory*. Englewood Cliffs, NJ: Prentice Hall.

Bohm, D. 1952. *Phys. Rev.* 85:66–179 and 180–193.

Bohm, D. 1965. Space, Time, and the Quantum Theory Understood in Terms of Discrete Structural Process. *Proc. Int. Conf. on Elementary Particles*, Kyoto, 252–287.

Bohm, D. 1980. *Wholeness and the Implicate Order*. London: Routledge.

Bohm, D. 1986a. A New Theory of the Relationship of Mind and Matter. *Journal of the American Society for Psychical Research* 80:113–135.

Bohm, D. 1986b. The implicate order and the super-implicate order. In (R. Weber, ed.), *Dialogues with Scientists and Sages: The Search for Unity*. London: Routledge

Bohm, D. 1989. Meaning and Information. In (P. Pylkkänen, ed.), *The Search for Meaning*. Wellingborough: Crucible.

Bohm, D. 1990. A New Theory of the Relationship of Mind and Matter. *Philosophical Psychology* 3(2):271–286.

Bohm, D. 2003. Soma-significance and the activity of meaning. In (L. Nichol, ed.), *The Essential David Bohm*. London: Routledge.

Bohm, D. and Hiley, B. J. 1987. An Ontological Interpretation of Quantum Theory. *Physics Reports* 144(6):321–375.

Bohm, D. and Hiley, B. J. 1993. *The Undivided Universe: An Ontological Interpretation of Quantum Theory*. London: Routledge.

Bohm, D. and Pylkkänen, P. 1991. Cognition as a movement toward coherence. Unpublished manuscripts B36-B38. The David Bohm papers, Birkbeck Library Archives and Special Collections, University of London.

Bohr, N. 1961. *Atomic Physics and Human Knowledge*. New York: Science Editions.

Bondi, H., van der Burg, M. G. J. and Metzner, A. W. K. 1962. Gravitational Waves in General Relativity. VII. Waves from Axi-Symmetric Isolated Systems. *Proc. Roy. Soc.* 269A:21–52.

Chalmers, D. J. 1995. Facing Up to the Problem of Consciousness. *Journal of Consciousness Studies* 2(3):200–219.

Chalmers, D. J. 1996. *The Conscious Mind: In Search of a Fundamental Theory*. Oxford: Oxford University Press.

Chalmers, D. J. 2015. Panpsychism and Panprotopsychism. In (T. Alter and Y. Nagasawa, eds.), *Consciousness in the Physical World: Perspectives on Russellian Monism*. Oxford: Oxford University Press.

Chalmers, D. J. 2020a. Idealism and the Mind-Body Problem. In (W. Seager, ed.), *The Routledge Handbook of Panpsychism*. London: Routledge.

Chalmers, D. J. 2020b. Is the Hard Problem of Consciousness Universal? *Journal of Consciousness Studies* 27(5–6):227–257.

Chrisley, R. C. 1997. Learning in Non-Superpositional Quantum Computers. In (P. Pylkkänen, P. Pylkkö and A. Hautamäki, eds.), *Brain, Mind and Physics*. Amsterdam: IOS Press.

Collini, E., Wong, C. Y., Wilk, K. E., Curmi, P. M. G., Brumer, P. and Scholes, G. D. 2010. Coherently wired light-harvesting in photosynthetic marine algae at ambient temperature. *Nature* 463:644–647.

Dirac, P. A. M. 1947. *The Principles of Quantum Mechanics*. Oxford: Oxford University Press.

Dretske, F. 1995. *Naturalizing the Mind*. Cambridge, MA: MIT Press.

Eccles, J. C. 1986. Do mental events cause neural events analogously to the probability fields of quantum mechanics? *Proc. Roy. Soc.* B277:411–428.

Einstein, A. 1924. Über den Äther. *Schweizerische Naturforschende Gesellschaft, Verhandlungen* 105:85–93. English translation in (S. Saunders and H. R. Brown, eds. 1991), *The Philosophy of the Vacuum*, Oxford: Clarendon Press, 13–20.

Einstein, A., Podolsky, B. and Rosen, N. 1935. *Phys. Rev.* 47:777–780.

Emlen, S., Wiltschko, W., Demong, N. and Wiltschko, R. 1976. Magnetic direction finding: Evidence for its use in migratory indigo buntings. *Science* 193:505–508.

Feynman, R. P. 1948. Space-time Approach to Non-Relativistic Quantum Mechanics. *Rev. Mod. Phys.* 20:367–387.

Finkelstein, D. 1969. Space-time code. *Phys. Rev.* 184:1261.

Flack, R. and Hiley, B.J. 2018. Feynman paths and weak values. *Entropy* 20:105.

Floridi, L. 2019. Semantic conceptions of information. In (E. N. Zalta, ed.), *The Stanford Encyclopedia of Philosophy* (Winter 2019 ed.), https://plato.stanford.edu/entries/information-semantic/

Gennaro, R. 2012. *The Consciousness Paradox*. Cambridge, MA: MIT Press.

Goel, A. 2008. Molecular Evolution: A Role for Quantum Mechanics on the Dynamics of Molecular Machines that Read and Write DNA. In (D. Abbott, O. C. W. Davies and A. K. Pati, eds.), *Quantum Aspects of Life*. Singapore: World Scientific.

Groenewold, H. J. 1946. The Principles of Elementary Quantum Mechanics. *Physica* 12:405–460.

Hameroff, S. and Penrose, R. 2014. Consciousness in the Universe: A Review of the Orch OR theory. *Physics of Life Reviews* 11(1):39–78.

Hiley, B. J. 1996. Mind and matter: Aspects of the implicate order described through algebra. In (K. H. Pribram and J. King, eds.), *Learning as Self-Organisation*. Mahwah, NJ: Lawrence Erlbaum, 569–586.

Hiley, B. J. 2004. Information, quantum theory and the brain. In (G. G. Globus, K. H. Pribram and G. Vitiello, eds.) ,*Brain and Being*. Amsterdam: Benjamins, 199–216.

Hiley, B. J. 2012. Weak Values: Approach through the Clifford and Moyal Algebras. *J. Phys.: Conference Series* 361.

Hiley, B. J. 2015. On the Relationship between the Moyal Algebra and the Quantum Operator Algebra of von Neumann. *Journal of Computational Electronics*. 14:869–878.

Hiley, B. J. 2016. The Algebraic Way. In (I. Licata, ed.), *Beyond Peaceful Coexistence: The Emergence of Space, Time and Quantum*. Singapore: World Scientific, 1–25.

Hiley, B. J. 2019. Stapp, Bohm and the Algebra of Process. *Activitas Nervosa Superior* 61(1):102–107.

Hiley, B. J. and Callaghan, R. E. 2010a. The Clifford Algebra approach to Quantum Mechanics A: The Schrödinger and Pauli Particles. arXiv: 1011.4031.

Hiley, B. J. and Callaghan, R. E. 2010b. The Clifford Algebra Approach to Quantum Mechanics B: The Dirac Particle and its relation to the Bohm Approach. arXiv: 1011.4033.

Hiley, B. J. and Callaghan, R. E. 2012. Clifford Algebras and the Dirac-Bohm Quantum Hamilton-Jacobi Equation. *Foundations of Physics* 42:192–208.

Hiley, B.J., Dennis, G. and de Gosson, A. M. 2021. Further Developments of the Dirac-Bohm Picture. To appear in 2021.

Hiley, B. J. and Pylkkänen, P. 2001. Naturalizing the Mind in a Quantum Framework. In (P. Pylkkänen and T. Vaden eds.), *Dimensions of Conscious Experience*. Amsterdam: John Benjamins.

Hiley, B. J. and Pylkkänen, P. 2005. Can mind affect matter via active information? *Mind and Matter* 3(2):7–26.

Hirshfeld, A. C. and Henselder, P. 2002. Deformation quantisation in the teaching of quantum mechanics, *Am. J. Phys.* 70:537–547.

Klinman, J.P. and Kohen, A. 2013. Hydrogen tunnelling links protein dynamics to enzyme catalysis. *Ann. Rev of Biochem.* 82:471–496.

Kolli, A, O'Reilly, E. J., Scholes, G. D. and Olaya-Castro, A. (2012) The fundamental role of quantized vibrations in coherent light harvesting by cryptophyte algae. *J. Chem. Phys.* 137:174109.

Levine, J. 2017. Anti-materialist arguments and influential replies. In (S. Schneider and M. Velmans, eds.), *The Blackwell Companion to Consciousness* (2nd ed.). Hoboken, NJ: John Wiley and Sons, Inc.

Miller, G.L. 1987. *Resonance, Information and the Primacy of Process: Ancient Light on Modern Information and Communication Theory and Technology*. Doctoral Thesis, Rutgers, New Brunswick, NJ.

Moyal, J. E. 1949. Quantum Mechanics as a Statistical Theory. *Proc. Camb. Phil. Soc.* 45:99–123.

Oriols, X. and Mompart, J. 2019. *Applied Bohmian Mechanics: From Nanoscale Systems to Cosmology* (2nd ed.). Singapore: Jenny Stanford Publishing Pte. Ltd.

Penrose, R. 1991. The Mass of the Classical Vacuum. In (S. Saunders and H. R. Brown, eds.), *The Philosophy of Vacuum*. Oxford: Clarendon Press, 21–26.

Penrose, R. and Rindler, W. 1984. *Spinors and space-time*. Vol. 2. Cambridge: Cambridge University Press.

Pribram, K.H. 1977. *Languages of the Brain: Experimental Paradoxes and Principles in Neuropsychology*. Monterey, CA: Brooks/Cole Publishing Company.

Pribram, K.H. 1991. *Brain and Perception: Holonomy and the Structure in Figural Processing*. Mahwah, NJ: Lawrence Erlbaum.

Pylkkänen, P. 1992. *Mind, Matter and Active Information: The Relevance of David Bohm's Interpretation of Quantum Theory to Cognitive Science*. Reports from the Department of Philosophy, University of Helsinki.

Pylkkänen, P. 1995. On Baking a Conscious Cake with Quantum Yeast and flour. In (B. Borstner and J. Shawe-Taylor, eds.), *Consciousness at a Crossroads of Cognitive Science and Phenomenology*. Thorverton: Imprint Academic.

Pylkkänen, P. 1999a. The Philosophical Implications of Quantum Level Active Information. In (D. Dubois, ed.), *Computing Anticipatory Systems. CASYS'98—Second International Conference*. Woodbury, NY: American Institute of Physics.

Pylkkänen, P. ed. 1999b. *Bohm-Biederman Correspondence: Creativity and Science*. London: Routledge.

Pylkkänen, P. 2007. *Mind, Matter and the Implicate Order*. Berlin and New York: Springer Frontiers Collection.

Pylkkänen, P. 2014. Can quantum analogies help us to understand the process of thought? *Mind and Matter* 12(1):61–92.

Pylkkänen, P. 2016. Can Bohmian quantum information help us to understand consciousness? In (H. Atmanspacher, T. Filk and E. Pothos, eds.), *Quantum Interaction 2015: 9th International Conference, QI 2015, selected papers*. Heidelberg: Springer.

Pylkkänen, P. 2017. Is there room in quantum ontology for a genuine causal role of consciousness? In (A. Khrennikov and E. Haven, eds.), *The Palgrave Handbook of Quantum Models in Social Science*. London: Palgrave Macmillan.

Pylkkänen, P. 2020. A Quantum Cure for Panphobia. In (W. Seager ed.), *The Routledge Handbook of Panpsychism*. London: Routledge.
Pylkkänen, P., Hiley, B.J. and Pättiniemi, I. 2016. Bohm's approach and individuality. In (A. Guay and T. Pradeu, eds.), *Individuals Across the Sciences*. Oxford: Oxford University Press.
Rédei, M. 1996. Why von Neumann did not like the Hilbert space formalism of quantum mechanics (and what he liked instead). *Stud. His. Phil. Mod. Phys.* 27B:493–510.
Rosenthal, D.M. 1997. A theory of consciousness. In (N. Block, O. Flanagan and G. Güzeldere, eds.), *The Nature of Consciousness*. Cambridge, MA: MIT Press.
Salari, V., Naeij, H. and Shafiee, A. 2017. Quantum interference and selectivity through biological ion channels. *Scientific Reports* 7:41625.
Schaffer, J. 2010. Monism: The Priority of the Whole. *The Philosophical Review* 119(1):31–76.
Schempp, W. 1984. Radar ambiguity functions, nilpotent harmonic analysis and holomorphic theta series. In (R. A. Askey, T. H. Koornwinder and W. Schempp, eds.), *Special Functions: Group Theoretical Aspects and Applications*. Dordrecht: Reidel Publishing Company.
Seager, W. ed. 2020. *The Routledge Handbook of Panpsychism*. London: Routledge.
Searle, J.R. 1980. Minds, Brains and Programs. *Behavioral and Brain Sciences* 3:417–457.
Shannon, C. E. 1948. *Bell Systems Tec. Journal* 27:379.
Skrbina, D. 2005. *Panpsychism in the West*. Cambridge, Mass: MIT Press,
Stan, M. 2021. Phoronomy: Space, construction, and mathematizing motion. In (M. B. McNulty, ed.), *Kant's Metaphysical Foundations of Natural Science: A Critical Guide*. Cambridge: Cambridge University Press.
Tononi, G., Boly, M., Massimini, M. and Koch, C. 2016. Integrated information theory: From consciousness to its physical substrate. *Nature Reviews Neuroscience* 17:451–461.
van Gelder, T. J. 1997. The Dynamical Alternative. In (D. M. Johnson and C. Erneling, eds.), *The Future of the Cognitive Revolution*. Oxford: Oxford University Press, 227–244.
von Neumann, J. 1931. Die Eindeutigkeit der Schrödingerschen Operatoren. *Math. Ann.* 104:570–587.
von Neumann, J. 1935. Letter to Garrett Birkoff dated Wednesday, November 13 (presumably 1935).
Weber, R. ed. 1986. *Dialogues with Scientists and Sages: The Search for Unity*. London: Routledge

16

Strange Trails: Science to Metaphysics

William Seager

What is ultimately most philosophically interesting about quantum theory is trying to understand what a quantum world is, or would be, really like. In the evocative words of David Mermin (1998), it is to get an answer to the question "what is quantum mechanics trying to tell us" about the nature of reality (always recalling that a possible answer is "nothing"). Every grand theory invites extension to a total picture of its world. It seems to be an essential aspect of human nature to seek a vision in which we can see "how things in the broadest possible sense of the term hang together in the broadest possible sense of the term" (Sellars 1963, p. 1). And so people have always been developing and sometimes believing in such grand frameworks.

But in the pre-scientific past, which still represents the vast majority of the extent of human thought, metaphysics sprang from ordinary observation and concepts drawn from immediate experience. Here are two obvious examples. The first is early atomism, drawing upon notions of commonplace material interactions extended by a very impressive feat of imagination to a hidden, simplified and purified realm of invisible constituents interacting in much the same way, like a domain of tiny marbles. The second is the basically Aristotelian idea of goal directed motion characterized by "striving activity." So far as they go, both are fairly reasonable conceptions of the world. The material world is evidently made of things which are composed of material parts. There doesn't seem to be any reason there cannot be a set of smallest constituents which fundamentally behave in much the same way as macroscopic material objects. It's then not very hard to get yourself to believe that composite objects are nothing more than conglomerates of these minuscule building blocks.

But in addition to brute material interaction we also see everywhere around us movement that is self-initiated and not caused by any visible material contact. Such movement is most typically the result of an effort of striving or will towards some goal. I would think it took no less of an

William Seager, *Strange Trails: Science to Metaphysics* In: *Consciousness and Quantum Mechanics.*
Edited by: Shan Gao, Oxford University Press. © Oxford University Press 2022.
DOI: 10.1093/oso/9780197501665.003.0017

imaginative feat to extend the basic notion of goal directedness to those aspects of the material world's behavior that are not—at least not obviously—contact dependent (although in this case, perhaps echoes of a still more ancient animism buttressed imagination).

When science finally arrived on the scene, very sophisticated metaphysical pictures of reality were already in place, centered upon a religious backdrop ripe for overturning, and still based upon a set of concepts closely related to ordinary experience. A scientific metaphysics is constructed not from a basket of such quotidian concepts but similarly is a vast extension of a certain initially quite limited picture of how things work in restricted domains of application.

It is very well known how early 17th century physics devised and devoted itself to the metaphysical world view that came to be called the "mechanical philosophy." According to a pure form of this view, everything that happens (in the material world at least, for most mechanists were not averse to there being an additional *immaterial* realm) can be and should be explained in strictly "mechanical" terms. Mechanism was thought of as just material contact interaction or collisions. This was regarded as both inherently intelligible and quite evidently something that actually occurs in nature. The mechanists' dream was that everything that happens is just the—undoubtedly vastly complex—ramification of microscopic collision interactions, the theoretically explicated details of which "explains everything in the nature of body through number, measure, weight, or size, shape and motion, [and] which teaches that nothing is moved naturally except through contact and motion, and so teaches that, in physics, everything happens mechanically, that is, intelligibly" (Leibniz 1710-16/1989).

This dream did not last very long, at least in its pure form. Its *aspirational* form lasted much longer and is probably not yet exhausted. But towards the end of the 17th century Isaac Newton's spectacularly successful account of motion and gravity infamously and somewhat reluctantly introduced a non-mechanical *force* into this picture of nature. This was met with immense skepticism, including from Newton himself.[1] The normally politic and conciliatory Leibniz was aghast. Writing in a paper of around 1712 he

[1] Newton disparaged his own view and those who took it metaphysically seriously: "... that one body may act upon another at a distance through a vacuum, without the mediation of anything else, by and through which their action and force may be conveyed from one to another, is to me so great an absurdity that I believe no man who has in philosophical matters a competent faculty of thinking can ever fall into it" (2004, p. 102).

entitled "Anti-barbarus physicus" Leibniz lamented that "people love to be returned to darkness" resulting in a "return to chimeras, to Archae, to certain plastic intelligences." Worst of all, because sometimes attended with impressive scientific and mathematical ability certain natural philosophers "return to *occult qualities* or to *Scholastic faculties*, but since . . . those [terms are] in bad repute, changing the name, they call them *forces*" (1710-16/1989, pp. 312–13). These supernatural entities destroy intelligibility. How, exactly, would the attractive force (which Leibniz archly notes "some in England" are trying to revive) between material bodies work? Are we, Leibniz wonders, supposed to "imagine that every body whatsoever is attracted by every other by virtue of a force in matter itself, whether it is as if a thing takes pleasure in another similar thing, and senses it even from a distance, or whether it is brought about by God, who takes care of this through perpetual miracle, so that bodies seek one another . . . (1710-16/1989, p. 317).

There were many efforts to give scientific, as opposed to philosophical or polemical, arguments against the introduction of these mysterious forces of nature. Leibniz himself, in very good company such as Christian Huygens, made some ultimately unsuccessful attempts to mathematically show that a Cartesian style vortex theory could explain in mechanical terms the known facts of planetary motion such as Kepler's laws (see Smeenk (forthcoming), Aiton (1972)). These were eventually shown to be provably inadequate. Other forces, such as that of electric attraction, were also found to be amenable to mathematical treatments which assumed instantaneous action, and physicists became inured to the absurdity of "matter acting where it was not."

The urge for mechanical intelligibility did not disappear however. The growth of electromagnetic field theory from Faraday to Maxwell testifies to this. Maxwell devoted great efforts to devising mechanical models of the electromagnetic field which at least initially he developed "with ontological intent" and even later he "continued in his allegiance to the core hypothesis of the model—that is, to the hypothesis of molecular vortices" (see Siegel 2003, p. 56). This kind of search for intelligibility was common. Maxwell was much influenced by earlier ether vortex models of William Thomson (aka Lord Kelvin) who resolutely cleaved to a mechanistic constraint on optical, electromagnetic and atomic theorizing. Maxwell's later distancing himself from the necessity of producing plausible (as opposed to heuristic) mechanical models of the electromagnetic field was not regarded fondly by Thomson who in some instances he lampooned as someone "trying to evolve [theory] out of his inner consciousness" (see Thomson 1911, p. 544). It's also worth noting that Thomson hoped for a mechanical theory of atoms based upon vortices in an all pervasive ether, promising a unified, mechanistically intelligible, account of matter and electromagnetism (see Thomson 1867; Silliman 1963).

It may be that the acceptance of the electromagnetic field as a bonafide physical entity in its own right and Einstein's efforts to provide a field theory of gravitation laid to rest 19th century worries about action-at-a-distance. But if they did so, they obviously did not thereby provide a mechanical account which made electricity, light and gravitation all as intelligible as, say, a steam engine. Instead, a certain kind of constraint on physical theorizing was finally given up.

Did that mean the end of intelligibility itself as an important scientific desideratum? Certainly not. But here we should distinguish two forms or types of intelligibility: (1) Mundane Intelligibility and (2) Mathematical Intelligibility. By the first, I mean to suggest the kind of intelligibility in which we come to understand a phenomenon by matching it to our intuitions or familiarity with ordinary sorts of interactions we encounter in everyday life. We, at least some of us, encounter billiard balls bouncing off each other in very precise and apparently fully predictable directions and speeds, or we see how sets of gears can engender complex motions. There is really a vast set of interactions we are familiar with and which seem inherently "reasonable."

The step towards mathematical intelligibility begins with the observation that mundane intelligibility can be represented mathematically. The speed, torque and direction of clockwork systems of gears can be calculated from a mathematical description of the device. Adding measures of velocity, mass and angles permits calculation of, for example, billiard ball interactions as innumerable computer pool games illustrate. It took the human race quite a long time to figure this out. Galileo stands as the icon of this insight, as he famously likened the universe to a book but one that cannot be read "unless one first learns to comprehend the language and recognize the letters in which it is composed. It is written in the language of mathematics, and its characters are triangles, circles and other geometric figures without which it is humanly impossible to understand a single word of it" (1623/1957, p. 238). Of course, a large part of the problem was that we had to *invent* this language, most especially the calculus.

In these first simple cases there is a satisfying correspondence between intuitive or so to speak qualitatively understood phenomena and a precise representation which strips away everything that is inessential. What is more surprising is that the new way of representing phenomena then suggests new phenomena, perhaps never before encountered. It's a deep philosophical question exactly why mathematical representations so often succeed in drawing forth new possibilities of nature that turn out actually to be there when we look, the more so when we recall that all mathematical theories prior to our current ones turned out false.[2] It is likely that our current theories will also eventually fail, but they nonetheless have succeeded magnificently to date.

As we develop a faith in mathematics, the need to coordinate the mathematical representation with some intuitively intelligible system falls away. What we seek then is pure mathematical intelligibility. Modern physical theory is in this sense intelligible to a truly remarkable degree. One of the most stunning and fruitful examples is Noether's theorem and the uses it is put to in modern physics. Most readers of this book will be familiar with her theorem (proven in 1915) and one can readily give a rough non-mathematical account of its content: physical symmetries entail conserved quantities (and vice versa), but there is no non-mathematical way to explain why it is true. For that one needs to define precisely what is meant by a "continuous symmetry" and one needs a lot of mathematical machinery to express the dynamics or the structure of the kinds of physical theories to which the theorem applies (e.g. the Lagrangian[3] of the system). Nonetheless, as a theorem, we have reached a kind of pinnacle of intelligibility insofar as the proof of the theorem completely "explains" why it is true. If we have theories which take forms meeting the constraints of the theorem, the theories will inherit this kind of mathematical intelligibility. Of course, it is another question why nature is such as to admit a description which meets the conditions for the theorem. There is something of an intuitive sense that some symmetries should be "invisible" to the physics of the world, especially things like the time or orientation of an experiment but these do not seem to be a priori knowable.

[2] Eugene Wigner famously asked why we find such an astounding and seemingly "Unreasonable Effectiveness of Mathematics in the Natural Sciences" (1960), a question which has spawned a large literature but no definitive answer (see Bangu 2012 for an overview).

[3] The Lagrangian is a function of what are called generalized coordinates, features which encode the state of the system, e.g. position, and derivatives thereof. In general terms the Lagrangian of a system is $L = T - U$, where T is kinetic energy and U is potential energy. As an example, the Lagrangian for an ideal pendulum is $L = \frac{1}{2}ml^2(d\theta/dt)^2 - (1 - \cos\theta)mgl$, where θ is the angle of the pendulum arm, m is the mass, g is the gravitational constant and l is the length of the pendulum. The equations of motion for the system can be derived from the Lagrangian.

Modern physics has taken this kind of intelligibility very far indeed. Here is one example (cribbed from Schwichtenberg 2015, pp. 129 ff.). Striving to describe interacting quantum fields and beginning with the Lagrangian of the free spin-½ field (from which the Dirac equation for the electron can be derived) it can be observed that multiplying the[4] field state, ψ, by a unit complex number, e^{ia}, makes no difference to the physics described by the field. Such a multiplication "changes" the field throughout space so this universal transformation which makes no physical difference is a global *symmetry*. The set of transformations encoded by unit complex numbers, e^{ia}, is the group called U(1) (also called the circle group as it represents rotations around a unit circle). The Lagrangian is thus said to be U(1) invariant.

However, such global transformations are regarded as "unphysical" insofar as they violate the constraint imposed by special relativity that nothing can move faster than the speed of light. That is, the proposed transformation instantly and thus illicitly changes the field everywhere all at once. One might wonder why this should matter if no detectable difference is the result of the transformation, but the consequence of replacing the global symmetry is remarkable. Schwichtenberg asks us to consider a local transformation instead, where we multiply the field state by something which depends on location: $\psi \Rightarrow e^{ia(x)}\psi$. What happens to the Lagrangian? Sadly, it is no longer invariant under this transformation (*local* U(1) symmetry) because of the appearance of an extra term in the Lagrangian with a coefficient of the form $-(\partial_\mu a(x))$.

It turns out that if we consider another field, the free massless spin-1 field we can find a way to restore local symmetry. This field too has a global symmetry (the massless nature of the field is important here to achieve this). But again, when we look for a local symmetry trouble arises. In this case a subtle change reveals a transformation which provides local symmetry: we must "add the derivative $\partial_\mu a(x)$ of some arbitrary function instead of an arbitrary function" (Schwichtenberg 2015, p. 132). Now something really remarkable happens. As the reader will note there is an obvious connection between the two situations. And indeed if we couple together the two fields the problematic term in the spin-½ field is canceled! Miraculously, the resulting combined Lagrangian is the correct Lagrangian for quantum electrodynamics. As Schwichtenberg puts it, "we are able to derive this

[4] The situation is more complicated since two fields are involved in this Lagrangian but I am oversimplifying for brevity.

Lagrangian simply by observing internal symmetries of the Lagrangians describing free spin-1/2 fields and free spin-1 fields" (p. 134).

Now, back to Noether's theorem. Since we found a Lagrangian which expresses a continuous symmetry (a U(1) symmetry) there must be an associated conserved quantity, which is the coupling constant which governs the strength of the interaction between the fields. This constant is proportional to electric charge so we have an account of why electric charge is conserved (as is, of course, empirically observed).

This is a case of rather magnificent mathematical intelligibility. Still more remarkable: if we reenact the search for local symmetries using this group theory methodology we find that the physics of the nuclear weak force (with a conserved quantity or charge called "isospin") and the nuclear strong force (with "color" conserved) can be derived. Sadly, if we try one more step on this ladder, perhaps looking for gravity or just some new fundamental force, it does not work. Schwichtenberg writes "Why is there no fundamental force following from SU(4)? Nobody knows!" (2015, p. 3).

Although the attainment of mathematical intelligibility is impressive and desirable, it comes at the cost of abandoning mundane intelligibility. If we are wondering what kind of world is described by these abstract mathematical structures, there is no anchor to any intuitive understanding of how they constitute the world as experienced, even as invisible underpinnings as in the mechanists' atomistic dreams. Instead, the link to experience is via a very long chain of calculation which allows us to make probabilistic predictions about experimental outcomes in highly contrived circumstances (e.g. inside the environment of the detectors at the Large Hadron Collider).

We might think to cut the distance between these highly abstract mathematical structures and the experienced world by applying them to a more specific physical theory. The core "application" is however quantum mechanics, notoriously unintuitive or positively at odds with customary ways of interacting with and understanding the material world. There are a great many examples of how strange quantum mechanics can seem. The case of non-interaction or interaction-free measurement is a familiar and simple but perennially mind boggling instance (see Elitzur and Vaidman 1993). For the vivacity of the example, imagine we have some bombs which are so sensitive that if exposed to any light, even a single photon, they will explode (let's say that normally they are packaged in a magic light proof box once they are verified as operational and they are later detonated by exposing them to light). But how to verify a bomb is not defective given that any verification procedure will seemingly have to expose it to light? Quantum mechanics permits this via the following arrangement:

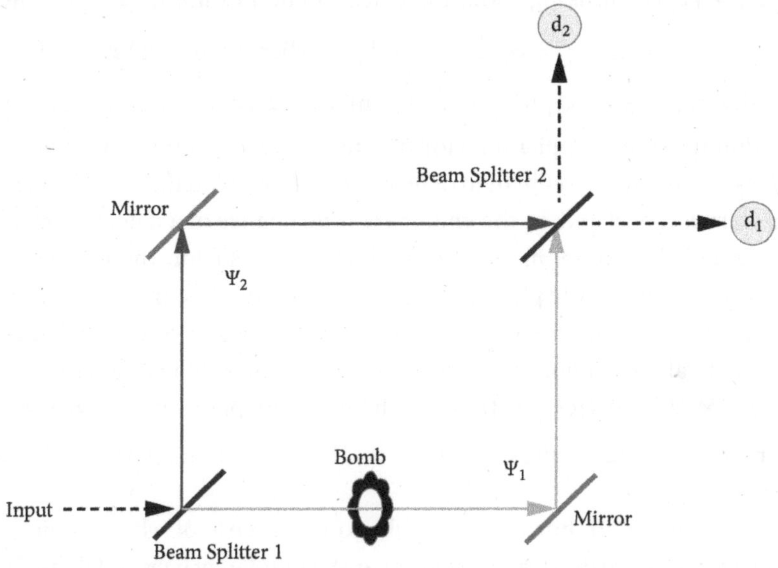

Here we have a Mach-Zehnder interferometer. We can input photons one by one and, if the bomb was not there, we will only ever measure an output photon at detector d_1. This is because the two paths are carefully arranged so that there will be destructive interference between the two possible paths to d_2. The Bomb then acts as a which-path detector IF it is an operational bomb; if not we set things up so the photon just continues on through the device. With such a detector in place the interference is eliminated and we have an equal chance of getting a detection at both d_1 and d_2. Since we are sending photons in one by one however, one quarter of the time a good bomb will not explode and we will get a result at detector d_2 whereas if the bomb is a dud we will never get such a result (clever adjustments to the system can increase our success rate substantially). That is, if whenever we get a result at d_2 we know we have a good bomb.

Familiarity with quantum mechanics should not make us forget that this is, on the face of it, absurd.[5] We have learned something about the Bomb without ever interacting with it (at least it certainly seems there has been no interaction because if there had been, the Bomb would have exploded).

[5] Another somewhat more complex example is the possibility of imaging an object by detecting the entangled partners of photons which interact with the object. The photons which do interact with the object need not be detected at all. See Lemos et al. (2014).

Needless to say, the mathematics of this situation is clear. We describe the the state of the system without the Bomb as a simple superposition: $\psi = \frac{1}{\sqrt{2}}(|\psi_1\rangle + |\psi_2\rangle)$. Thinking about the detector end of things, we can express ψ_1 as evolving thus: $\psi_1 \Longrightarrow \frac{1}{\sqrt{2}}(|d_1\rangle + |d_2\rangle)$ whereas the evolution of ψ_2 is instead this: $\psi_2 \Longrightarrow \frac{1}{\sqrt{2}}(|d_1\rangle - |d_2\rangle)$. The minus sign here arises because each reflection introduces a phase factor of i and there are two extra reflections (hence $i^2 = -1$) in the ψ_2 path to detector d_2 compared to the ψ_1 path. Hence the d_2 terms cancel and, in the end, there is no chance of a detection at d_2.

If we add the Bomb, in essence a which-way detector, the interference disappears. In effect, the photon is now going to either be in state $|\psi_1\rangle$ or in state $|\psi_2\rangle$ and each path has a probability of ½ to arrive at either detector. Mathematically, the photon + Bomb superposition, which is something like $\frac{1}{\sqrt{2}}((|\psi_t\rangle \otimes |B^t\rangle) + (|\psi_b\rangle \otimes |B^b\rangle))$, will couple its path interference terms to terms like $\langle B^t|B^b\rangle$ whose value is 0. The interference terms will thus disappear.

All well and good, but we would like to get a sense of what is going on *physically*. Sadly, we don't really have an answer and may have to fall back on the wisdom of Richard Feynman:

> I can safely say that nobody understands quantum mechanics. . . . I am going to tell you what nature behaves like. If you will simply admit that maybe she does behave like this, you will find her a delightful, entrancing thing. Do not keep saying to yourself, if you can possible avoid it, "But how can it be like that?" because you will get "down the drain," into a blind alley from which nobody has escaped. Nobody knows how it can be like that.
>
> (Feynman 1965, p. 129)

There does not seem to be an intuitive physical model of what is going on here, and in innumerable other of the puzzles which quantum mechanics poses to our common sense understanding of the world. It is, of course, this "disconnect" between mundane intelligibility and mathematical intelligibility that explains the plethora of interpretations of quantum mechanics.

There is another form of intelligibility which philosophers in particular have often appealed to and which might ameliorate this disconnect and help reveal the philosophical place of interpretations of quantum mechanics: *metaphysical* intelligibility. Leibniz is a perfect example. We have seen that Leibniz strongly endorsed the need for mundane intelligibility in the scientific endeavor. He firmly believed that everything that happens in the physical world has an intelligible mechanical explanation. But he certainly did not think that accepting the mechanical world view completed the metaphysical job of understanding the world as a whole (recall Sellars's dictum: "how things in the broadest possible sense of the term hang together in the broadest possible sense of the term").

Leibniz's views are complex and evolved significantly across his life. In what we might call his "middle period" Leibniz seems to have regarded the material world as something which existed in its own right, and within which mechanical explanations reigned supreme and were complete (for a discussion of Leibniz's evolving conceptions of matter see Garber and Rauzy 2004). In fact, in 1678 Leibniz could write that

> I recognize nothing in the world but bodies and minds, and nothing ... in bodies insofar as they are separated from mind but magnitude, figure, situation, and changes in these, either partial or total. Everything else is merely said, not understood; it is sounds without meaning.
>
> (Leibniz 1678/1976)

However, he soon came to see that the mechanical view was metaphysically puzzling and required an interpretation to render it fully intelligible. In the first instance, this was not couched in terms of his trademark monadic idealism but rather in terms of matter requiring an inward substantial being. Otherwise, infinitely divisible matter would resolve itself into nothing, or, as Brandon Look puts it: "if matter is infinitely divisible, then one can never arrive at the simple unities that must exist at some ontological ground level" (2017). Furthermore, Leibniz regards the pure passivity of mechanical matter as metaphysically objectionable. Matter requires a principle of activity to become real. In a remarkable passage, Leibniz writes:

> Primitive force, which is nothing but the first *entelechy*, corresponds to the soul or substantial form, but for this very reason it relates only to general causes which cannot suffice to explain phenomena,

and he says of this kind of force, irresistibly reminding us of the uncertainty principle:

> Primitive force, which is in all corporeal substance as such, since I believe that a body entirely at rest is contrary to the nature of things. (1695/1976)

Here we see Leibniz maintaining that mechanical explanations will suffice for empirical adequacy, but leave mysterious the deeper reason for things being the way they are in general. Lastly, Leibniz recognized that the laws of nature did not and could not follow from the mechanical idea of passive matter but must in some way be "imposed" upon it. Following general scientific opinion of the day, Leibniz naturally credited the laws to divine power.

As Leibniz's thought developed he moved towards a more overtly idealistic metaphysics in which minds alone were ontologically fundamental.[6] Perhaps it is a natural line of argument: if all the appearances remain the same, what reason is there to posit a material basis in addition to minds? Still more, if one can show that matter itself requires mind-like entities to underlie it, what purpose does matter serve in one's metaphysics? Leibniz had an additional argument however. He held the view that all relations can be reduced to, or deduced from, the intrinsic features of things. Relations are not exactly unreal on this view, but they are derivative—they supervene upon the intrinsic properties of genuine substances and form part of the total universe which monadic perceptions share from different viewpoints. As Benson Mates puts it, Leibniz's "nominalistic metaphysics provides no place in the real world for anything other than individual substances-with-accidents" (1989, p. 209). Once God puts into place these substances, everything else follows.

Such a metaphysics has a hard time dealing with space and time, which seem to support "pure relations" which are not fully dependent on the intrinsic nature of individual substances. Nothing about my intrinsic properties, for example, seems to reveal or entail that I am fifty miles from a burning barn (and adding the barn's intrinsic properties does not seem to help). Monadic idealism provides a beautiful answer to this problem: if all that exists are the monads and their coordinated perceptions of a spatial world, it will be possible to infer spatial relations in that world from the system of perceptions—that is, the intrinsic properties of the monads—after all.

But the point here is not to explore Leibniz's philosophy but to point out that, for him at least, the need for metaphysical intelligibility went beyond both mundane intelligibility (which the mechanical philosophy provided in abundance) and mathematical intelligibility (which the efforts of Galileo, Descartes, Newton, Leibniz himself and many others were revealing as a

[6] It is controversial to what extent Leibniz was always an idealist about the material world. See Adams (1994) for the case for a consistently idealist Leibniz; Garber (2009) for an evolving Leibniz.

deep abstract ground for the mechanical philosophy). Leibniz could have dreamt of the possibility of identifying these two forms of intelligibility: the mathematics would map exactly onto a mundanely intelligible picture of a world of matter characterized solely by motion and contact interaction. Subsequent developments dashed the hope for a pure mechanistic account, but not the general possibility of the identification of the worlds of mathematical and mundane intelligibility.

I have argued that in quantum mechanics and modern physics more generally we have lost any hope that mundane intelligibility will be preserved but that mathematical intelligibility has deepened to an almost unbelievable extent. It seems to me however that the need for metaphysical intelligibility is greatly exacerbated by this situation. In fact, the existence of the large number of interpretations of quantum mechanics (Wikipedia currently lists some eighteen) testifies to the desire for some way to make quantum mechanics intelligible. Obviously, mundane intelligibility is not going to be forthcoming so it looks like metaphysical intelligibility will have to do the job.

But the interpretations of quantum mechanics vary greatly on how metaphysical they are, ranging all the way from an anti-realism that denies that quantum mechanics describes nature at all to some kind of mind dependent idealistic account. A natural way to grade them is on a score of empirical testability: the less testable the more metaphysical and vice versa. For example, the dynamic collapse interpretation (see, e.g., Ghirardi et. al. 1986) is really more of a new scientific theory than a metaphysics of quantum mechanics which, in principle at least, would permit an empirical test of the spontaneous collapse hypothesis. The mysteries of quantum mechanics remain even if they are hidden in the "tails" (i.e. the low amplitude remnants of the wave function after induced collapse) the proper treatment of which remains unresolved (see McQueen 2015).

The Everett, or Many Worlds, interpretation perhaps presents a more viable effort at a genuine kind of metaphysical intelligibility. Very far from mundane intuition, it postulates that at every instant innumerable "worlds" or world-copies are being formed in exact correspondence to the forever foliating wave function of the universe. Every quantum mechanically possible evolution (from the presumed initial state of the universe) is equally real on this view. Two internal technical problems have dogged the Many Worlds interpretation since its inception: probability and classicality. The latter is the question: why does the world appear to be more or less as classical physics describes it? The theory of decoherence now seems likely to provide a viable answer to this problem (see Joos et. al. 2003; Wallace 2012). Given that the environment of any quantum system has a huge number of components ("degrees of freedom") it follows that large scale environment states, E_n, which register or measure the components of a superposition are orthogonal (or almost so). Somewhat as in the discussion of the Bomb detector above, the fact that $\langle E_n|E_m\rangle = 0$ will lead to the elimination (or at least suppression) of distinctively quantum interference phenomena.

The former, probability, problem is that of justifying the use of the Born rule to assign probabilities to experimental outcomes despite the presumed reality of *all* of the outcomes. Consider a quantum coin toss with outcome: $\sqrt{1/1000}|Heads\rangle + \sqrt{999/1000}|Tails\rangle$. Given that both *Heads* and *Tails* will occur, why should I assign one a greater probability than the other? The amplitudes appear otiose, a meaningless addition which make no difference to the way the world actually evolves. No approach to the probability problem has reached the level of consensus attained by the decoherence approach to the classicality problem. But efforts have been made to show that a decision theoretic analysis can support use of the Born interpretation as a guide to fixing one's own subjective probability assignments (see Deutsch 1999; Wallace 2007).

But a fundamental metaphysical problem remains: What is the nature of the world itself? The mundane intelligibility of the mechanical world view simply took at face value the quotidian perception of matter as massy, impenetrable, movable *stuff* and extended that into a mathematical elaboration. But the nature of matter now seems quite unintelligible. Instead, we have an extremely successful and precise recipe for determining what we are likely to observe (experience) in any situation to which we can apply a viable quantum description. These descriptions—for any moderately realistic case—are couched in terms of extremely or infinitely high dimensional spaces and seem to assign something like potentials for probabilities for measurements of various properties. If we take the description in full ontological seriousness, maybe we should accept that reality is exactly as thus portrayed (see Albert 2013, and many of the other essays in Ney and Albert 2013). But what exactly we are being asked to believe about the nature of the world is not clear, save that it generates correct expectations about forthcoming experiences, whatever those are exactly.

Could more radical metaphysical speculations increase intelligibility? What if we take experience itself as in some way fundamental? Certain interpretations of what is called QBism might suggest something like this.

QBism is the view that quantum mechanics is nothing but a "mathematical machine" for assigning subjective probabilities in a rational and empirically warranted way. On this view, quantum mechanics is not a description of the world at all. The nature of the world will be approached in some way that permits this view of the role of quantum mechanics. Chris Fuchs, one the chief proponents of QBism, draws on and quotes the philosophy of William James here:

> The case QBism makes before the forum is this. What the quantum agent—the protagonist in the drama of any application of quantum theory—is ultimately doing is hitching a ride with a new kind of ontology or metaphysics. An ontology of all-pervasive, pan-creative experience: "My thesis is that if we start with the supposition that there is only one primal stuff or material in the world, a stuff of which everything is composed, and if we call that stuff "pure experience," then knowing can easily be explained as a particular sort of relation towards one another into which portions of pure experience may enter. The relation itself is a part of pure experience; one of its "terms" becomes the subject or bearer of the knowledge, the knower, the other becomes the object known". (2017, p. 290; the James quote is from "Does Consciousness Exist?" (1904)).

It is vital to recognize here that James's term "pure experience" does not refer, in the first instance, to mentalistic phenomena. Rather, roughly speaking, it refers to what is "present to us" or what confronts us in experience independent of our ways of conceptualizing it. As James would have it, both matter and mind are "constructed" of the same stuff which is non-mental but directly accessible, which is why his view came to be known as "neutral monism".[7] Given the mystery of the nature of the world that quantum mechanics leaves us with, a view that couples subjective Bayesianism with a metaphysics of "pure experience" seems a promising route towards increased metaphysical intelligibility and one that naturally conjoins with the tremendous mathematical intelligibility already in place.

But instead of exploring this further, I want to conclude with a brief look at the views of David Bohm because in this context they are fascinating and

[7] An important follower of James was Bertrand Russell (see 1927). Neutral monism is having something of a comeback in philosophy these days as evidenced by Alter and Nagasawa (2015). An interesting application of Russellian neutral monism to quantum mechanics can be found in Lockwood (1989).

somewhat peculiar. Unlike QBism, the Bohmian interpretation of quantum mechanics is one that is staunchly ontological and can be seen—initially—as an attempt to retrieve some mundane intelligibility for the quantum realm. As is very well known, Bohm, following Louis de Broglie's pilot-wave theory of 1925, managed to show there was a version of (at least non-relativistic) quantum theory that permitted particles of matter to have both position and momentum, apparently contrary to orthodox views (the view was especially unorthodox at the time Bohm was beginning to explore his theory, the early 1950s).[8]

So we might expect the Bohmian picture is one of little bits of matter flying about interacting with each other, an obvious echo—with some extra forces thrown in—of the mechanical world view. I guess something like this picture underlies David Albert's remark that "the metaphysics of [Bohm"s] theory is exactly the same as the metaphysics of classical mechanics" (1992, p. 134) and, perhaps, Fuchs's dismissive: "Bohmism" represents a hopeless "return to the womb of classical physics... yuck!" (2011, p. 15). But that can't be right since the linchpin idea of the mechanical view is that given the position and velocities of all the particles at one time, the history of the universe is completely determined for all other times, past and future. The Bohmian addition of the "quantum potential" or the "guiding wave" is an ineliminable and radical difference from classical conceptions:

> 1. The quantum potential does not produce, in general, a vanishing interaction between two particles as the distance between those particles becomes very large. Thus two distant systems may still be strongly and directly connected....
> 2. [T]he quantum potential cannot be expressed as a universally determined function of all the co- ordinates of the particles. Rather it depends on the "quantum state" $\psi(r_1 \ldots r_n)$ of the system as a whole. This means that even if at some time the positions and momenta of two sets of particles are the same, but they are in different quantum states, then their subsequent evolution can be very different. (Hiley and Peat 2012)

[8] At that time, there were "proofs," such as that of John von Neumann which seemed to show that no hidden variable theory could be consistently developed. Hence we have John Bell's famous reaction to Bohm: "in 1952 I saw the impossible done. It was in papers by David Bohm" (1987, p. 160). An interesting counterpoint to the contemporary denigration of von Neumann's proof is Dieks (2017).

I don't think the Bohm ever thought that his project was a return to some classical-like picture of reality; his 1952 paper offered a "proof of concept." His attempt at metaphysical intelligibility went in quite a different direction.

One core metaphysical idea of Bohm is universal holism which we can express with the slogan that the whole is prior to the parts, rather than vice versa.[9] Very early on, Bohm writes that the "entire universe must, on a very accurate level, be regarded as a single indivisible unit in which separate parts appear as idealizations permissible only on a classical level of accuracy of description" (1951, p. 167). Much later he wrote, more exotically: "We may regard each of the "particles" constituting a system as a projection of a "higher-dimensional" reality, rather than as a separate particle, existing together with all the others in a common three-dimensional space" (1980, p. 238) and, again:

> Ultimately, the entire universe (with all its "particles", including those constituting human beings, their laboratories, observing instruments, etc.) has to be understood as a single undivided whole, in which analysis into separately and independently existent parts has no fundamental status. (1980, p. 221)

This is more than "mere" non-locality which is compatible with a picture of separable individuals that have some kind of mysterious communicative link between them. Instead of thinking of the components as independent existents interacting in various ways, we find that "each element participates irreducibly in all the others" (Bohm and Hiley 1993, p. 177)—a symptom that their apparent individuality is an illusion.

This is not to say that the particle picture should be jettisoned. Under most conditions, it is practically effective. Even if the particles are not fundamentally real but rather a "projection" of a deeper reality:

> Under the ordinary conditions of our experience, these projections will be close enough to independence so that it will be a good approximation to treat them in the way that we usually do, as a set of separately existing particles all in the same three-dimensional space. (Bohm (1980), p. 239)

9 Recent philosophical work has clarified the nature of so-called priority monism: the idea that the whole is metaphysically prior to the parts and linked it to quantum mechanics as well (see Schaffer 2010; Ismael and Schaffer 2020).

Bohm had an idea about what underlies universal holism. Obviously, it cannot be mere causal inter-relatedness. Causation is naturally conceived as a relation between independently existing events. Furthermore, we already know that quantum mechanics demands connections which are non-causal. Bohm postulated that *information* was a fundamental feature of reality supporting and/or expressing holism. The quantum wave function is thus a field of mutual information.

This is not the kind of information now familiar from information theory, which Bohm and his long time collaborator Basil Hiley call "Shannon-information" in contrast to the metaphysically significant "active information" which underpins quantum phenomena and perhaps much else. Shannon information is a mere relational structure which requires interpretation to reveal the meaning or significance of the carried messages. Modern information theory is a theory of information transmission, not semantic content. Active information by contrast is semantically significant; it carries intrinsic meaning that does not demand any "outside" interpretation. Bohm and Hiley frequently put this point in terms of the information being 'for the particle' (objective) versus "for us" (subjective, interpretation based).

Bohm and Hiley also use the locution of form and matter to explicate this, speaking of active information as able to "actively... put form into something or to imbue something with form" (Bohm and Hiley 1993, p. 35). Although Bohm and Hiley typically use causal metaphors to explicate the notion of active information, as in the example of weak radio waves modifying the course of a gigantic cargo ship, active information is not a kind of causal relation, since, as noted, it can act independent of distance and its influence need not fall off with distance. The amplitude, *R*, of the wave function does figure in the quantum potential: $Q = -(\hbar^2/2m)\nabla^2 R/R$, but clearly its magnitude is irrelevant; only its "form" matters.

If we recall the Elitzur-Vaidman Bomb tester, it is the presence of information about the possible paths within the apparatus and its availability throughout the system which accounts for the otherwise highly peculiar apparent fact that, so to speak, what happens in one path depends on what *could* happen in the other. It is a quite natural metaphor to say that the photon in one path "knows" the situation in the other path. This is not explicable in terms of a "bit channel" or a causal link: the arms of the interferometer could be widely (space like) separated and the Bomb could—in principle—be added *after* the "particle" is in the one path or the other.

If we find the idea of intrinsic information acceptable, we have at least gone some way toward providing a picture which provides metaphysical intelligibility to the quantum world. That is, although very far from our ordinary picture of material reality, a pervasive holistic system of active information does go some way to explain why the quantum world acts as it does.

However, the idea of brute intrinsic information remains mysterious. I think this mystery did prod Bohm, very much in collaboration with Hiley, to extend his metaphysical system further, and, in a way, toward a more Leibnizian view. If we ask where in the world do we find intrinsic semantic information there is an obvious, perhaps unique, answer: the conscious mind. Arguably, the assignment of some information to any non-mental state requires interpretation, or the possibility of interpretation, by a conscious subject. There is nothing intrinsic to a bit string that reveals whether it is a system of missile guidance or a video representation of a Gilligan's Island rerun.[10]

It might thus be natural to seek some primitive mental aspect, some basic kind of proto-consciousness, which imbues nature with the semantically significant and intrinsic information which quantum holism suggests lurks behind the appearance of the physical world. And, indeed, we do find Bohm heading, more or less tentatively, in that direction. In a rather cryptic passage which we can read in terms of the present hypothesis, Bohm writes that

> Reality can be considered as in essence a set of forms in an underlying universal movement or process... Thus, the way could be opened for a world view in which consciousness and reality would not be fragmented from each other. (1980, p. xiv)

But if we want a more forthright endorsement of universal mentality, we can find it: "the particles of physics have certain primitive mind-like qualities" (1990, p. 272). About active information, Hiley and Pylkkänen explicit link it to some form of consciousness when they write that "if we now interpret active information... by assuming that it has both a phenomenal and a

[10] Here we approach the deep technical problem of intensionality: although we can think the difference between "equiangular triangle" and "equilateral triangle" these two concepts are logically equivalent, so it is hard to see how nature could manage to assign one meaning rather than the other to distinct physical systems. This problem was notoriously explored in Fodor and Piattelli-Palmarini (2010); see also Hornstein (2010).

physical aspect, we have a concept of information which is both concrete and fundamental" (2001, p. 134).[11]

Now we have circled back again to Leibniz. He found that adding a fundamental mentalistic aspect to reality was required to secure metaphysical intelligibility. His reasons included the worry that matter, as it had to be within the mechanical world view (passive, intrinsically inert, infinitely divisible) lost its claim to be an intelligible ground of nature. Mind's activity and intrinsic representational capacity (something very close if not identical to intrinsic semantically significant information) restored intelligibility to his picture of nature. Similarly, Bohm finds that matter, as it turns out to be in the quantum mechanical picture (lacking individuality, universally interconnected, of an unknowable nature) similarly is inadequate to stand as a metaphysically intelligible underpinning for reality. Like Leibniz, Bohm moves towards a view which puts mentality, as the bearer of intrinsic information, as a fundamental feature of the world, whose attributes go at least some way towards providing the metaphysical intelligibility needed to buttress and complete the mathematical intelligibility so evident in modern physics. Whether Bohm's vision is the best way to provide a metaphysically intelligible basis for quantum theory is unclear—as unclear as the ongoing struggle among the plethora of interpretation of quantum mechanics.

References

Adams, Robert (1994). *Leibniz: Determinist, Theist, Idealist*. Oxford: Oxford University Press.

Aiton, Eric (1972). *The Vortex Theory of Planetary Motions*. London: Macdonald.

Albert, David (1992). *Quantum Mechanics and Experience*. Cambridge, MA: Harvard University Press.

Albert, David Z (2013). "Wave function realism." In A. Ney and D. Albert (eds.), *The Wave Function: Essays on the Metaphysics of Quantum Mechanics*, pp. 52–7. Oxford: Oxford University Press Oxford.

Alter, T. and Y. Nagasawa (eds.) (2015). *Consciousness in the Physical World: Perspectives on Russellian Monism*. Oxford: Oxford University Press.

Bangu, Sorin (2012). *The Applicability of Mathematics in Science: Indispensability and Ontology*. Houndmills: Palgrave Macmillan.

[11] It should be noted that Bohm did not want to say that the particles were conscious (see 1990). I think his worry stems from having a very demanding conception of consciousness as a highly sophisticated mental process. Something like Hiley and Pylkkänen's notion of "phenomenality" is a suitably weaker form of mentality which retains the essential link to consciousness (for more on this see Seager 2018).

Bell, John (1987). *Speakable and Unspeakable in Quantum Mechanics*. Cambridge: Cambridge University Press.

Bohm, David (1951). *Quantum Theory*. Englewood Cliffs, NJ: Prentice-Hall.

Bohm, David (1980). *Wholeness and the Implicate Order*. London: Routledge and Kegan Paul.

Bohm, David (1990). "A new theory of the relationship of mind and matter." *Philosophical Psychology*, 3 (2): pp. 271–86.

Bohm, David and Basil Hiley (1993). *The Undivided Universe: An Ontological Interpretation of Quantum Mechanics*. London: Routledge.

Deutsch, David (1999). "Quantum Theory of Probability and Decisions." *Proceedings of the Royal Society of London A*, 455 (1988): pp. 3129–37.

Dieks, Dennis (2017). "Von Neumann's impossibility proof: Mathematics in the service of rhetorics." *Studies in History and Philosophy of Science Part B: Studies in History and Philosophy of Modern Physics*, 60: pp. 136–48.

Elitzur, Avshalom C and Lev Vaidman (1993). "Quantum mechanical interaction-free measurements." *Foundations of Physics*, 23 (7): pp. 987–97.

Feynman, Richard (1965). *The Character of Physical Law.* Cambridge, MA: MIT Press.

Fodor, Jerry and Massimo Piattelli-Palmarini (2010). *What Darwin Got Wrong.* New York: Farrar, Straus & Giroux.

Fuchs, C.A. (2017). "Notwithstanding Bohr, the Reasons for Qbism." *Mind and Matter*, 15 (2): pp. 245–300.

Fuchs, Christopher A. (2011). *Coming of Age With Quantum Information: Notes on a Paulian Idea.* Cambridge: Cambridge University Press.

Galilei, Galileo (1623/1957). "The Assayer." In D. Stillman (ed.), *Discoveries and Opinions of Galileo*, pp. 229–80. New York: Anchor Books. D. Stillman (ed. and trans.).

Garber, Daniel (2009). *Leibniz: Body, Substance, Monad.* Oxford: Oxford University Press.

Garber, Daniel and Jean Baptiste Rauzy (2004). "Leibniz on Body, Matter and Extension." *Proceedings of the Aristotelian Society, Supplementary Volumes*, 78: pp. 23–40.

Ghirardi, Gian Carlo, Alberto Rimini et al. (1986). "Unified dynamics for microscopic and macroscopic systems." *Physical Review D*, 34 (2): pp. 470–91.

Hiley, Basil and F. David Peat (eds.) (2012). *Quantum Implications: Essays in Honour of David Bohm.* London: Routledge.

Hiley, Basil and Paavo Pylkkänen (2001). "Naturalizing the Mind in a Quantum Framework." In P. Pylkkänen and T. Vadén (eds.), *Dimensions of Conscious Experience*, pp. 119–144. Amsterdam: John Benjamins.

Hornstein, Norbert (2010). "An Outline of the Fodor and Piattelli-Palmarini Argument against Natural Selection." *Biolinguistics*, 4 (4): pp. 382–4.

Ismael, Jenann and Jonathan Schaffer (2020). "Quantum holism: nonseparability as common ground." *Synthese*, 197: pp. 4131–60.

James, William (1904). "Does "Consciousness" Exist?" *Journal of Philosophy, Psychology, and Scientific Methods*, 1: pp. 477–91.

Joos, Erich, H. Dieter Zeh et al. (2003). *Decoherence and the Appearance of a Classical World in Quantum Theory.* Berlin: Springer-Verlag, 2nd ed.

Leibniz, G. W. (1678/1976). "Letter to Herman Conring." In L. Loemker (ed.), *Gottfried Wilhelm Leibniz: Philosophical Papers and Letters* (2nd ed.), pp. 186–91. Dordrecht: Reidel.

Leibniz, G. W. (1695/1976). "Specimen Dynamicum." In L. Loemker (ed.), *Gottfried Wilhelm Leibniz: Philosophical Papers and Letters* (2nd ed.), pp. 435–52. Dordrecht: Reidel.

Leibniz, Gottfried Wilhelm (1710-16/1989). "Against Barbaric Physics: Toward a Philosophy of What There Actually Is and Against the Revival of the Qualities of the Scholastics and Chimerical Intelligences." In R. Ariew and D. Garber (eds.), *G. W. Leibniz: Philosophical Essays*, pp. 312–20. Indianapolis: Hackett. (Written sometime between 1710-16.).

Lemos, G.B., V. Borish et al. (2014). "Quantum imaging with undetected photons." *Nature*, 512 (7515): p. 409.

Lockwood, Michael (1989). *Mind, Brain and the Quantum*. Oxford: Blackwell.

Look, Brandon (2017). "Gottfried Wilhelm Leibniz." In Edward N. Zalta (ed.), *The Stanford Encyclopedia of Philosophy*. Metaphysics Research Lab, Stanford University. https://plato.stanford.edu/archives/sum2017/entries/leibniz/.

Mates, Benson (1989). *The Philosophy of Leibniz: Metaphysics and Language*. Oxford: Oxford University Press.

McQueen, Kelvin J (2015). "Four tails problems for dynamical collapse theories." *Studies in History and Philosophy of Science Part B: Studies in History and Philosophy of Modern Physics*, 49: pp. 10–18.

Mermin, David (1998). "What is Quantum Mechanics Trying to Tell Us?" *American Journal of Physics*, 66 (9): pp. 753–67.

Newton, Isaac (2004). *Isaac Newton: Philosophical Writings*. (Andrew Janiak, ed.). Cambridge, UK: Cambridge University Press.

Ney, Alyssa and David Z Albert (2013). *The Wave Function: Essays on the Metaphysics of Quantum Mechanics*. Oxford: Oxford University Press.

Russell, Bertrand (1927). *The Analysis of Matter*. London: K. Paul, Trench, Trubner.

Schaffer, Jonathan (2010). "Monism: The priority of the whole." *Philosophical Review*, 119 (1): pp. 31–76.

Schwichtenberg, Jakob (2015). *Physics From Symmetry*. New York: Springer.

Seager, William (2018). "The Philosophical and Scientific Metaphysics of David Bohm." *Entropy*, 20 (7).

Sellars, Wilfrid (1963). "Philosophy and the Scientific Image of Man." In *Science, Perception and Reality*, pp. 1–40. London: Routledge and Kegan Paul.

Siegel, Daniel M (2003). *Innovation in Maxwell's Electromagnetic Theory: Molecular Vortices, Displacement Current and Light*. Cambridge, UK: Cambridge University Press.

Silliman, Robert H. (1963). "William Thomson: Smoke Rings and Nineteenth-Century Atomism." *Isis*, 54 (4): pp. 461–74.

Smeenk, Christopher (forthcoming). "Cosmology and Physical Astronomy in Newton's General Scholium." In Forthcoming S. Snobelen, S. Ducheyne and S. Mandelbrote (eds.), *Isaac Newton's General Scholium: Science, Religion and Metaphysics*.

Thomson, William (1911). In *Mathematical and Physical Papers, Vol. 4*, pp. 539–44. Cambridge University Press.

Thomson, William (Lord Kelvin) (1867). "On Vortex Atoms." *Proceedings of the Royal Society of Edinburgh*, 6: pp. 94–105.

Wallace, David (2007). "Quantum Probability from Subjective Likelihood: Improving on Deutsch's Proof of the Probability Rule." *Studies In History and Philosophy of Science Part B: Studies In History and Philosophy of Modern Physics*, 38 (2): pp. 311–32.

Wallace, David (2012). *The Emergent Multiverse: Quantum theory According to the Everett Interpretation.* Oxford: Oxford University Press.

Wigner, Eugene (1960). "The Unreasonable Effectiveness of Mathematics in the Natural Sciences." *Communications on Pure and Applied Mathematics*, 13: pp. 1–14. Reprinted in Wigner's *Symmetries and Reflections*, Bloomington: Indiana University Press, 1967.

17

On the Place of Qualia in a Relational Universe

Lee Smolin

1. Introduction

This paper addresses the mind-body problem. More specifically, what Chalmers has called "The Hard Problem" of consciousness [1]. The problem is why conscious perceptions, such as the experiences of the colors red or blue, are associated with particular physical processes—namely excitations of neurons in the cerebral cortex—whose complete description in terms of the laws of physics does not and indeed cannot reference these subjective aspects of experience. The languages the neurosciences give us so far, such as synaptic potentials, electric currents, neurotransmitters, etc. explain a lot about how neurons work, but do not seem to have the capability of either describing or explaining, why or how these relate to the subjective mental experiences we all perceive.

The point has been put several different ways, from Leibniz's mill [2], to Jackson's knowledge argument [3], featuring Mary the color-blind neuroscientist to Chalmer's zombie argument [1]

Leibniz, in his Monadology, put the point as well as any:

> It must be confessed, moreover, that perception, and that which depends on it, are inexplicable by mechanical causes, that is, by figures and motions, And, supposing that there were a mechanism so constructed as to think, feel and have perception, we might enter it as into a mill. And this granted, we should only find on visiting it, pieces which push one against another, but never anything by which to explain a perception. This must be sought, therefore, in the simple substance, and not in the composite or in the machine.

Lee Smolin, *On the Place of Qualia in a Relational Universe* In: *Consciousness and Quantum Mechanics.*
Edited by: Shan Gao, Oxford University Press.
DOI: 10.1093/oso/9780197501665.003.0018

But I should hasten to add that while I address the "hard problem," I do not solve it. What I aim to do is to solve the "next hardest problem," which is 1) to identify those *physical processes, events or states* that are "associated with" or "correspond to" instances of conscious perception (we will call these PCC for physical correlates of consciousness), and 2) explain some of the characteristic features of conscious perception, to be described below, in terms of the features or structure of these correlates of consciousness. While I do not solve this either, I have some specific proposals and hypotheses to make concerning the physical correlates of consciousness.

These proposals situate the search for an approach to the mind-body problem within the search for what I call a *completion of quantum mechanics.* This is what is sometimes called a *non-local hidden variables theory*: it is a realist description of precisely what goes on in each individual event or process, which reduces to quantum mechanics in a certain limit and averaging procedure. For reasons I described elsewhere [21, 22, 30], I confine the search to completions of quantum mechanics which are relational, so that they address also issues in quantum gravity.

1.1 Brief Summary of the Proposals Made Here

In this chapter I look to find the correlates of consciousness within the framework of a recently proposed completion of quantum theory, called *the causal theory of views* [27]. This incorporates elements of two other frameworks, the real ensemble formulation of QM [27, 48, 49], and the energetic causal set framework, which we developed with Marina Cortes [51–54, 56]. I also look to another proposed completion of quantum mechanics, called the principle of precedents [28], and find there another suggestion for the PCC.

To anticipate in the simplest possible terms, the key ideas are first, that the universe is constructed from nothing but a collection of *views* of events, where the view of an event is what can be known about that event's place in the universe from what can be seen from that event [27]. In other words, the beables of this theory are views from events, comprised of the energy, momentum, and other conserved charges that combine to create the event, from its causal predecessors. Within such an ontology of views, it seems natural to propose that instances or moments of conscious experience are aspects of some views. That is, an elementary unit of consciousness is not a single qualia, but the entire of a partial view of the universe, as seen from one event.

The second proposal restricts the views that are associated with consciousness to within a very small set. Most events and their views are common, in that they have many near copies in the universe. Most events could also be called routine, in that nearly identical views appear numerous times within their causal pasts. I then propose that these common and routine views have no conscious perceptions. Then, there are a few, very rare views which are unprecedented, which are having their first instance, or are unique, in that they have no copies in the universe. I propose it is those few views of events, which are unprecedented, and/or unique, and are hence *novel*, which are the physical correlates of conscious perceptions.

Elsewhere [30, 59, 64], I have argued in detail for the view that quantum mechanics is incomplete, in that it does not give a full description of, nor explanation for, all individual physical processes, independent of our knowledge, or interventions. This implies the need for a "realist" completion of quantum mechanics. I will not repeat these arguments here. Of course, a completion is also required for unifying gravity, spacetime and cosmology into the rest of physics. The causal theory of views is intended to be a completion in both senses.

The causal theory of views is the present stage of a research program I have been carrying out since the 1980s to construct such a realist double completion of quantum mechanics and general relativity [59–64]. The common theme, motivating all these attempts, has been that of a *relational hidden variables theory* [59–64]. This situates it within the relational tradition developed by Leibniz, Mach, Einstein, and others, according to which physical degrees of freedom describe dynamically evolving relationships amongst particles, fields or other dynamical actors, rather than being defined against a fixed, non-dynamical background structure such as absolute space. My aim has been that this would be the case also for the additional "hidden variables" needed to complete the quantum dynamics.

I simultaneously developed the idea that the relational variables which completed quantum mechanics would also underlie a discrete formulation of spacetime, so that the full theory would be a completion of both quantum mechanics and general relativity, i.e., it would also be the quantum theory of gravity. So, I like to think that it is no coincidence that the basic tools for describing a system of relations, namely, networks, and matrices, occur in all background independent formulations of quantum gravity [20–26] as well as in the various relational hidden variable theories I and others introduced [59, 64, 66, 67].

All of these developments, beginning with several formulations of relational hidden variables theories [59–64], and developing into the real ensemble formulation [27, 48, 49], the principle of precedence [28] and the causal theory of views [27] are each described in detail in previous publications [30]. The causal theory of views, is also a development of a class of models of causal spacetime, called Energetic Causal Sets, developed in a series of papers with Marina Cortes [51–54, 56].

To summarize briefly this already brief summary, taking the relational philosophy seriously—in a way that lets us read the Principle of the Identity of the Indiscernible as a dynamical principle (see below)—makes possible two concepts on which to base a theory: First is the proposal of taking the beables of a relational theory to be the views of events. Second is the possibility of making a physical distinction between common and routine states, on the one hand, and novel and unique states, on the other. The whole point of this chapter is that a relational theory that incorporates both ideas offers a possible setting for bringing qualia and consciousness into physics. The physical correlates of consciousness are the novel or unique views of events.

The proposals I describe here are made within the language and frameworks of the specific theories, just mentioned. This does not rule out the possibility of implementing them in the context of another approach to quantum foundations. But at present I only know how to express them within these, closely related frameworks.

1.2 Some Comments on the Mind-body Problem

Before getting to my specific proposals, I make some general comments on the mind body problem.

First of all, I take the existence of our conscious perceptions such as colors and tones-which are often called qualia, to be a fact about the natural world [6], to be naturally incorporated into, and even explained by, any fully complete theory of the natural world [4–6]. Further, I take various features and characteristics of our experience (to be discussed below) as phenomena to be explained by the theory.

I should hasten to add that the suggestions I make here are all within the paradigm of naturalism, so that I assume that qualia and other aspects of our conscious experiences are aspects of the natural world, alongside energy and charge. Here I follow a number of philosophers,[1] including William James [10], Russell [11], Eddington [12], Strawson [6] and Goff [4, 5].

I would also consider these suggestions to be within the constraints of physicalism, if by that we mean the assumption that all our understanding of the natural world is eventually to be grounded in (which is to say, possibly emergent from) the final, correct complete theory.[2] of physics [4].

My thinking on these questions was to a large extent inspired by the panpsychists, or Russellian neutral monists, whose proposal that qualia are universal aspects of the intrinsic or fundamental nature of matter, seemed to me, for a long time, to be the most plausible view [13]. The neat trick of making qualia causally null, by making them "inner aspects" of all physical processes, and so not conflict with the causal completeness of the standard laws of physics, in spite of being real, seemed to me too perfect a solution to the body-mind problem not to take seriously.[3]

But when you take it seriously, it raises a number of puzzles which seem hard to address, given the assumptions made. Among them, if all physical processes have intrinsic aspects which are qualia, what distinguishes those bundles of qualia we experience from the qualia associated to the myriads of processes that unconsciously fill our heads? Following this are a number of questions about the particular ways our conscious experience appears to be structured which seem hard to address, given the claimed universality of the merging of physical and experienced aspects of nature.

The monist step of seeing qualia as aspects of physical processes seems promising. It is the assumption that the correspondence is universal that seems to land us in trouble.

The question I want to raise in this chapter is then whether we might find a more explanatory theory if we keep the monist-dual aspect ontology, but restrict the association of qualia and consciousness to a restricted subset of physical processes, events or states, where the restriction was based on purely physical properties? The proposals I make here are intended as first steps in this direction.

The view I propose here might be called a kind of *restricted panpsychism*. It departs from the usual formulation of panpsychism on several points.

[1] A very clear introduction to the philosophy of "neutral monism" espoused by these philosophers is in the book of Rodolfo Gambini [9].

[2] Note that I say theory here, rather than laws, because I believe that the laws must be restricted to subsystems and evolve on cosmological scales [31–33]. Indeed, will suggest below that the physical correlates of consciousness may be related to the evolvability of the laws.

[3] For a good discussion, see [14].

It does not propose that qualia are associated with all material states or processes. Instead, I propose that there are specific, generally rare, aspects or features of certain physical processes, which allows them to be correlates of consciousness.

We are thus challenged to discover a *selection principle* which is a physically based criteria which distinguishes the small minority of events, or states, or processes, to which correspond conscious perceptions. These will be **physical** *correlates of consciousness* (PCC).

This strategy is not new; a prior example is *integrated information theory*, which proposes a certain physical quantity, *the integrated information*, as a measure of, or selection principle for, consciousness [43].

Note here something important: Neuroscientists look for **neural** *correlates of consciousness* (NCC) [34–37], and our program will gain a lot by their success. But, we should expect that neural correlates of consciousness should be such because they are composed partly of physical processes which are physical correlates of consciousness!

Among the things we would like to explain are structural questions about consciousness, some of which are the following:

1. **The "scenef" or "bundling" problem:** Why are qualia never perceived singly, but only bundled together with others into a frame or scene?
2. **The "viewpoint" or "self" problem:** Why is our conscious experience shaped in ways that appear to depend on our prior beliefs and intensions? That is, why does our consciousness seem to reflect a point of view of the world around us, more an active probe than passive recipient of "raw sensations."
3. **The "presentism" problem:** Why does that scene bundled together represent, approximately to be sure, a thickened (i.e., of some small duration) moment of time.
4. **The "unique self" problem:** Why does there seem to be only a single unique framed or bundled scene for every human brain, at each time? Why not many? This is especially puzzling given that the brain is simultaneously running a great many parallel processes, most of which we are unaware.
5. **The structural mismatch problem** [18]: Goff makes the point clearly [4, p. 203]:

   On the face of it, the structure of the brain seems radically unlike the structure of the.... conscious mind, as revealed through introspection. But if the mind and the brain are identical, or at least grounded in the same micro level base, one might expect them to have the same structure

Or, as Edelman and Tononi emphasize, the majority of neural processing in the brain is never experienced or perceived [34]. It is the job of a physical theory to explain why some are and most aren't, not just by describing the NCC, but by explaining on the basis of a physical selection principle why the NCC are, such, because they incorporate the PCC.

6. **The modality problem**[4]: Why are different modalities, such as sight, sound, pain, smell, etc. experienced as qualitatively quite different, in spite of their corresponding NCC being quite similar?
7. **The "linear scale" problem:** Why do several of the modalities, while qualitatively distinct, distinguish within themselves by a linear scale? Why is this true of somemodalities (vision and hearing), while not true of others (olfaction, touch)?

Panpsychists are exploring ways to address these questions, this has led to a large literature, mostly in philosophy of mind, that I am not sufficiently a scholar of to summarize.[5] But it does seem to me that the obstacles are challenging. Because they must assume that all physical processes are PCC, they have a hard time explaining why all neuronal processes are not experienced, or why the tiny fraction that are experienced, are those that are structured as they are.

That is—and this seems to me an important fact—there is nothing easier than imaging instances of conscious experience which are never experienced by us, to start with experiences of pure color or sound, or experiences of higher level organizations like species or families. Pan-psychism starts with a disadvantage, because they are committed to believing all forms and organizations of matter have or manifest internal aspects which are instances of consciousness. There is, if they are right, a universe of experience that they must explain how and why is disconnected from ours.

Any proposal for such a selection principle is bound at this point to be highly speculative and the ones that follows are no exception.[6] However, even

[4] Also called the "Palette problem [4, pp. 193–202].

[5] A good introduction is the recent book [4].

[6] One that comes to mind as a partial, if not yet successful, precedent for what I am about to propose is the Penrose-Hammeroff proposal [39], that consciousness has something to do with

if, as is likely at the present state of knowledge, any particular proposal for the selection principle or PCC turns out to be wrong, I would suggest that the search for them is likely eventually to lead to success.

It seems to me unlikely that qualia can be grounded in the classical, i.e., non-quantum side of physics. For one thing, I don't see that they have anything to offer for a selection principle for addressing the structural questions just mentioned. Consider the structural mismatch problem, applied to proposals for a PCC. If our conscious experience is not structured anything like the brain, it is even less structured like a classical physical system. We know how to reduce the description of a metal to the physics of its atoms, but as we have remarked (the self problem) conscious experience has no atoms of experience it can be reduced to.

But as pointed out by several authors, most recently by Chalmers and McQueen [46] and Gambini and Pullin [47], the ontology of quantum states does seem to be structured in some aspects like a conscious experience. Due to the entanglement, or non-separability of quantum states, such states have properties attributable only to the entire of a composite system, which are not reducible to properties of the parts. The way in which single colors, motions and sounds are woven together into a single framed conscious moment, seems to have an at least partly similar character. It has also been pointed out that both conscious states and quantum states seem vulnerable to an instability of observation, in that measuring or observing them necessarily changes them.

On the other hand, if conscious states may seem in some aspects like non-separable or entangled states, we seem never to experience a conscious state which is a superpositions of two or more conscious states. It seems true that in all the conscious states we experience, everything which is part of our view has a definite value. This has led to a long tradition according to which consciousness is proposed to be responsible for, or in some way associated with collapse of the wave-function, so that we always experience the results of our observations to be single, definite values [44–47].

Indeed, Penrose and Hammeroff [39], Chalmers and McQueen [46], Gambini and Pullin [47], and others have already put forward proposals for how consciousness may play a role in the domain of possible completions of quantum phenomena. The proposals I make here are similar.

dynamical collapses of the wavefunction in a theory where that has become a physical process, where those events are taking place within microtubules of neurons in the brain. But see also [40] for criticism of specifics of this proposal.

2. The Context: Relational Physics

The relational tradition, which motivates the theories within which my proposals concerning PCC are made, is based on a few simple ideas. To put what follows in context, I state the main ideas in this short section.

2.1 Basic Principles of Relational Physics

The following are all different ways to state the idea of relationalism [15–17, 20–22].

- *The principle of the identity of indiscernible* (PII) asserts that two individuals (events) with exactly the same view of the rest of the universe are to be identified [15, 16].

  What is particular about the following is that I aim to achieve this dynamically, through a term in the action that seeks to maximize a measure of the diversity of views [27, 49], which we call the variety [50].

  Thus, one can say this theory is one that takes the PII seriously, and that this makes possible the proposals concerning consciousness I make here.

  A consequence is the following.
- *Individuals are identified through their relations.* Individuals, whether events, particles, or subsystems, don't have intrinsic names or labels or coordinates. They are distinguished and named only by what they know about the rest of the system through their interactions or shared relationships, i.e., by what they see when they look around This is expressed as the *view of that individual* [15, 16].

  So we never think of an individual particle as placed or moving in a background space. We think of an individual only in terms of its view of the rest.
- *Background independence* [20]. Most theories have non-dynamical fixed elements with respect to which the dynamical variables are defined. We call these the background, because it usually refers to a part of the universe that is not being modelled that is used as a fixed reference point against which to measure changes in the dynamical variables we are modeling. An example is Newton's absolute space, which Mach

understood is a stand in for the "fixed stars," i.e., the rest of the universe. A theory that has a chance to make a complete description of nature must hence be background independent; and depend on no fixed background structure.

- *No view of the whole universe as if from outside of it* [57, 58, 70]. We seek a theory that could be applied to the universe as a whole. Most theories we work with, including quantum mechanics, Newtonian mechanics, special relativistic field theory, cannot be extended to the entire universe because they rely on fixed background structure created by splitting the universe into two parts, one of which serves as the fixed, non-dynamical reference. By this slogan we remind ourselves that a theory of a whole universe—a cosmological theory—must be structurally very different from the theories we use.

  One structure that is common to theories suited for subsystems of the universe is a fixed, timeless configuration or phase space, on which acts an Hamiltonian which generates ahead of time all the possible lawful histories. We call theories like this the Newtonian paradigm [57, 58]. The ability to determine all the possible configurations ahead of the evolution of the system corresponds to the existence of a clean separation between the effects of laws and initial conditions. This requires that we are studying a class of subsystems of the universe, so we can operationalize the split in the functions of laws and initial conditions.

  Systems outside the Newtonian paradigm include those that, by their intrinsic complexity, make it impossible to work out the set of possible configurations ahead of letting the system evolve dynamically. Because there is only one universe, cosmological theories are cases where one cannot make a clean separation of the effects of laws and initial conditions; this is one of several reasons they fall outside the Newtonian paradigm.

  So the slogan here is *one universe, described by many partial views.*
- *No causes from outside the universe.* Similarly, all chains of causation must stay within the universe.
- *No global symmetries, hence no global conservation laws.* A symmetry is a way to move the subsystem whose dynamics you are modelling relative to the fixed background without costing energy. Because of locality, the dynamical system can be assumed to be weakly coupled to the background—whose own dynamics can then be neglected.

Gauge invariances, such as diffeomorphism invariance in general relativity, are another story, they are a strategy to expunge non-dynamical backgrounds and transform a background dependent theory into a more relational, background independent theory.

- There are two very different notions of time within relational theories, giving rise to two different versions of relationalism, and hence, two versions of naturalism [21]. Some hold that time is not fundamental, but can be eliminated in favor of timeless law, this is called timeless naturalism. On this view there is no objective distinction between past, present and future; these are just differences of perspective. One version of timeless naturalism is the block universe of general relativity. Opposed to this is the temporal version of naturalism [21, 22, 57, 58], that embraces an objective distinction between past and the future, separated by an ever moving present moment. In these ontologies time is fundamental and irreversible, space is often emergent and laws evolve.[7]

  Temporal naturalism is similar to presentism, but they are not the same thing.

  It has been noted before that if one takes a realist attitude towards qualia and conscious experience one is pushed towards embracing time and the present moment as real. One way to say why is that there are two aspects of nature that are not capturable in a purported identification of the history of the universe with a timeless[8] mathematical object. One is the present moment, the other is a conscious sensation. And indeed they seem related because every sensation is contained in a present moment.

2.2 How to Describe a Cosmological Theory? The General Picture and Vocabulary

So how are we to go about constructing a cosmological theory within a relational framework?

[7] Very briefly, because timeless laws can be used to express variables at any time as functions of initial conditions, and hence eliminate time from the description of nature [57, 58].

[8] Timeless because temporal causation is modeled by logical inference.

The key thing we learned from the previous subsection is that if the completion we seek of quantum mechanics is to be at least the double completion including the requirement that it make sense as a theory of cosmology, it has several important features not shared by theories of subsystems. These will all be useful for identifying the PCC.

In a completely relational theory, a universe with no fixed background can be described as an inter-nested complex of dynamical subsystems, each one of which can be thought of as being described in terms of its interactions with the rest. We talk about this in terms of the *view* a subsystem has of its surroundings, by which we mean the information available to it of those other subsystems it interacts with most strongly. We call them the subsystem's *neighbors*, by virtue of its interaction with them. The view of a subsystem, I, will be denoted V_I.

In different theories in this class, the views of the subsystems will have diverse mathematical descriptions. This may be given by labeled graphs, or by vectors in a vector space, or even the space of functions of a certain kind on a sphere. There is in each case, a set or space of possible views, S.

In a background independent theory, the notions of space, position, relative distance in space are no longer fundamental, We posit that space is not present at the fundamental level of description, but is emergent, together with the relative positions of the various subsystems.

In this world a fundamental role is played, not by distances or coordinates in a background space, but by differences between pairs of views.[9] As a result, the principle of locality—that you interact most strongly and directly with that which is closest to you (in a background space) is replaced by a principle of similarity, according to which you interact with those whose views of the universe are most similar to yours. Note that this interaction will need to be repulsive to satisfy the PII as well as maximize diversity of views.

In many situations locality will track similarity of views, because you and your neighbor will have similar views of the surroundings. This is especially likely for the views of massively composite systems, which we expect to be both complex and unique. This will be the basis for the recovery of standard physics. An example we find in the CTV is the emergence from an energetic causal set of a Minkowski spacetime, together with an embedding of the network of events and causes in it [27].

[9] For a fully worked out model exhibitiing these principles, please see [51–54, 56].

But not always: sometimes similarity of views will conflict with emergent notions of spatial distance. This is likely to be the case for views of microscopic or funamental events, as their views will have few degrees of freedom, and hence, in a large universe will have many accidental near copies. This, we show in the real ensemble theory and the CTV, is the origin of quantum phenomena, including nonlocal entanglement [27, 48, 49].

To study these ideas concretely we will need to formulate a space of views, *S*, on which we will need to define a measure of difference of two views, *I* and *J*, i.e., a distance function on *S*, which we will call $D(I, J)$ [27, 51].

In the absence of any notion of space, how do we define dynamics? We propose in [27, 49, 51] that the dynamics of a closed, relational system is to be expressed in terms of differences between views of its subsystems.

The other big idea is that the most important relationship involved in physical systems is causation: event E is the partial cause of event F [23, 24, 51, 70]. Indeed, nine of the ten functions that define the metric of spacetime go to describing the causal structure, that is which events were causally prior to which events. If we know the causal relations amongst events we have a very physical definition of a neighborhood: a causal neighborhood .of event *E* consists of those events causally prior to *E*.

Rafael Sorkin and collaborators have been studying models of discrete analogues of spacetimes where the only property an event may have is who is in its causal past [23]. These are called causal set models. Cortes and I made their structures slightly more complicated by giving the events and causal relations quantities of energy and momentum, as internal or intrinsic quantities; we call these energetic causal sets [51–54, 56]. The causal theory of views [27] joins this with the real ensemble framework for quantum realism [48, 49].

3. Hypotheses about Physical Correlates of Consciousness (PCC)

There is a great deal more to say to explain the motivation for the theories I will now discuss. However, as these are available in various books [30, 57, 58] and papers (cited above), I want to jump right in and describe in as simple language as possible the basic ideas as well as how these may work to generate hypotheses as to the physical correlates of consciousness.

3.1 The Causal Theory of Views [27]

This approach is based on the application of the real ensemble formulation [48, 49] to a class of models of discrete or quantum spacetime we developed with Marina Cortes, called energetic causal sets (ECS) [51–54, 56].

We begin with the latter.

3.1.1 The basics of energetic causal sets [51–54, 56].

We work within an ontology of events according to which the history of the universe is composed of events, connected by causal relations. We also hypothesize that energy and momentum are fundamental, and are transmitted from parent to child events, by the causal processes that create new events from present events, in such a way as to conserve them. Thus, while there is no space, and no spacetime fundamentally, there is a momentum space, whose geometry dictates how energy-momentum vectors are to be combined [68, 69].

Space is not part of the fundamental description, the events do not live in a space or a spacetime. At a later stage, a spacetime may emerge as a consequence of solving the equations of motion that guide the transmission of energy and momentum through the dynamically generated causal structure [51, 52].

The fundamental causal processes, i.e., the activity of time, is the continual creation of new events, by means of the selection of their parent events, from a reservour of present events. At each step in the construction, parents events are chosen and an event is added, together with its causal connections to its parent events, following a rule which is part of the definition of the model. The details are in [51, 52], but the important thing to say is that the rule aims to increase the diversity of the views of the present events. This event creation process creates a discrete partial order, or causal structure. Each event has a specified number of parents, N_p, and can have at most N_c children.

After each step, the set of events can be divided into present and past events. A past event has had its limit, N_c, of children and can no longer directly influence the future, by the creation of new events. So the particular history of the system defines, at each step in the event creation process, a physically defined, thick present, which are those events which have after their creation, not yet been parents to N_c events. By thick we mean that two present events may nonetheless be causally related.

The future does not yet exist, and indeed may be undeteremined, e.g., if the event creation rule has a stochastic element. This is important for what follows.

3.1.2 Connecting to the real ensemble formulation

In order to join this theory of events to the real ensemble formulation, which is a realist formulation of quantum physicis, we need to specify what are the beables. What are the beables of an event?

The beable of an event is defined as its momentary *view* of the rest of the universe, i.e., it is the information available at that event of its near causal past (or past causal neighborhood) through the causal processes that created it. Hence the view of an event is a snapshot of information about its ancestry, transmitted by its parent events. This includes the set of immediate past causal links and the energy and momentum they carry.

No elementary event has only one cause. Therefor there are no isolated elements of views, each view contains a number of elements unified by being part of the view of a particular event.

At an elementary level, the view consists of a framing, which is a two-sphere, marked with labeled points, each representing the incoming direction of each of its N_p parental causal processes (gotten from the direction of the incoming momentum), with the label taken to be a measure of the energy transmitted by that causal process.[10] There is a space of possible views, V, of an event.

We can call this description of the view of an event its *sky*; it is literally what is seen looking outward (and hence back into its causal past) from the event. The views are sufficient to reconstruct the events and their causal relations as well as flows of energy-momentum. So, while we started thinking of events and causal processes, we end up with just the views, as the only beables.

Events may be combined into sets of events, which have a joint causal past, hence each set of events also has a joint view. There is a natural algebra for combining views. This remind us of surgery joining topological two surfaces. Some laws act together on sets of events and their views, we call these law-bound. Entangled states and coherent states will require this treatment.These must be dealt with irreducibly.

[10] Among the classical constraints satisfied is the energy-momentum relation.

The event creation dynamics knows about the views only through their differences, $D(I, J)$. It is specified by requiring that the diversities of views is maximal.

In a "continuum limit" in which a spacetime emerges, together with an embedding of the discrete causal structure into it, a view must become approximated by a cross-section of a backwards light cone (i.e., a sphere of some dimension), on which are found punctures or (coarse grained) fields, which carry energy-momentum and other conserved quantities [27]. In our universe, $d = 2$. Thus a view physically is a full or partial S^2, on which live, depending on the level of description, punctures labeled by energy, momentum and other conserved charges, or pull backs of fields (into the light cone's cross-section) carrying those conserved quantities.

The real ensemble theory is a realist completion of QM, which, in its most general formulation, is based on the principle that the dynamical variables of the theory are differences of views of subsystems. This can be thought of as a norm or distance function $D(I, J)$.

On a theory with many subsystems, S_I, each of which has a view of the others: v_I^J represents the view of the subsystem J from the subsystem I. The subsystems may also be composites and have internal degrees of freedom, y_I^α, that also are seen in the views.

In this world a fundamental role is played, not by distances in a background space, but by differences between pairs of views. We will then employ $D(I, J)$ as a measure of difference of two views, I and J, i.e., a distance function on a space of views, S.

The dynamics has two parts. There is the event generator, which picks the parents of each newly create event. Then there is an action principle which determines how the energy and momentum are distributed to the recently formed events. This is of the rough form

$$S = K[\Delta D(I, J)] - U[D(I, J)], \tag{1}$$

where the potential energy is a functional of difference between views and the kinetic energy, K is a functional of the rates of change of the differences between views, $\Delta D(I, J)$. The kinetic energy comes from the energetic causal set models, while the potential energy is taken from the real ensemble models [49], and is chosen so that there is a strong repulsive interaction when two views are very similar. A good choice is a measure of complexity called the variety, V.

The variety, V of a system of relations is a useful measure of the diversity present in the set of views [27, 49, 50], and is a function of the $D(I, J)$:[11]

$$V = \frac{1}{N^2} \sum_{J<K} \frac{1}{D(J, K)}, \tag{2}$$

where N is the number of subsystems. The potential energy U will be proportional to the negative of V:

$$U = -\alpha V. \tag{3}$$

This has the consequence that the forces are strongest—and are repulsive—between two views whose difference is small.

The constant α will turn out to be related to $(\hbar/m)$.

We intend something like an ensemble interpretation of quantum mechanics, but rather than being an interpretation of the usual formulas this is a construction of Schrödinger quantum dynamics for a function of variables which describes a real ensemble, by which I mean that, rather than being a mathematical description of an imaginary ensemble, each and every member of this ensemble exists somewhere in the universe. That is, the wavefunction of a water molecule at the tip of my cat's left whisker is built from information about an ensemble of water molecules in similar conditions scattered through out the universe.

This general framework is presented in [48, 49]. We are interested here in the application to the energetic causal sets, which is the main subjects of [27]. Then the whole system is an energetic causal set, constructed up to some step. The elementary subsystems are the events, and we have already said that their views are extracted from the information in their past causal neighborhoods; these will generally be truncated a finite number of causal links. Thus, we have a graded structure in which the p'th past causal neighborhood of the event E, which we will also call the p'th view, is the past of E truncated p steps.

As we sketched above for the first view, the information in these views can be represented as a 2-sphere with labeled punctures.

At each step in the construction, there is a present set of events and their views.

[11] To eliminate coordinate dependent effects, which arise from fictional "outside observers," the distance functional, $D(I, J)$., is sometimes defined through a best matching procedure (see [72]).

For each event, we ask if there are other events in the present which are similar to it. We make the arbitrary distinction that a set of M events E_α $\{E)$ have views which are all within some small number ϵ of each other:

$$\forall \alpha, \beta \quad D(E_\alpha, E_\beta) < \epsilon, \tag{4}$$

We note that these are, by the assumptions stated above, all strongly interacting with each other. We hypothesize that the result is that the views of the members of the ensemble may be changing their values often, but in a way that can be expressed by a static probability function:

$$\rho(V_I) \tag{5}$$

Now let us consider two opposite cases, first a set of many similar events, so $M >> 1$. The individual values are rapidly changing within a narrow range, due to the strong interactions within the ensemble, but there can be an equilibrium, in which the probability distribution is slowly changing, and becomes uniform on the ensemble. We are able to show that the probability distribution represents the modulus of a wave-function that satisfies the Shrodinger equation.

This quick summary leaves out many steps, to mention just one—we are able to extract from the information in the views, phases, which are correlated with the energy in the views and which become the phases of the wavefunction.

Thus, events whose views have many near copies define ensembles of views, whose evolution is in a certain limit defined to leading order (in $1/M$) desscribed by the Schrödinger equation.

But what of events whose views have no near copies? Most likely these will be composite events of sufficient complexity and numbers of degrees of freedom that they would be associated with a history made up by a mesoscopic or macroscopic process or system.

When a subsystem is too complex or large, it will not have any copies in the universe, so it is not a part of such a large ensemble. Hence they do not correspond to pure quantum states. These are the novel states. Their evolution law is not the Schrodinger equation, but a more complex non-linear equation, governed by the full dynamics of the completion. The dynamics is specified by the same theory, but without taking the large M limit or constructing an ensemble and its probability distribution.

Some first steps towards exploring this anti-quantum limit are taken in [48, 49]. To leading order in the inverse of the number of copies, this can be expressed as a non-linear extension of the Schrödinger equation, that conserves probability but is not unitary. This then falls into a class of modifications of quantum mechanics that is very vulnerable to experimental tests.

This is we may note a solution to the measurement problem, because macroscopic devices, clocks, ourselves and our cats all are unique and have no copies, hence are not described by a wavefunction.

More generally, we see that there are naturally three classes of views: views which have many near copies, views which have a small number of near copies and views that are unique. Thus, this completion of quantum mechanics is also an extension of quantum physics. In the first class we find a derivation of ordinary quantum mechanics, in the second, a new class of nonlinear quantum phenomena. The third, unique views, will play a special role in the following.

3.1.3 From views to correlates of consciousness

We are now ready to state my first hypothesis about the physical correlates of consciousness. This is related to what we will call the first observation regarding consciousness.

- *First observation:* There are no "atoms of experience," i.e., experiences of nothing but a shade of red or blue or a pure high c. *Each conscious perception comes as a complex but irreducible unity, which may contain a number of qualia, thoughts, sounds, smells, all together, defining a (thick) moment of time, always in a frame, which is often experienced as a two-sphere, or a piece of one.* We call this a *framed conscious perception.*
- *First hypothesis: Each framed conscious perception corresponds to the view of a physical event or law-bound sets of events.*

  The framing of a conscious perception is another way of stating the self problem: there are no elementary qualia which are pure colors or pure tones. Each view is experienced as a whole, with colors or tones bound into a frame defined by the two-sphere. That is it suggests there is something or someone whose bundled or framed complex perception this is.

  The first hypothesis answers the self-problem: it is the conscious perception correlated to the view of a particular event or law-bound set of events. This addresses how qualia are bound into a single frame.

The "presentist problem" is addressed because each framed conscious perception is associated to a single event or, to a law-bound set of events. In the latter case, the moment may be thick.

The uniqueness of the self-problem is not yet addressed.[12]

- We notice that the bundled qualia are each associated with beables, so there is no ambiguity, as beables always have definite values. This explains the fact that we experience a world in which all variables have definite values.

  If we look ahead at how the theory we are sketching may, we hope, be a completion of QM in two senses-that it resolves the foundational issues of quantum theory, and that it gives a home to qualia and framed perceptions in a physical theory. These two issues must relate, because we want the perceived (i.e., selected) events to arise at the level where we experience beables, which take on definite values, not observables. We do not perceive superpositions, hence the events whose views are perceived must be limited to those events where the superpositions are resolved. These events must be rare, within a framework in which possible or potential histories are summed over. This then underlies the hypothesis that the rare events associated with conscious experience are expressions of the mechanism by which the real beables are defined, in a completion of quantum mechanics.

3.2 Common and Unique Views

I have proposed that the PCC are views, and that this addresses the "self" or bundled problem.

In the last section, we saw that the real ensemble formulation gives a special role to unique views, which have no present near copies. We also saw that these are sometimes the views of large or complex composite events. This makes it natural to propose the following two hypotheses:

- *Second hypothesis: Common events, which are those whose views have many near copies, are those described to good approximation by quantum*

[12] To do that we need to be able to define a hierarchy of inclusions of events, which in turn dictates an algebra of views. The problem may be solved if there is a property like entanglement, which can serve as a selection principle, which will be present up to a top level. [See Markopoulou et al. [70, 71].]

mechanics as formulated presently. These are not correlates of conscious perceptions.

- *Third hypothesis: Only views of unique events or unique law-bound sets of events are correlates of conscious perceptions.*

  Note that a state of a quantum many body system may be unique, in the sense that it has no copies in the universe, but it is composed of subsystems which have many copies (think of a neuron in your brain, and the atoms it is made of). There will be no copies for that neuron, but its component atoms have copies, and hence evolve normally.

  This gives us an important tool to incorporate into the selection criteria. We now may propose:

- *Fourth hypothesis: Only views of unique views or unique law-bound composite views, which are the smallest composite views, in a hierarchy of composite views, which have no copies, are correlates of conscious perceptions.*

3.3 Routine and Novel Events

I now want to introduce an alternative selection principle which also vastly reduced the set of views that will be candidates for PCC. This is connected with the principle of precedence I introduced in [28]. This leads to a similar second and third hypothesis.

Let me begin with the idea that the laws of physics, or at least their dimensionless parameters, must change and evolve, if they are to be explainable [31, 32]. I have made the case for this at length in several books and papers [21, 31–33, 57, 58], here I will simply work with the simplest proposal so far made for laws to evolve, which I called the principle of precedence [28].

The principle of precedence is a mechanism I invented by which the dynamics of a quantum system (i.e., the unitary evolution operator, U) could be seen to evolve in time. More specifically the dynamical law is replaced by a principle of precedence, which I describe below [28].

But first I want to introduce some language to frame this and other proposals.

I propose we divide events and law-bound sets of events into *routine events* and *novel events.*

Routine events are the vast bulk of events where the laws act as on those in the past, so that no mechanism of law-change is acting.

Novel events are those that have no near[13] precedents in their causal past. Similarly, we can speak of routine states and novel states, a novel state is one that has no precedents in its causal past, Those that have many precedents are routine.

I now make two alternative hypotheses:

- *Second' hypothesis: Routine events are those described to good approximation by quantum mechanics as formulated presently. Novel events will require a realist completion of quantum physics. Thus, the PCC will be associated with the completion of QM.*
- *Third' hypothesis: Only views of novel events or novel law-bound sets of events are correlates of conscious perceptions.*

  Note that a state of a quantum many body system may be novel, in the sense that it has no precedents, but it is composed of subsystems which have many precedents (think of a neuron in your brain, and the atoms it is made of). There will be no precedent for that neuron, but its component atoms have precedents, and hence evolve normally.

  Thus the distinction between novel and routine states will often be scale dependent, and may arise as we coarse grain the description. This gives us an important tool to incorporate into the selection criteria.

  We now may propose:
- *Fourth' hypothesis: Only views of novel (or unique) events or novel (or unique) law-bound sets of events, which are the smallest set of views, in a hierarchy of views, which have no precedents or copies, are correlates of conscious perceptions.*

This addresses the unique self problem.

Let me now describe the principle of precedence in a bit more detail, after which I will give a second example of a completion of quantum mechanics where a distinction between routine and novel events or states could arise.

3.4 Principle of Precedence [28]

The *Principle of Precedent* is an idea about the origin of laws, or rather how the notion of dynamical law could be replaced by a simpler hypothesis.

[13] We say near-precedents because complete precedents are we hope ruled out in respect for the PII.

One context in which it has an especially clear presentation is an operational formulation of quantum mechanics. In such a formulation, each quantum process is broken up into three stages: i) a preparation, whose output is a steady stream of quantum states in specified states $\rho_{initial}$, ii) a unitary evolution generated by a Hermitian hamiltonian, U, and iii) a measurement. Given a set of possible output states, the result is a set of numbers, $p(output, input) = Tr\rho_{final}U\rho_{initial}$. From the time independence of the U we can deduce the following.

The various probabilities $p(output, input)$ for the different possible outcomes are constant, so that the probabilities measured in the next year will converge with those measured over the last N years. Given this we could posit a precedence law:

Law of precedence: *Given a preparation for a physical system, chose the output state by pulling it randomly from the set of past precedents.*

The routine states are those that have a large number of precedence. The novel states are those without precedents.

How does the universe choose the outcomes of preparations which have no or few precedents?

This question might require a completion of quantum mechanics to answer. The causal theory of views and the real ensemble theories are candidates.

But so far as the question of consciousness is concerned we have here a striking suggestion:

The novel states are the physical correlates of conscious events!

That is, experience is made up of those moments where the universe is not guided by the law of precedents. At these moments, the universe has perhaps some degree of freedom to choose what happens next. It is these moments of freedom which make up conscious experience.

Those unprecedented moments are presumably common near the universe's origin, and spread throughout the universe. As the universe ages, it takes a higher degree of complexity for a state to be unprecedented. But we can wonder whether complex biomolecules might serve as a reservoir of novel states. Might the biosphere and the brain have evolved, to make use of the special properties of novel states, including the freedom present at those moments to choose a small part of the future. It is not difficult to see that this access to novel states might be a selective advantage.

Note that large molecules are made up of smaller subsystem, such as atoms. The component atoms will not be novel. What I want to suggest is that if there are entangled or coherent states which are made of many atoms which are sufficiently large and complex to be without precedent, these may serve as novel states.

The freedom in choosing the unitary evolution operator acting on such states will not impinge on the microscopic local dynamics governing each routine component, it will have to act non-locally, on the whole molecule, and be sufficiently weak so as to not have been discovered. Such a term might, e.g., favor one folding of a protein over others.

3.5 The principle of Precedence Applied to an Energetic Causal Set Theory

Ideally, we would like to derive relativistic quantum mechanics from the causal theory of views, along the lines of the derivation of non-relativistic QM we give in [27]. But in the absence of such a derivation we can anticipate the outcome. This is a quantum theory built on a causal set, somewhat analogous to a discrete version of a quantum field theory built over a fixed classical spacetime.

This is an example of a class of theories invented by Fotini Markopoulou called *quantum causal histories* [70, 71] They are defined as follows.

- Generate a causal set. There is then a map which assIgns to the causal link connecting event K with an event I in its immediate future, a Hilbert space H_I^K.
  Associated with the sky of I is a joint HIlbert space constructed by direct product:

$$H_I = \otimes_{K<I} H_I^K. \tag{6}$$

- Associated with the same event I there also is the anti-view, which consists of all the information sent out by I. In classical general relativity the anti-view of an event I is a cross section through the future light cone of I. If $L > I$ is in the immediate future of I then

$$\bar{H}_I = \otimes_{I<L} H_L^I. \tag{7}$$

- Let us assume that the view and anti-view have the same dimension. The quantum dynamics is constructed to give a unitary evolution from the quantum view to the quantum pre-view:

$$U_I : H_I \rightarrow \bar{H}_I. \tag{8}$$

- To construct the quantum casual histories associated to an energetic causal set (which we may call an energetic quantum causal history (EQCH)), replace the energy-mometum vector p_I^K associated to the link from K to I by the free-particle Hilbert space:

$$p_I^K \rightarrow H_I^K \in \psi(p_I^K). \tag{9}$$

subject to the usual mass shell constraint and norm:

$$C(p) = 0 \rightarrow C(p)\psi(p) = 0. \tag{10}$$

Now we can apply the principle of precedence, as follows.

- Now let us choose a basis for the space of views, H_I given by $|a>$. These are the preparations. A dual basis for the anti-views is given by $< w|$. Then as usual

$$T(I, < w|, |a >) =< w|U_I|b > \tag{11}$$

is the probability that the sky preparation $|a>$ evolves by the event I and yields the output state $< w|$ (in a fixed basis).
- Now, let $D(I, J)$ be the distant operators on the past causal sets of events I and J.
- Let us consider for an event I, and preparation, $|a>$, the set of causally past events J which share the same preparation, $|a>$, and which are close in the space of views. We will define $P^\epsilon(I, |a>)$ to be the past set of event I consisting of events $J << I$, such that

$$D(I, J) < \epsilon. \tag{12}$$

We call the members of this set the precedents of $(I, |a>)$. Let there be $N(I, |a>)$ elements in the set of precedents and let this number be very large. Each event in this set has an output which is one of the $< w|$. This gives us an ensemble of outputs, called $O(P^\epsilon(I, |a>))$.

Then we apply the principle of precedence, which says that to find the output of I pick a random element of the ensemble of outputs.

This will agree with the quantum prediction given by (11).

- The interesting question is what to do if the event generator builds an event N, that has no precedent; the set $O(P^{\epsilon}(I, |a>))$ is empty.

  In the case of no precedent, the universe must make a choice of output that is not determined by past data; in a certain minimal sense of the word freedom, the universe makes a free choice.
- I now propose the following: *the correlates of conscious perceptions are the smallest views which are unprecedented.*

3.6 Energy and Qualia

- *Fifth hypothesis: Different qualia of the same modality (i.e., colors, tones) correspond to differences in energy.*

4. Conclusions

I close with a brief summary of the main assumptions and hypotheses.

- We propose a new ontology according to which the universe consists of a dynamically evolving collection of partial views of itself.
- The view of an event contains information about its recent causal past neighborhood. The view also represents flows of energy, momentum and other conserved quantities. The views are the only beables of this theory.
- The dynamics which creates the events and guides the flow of energy on the causal links depends only on differences amongst views, and expresses a principle of maximizing the diversity, or variety of views [50]. There is no fixed background space or spacetime.
- The replacement of locality (in a background space) with similarity (in a space of views) has striking consequences. Since space is emergent, so is locality, and the mechanism by which that happens is that much of the time the world is arranged so that locality in the emergent space tracks similarity of views. (Come here and look at what I see!) But locality, being emergent, will have defects, where two very similar views represent two events which are very far from each other in the emergent spacetime geometry [62]. In [27, 49, 61] I show that this leads to the

recovery of quantum mechanics. The key point is that a small composite system, like an atom or a small molecule, will have *copies* which are scattered across the universe. These nonetheless interact strongly with each other. When there are many copies, the evolution develops sufficient coherence and the result is unitary Schrodinger dynamics.

But what of the subsystems that have no copies? How does it evolve? This is addressed in detail by the real ensemble theory [48], showing that it is indeed a good candidate for a completion of quantum mechanics.

This suggests a new picture of evolution in quantum physics, which I called the principle of precedence [28]. A quantum state evolves because the underlying dynamics being local in the space of views, it is coupled to the members of an ensemble of similar states in its causal past. These are its precedents, and the proposal is that quantum dynamics is simply the copying of random precedents from a state's causal past.

A state without precedents can be called a novel state—one that has not so far existed in the history of the universe. A key question to be addressed by a completion of quantum theory is then, how is an unprecedented state to chose what next to do?

Within this framework for a relational physics, we make five hypotheses about the physical correlates of consciousness.[14]

1. *First hypothesis: Each framed conscious perception corresponds to the view of a physical event or law-bound sets of events.*
2. *Second hypothesis: Common events, which are those whose views have many near copies, are those described to good approximation by quantum mechanics as formulated presently. These are not correlates of conscious perceptions.*

[14] Or, if we frame the dynamics in terms of the Principle of Precedence rather than the real ensemble view, the middle three are replaced by the alternative hypotheses:

1. *Second' hypothesis: Routine events are those described to good approximation by quantum mechanics as formulated presently. Novel events will require a realist completion of quantum physics. The PCC will then be associated with the completion of QM.*
2. *Third' hypothesis: Only views of novel events or novel law-bound sets of events are correlates of conscious perceptions.*
3. *Fourth' hypothesis: Only views of novel events or novel law-bound sets of events, which are the smallest set of views, in a hierarchy of views, which have no precedents, are correlates of conscious perceptions*

3. *Third hypothesis: Only views of unique events or unique law-bound sets of events are correlates of conscious perceptions.*
4. *Fourth hypothesis: Only views of unique views or unique law-bound composite views, which are the smallest composite views, in a hierarchy of composite views, which have no copies, are correlates of conscious perceptions.*
5. *Fifth hypothesis: Different qualia of the same modality (i.e., colors, tones) correspond to differences in energy.*

Thus, while standard panpsychism proposes that there are qualia (or proto-qualia) associated with all physical states or processes, this new view proposes that there are framed or bundled conscious perceptions, associated to a very restricted subset of views-those that are novel, and maximal in the sense described.

These first four hypotheses explain together the self problem, the presentist self problem and the unique self problem.

How do we proceed from here?

Of course, the most urgent question is to make contact with neurobiology. To do that we need a suggestion as to actual physical processes acting in specific neural tissues that have the needed characteristics. Principally, they have to be novel in the sense of having no or few precedents or copies. One way to build such unprecedented states would be to entangle a large enough number of qubits that simple combinatorial complexity could guarantee uniqueness. But this seems like pure fantasy, where in the brain's warm environment arc wc to find large sets of entangled qubits? Under present evidence, there would seem to be little chance the brain constructs protected channels, topological or otherwise?

One way to look for coherently entangled qubits in the brain is to make use of the suggestion of Mathew Fisher that sets of the nuclear spins of phosphorus shielded in phosphate and randomized to project out protected noise free channels might provide a source of qubits in biological systems [76]. One place they are found is in the bilayers of phospholipid molecules, which form the membrane of the neuron. Each such molecule has a "head," which is composed of a phosphate group, possibly linked to other groups. Two chains or tails extend downward, each composed of carbon-hydrogen units. The prospects for there being significant quantum effects involving these spins will be discussed separately [77].

Acknowledgments

I would like to thank Shan Gao first of all for the invitation to contribute a chapter to this book. I am grateful for correspondence and conversations with Sean Carroll, David Chalmers, Philip Goff, Jenann Ismael, Jaron Lanier, Stuart Kauffman, Kristoff Koch, George Northoff and Galen Strawson. I am especially grateful to Marina Cortes, Rodolfo Gambini, Philip Goff, Jaron Lanier, Andrew Liddle, Galen Strawson and Steve Weinstein for detailed critical comments on drafts. Jaron Lanier has been insisting to me for many years—until I finally got it—that there are no isolated qualia.

This research was supported in part by Perimeter Institute for Theoretical Physics. Research at Perimeter Institute is supported by the Government of Canada through Industry Canada and by the Province of Ontario through the Ministry of Research and Innovation. This research was also partly supported by grants from NSERC and FQXi. I am especially thankful to the John Templeton Foundation for their generous support of this project.

References

[1] David Chalmers, *The Conscious Mind: In Search of a Fundamental Theory* (1996). Oxford: Oxford University Press.

[2] Leibniz, in *The Monadology*.

[3] F. Jackson, Epiphenomenal qualia, *Philosophical Quarterly*, 32 (1982), 127–136.

[4] Philip Goff, *Consciousness and Fundamental Reality* (2017). Oxford: Oxford University Press.

[5] Philip Goff, *Galileo's Error: Foundations for a New Science of Consciousness* (2019). London: Rider Books, Penguin Random House.

[6] Galen Strawson, *Consciousness and Its Place in Nature: Does physicalism entail panpsychism?* (2006). Andrews, UK: Imprint Academic. Realistic Monism: Why Physicalism Entails Panpsychism, *Journal of Consciousness Studies*, 13, no. 10/11 (2006), 3–31.

[7] G. Strawson, *New York Review of Books* (March 13, 2018), https://www.nybooks.com/daily/2018/03/13/the-consciousness-deniers/

[8] Tom Nagles, *Mind and Cosmos* (2012). Oxford: Oxford University Press.

[9] Rodolfo Gambini (in collaboration with Jorge Pullin), *A hospitable universe: Addressing Ethical and Spiritual Concerns in Light of Recent Scientific Discoveries* (February 1, 2018). Andrews, UK: Imprint Academic.

[10] William James, *Essays in Radical Empiricism* (1912). New York: Longmans, Green, and Co.

[11] Bertrand Russell, *The Analysis of Mind* (1921). London: G. Allen and Unwin; New York, Macmillan.

[12] A. Eddington, *The Nature of the Physical World* (1928). Cambridge, UK: Cambridge University Press.
[13] L. Smolin, *No Matter, Never Mind, II* (1974). Hampshire College, Division II paper, http://leesmolin.com/wp-content/uploads/2019/07/No-Matter-Never-Mind-1.pdf
[14] Adrian Kent, Quanta and qualia, *Found. Phys*, 48 (2018), 1021.
[15] Leibniz G. W., Monadology.
[16] G. W. Leibniz, *Leibniz's letters to Samuel Clarke.*
[17] Albert Einstein, *Preussische Akademie der Wissenschaften, Sitzungsberichte* (part 1, 1916), 688–696.
[18] D. Chalmers, Consciousness and its place in nature. In his *Philosophy of Mind: Classical and Contemperary Readings* (2022). New York: Oxford University Press.
[19] Mauricio Gonzlez-Forero and Andy Gardner, Inference of ecological and social drivers of human brain-size evolution, *Nature*, 557 (2018), 554–557.
[20] Lee Smolin, The case for background independence, arXiv: hep-th/0507235.
[21] Lee Smolin, Temporal naturalism, of Studies in History and Philosophy of Modern Physics, Special Issue on Time and Cosmology, arXiv:1310.8539.
[22] L. Smolin, Temporal relationalism. In *Beyond Spacetime*, edited by Nick Huggett, Keizo Matsubara and Christian Wuthrich (2020), arXiv:1805.12468.
[23] Luca Bombelli, Joohan Lee, David Meyer, and Rafael D. Sorkin, Spacetime as a causal set, *Phys. Rev. Lett.*, 59 (1987), 521–524.
[24] J. Ambjorn, S. Jordan, J. Jurkiewicz, and R. Loll, Quantum spacetime, from a Practitioner's point of view, *AIP Conference Proceedings*, 60 (2013), 1514, arXiv:1302.2181, https://doi.org/10.1063/1.4791726
[25] Daniele Oriti, Spacetime as a quantum many-body system. In *Many-Body Approaches at Different Scales: A Tribute to Norman H. March on the Occasion of His 90th Birthday*, edited by G. G. N. Angilella and C. Amovilli (2017), arXiv: 1710.02807. New York, Springer.
[26] Carlo Rovelli and Francesca Vidotto, *Covariant Loop Quantum Gravity: An Elementary Introduction to Quantum Gravity and Spin Foam Theory* (2015). Cambridge, UK: Cambridge University Press.
[27] Lee Smolin, The dynamics of difference, *Foundations of Physics*, arXiv:1712.04799.
[28] Lee Smolin, *Precedence and Freedom in Quantum Physics*, arXiv:1205.3707
[29] Roger Penrose, Conversation Chapter 14: Consciousness Involves Noncomputable Ingredients. Edge.com.
[30] Lee Smolin, *Einstein's Unfinished Revolution* (April 2019). New York: Penguin Press.
[31] Lee Smolin, Did the universe evolve? *Classical and Quantum Gravity*, 9 (1992), 173–191.
[32] Lee Smolin, *The Life of the Cosmos* (1997). New York; Oxford University Press; London, UK: Weidenfeld and Nicolson.
[33] Lee Smolin, A perspective on the landscape problem, Foundations of Physics, *Special Issue on Forty Years of String Theory: Reflecting on the Foundations*, arXiv:1202.3373.
[34] Gerald Edelman and Giulio Tononi, *A Universe of Consciousness: How Matter Becomes Imagination* (2000). New York: Basic Books.
[35] Gerald Edelman, Bright Air, Brilliant Fire: *On the Matter of the Mind* (1992). New York: Basic Books.
[36] Christof Koch, *The Quest for Consciousness: A Neurobiological Approach* (2004). Englewood, Colorado: Roberts and Co.

[37] Christof Koch, *Consciousness: Confessions of a Romantic Reductionist* (2012). Cambridge, MA: MIT Press.

[38] P. Pearle, Reduction of the state vector by a nonlinear Schrödinger equation, *Phys. Rev. A* 113 (1976), 857; G. C. Ghirardi, R. Grassi and A. Rimini, Continuous-spontaneous-reduction model involving gravity, *Phys. Rev. A* 42 (1990), 1057.

[39] Roger Penrose and Stuart Hameroff, Consciousness in the Universe: Neuroscience, Quantum Space-Time Geometry and Orch OR Theory, Journal of Cosmology; Consciousness in the universe: A review of the 'Orch OR' theory, *Physics of Life Reviews*, 11, no. 1 (2014), 39–78.

[40] Max, Tegmark, Importance of quantum decoherence in brain processes, *Physical Review E*, 61, no. 4 (2000) 4194–4206. arXiv:quant-ph/9907009; McKemmish, Laura K.; Reimers, Jeffrey R.; McKenzie, Ross H.; Mark, Alan E.; Hush, Noel S. (2009). "Penrose-Hameroff orchestrated objective-reduction proposal for human consciousness is not biologically feasible" (PDF). *Physical Review E*. 80 (2): 021912. Bibcode:2009PhRvE..80b1912M. doi:10.1103/PhysRevE.80.021912. PMID 19792156; Georgiev, D.D. (2007). "Falsifications of Hameroff-Penrose Orch OR model of consciousness and novel avenues for development of quantum mind theory". *NeuroQuantology*. 5 (1): 145–174. doi:10.14704/nq.2007.5.1.121; Koch, Christof; Hepp, Klaus (2006). "Quantum mechanics in the brain". *Nature*. 440 (7084): 611. Bibcode:2006Natur.440..611K. doi:10.1038/440611a. PMID 16572152; Hepp, K. (27 September 2012). "Coherence and decoherence in the brain". *J. Math. Phys.* 53 (9): 095222.

[41] R. Gambini and J. Pullin, The Montevideo interpretation of quantum mechanics: A short review, *Entropy*, 20, no. 6 (2018), 413, arXiv:1502.03410.

[42] D. Dennett, *Consciousness Explained* (1992). Boston: Back Bay Books.

[43] Giulio Tononi, Melanie Boly, Marcello Massimini and Christof Koch, Integrated information theory: From consciousness to its physical substrate, *Nature*, 17 (July 2016), 450; Masafumi Oizumi, Larissa Albantakis, and Giulio Tononi, From the Phenomenology to the Mechanisms of Consciousness: Integrated Information Theory 3.0, *PLoS Comput. Biol.*, 10, no. 5 (May 8, 2014), e1003588.

[44] E. P. Wigner, Remarks on the mind-body question. In *The Scientist Speculates*, edited by I. J. Good (1961).

[45] H. Stapp, *Mind, Matter and Quantum Mechanics* (1993). New York: Springer.

[46] D. J. Chalmers and K. J. McQueen, Consciousness and collapse of the wave function. In *Consciousness and Quantum Mechanics*, edited by Shan Gao (2022), 11–63. New York: Oxford University Press.

[47] R. Gambini and J. Pullin, *Physical Requirements for Models of Consciousness*, http://philsci-archive.pitt.edu/16405/

[48] L. Smolin, A real ensemble interpretation of quantum mechanics, *Foundations of Physics* 42, (2012), 1239–1261, arXiv:1104.2822.

[49] L. Smolin, Quantum mechanics and the principle of maximal variety, *Foundations of Physics*, 46, (2016), 736–758, arXiv:1506.02938.

[50] J. Barbour and L. Smolin, *Variety, Complexity and Cosmology*, hep-th/9203041.

[51] M. Cortês and L. Smolin, The universe as a process of unique events, *Phys. Rev. D*, 90 (2014), 084007, arXiv:1307.6167.

[52] M. Cortês and L. Smolin, Energetic Causal Sets, *Physical Review D*, 90 (2014), 044035.

[53] M. Cortês and L. Smolin, Spin foam models as energetic causal sets, *Phys. Rev. D*, 93 (2016), 084039, arXiv:1407.0032.
[54] M. Cortês and L. Smolin, *Reversing the Irreversible: From Limit Cycles to Emergent Time Symmetry*, arXiv:1703.09696.
[55] W. M. Wieland, New action for simplicial gravity in four dimensions, *Class. Quantum Grav.*, 32 (2015), 015016, arXiv:1407.0025.
[56] Eliahu Cohen, Marina Corts, Avshalom C. Elitzur, and Lee Smolin, *Realism and causality II: Retrocausality in Energetic Causal Sets*, arXiv:1902.05082
[57] L. Smolin, *Time Reborn* (2013). New York: Houghton Mifflin Harcourt, Penguin and Random House Canada.
[58] Roberto Mangabeira Unger and Lee Smolin, *The Singular Universe and the Reality of Time: An Essay in Natural Philosophy* (November 2014). Cambridge, UK: Cambridge University Press.
[59] Lee Smolin, Derivation of quantum mechanics from a deterministic non-local hidden variable theory, I. The two dimensional theory. In *Stochastic mechanics, hidden variables and gravity in Quantum Concepts in Space and Time*, edited by C.J. Isham and R. Penrose, 1085). Oxford: Oxford University Press.
[60] Lee Smolin, *Matrix Models as Non-Local Hidden Variables Theories*, hepth/0201031.
[61] Fotini Markopoulou and Lee Smolin, Quantum theory from quantum gravity, *Phys. Rev. D* 70, (2004), 124029, arXiv:grqc/0311059.
[62] Fotini Markopoulou and Lee Smolin, Disordered locality in loop quantum gravity states, *Class. Quant. Grav.*, 24 (2007), 3813–3824.
[63] Lee Smolin, *Could Quantum Mechanics be an Approximation to Another Theory?* arXiv:quant-ph/0609109.
[64] Lee Smolin, Nonlocal beables, *International Journal of Quantum Foundations*, 1 (2015), 100–106, arXiv:1507.08576.
[65] E. Nelson, *Phy. Rev.*, 150 (1969), 1079; E. Nelson, *Quantum Fluctuations* (1985). Princeton, NJ: Princeton University Press.
[66] S. Alder,*Quantum Theory as an Emergent Phenomenon* (2004). New York: Cambridge University Press; *Statistical Dynamics of Global Unitary Invariant Matrix Models as Pre-Quantum Mechanics*, hep-th/0206120.
[67] Artem Starodubtsev, A note on quantization of matrix models, *Nucl. Phys.*, B674 (2003), 533–552, arXiv:hep-th/0206097.
[68] G. Amelino-Camelia, L. Freidel, J. Kowalski-Glikman and L. Smolin, The principle of relative locality, *Phys. Rev. D*, 84 (2011), 084010, arXiv:1101.0931.
[69] L. Freidel and L. Smolin, *Gamma Ray Burst Delay Times Probe the Geometry of Momentum Space*, arxiv:hep-th/arXiv:1103.5626
[70] Fotini Markopoulou, Quantum causal histories, *Class. Quant. Grav.*, 17 (2000), 2059–2072.
[71] Eli Hawkins, Fotini Markopoulou, and Hanno Sahlmann, Evolution in Quantum Causal Histories, *Class. Quant. Grav.*, 20 (2003), 3839.
[72] J. Barbour and B. Bertotti, Mach principle and the structure of dynamical theories, *Proceedings of the Royal Society* (London), A 382 (1982), 295.
[73] A. Criscuolo and H. Waelbroeck, Causal Set Dynamics: A Toy Model?, *Class. Quant. Grav.*, 16 (1999), 1817–1832, arXiv:grqc/9811088.
[74] Cohl Furey, *Notes on algebraic causal sets*, unpublished notes (2011); *Cambridge Part III research essay* (2006).

[75] D. Bohm, A suggested interpretation of quantum theory in terms of hidden variables, I, *Phys. Rev.*, 85 (1952), 166179.
[76] M. P. A. Fisher, Quantum cognition: The possibility of processing with nuclear spins in the brains, *Annals of Physics*, 362 (2015), 593–602, arXiv:1508.05929.
[77] Lee Smolin, *Design for a Bionic, Hybrid Silicon and Biological Quantum Device*, unpublished draft (2016).

Index